PHOTOGRAVURE

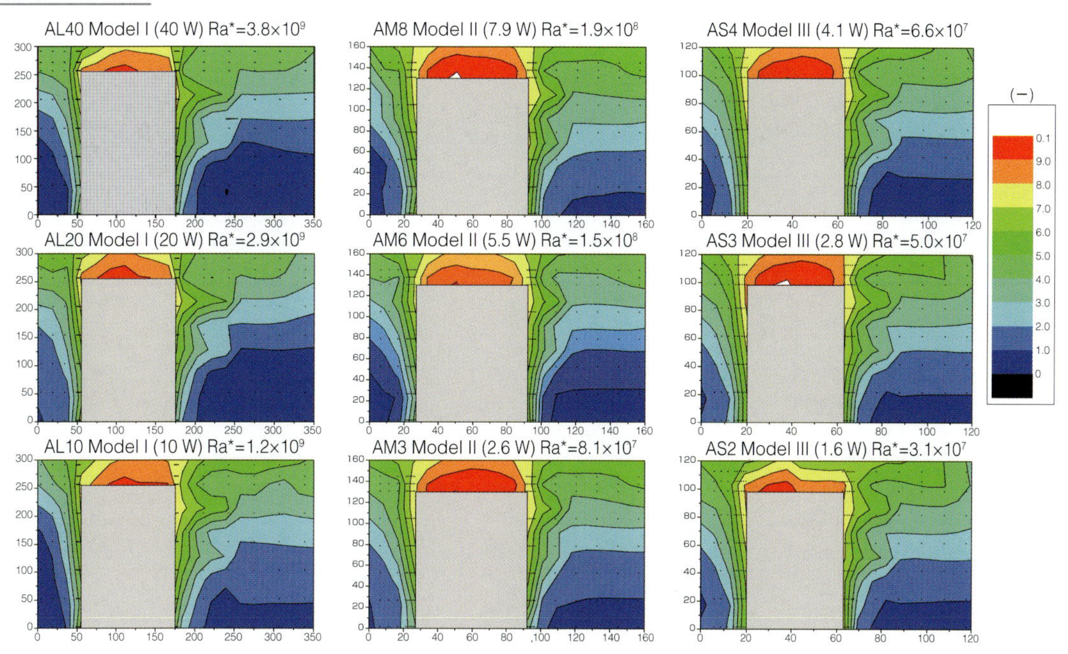

Fig. 3.2.2(1)-4 Non-dimensional temperature distribution (Air)

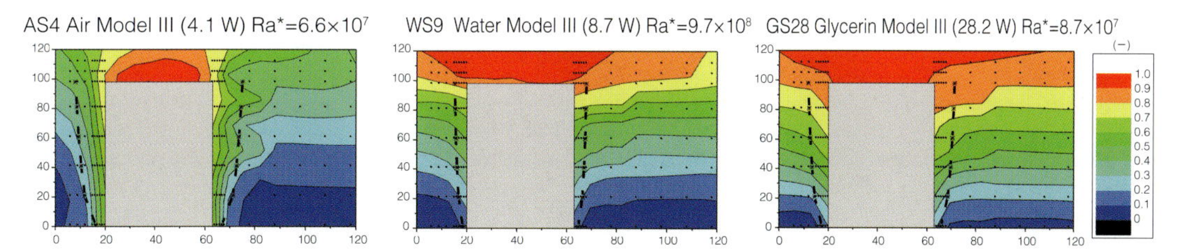

Fig. 3.2.2(1)-5 Non-dimensional temperature distributions and boundary layer (Air, Water, Glycerin)

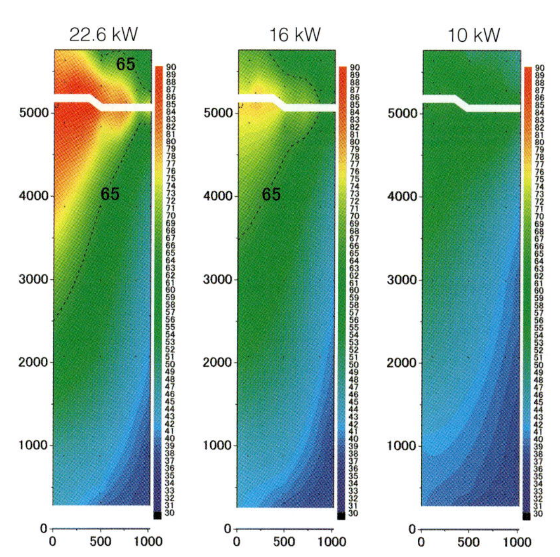

Fig. 4.3.2-6 Concrete temperature (RC cask)

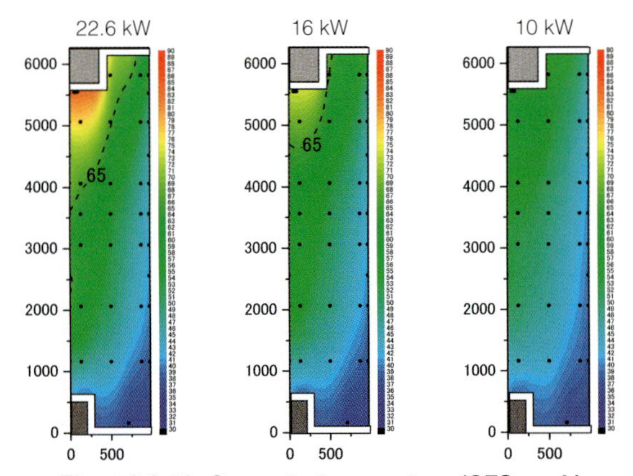

Fig. 4.3.2-10 Concrete temperature (CFS cask)

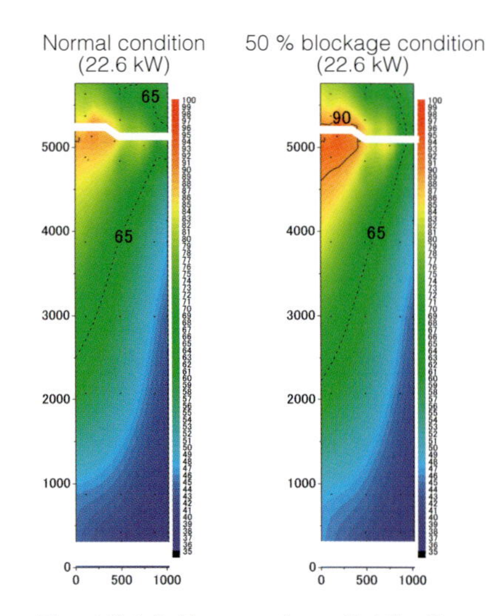

Fig. 4.3.4-2 Temperature distribution of concrete body (Cross section of RC cask at 90 °)

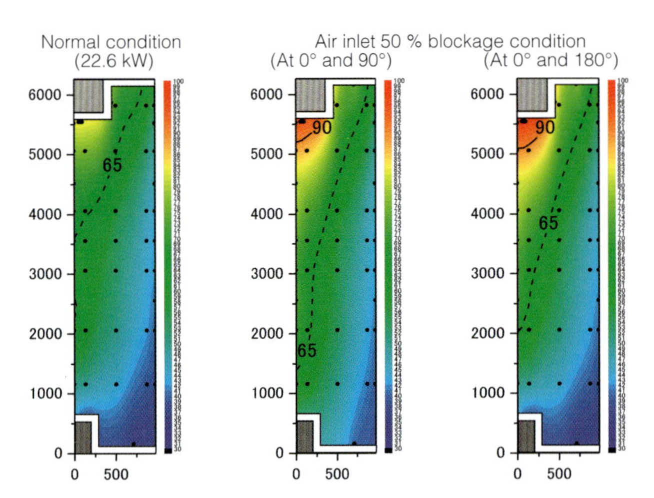

Fig. 4.3.4-3 Temperature distribution of concrete body (Cross section of CFS cask at 90 °)

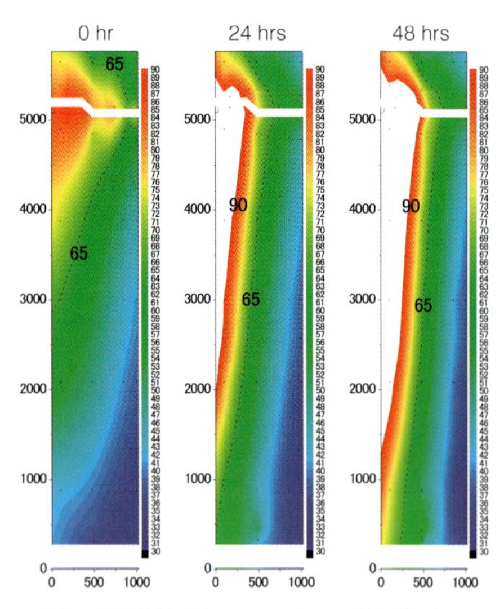

Fig. 4.3.4-7 Temperature distribution change of concrete body (Cross section of RC cask at 90 °)

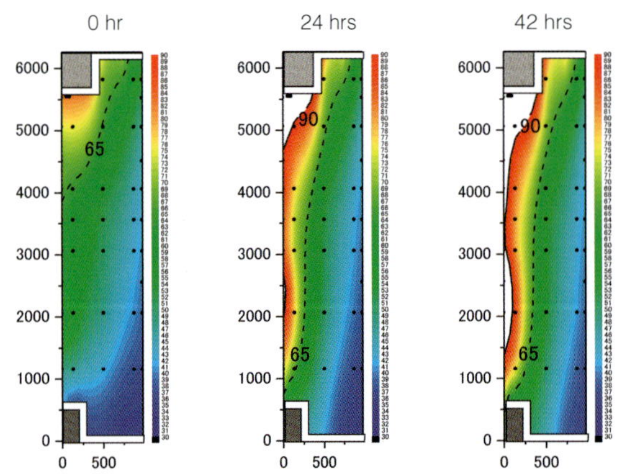

Fig. 4.3.4-11 Temperature distribution change of concrete body (Cross section of CFS cask at 90 °)

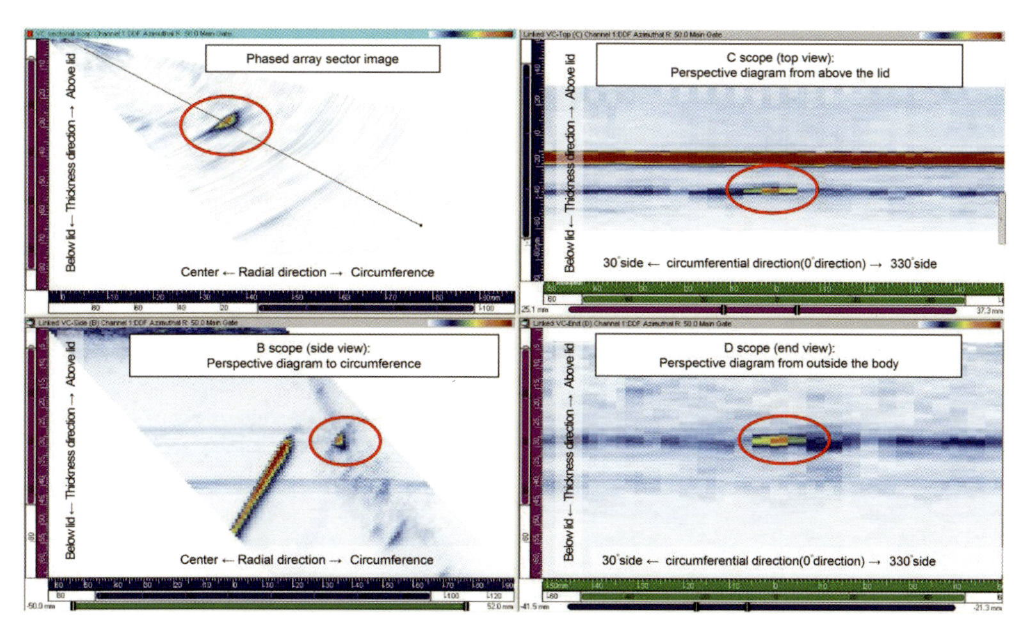

Fig. 4.5.2-7 Example of detection of defect (i) on the welding root (transverse wave, 2 MHz, reflecting angle of around 45°)

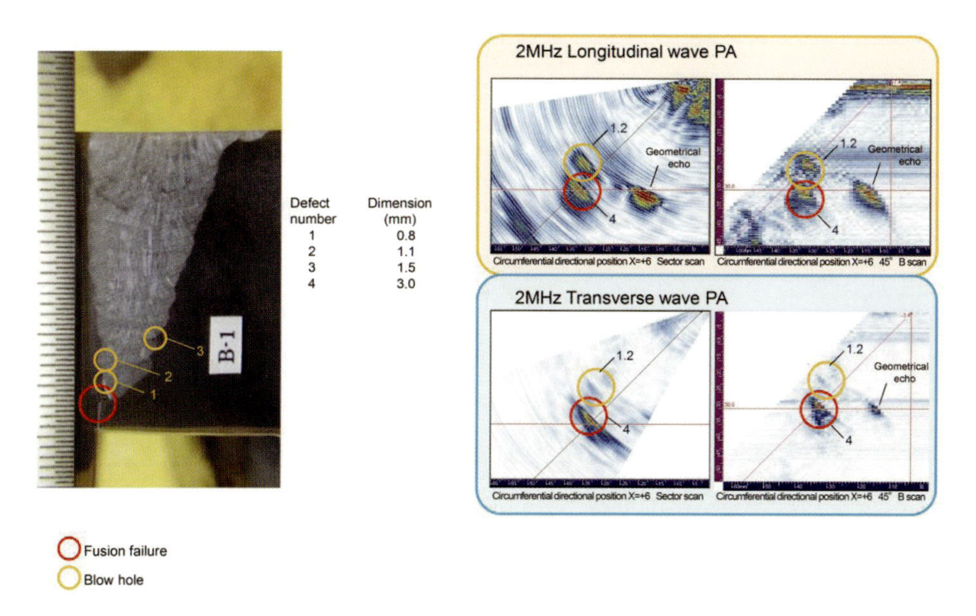

Fig. 4.5.2-9 Macroscopic picture of cross section of fusion failure area and results of UT

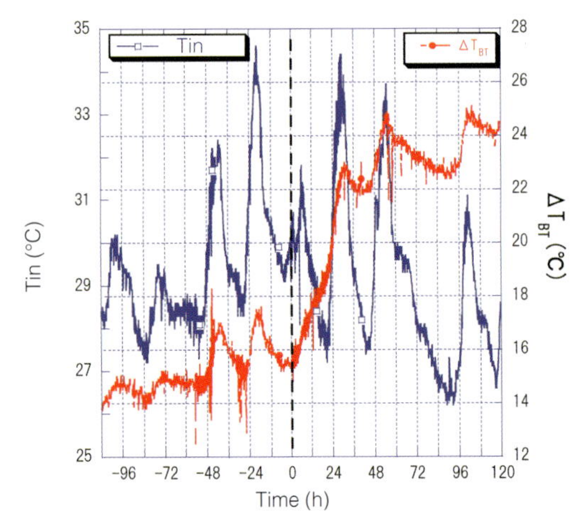

Fig. 4.7.3-8 Change of ΔT_{BT} and Tin (CASE 1)

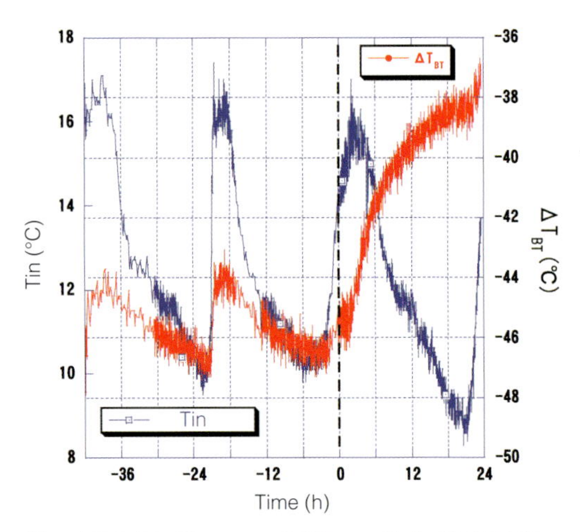

Fig. 4.7.3-14 Change of ΔT_{BT} and Tin (CASE 3)

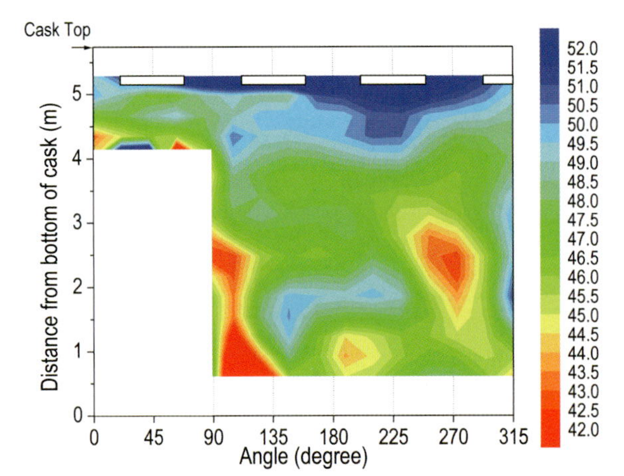

Fig. 4.7.4(4)-2 Concrete strength distribution converted from scleroscope hardness

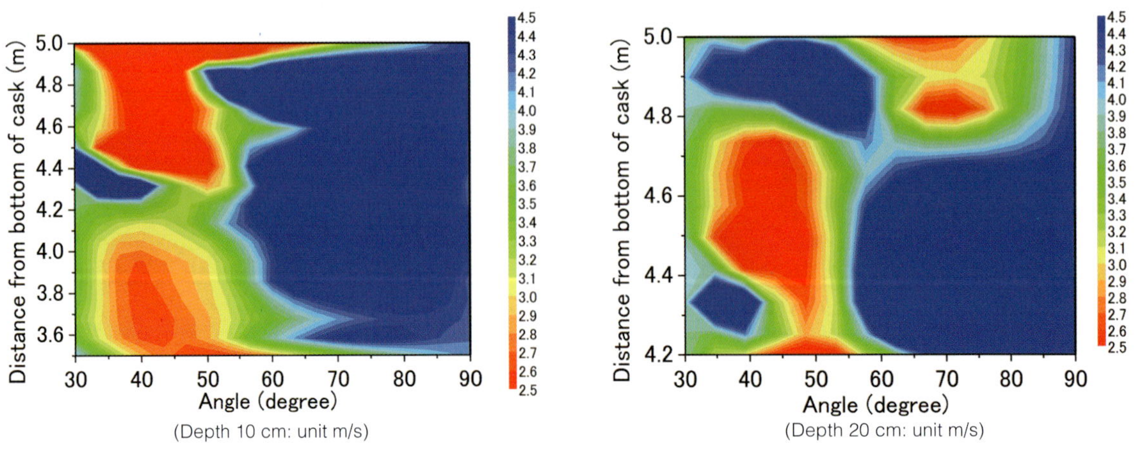

Fig. 4.7.4(4)-5 Elastic wave velocity distribution in depth direction

Basis of Spent Nuclear Fuel Storage

Central Research Institute of Electric Power Industry

ERC PUBLISHING CO., LTD

Recommendation 1

Masanori Aritomi, Professor Emeritus, Tokyo Institute of Technology

"Regarding the situation of spent fuels, even when we only consider those of OECD member states, there are approximately 185,000 tons of spent fuels as of 2011, and how to manage spent fuels is a global challenge. Japan currently stores about 17,000 tons of spent fuels as of 2014. Spent fuels produced by nuclear power generation must be safely managed... It is therefore necessary to expand the capacity for storing the spent fuels and is urgently important to broaden the range of choices for managing the spent fuels while ensuring safety… While studying a wide range of locations as possible sites, regardless of whether they are inside or outside the premises of a power plant, Government of Japan will strengthen its effort for facilitating construction and utilization of new intermediate storage facilities and dry storage facilities." (Strategic Energy Plan, April, 2014, Japan)

This book introduces results of experiments, researches and developments on metal cask storage, concrete cask storage, and other storage technologies with their evidence of safety. Those were carried out by CRIEPI's in-house researches, contractual researches from Japanese governments, etc. from 1980's. These results include fundamental knowledge extending to the most advanced technology that have been reported and discussed at international (IAEA, PATRAM, etc.) and domestic journals as well as conferences.

This book provides knowledge answering to issues on spent fuel storage safety and explains their basis kindly. This book is recommended to wide range of readers including persons in charge of siting, design, construction, safety evaluation, licensing, operation, and communication with public.

(April 2014)

Recommendation 2

Dr. Klaus Janberg, German Engineer, French Physicist, retired.

It was in summer 2014 that I could have a close look at the Japanese version of this book, which in my opinion will become a "vade mecum" for everybody engaged in transports and / or storage of spent nuclear fuel and my first and natural reaction was to acquire an English version of it as I have not been able to deepen my interest in the Japanese culture and language to the point of reading. I am happy to learn that this point is coming closer.

Never, ever, have I seen such a complete collection of research and development results which will help the utilities to extend their strategic thinking with respect to storage and I suppose that this goal is part of the basic idea behind the foundation of CRIEPI as a service organization of the Japanese utilities.

My first encounter with CRIEPI in the persons of Drs. Fukuda (at that time already widely known internationally) and Saegusa (then his futokoro-gatana) took place in 1983.

This encounter started a long-lasting exchange involving also licensing authorities and with a major mile-stone in 1989 in Tokyo: a full-scale big cask regulatory drop test with an artificial flaw and an international seminar where the different developers presented their solutions to dry spent fuel storage. This event made CRIEPI a leading international research organization in this area and Dr. Saegusa pushed the engagement of CRIEPI further, soon assisted by the also highly respected Dr. Shirai, who became known as a leading expert in fracture toughness questions, flaw detection, etc.

The book addresses now all storage systems as Prof. Aritomi already pointed out but with the increasing acknowledgment that many fuel elements would have to be stored for much extended periods. CRIEPI's research started also to cover fuel behavior questions together with other Japanese research organizations or at least giving them a representative place in this book. CRIEPI also cooperated many years with EPRI (American utility-owned counterpart).

I highly recommend this book to all engineers / physicists working in these areas, developers, fabricators or members of the licensing authorities.

I sincerely hope to become in 2015 one of the happy owners of this very respectable book.

Dr. Klaus Janberg, German Engineer, French Physicist, retired.
1976-1980: Department Head of the German Reprocessing Comp. DWK. And as such in charge of storage projects and negotiation of Reprocessing contracts with COGEMA and the Development of Double Purpose Casks by GNS and TN.
1980-2000: General Manager of GNS, President of GNB, the CASTOR Fabricator.

(April 2015)

Acknowledgement

Parts of this book were results of researches contracted from Agency for Natural Resource and Energy of the Ministry of Economy, Trade and Industry, and Nuclear and Industrial Safety Agency (now, Nuclear Regulatory Authority), and the former Science and Technology Agency of the Japanese Government to CRIEPI.

We were encouraged and supported by many people including Mr. Kiyoshi Sato, General Manager of Public Communications Group, Mr. Shigeya Nozaki, senior staff member of the Center of Intellectual Property & Technology Licensing, and Ms. Motoyo Yamamoto and Ms. Miki Takano, Backend Research Center, Civil Engineering Laboratory of CRIEPI. Special thanks to Ms. Ryuko Takeda and Ms. Kumiko Saegusa for their continuous support. Without the support of those people, this book has not been published. We heartily appreciate their support.

Finally, Mr. Takashi Osada, the president of ERC publisher kindly advised and led the editing this book. On top of that, Ms. Megumi Sawatari rapidly, precisely, and patiently performed the editing work. We deeply appreciate their work.

Table of contents - Basis of Spent Nuclear Fuel Storage

CHAPTER 1 CHARACTERISTICS OF SPENT FUEL AND DEMAND OF STORAGE

1.1 What is spent fuel?

Spent fuel means nuclear fuel generated after nuclear power generation. According to the definition of the Safety Glossary (2007) of the International Atomic Energy Agency (IAEA), it means nuclear fuel which cannot be used anymore as it is after being irradiated in a nuclear reactor because a fissile product is consumed, a reactive poison is accumulated, and radiation damage is caused.

Uranium fuel for nuclear power generation is used in a form of a fuel assembly in which fuel rods are assembled in a lattice shape. This fuel assembly has two types: for a boiling water reactor (BWR) and for a pressurized water reactor (PWR). The both types of assemblies are formed by arranging fuel rods containing a number of fuel pellets in a lattice shape.

(1) Specification of fuel assembly

An example of specifications of main spent fuel assemblies for BWR and PWR is shown in Table 1.1 and 1.2[1-4].

Table 1.1 Example of specifications of main spent fuel assemblies (BWR)

<table>
<tr><td colspan="2">Item</td><td colspan="5">Specification</td></tr>
<tr><td colspan="2">Fuel type</td><td>8x8 fuel</td><td colspan="2">New-type 8x8 zirconium liner fuel</td><td>High-burnup 8x8 fuel</td><td>New-type 8x8 zirconium liner fuel</td></tr>
<tr><td rowspan="2">Shape</td><td>Maximum width: mm</td><td colspan="2">133</td><td colspan="2">134</td><td>133</td></tr>
<tr><td>Total length: mm</td><td colspan="5">4,470</td></tr>
<tr><td colspan="2">Weight: kg</td><td>280</td><td colspan="3">270</td><td>270</td></tr>
<tr><td colspan="2">Initial enrichment: wt%</td><td>2.8</td><td colspan="2">3.1</td><td>3.6</td><td>3.1</td></tr>
<tr><td rowspan="2">Burnup</td><td>Average: MWd/t</td><td>26,000</td><td>34,000</td><td>38,000</td><td>44,000</td><td>34,000</td></tr>
<tr><td>Maximum: MWd/t</td><td>29,000</td><td colspan="2">40,000</td><td>50,000</td><td>40,000</td></tr>
<tr><td colspan="2">Cooling period: year</td><td colspan="2">17</td><td colspan="2">8</td><td>17</td></tr>
<tr><td colspan="2">Heat generation kW/assembly</td><td colspan="2">0.143</td><td>0.282</td><td>0.177</td><td>0.177</td></tr>
<tr><td colspan="2">Name of transport/storage cask</td><td colspan="2">NEO-2569CB</td><td colspan="2">NEO-2552CB</td><td>HDP-69B</td></tr>
</table>

Table 1.2 Example of specifications of main spent fuel assemblies (PWR)

<table>
<tr><td colspan="2">Item</td><td colspan="2">Specification</td></tr>
<tr><td colspan="2">Fuel type</td><td>R4-6
17x17 fuel</td><td>R7
17x17 fuel</td></tr>
<tr><td rowspan="2">Shape</td><td>Maximum width: mm</td><td colspan="2">214</td></tr>
<tr><td>Total length: mm</td><td colspan="2">4,100</td></tr>
<tr><td colspan="2">Weight: kg</td><td colspan="2">680</td></tr>
<tr><td colspan="2">Initial enrichment: wt%</td><td>3.5</td><td>4.2</td></tr>
<tr><td rowspan="2">Burnup</td><td>Average: MWd/t</td><td>32,000</td><td>44,000</td></tr>
<tr><td>Maximum: MWd/t</td><td>36,000</td><td>48,000</td></tr>
<tr><td colspan="2">Cooling period: year</td><td colspan="2">15</td></tr>
<tr><td colspan="2">Heat generation kW/assembly</td><td>0.46</td><td>0.66</td></tr>
<tr><td colspan="2">Name of transport/storage cask</td><td colspan="2">MSF-26PJ</td></tr>
</table>

(2) Structure, shape and color of spent fuel assembly

The structure and shapes of spent fuel assemblies are the same as those of fresh fuel assemblies. Fig. 1.1 shows the structure and shape of a fresh fuel assembly. How does fuel change after being used for nuclear power generation?

There is no change in structure and shape of a fuel assembly in appearance after use. The color is light gray with metallic luster before use, and turns into dark gray with no metallic luster after use (Fig. 1.2). The following changes are caused to the composition, radioactivity, and a heat generation of fuel contained in fuel assemblies through the power generation.

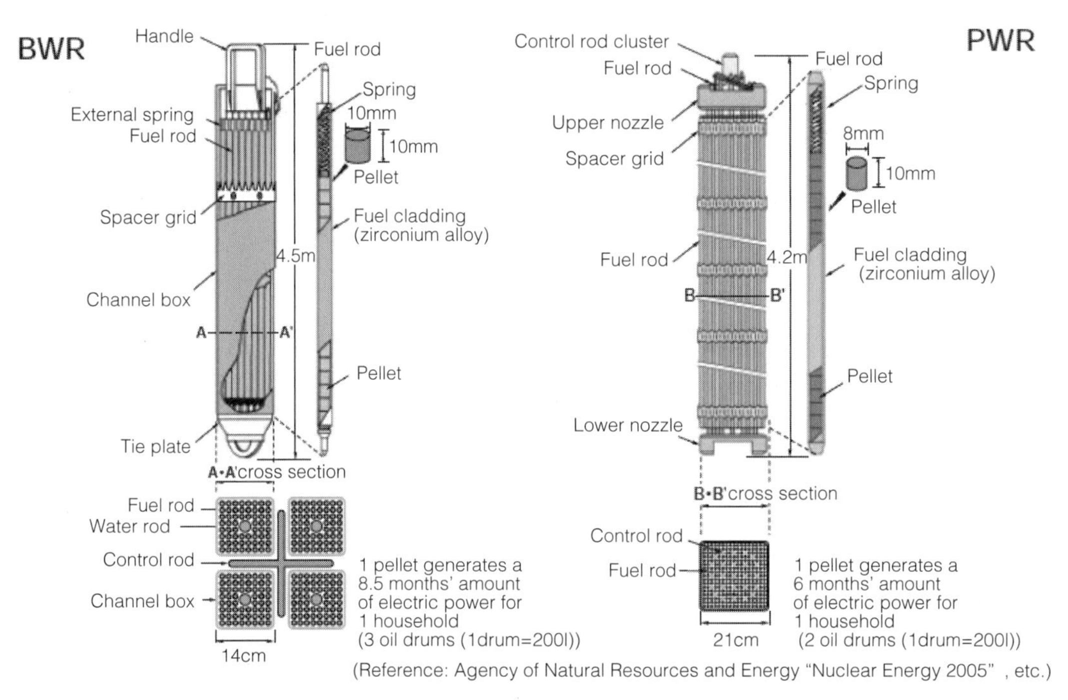

Fig. 1.1 Structure and shape of fuel assembly

Fig. 1.2 Change in appearance of fuel after nuclear power generation

(3) Inclusion of spent fuel

Fig. 1.3 shows a change in composition of uranium fuel due to nuclear power generation. After use, 3-6 % of a fission product and 1 % of plutonium are produced from uranium 235 and 238 that comprise fuel before use. 96 % of the total including the produced plutonium can be reused.

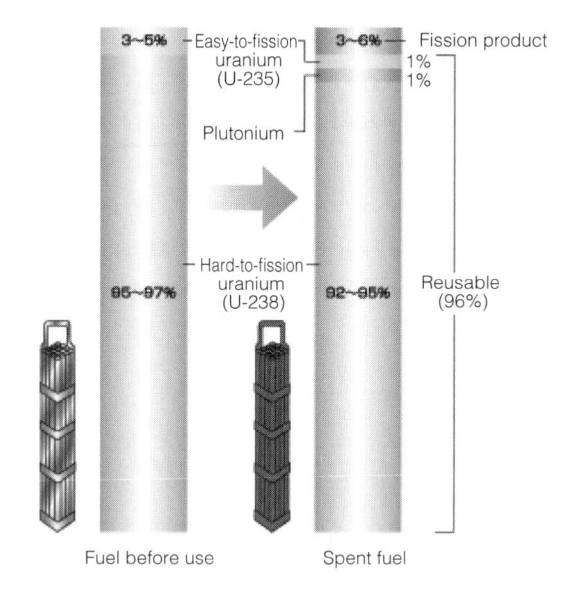

Fig. 1.3 Change in composition of uranium fuel due to nuclear power generation (example)

Nuclides largely affecting safe transport and storage of spent fuel include the following: tritium (^{3}H) and krypton (^{85}Kr) as gaseous fissionable nuclides, which largely affect sealing evaluation; iodine (^{131}I), cesium (^{237}Ce), and strontium (^{239}Sr) which largely affect shielding; cobalt (^{60}Co) produced due to radioactivation of the components of fuel assemblies; and fissionable uranium (^{235}U) and plutonium (^{239}Pu) that largely affect criticality prevention.

(4) Radioactivity and heat generation of spent fuel

Radiation of spent fuel is generated mostly from a "radioactive material" such as a fission product. There are the following types of radiation: alpha, beta, gamma, and neutron rays. The radioactive material turns into stable material (material not emitting radiation) as emitting radiation. The radiation decreases over time along with the change of the radioactive material in spent fuel. Heat is generated from spent fuel, and its quantity decreases over time as well as the radiation (Fig. 1.4).

As just described, spent fuel includes a fission product produced during power generation, so as to emit radiation and generate heat. Thus, sealing, shielding, and heat removal are required for safe storage. Furthermore, criticality prevention is required because spent fuel contains fissionable uranium as an ember and newly-produced plutonium. These four safety functions are required on spent fuel facilities, and the verification (safeguards) of the IAEA is also required on the facilities in order to secure the utilization of nuclear material only for peaceful purposes and no diversion of the material to nuclear weapons.

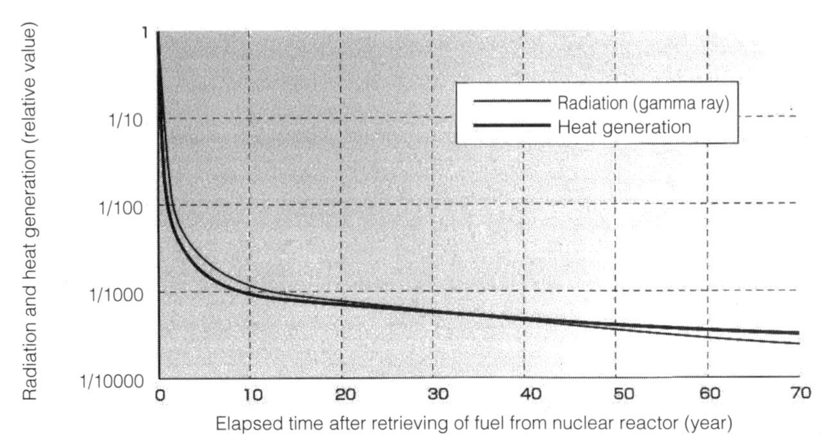

Fig. 1.4 Change in radiation and heat generation of spent fuel with time (example)

(5) High-burnup spent fuel and MOX spent fuel

High-burnup spent fuel and MOX spent fuel have such characteristics that the initial enrichment of U-235 and the concentration of Pu are comparatively high, and radioactive intensity and decay heat are high. The main specifications of these kinds of spent fuel and the calculation results of radioactive intensity and decay heat by the ORIGEN2 code (case of PWR) are shown in Table 1.3[5].

Table 1.3 Characteristics of high-burnup spent fuel and MOX spent fuel (PWR)

Fuel type	Conventional fuel 4.1 wt% U-235fuel (Max.48 GWd/t) (Avg. 40 GWd/t)	High-burnup fuel 4.7 wt% U-235fuel (Max. 55 GWd/t) (Avg. 49 GWd/t)	MOX spent fuel (Max.48 GWd/t) (Avg. 43 GWd/t)	Cooling period
Radioactive intensity (Ci/t)[a]	6.82×10^5 4.75×10^5	8.00×10^5 (1.17)[b] 5.64×10^5 (1.19)[b]	1.45×10^6 (2.13)[b] 1.06×10^6 (2.24)[b]	5 10
Decay heat (W/t)	2.16×10^3 1.38×10^3	2.69×10^3 (1.25)[b] 1.73×10^3 (1.26)[b]	5.09×10^3 (2.36)[b] 4.11×10^3 (2.99)[b]	5 10

[a] 1 curie (Ci) = 3.70×10^{10} Bq
[b] Figures in parentheses represent the ratio of fuel concerned to conventional fuel.

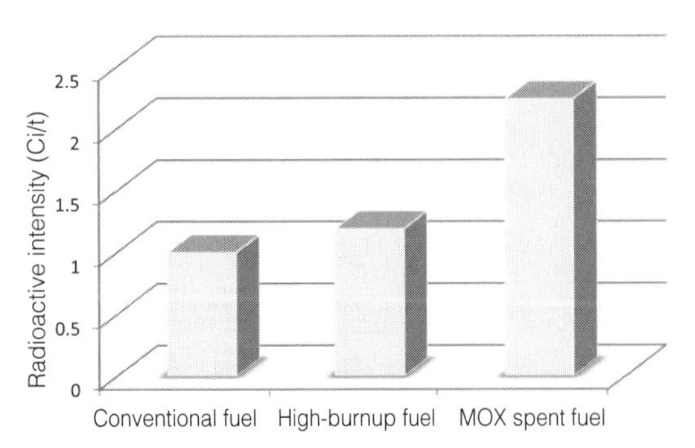

Fig. 1.5(a) Comparative example of radioactive intensity of high-burnup spent fuel and MOX spent fuel (10-year cooling period)

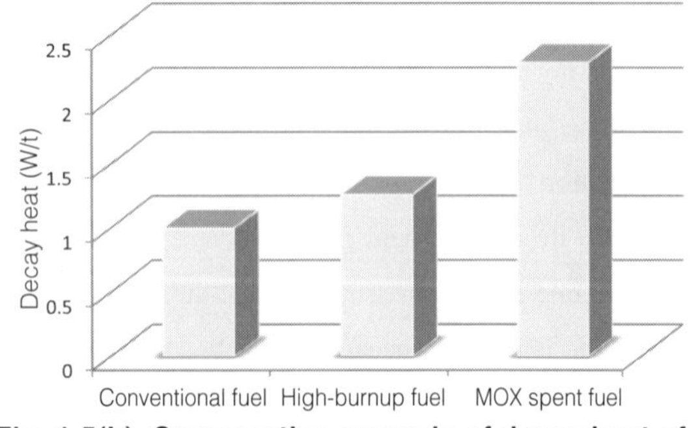

Fig. 1.5(b) Comparative example of decay heat of high-burnup spent fuel and MOX spent fuel (10-year cooling period)

A relative comparison between the radioactive intensity and decay heat of these kinds of spent fuel cooled for 10 years is shown in Fig. 1.5 (a), (b). These characteristics should be noted in regard to the design for subcriticality, shielding, and heat removal of transport/storage casks (fuel baskets, etc.) for high-burnup spent fuel and MOX spent fuel. Additionally, it should also be noted that long-term ageing is small in comparison with conventional fuel because these characteristics depend on the existence of long half-life nuclides.

1.2 Necessity of storage

In Japan, spent fuel is supposed to be stored for about 40-60 years until reprocessing, and there are safety review guidelines for the interim storage. Many discussions on the future of nuclear power generation were made after the Great East Japan Earthquake on March 11, 2011. The Japan Atomic Energy Commission set the proportion of nuclear power generation to the total power generation in four cases: 35, 20, 15, 0 % (an installed capacity will be 50, 30, 20, 0 GW for each case in 2030), and performed evaluation of the amount and economy of a nuclear fuel cycle[6]. The evaluation was performed on three scenarios: Scenario 1/Full reprocessing, Scenario 2/Coexistence of reprocessing and disposal, and Scenario 3/Full direct disposal. The following is the results.

(1) Evaluation scenario of Japan Atomic Energy Commission

1) Common items

- The total amount of spent fuel was about 17,000 tU as of the end of 2010 (as shown in Table 1.4[7]). The amount of spent fuel additionally generated until 2030 will be about 16,000 tU. Thus, the total will come up to 32,000 tU.
- The storage capacity of spent fuel pools in power plant sites was about 20,000 tU (as of 2010). In the case II of the nuclear power proportion (20 %), a management capacity (calculated by an expression: a management capacity = a storage capacity - (1 full core reserve + 1 refueling)) of spent fuel pools will gradually decrease because an installed capacity will decrease to 30,000,000 kw.
- The storage capacity of the Rokkasho Reprocessing Plant is 3,000 tU. The Recyclable-Fuel Storage Center in Mutsu (Mutsu RFS) under construction has a storage capacity of 5,000 tU.

Securing further storage capacity is an issue in the future, regardless of at-reactor or away-from-reactor storage.

2) Scenario 1 (Full reprocessing)

- The total amount of spent fuel will be about 19,000 tU when reprocessing is performed until 2030.
- An increase in storage capacity of spent fuel is necessary because there is a possibility that the storage capacity might run out depending on the operational status of the reprocessing plants.

3) Scenario 2 (Coexistence of reprocessing and disposal)

- The storage capacity and total amount of generated spent fuel are the same as Scenario 1.
- Mutsu RFS stores spent fuel to be reprocessed because it is a storage facility aimed at reprocessing.
- The increase in storage capacity of spent fuel is necessary because there is a possibility that the storage capacity might run out depending on the operational status of the reprocessing plants.

Table 1.4 Amount of spent fuel storage (ref. 2011 Report of the Japanese Government to the IAEA)

Plant, etc.		Storage amount (t)	Type of spent fuel
Japan Atomic Power Co.	Tokai No.2 Power Station	370	Uranium oxide fuel assembly
	Tsuruga Power Station	580	
Hokkaido Electric Power Co, Inc.	Tomari Power Plant	370	
Tohoku Electric Power Co, Inc.	Higashidori Nuclear Power Station	60	
	Onagawa Nuclear Power Station	420	
Tokyo Electric Power Co, Inc.	Fukushima Daiichi Nuclear Power Station	1,860	
	Fukushima Daini Nuclear Power Station	1,120	
	Kashiwazaki Kariwa Nuclear Power Station	2,270	
Chubu Electric Power Co, Inc.	Hamaoka Nuclear Power Station	1,140	
Hokuriku Electric Power Co, Inc.	Shiga Nuclear Power Plant	120	
Kansai Electric Power Co, Inc.	Mihama Nuclear Power Plant	370	
	Oi Nuclear Power Plant	1,370	
	Takahama Nuclear Power Plant	1,200	
Chugoku Electric Power Co, Inc.	Shimane Nuclear Power Plant	390	
Shikoku Electric Power Co, Inc.	Ikata Nuclear Power Plant	560	
Kyushu Electric Power Co, Inc.	Genkai Nuclear Power Station	840	
	Sendai Nuclear Power Station	850	
Japan Atomic Energy Agency	Reactor Decommissioning Research and Development Center	70	Uranium oxide fuel assembly, MOX fuel assembly
	FBR Research and Development Center	0	
	Reprocessing Plant of Nuclear Fuel Cycle Engineering Laboratories, Tokai Research and Development Center	41	Uranium oxide fuel assembly, MOX fuel assembly
	Nuclear Science Research Institute, Tokai Research and Development Center	18	Uranium oxide fuel assembly
	Oarai Researchand Development Center	16	Uranium oxide fuel assembly, MOX fuel assembly
Japan Nuclear Fuel, Inc.	Reprocessing Plant	2,834	Uranium oxide fuel assembly
Total		16,869	

4) Scenario 3 (Full direct disposal)

- The amount of spent fuel generated as waste until 2030 will be 32,000 tU, which exceeds a current storage capacity. Thus, the increase in storage capacity is a pressing issue.
- Mutsu RFS cannot be used because it is a storage facility aimed at reprocessing. The Rokkasho Reprocessing Plant cannot be used for storage.

Fig. 1.6 shows the analysis result of a storage amount of spent fuel in the case of the 20 % proportion of nuclear power generation.

In the case of Ⅱ-3/Full direct disposal, a storage amount of spent fuel keeps increasing in a linear manner; however, in the case of Ⅱ-1 and Ⅱ-2 that the Rokkasho Reprocessing Plant is operated, the increase can be suppressed because spent fuel is recycled.

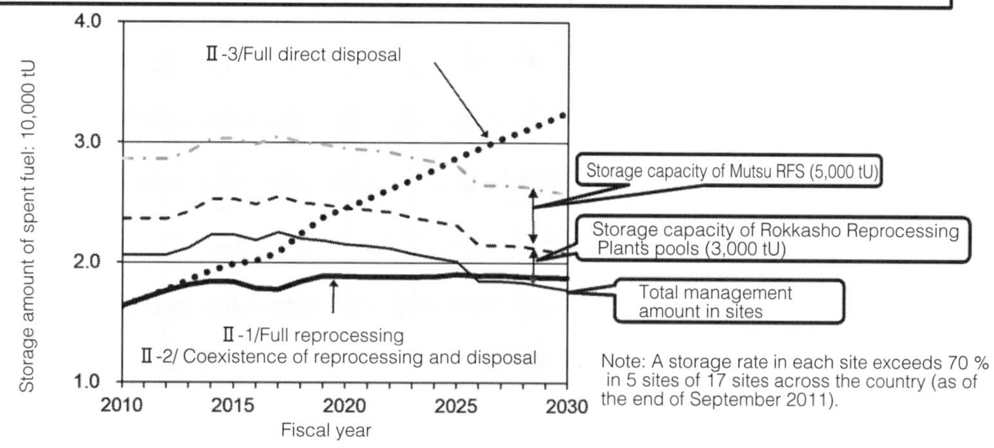

Fig. 1.6 Comparison of spent-fuel storage amount among scenarios

The increase in storage capacity is necessary in any scenario.

(2) Government policy principle

The Japan Atomic Energy Commission decided FY 2014 basic guidelines on a budget request for research, development, and utilization of nuclear [8]. The guidelines state that efforts to increase a capacity to store spent fuel outside reactor buildings in a manner of dry storage are required, and also cite the confirmation of long-term reliability of the dry storage technique of spent fuel as an effort for nuclear research and development that are of upmost importance at the moment.

On the other hand, the Fundamental Issues Subcommittee of the Advisory Committee for Energy of the Agency of Natural Resources and Energy discussed a future nuclear energy policy. The recent status of spent fuel storage in each nuclear power plant is shown in Table 1.5 [9]. Each nuclear power plant stores spent fuel in its spent fuel pools and dry casks. Spent fuel of about 14,000 tU was stored as of the end of September 2013 while the total storage capacity is about 20,000 tU. At present, a certain acceptable amount for storage is secured as a whole; however, some sites have only a small acceptable amount. It is considered that the improvement and enhancement of spent fuel storage measures are one of the important issues.

The Strategic Energy Plan decided by the Cabinet of the Japan on April 2014 described as follows in the section 4 of the chapter 3. “As spent fuels are sure to be produced through the use of nuclear energy, it is essential to implement measures to resolve this challenge as a responsibility of the current generation so that the burden is not passed on to future generations. Therefore, Japan will drastically reinforce and comprehensively promote efforts to resolve the challenge of how to manage and dispose of spent fuels. ... Japan currently stores about 17,000 tons of spent fuels. ... Spent fuels produced by nuclear power generation must be safely managed. It is therefore necessary to

Table 1.5 Status of spent fuel storage in each nuclear power plant (LWR)

(As of the end of March 2014)(Unit:tU)

Power plant		1 full core reserve	1 refueling (A)	Spent-fuel storage amount (B)	Management capacity (C)	Management allowance (C)–(B)	Time to exceed management capacity (year) (C)–(B)/(A)*12/16
Hokkaido	Tomari	170	50	400	1,020	620	16.5
Tohoku	Onagawa	260	60	420	790	370	8.2
	Higashidori	130	30	100	440	340	15.1
Tokyo	Fukushima Daiichi	580	140	1,960	2,100	–	–
	Fukushima Daini	520	120	1,120	1,360	–	–
	Kashiwazaki Kariwa	960	230	2,370	2,910	540	3.1
Chubu	Hamaoka	410	100	1,140	1,740	600	8.0
Hokuriku	Shiga	210	50	150	690	540	14.4
Kansai	Mihama	160	50	390	670	280	7.5
	Takahama	290	100	1,160	1,730	570	7.6
	Oi	360	110	1,420	2,020	600	7.3
Chugoku	Shimane	170	40	390	600	210	7.0
Shikoku	Ikata	170	50	610	940	330	8.8
Kyushu	Genkai	270	90	870	1,070	200	3.0
	Sendai	140	50	890	1,290	400	10.7
JAPC	Tsuruga	140	40	580	860	280	9.3
	Tokai No.2	130	30	370	440	70	3.1
Total		5,070	1,340	14,340	20,640	6,300	–

Note: A management capacity is calculated by the following expression in principle: a management capacity = a storage capacity - (1 full core reserve + 1 refueling). The management capacity of Hamaoka No.1 and No.2 reactors of Chubu Electric Company is equal to the storage capacity because of shutdown.

Note: The time to exceed a management capacity is obtained by a trial calculation in the case of assuming that all reactors in a plant are operated simultaneously, the replacement of fuel is performed every 16 months, and spent fuel is not transported to reprocessing plants. (Ref. the Agency of Natural Resources and Energy)

Ref.: The spent-fuel storage capacity of the Rokkasho Reprocessing Plant/2,945 tU (maximum storage capacity/ 3.000 tU); the spent-fuel storage capacity of Mutsu RFS/0 tU (maximum storage capacity/ 3.000 tU, schedule to start operation Oct. 2013 and to increase the maximum capacity to 5,000 tU)

expand the capacity for storing the spent fuels and is urgently important to broaden the range of choices for managing the spent fuels while ensuring safety. Based on this concept, the storage capacity of spent fuels will be expanded. Specifically, while studying a wide range of locations as possible sites, regardless of whether they are inside or outside the premises of a power plant, the Government of Japan will strengthen its effort for facilitating construction and utilization of new intermediate storage facilities and dry storage facilities."

References

1) OCL Corporation (March 2007): "Metal Dry Cask in Interim Storage Facility for Spent Fuel (NEO-2569CB)".

2) OCL Corporation (March 2007): "Metal Dry Cask in Interim Storage Facility for Spent Fuel (NEO-2552CB)".

3) Hitachi, Ltd. (March 2007): "Metal Dry Cask in Interim Storage Facility for Spent Fuel".

4) Mitsubishi Heavy Industries, Ltd. (March 2007): "Metal Dry Cask in Interim Storage Facility for Spent Fuel (MSF-26PJ)".

5) N.Takahashi, C.Ito, T.Matsumura, et al.: "Development of Advanced Spent Fuel Storage Technology at Reactor", Proc. Int'l Sympo.on Safety and Engineering Aspects of Spent Fuel Storage, IAEA and OECD, Vienna, 10-14 Oct. 1994.

6) Nuclear Power/Nuclear Fuel Cycle Technology Subcommittee of Cabinet Nuclear Energy Policy Office: "Evaluation of Step 3: Until 2030 (Case II of Nuclear Power Generation Proportion)" Document No.1-2 of the 11th meeting.
7) Government of Japan (October 2011): "Joint Convention on Safety of Spent Fuel Management and on Safety of Radioactive Waste Management" The Fourth National Report of Japan.
8) Atomic Energy Commission of Japan (July 17, 2013): "FY 2014 Basic Guidelines on Budget Request for Research, Development, and Utilization of Nuclear".
9) Fundamental Issues Subcommittee of Advisory Committee for Natural Resources and Energy (Oct 16, 2013): "Future Nuclear Energy Policy" Document No.1 of the 7th meeting.

CHAPTER 2 SAFETY STANDARDS AND CODES FOR SPENT FUEL STORAGE

The Nuclear Regulation Authority (NRA) was established in September 2012 in the light of the accident occurring at Tokyo Electric Power Company's Fukushima No.1 Nuclear Power Plant due to the Great East Japan Earthquake in March 2011. The NRA was established as an extra-ministerial bureau of the Ministry of the Environment, which is independent of nuclear-propelled administrative organizations so as to exercise its authority on a neutral and fair ground. In the Act for Establishment of the Nuclear Regulation Authority legislated in June 2012, the Act on the Regulation of Nuclear Source Material, Nuclear Fuel Material and Reactors (hereinafter referred to as the Nuclear Reactor Regulation Law), etc. was amended, so that the institution of safety regulations on power reactor facilities was reviewed.

In concrete terms, (i) the regulations based on the latest knowledge were enforced, (ii) the operators themselves have engaged in safety improvement, and (iii) the system of safety regulations on the power reactor facilities was amended for reorganization. The new safety standards for the power reactor facilities and the new regulatory standards for nuclear fuel facilities including spent fuel storage facilities were promulgated and enforced in July 2013 and December 2013, respectively. In the new safety standards for spent fuel storage, it is considered that the safety review guidelines of the former Nuclear Safety Commission and the technical requirements of the former Nuclear and Industrial Safety Agency will be basically followed. The following is an outline of these guidelines and requirements in relation to the safety researches mentioned in this book.

2.1 Metal cask storage

The new regulatory standards and guidelines for metal cask storage are outlined as follows.

2.1.1 New regulatory standards for spent fuel storage facilities

(1) Outline

The Nuclear Regulation Authority enforced a series of regulations, notices, and review standards related to the new regulatory standards in December 2013. The enforcement plan was presented at "The 33rd the Nuclear Regulatory Authority Meeting". (http://www.nsr.go.jp/disclosure/committee/kisei/h25fy/20131127.html)

It mainly includes the following with regard to the interim storage.

· Appendix (2/5) Regulations

"Regulations Concerning Spent Fuel Storage Activities" P.273-P.303

"Ministerial Ordinance on Technical Standards for Design and Construction Method of Spent Fuel Storage Facilities" P.303-P.311

"Regulations Concerning Standards for Location, Structure, and Equipment of Spent Fuel Storage Facilities" P.686-P.696

"Regulations Concerning Design, Method of Quality Control for Construction, and Technical Standards for Inspectors by Spent Fuel Storage Operators in Regard to Spent Fuel Storage Facilities" P.696-P.736

"Regulations Concerning Technical Standards for Performance of Spent Fuel Storage Facilities"

· Appendix (4/5) Review standards

"Interpretation of Regulation Concerning Standards for Location, Structure, and Equipment of Spent Fuel Storage Facilities" P.320-P.350

Here, "Framework Plan for New Regulatory Standards for Spent Fuel Storage Facilities (revision)" [1)]released by the NRA is outlined as follows. The constitution of contents is shown in Fig. 2.1. The details of requirements are clarified as bylaws.

1. General provisions: (1) Scope of application, (2) Definitions of terms, (3) Compliance standards and criteria 2. Basic safety functions: (1) Confinement function, (2) Radiation protection, (3) Criticality prevention, (4) Heat removal function 3. Radiation control/Environmental safety: (1) Radiation exposure control, (2) Disposal or storage of radioactive waste, (3) Radiation monitoring, (4) Consideration of ageing	4. Other safety measures : (1) Consideration of natural phenomena, (2) Consideration of external man induced events, (3) Consideration of fire and explosion, (4) Consideration of power loss, (5) Consideration of transport of metal casks, (6) Communication facilities, etc., (7) Consideration of common use, (8) Consideration of inspection, repair, etc. 5. Safety evaluation : (1) Evaluation during normal periods, (2)Evaluation at accidents

Fig. 2.1 Constitution of "Framework Plan for New Regulatory Standards for Spent Fuel Storage Facilities (revision)"

Compared with the safety review guidelines of the former Nuclear Safety Commission (next section 2.1.2), the heat removal only by natural convection was regulated in (4) Heat removal function in the second clause, and the protection design for accidental fall of aircraft, etc. (10^{-7} fall x facility/year or more) was documented in (2) Consideration for external man induced events in the fourth clause. Substantially, there is no major change.

The public examination was performed on this framework plan (revision). Comments and responses[2)]were made, so that a part of the framework plan was amended. It mainly includes the following.

a. "Storage period" and "Design storage period"

"Storage period" in a definition means an in-service period of metal casks used in Japan and overseas. "Design storage period" means "the longest period of time when a metal cask is assumed to be stored in a spent fuel storage facility at the time of designing" as stated in 1(2) Definition of terms in this framework plan, and differs from the "storage period" in concept. A concrete storage period itself is not required as a regulation.

b. "Design basis accident" and " Design evaluation accident"

As to the "Design basis accident" in a definition, the evaluation of validity of design for the escalation prevention and the impact mitigation function at the time of accidents is required. On the other hand, in the spent fuel storage facilities, the possibility of the occurrence of the accidents is considered to be quite low because the spent fuel is stored in the metal casks which are strong transport containers and also dynamic components for cooling the fuel are not required. However, the accidents assumed to technically occur are considered here. The one releasing the radiation with the largest radiological dose to general person among such accidents is evaluated as the "design evaluation accident", and the confirmation that there is no risk of significant radiation exposure to the public is required.

c. "2 (1) Confinement function" Spent fuel and adherent radioactive materials are stated as containment objects.

d. "2 (3) Criticality prevention" Cask sliding due to earthquake is not excluded.

e. "4 (2) Consideration for external man induced events" The spent fuel storage facilities should be designed not to impair the safety against postulated accidental externally man-caused events. It is considered that the measures for severe accidents such as intentional aircraft collisions with nuclear power plants, etc. are not needed for the spent fuel facilities because the spent fuel is stored in the metal casks which are the strong transport containers and also the dynamic components for cooling the fuel are not required.

Meanwhile, the measures for armed attacks are supposed to be taken under the law concerning measures to protect the people in the case of the armed attack, etc.

(2) Comparison with international standards

As to the spent fuel storage facilities, the items required in the General Safety Requirements GSR Part5 "Predisposal Management of Radioactive Waste" by IAEA and the Specific Safety Guide SSG-15 "Storage of Spent Nuclear Fuel" by IAEA were incorporated. Especially, the heat removal function by natural convection is required in reference to the regulatory standards of Germany, "Safety Guidelines for Dry Interim Storage of Irradiated Fuel Assemblies in Storage Casks".

2.1.2 Safety review guidelines

The former Nuclear Safety Commission enacted two safety review guidelines as follows.

- "Dry Cask Storage of Spent Fuel in Nuclear Power Plants" (approved by the Nuclear Safety Commission on August 27, 1992, partially revised in 2001 and 2006)
- "Safety Review Guidelines for Spent Fuel Interim Storage Facilities Using Metal Dry Casks" (decided by the Nuclear Safety Commission on October 3, 2002, partially revised in 2006 and 2010)

(1) "Dry Cask Storage of Spent Fuel in Nuclear Power Plants" (at reactor storage)

The constitution of this document is shown in Fig. 2.2.

1. Introduction

2. Definitions of terms

3. Basic concept of safety design
   (1) Safety functions
   (2) Structural strength
   (3) Management/Operation

4. Items to be confirmed in safety functions and safety review of dry cask storage facilities
   4.1 Items pertaining to safety design
      (1) Heat removal
      (2) Containment
      (3) Shielding
      (4) Criticality prevention
      (5) Structural strength
   4.2 Items pertaining to safety review
      (1) Abnormal events to be assumed in safety review
      (2) Safety review standards

5. Conformance to guidelines

6. Summary

Fig. 2.2 Constitution of "Dry Cask Storage of Spent Fuel in Nuclear Power Plants"

The following is main contents of this document. The titles of items intimately related to this textbook are enclosed in boxes, and the notable contents are underlined.

1) Scope

The guidelines examined items to be confirmed in the safety review in the case of storing spent fuel in dry cask storage facilities in nuclear power plants, and were actually used for the safety review of the dry cask storage in the Fukushima No.1 Nuclear Power Plant of Tokyo Electric Power Company and the Tokai No.2 Power Station of Japan Atomic Power Company. The knowledge of qualification tests conducted by Central Research Institute of Electric Power Industry, etc. was consulted during the establishment of the guidelines. Each guideline item includes interpretation and concepts.

2) Basic concept of safety design

· The four functions are specified as safety functions of the dry cask storage of spent fuel: a heat removal function, a containment function, a shielding function, and a criticality prevention function.
· Structural strength design for maintaining these safety functions is required.
. Design for transport is required because dry storage casks are transported in the power plants.
· Sound spent fuel is loaded.
· When transported to the outside of the power plants after storage, the spent fuel is repacked in transport casks.
· On the grounds that the storage facilities have not been practically operated in Japan yet, it is desired that sampling inspection in regard to the containment performance of seal of metal gaskets, etc. and the integrity of spent fuel cladding is conducted at the appropriate time during the storage period by reactor establishers.

3) Safety function of dry cask storage facilities and items to be confirmed in safety review

. Reliable heat transfer analysis codes shall be used for the confirmation of the heat removal function.
· The fuel cladding shall be set at the temperature of allowing its accumulative creep to be 1 % or less.
· The containment function shall be designed to maintain a negative pressure inside casks during a design storage period.
· As shielding standards, a dose rate on a cask surface shall be 2mSv/h or less, and a dose rate at 1m from the surface shall be 100μSv/h or less. An air dose rate outside a property boundary shall be an air kerma of 50μGy/y or less.
· In criticality prevention, the depletion of neutron absorbers during the design storage period shall be considered. A calculation result of a neutron effective multiplication constant shall not exceed 0.95 even if calculation errors occur.
· In structural strength, low temperature brittleness at the lowest service temperature of a cask shall be considered. Structural strength evaluation shall be performed on baskets from the point of view of maintaining the criticality prevention function.
· In seismic design classification, dry storage casks and support structures shall be classified as S Class. Storage

buildings store the S-class dry storage casks and serve as support structures during the time of storage, so that it shall be confirmed that there is no problem for safety against basic earthquake ground motion Ss. The storage buildings shall be classified as C Class.

· In safety evaluation, the abnormal temperature rise at each part of dry storage casks may be prevented at assumed abnormal events. A required leakage rate may be maintained, which shall be 10mSv/h or less at 1m above a cask surface. A neutron effective multiplication constant shall not exceed 0.95 even if the calculation errors occur.

(2) "Safety Review Guidelines for Spent Fuel Interim Storage Facilities Using Metal Dry Casks" (away from reactor)

The constitution of the guidelines is shown in Fig. 2.3.

Introduction I.Scope II.Definitions of terms III.Site conditions (Guideline 1-3) IV.Radiation control (Guideline 4-6)	V. Environmental safety, etc. (Guideline 7-9) VI.Criticality safety (Guideline 10-12) VII.Other safety measures (Guideline 13-21) (Interpretation)

Fig. 2.3 Constitution of "Safety Review Guidelines for Spent Fuel Interim Storage Facilities Using Metal Dry Casks"

The following is main contents of the guidelines. The titles of items intimately related to this book are enclosed in boxes, and the notable contents are underlined.

Introduction

· A period of spent fuel storage is 40 to 60 years.

I . Scope

· The guidelines shall be applied to spent fuel interim storage facilities that are located independently of nuclear power plants, into which spent fuel assemblies are carried by metal dry casks as transport containers used for transport out of the power plants, where the assemblies are stored without being repacked in other containers, and out of which the assemblies are carried after the storage.

· Lids etc. of the metal dry casks shall not be opened for inspecting stored spent fuel assemblies, etc. during and after a storage period in the spent fuel interim storage facilities.

· Sound uranium dioxide fuel and MOX fuel shall be stored.

III. Site conditions (Guideline 1-3)

· The natural events such as earthquakes, tsunamis, etc. and social environment-related events such as a swoop of fragments due to an aircraft accident, etc. shall be examined, and it shall be confirmed that there is no interference with safety.

· On the basis of ageing of structural members of metal casks in association with long-term storage, excessive exposure to the general public shall not be caused at the maximum credible accidents. However, technically-inconceivable accidents shall not be assumed here.

IV. Radiation control (Guideline 4-6)

· For possible abnormality of the containment function, repairability such as additional installation of lids, etc. shall be considered.

V. Environmental safety, etc. (Guideline 7-9)

· Ageing in association with long-term storage shall be considered. The accumulation of knowledge and pre-shipment inspection after interim storage will be described hereinafter.
· Metal casks shall be stored in such a state that spent fuel assemblies are enclosed (loaded) together with inert gas.

VI. Criticality safety (Guideline 10-12)

· In the case of using neutron absorbers in baskets, their concentration, non-homogeneity, depletion during a design storage period, and the like shall be reasonably assumed.

VII. Other safety measures (Guideline 13-21)

· Guideline 13 "Consideration for earthquakes" does not require " design for preventing metal casks from tip over" that is required by after-mentioned technical requirements. Instead, the design for maintaining the basic safety functions relative to design seismic force is required. In extreme terms, it is estimated that the verification that the basic safety functions are maintained even if the casks tip over is only necessary. If this verification is difficult, the design for preventing the casks from tip over (allowing certain sliding or fixing the casks) is required.
· Facility design, material selection, manufacture, construction, and inspection important to safety are performed on the basis of the standards and criteria in Japan, which are admitted to be proper (at present, the Atomic Energy Society of Japan Standards, the Japan Society of Mechanical Engineers Codes, and the Japan Electric Association Code are being evaluated by the government, and are now used).

The documents (written decision by the Nuclear Safety Commission, on October 3, 2002) titled "Safety Review Guidelines for Spent Fuel Interim Storage Facilities Using Metal Dry Casks" and "Long-Term integrity of Metal Dry Casks and Spent Fuel in Spent Fuel Interim Storage Facilities" are annexed to the guidelines. Here, the operators and the administrative agencies are required to correspond as follows.

1. Operators shall continuously conduct the investigation on a state of dry storage in nuclear power plants and accumulate the knowledge about long-term integrity, in terms of the confirmation of the integrity of metal casks and spent fuel for the post-storage transportation.
2. Administrative agencies shall establish rational inspection methods considering the characteristics of interim storage facilities as pre- shipment inspection for securing the safety in regard to the transport of metal casks and spent fuel after interim storage, on the basis of the clause 1.

Until now, in reaction to the clause 1, the operators have conducted and reported the circumstantial investigation in the dry cask storage facilities in the Fukushima No.1 Nuclear Power Plant of Tokyo Electric Power Company and the Tokai No.2 Power Station of Japan Atomic Power Company. In reaction to the clause 2, the former Nuclear and Industrial Safety Agency proposed comprehensive methods as the rational methods, which are detailed in "Long-Term integrity of Metal Dry Casks and Spent Fuel in Spent Fuel Interim Storage Facilities Using Metal Dry Casks" (Nuclear Fuel Cycle Safety Subcommittee, Interim Storage WG and Transport WG, June 25, 2009).

http://www.meti.go.jp/report/downloadfiles/g90924a01j.pdf

2.1.3 Technical requirements (former Nuclear and Industrial Safety Agency)

The former Nuclear and Industrial Safety Agency compiled the items considered to be technically important to safety review in regard to the license for spent fuel storage activities using the metal casks, on the basis of "Technical Examination Report Concerning Spent Fuel Storage Facilities (Interim Storage Facilities) Using Concrete Casks" compiled in June 2004. Then, "Technical Requirements Concerning Spent Fuel Storage Facilities (Interim Storage Facilities) Using Metal Casks" was established. The constitution of the technical requirements is shown in Fig. 2.4.

I. Roles and scope of the technical requirements II. Definitions of terms III. Site conditions (Requirements 1-3) IV. Basic safety functions (Requirements 4-9) V. Radiation control and environmental safety (Requirements 8-10)	VI. Other safety measures (Requirements 11-21) Matters to be considered(with respect to Requirements 2,3,6,7,11,12,17)

Fig. 2.4 Constitution of "Technical Requirements Concerning Spent Fuel Storage Facilities (Interim Storage Facilities) Using Metal Dry Casks"

The following is main contents of the technical requirements.

- " Technical Requirements Concerning Spent Fuel Storage Facilities (Interim Storage Facilities) Using Metal Casks" (hereinafter referred to as "technical requirements") had approximately the same contents as "Safety Review Guidelines for Spent Fuel Interim Storage Facilities Using Metal Dry Casks" (hereinafter referred to as "safety review guidelines"); however, there was some difference in a part of the contents related to earthquake resistance. Then, the Nuclear Safety Commission conducted the examination on the clarification of how to apply the "safety review guidelines" in December 2008, and newly established a document clarifying its concrete application. As a result, the reorganization was performed by using the latest knowledge in a more fulfilling manner. For this reason, the former Nuclear and Industrial Safety Agency established the "safety review guidelines" and a series of documents established as the clarification of the application methods as new review standards for the interim storage facilities.
- On the other hand, the "technical requirements" which had been set as screening standards until then was excluded from the review standards established by the Minister of Economy, Trade and Industry because their items were included in the "safety review guidelines", etc. Ultimately, although the "technical requirements" were excluded from the review standards established by the minister, they have existed as the bylaw of the former Nuclear and Industrial Safety Agency since their enactment. They can be read through the following address.

http://www.nsr.go.jp/archive/nisa/oshirase/2006/files/181011-4.pdf

· Among them, the contents which were changed compared with the revision of the earthquake resistant design review guidelines are indicated as followed.

http://www.nsr.go.jp/archive/nisa/oshirase/2006/files/181011-2.pdf

"Consideration for earthquakes (Technical requirement 17)" requires "design for not allowing metal casks during storage to fall by assumed seismic force", which is not required by the guideline 13 of the safety review guidelines. Here, it is interpreted as meaning that the fixing is not necessarily required and the sliding is permissible. However, the interference of the casks due to large sliding requires the evaluation.

2.1.4 Safety design standards (Atomic Energy Society of Japan)

The Atomic Energy Society of Japan (AESJ) legislated and revised "Standards for Safety Design and Inspection of Metal Casks for Spent Fuel Interim Storage Facilities" in preparation for the start of the operation of the interim storage facilities of spent fuel. Until now, the standards were issued in June 2002, and revised in January 2004, May 2008, and July 2010.

The constitution of the standards is shown in Fig. 2.5.

1. Scope of application
2. Reference standards
3. Terms and definitions
4. Safety design of metal casks
   4.1 Basic requirements
      4.1.1 Storage conditions for metal casks
      4.1.2 Conditions for spent fuel to be stored in metal casks
      4.1.3 Consideration for double use for transport and storage
      4.1.4 Consideration for ageing
   4.2 Method of safety design
      4.2.1 Basic specifications
      4.2.2 Containment design
      4.2.3 Shielding design
      4.2.4 Criticality prevention design
      4.2.5 Heat removal design
      4.2.6 Structural strength design
5. Inspection on metal casks
   5.1 Inspection steps and inspection items
   5.2 Inspection guidelines

Annex A-V
Interpretation

Fig. 2.5 Constitution of "Standards for Safety Design and Inspection of Metal Casks for Spent Fuel Interim Storage Facilities"

The following is main contents of the standards. The titles of items intimately related to this book are enclosed in boxes, and the notable contents are underlined. In what follows, the numbers of sections and divisions correspond to those of the AESJ standards.

4.1.1 Storage conditions for metal casks

· Metal casks shall be stored in buildings

· Natural cooling shall be adopted.

· Evaluation shall be performed when the occurrence of accidents exceeding abnormal events is assumed, and metal casks shall be stored in facilities equipped in a manner to fulfill the following conditions even when damage is assumed.

- Radioactive material shall not be released from metal casks.
- Spent fuel cladding shall maintain integrity.

- Metal casks can be transported outside the facilities as packages by taking required repair measures.

· Spent fuel shall not be repacked in interim storage facilities.

4.1.2 Conditions for spent fuel to be stored in metal casks

· Spent fuel which is generated in BWR or PWR and fulfills the following:

  - BWR spent fuel – Burnup of 50,000 MWd/t or less
  - PWR spent fuel – Burnup of 48,000 MWd/t or less

· Spent fuel whose cladding is confirmed as sound

· Spent fuel cooled in spent fuel pools, etc. for one year or more

· The securing of integrity of fuel cladding under the conditions during transport and storage may be evaluated with accumulated data.

Meanwhile, the spent fuel is antecedently stored in the power plants or other facilities in Japan and overseas until when it can be confirmed that it is not systematically damaged due to the occurrence of unpredictable deterioration phenomena.

4.1.3 Consideration for double use for transport and storage

· Meeting the technical standards prescribed by the transport law

· Designed according to AESJ-SC-F006:2006 (standards for design and inspection of transport containers)

· Designed to be able to maintain the integrity of spent fuel by maintaining the basic safety functions even during storage

4.1.4 Consideration for ageing

· Materials and structure shall be selected in consideration of aged deterioration, and anticorrosion and usage environmental mitigation measures shall be provided. Metal casks shall be designed to reduce or prevent the influence on spent fuel by maintaining inert environment in their inside and controlling temperature environment.

· In structural members, it shall be antecedently confirmed that functions required for the design can be maintained over a design evaluation period in consideration of the above.

4.2.2 Containment design

· Internal environment of metal casks

  - Inert gas shall be filled, and its purity shall be managed.
  - Residual water shall be 10 % (mass) or less.
  - Negative pressure shall be maintained except under special test conditions prescribed by the transport regulations.

· Containment structure during transport (Fig. 2.6)

  - A secondary lid and a tertiary lid shall be arranged.

· Containment structure during storage

  - Double structure is constituted by a primary lid and a secondary lid, and a pressure barrier by inter-lid pressure

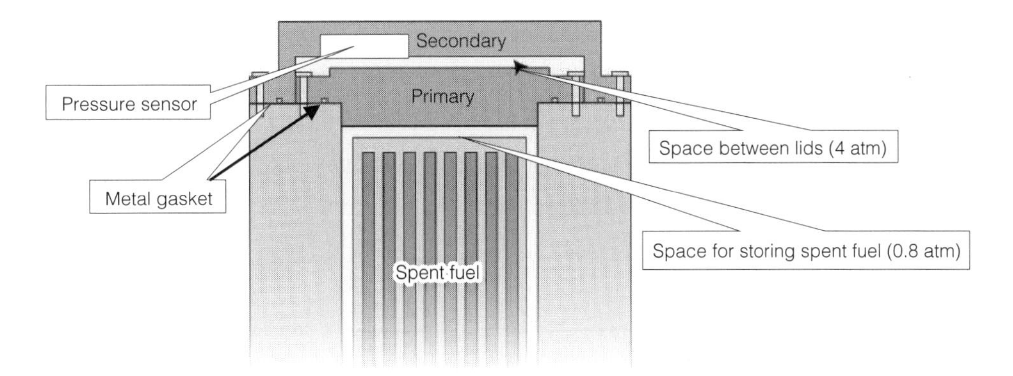

Fig. 2.6 Containment structure of lids of metal cask (forged steel type)

shall be adopted.

· Monitoring of containment structure during storage
  - Pressure sensors shall be installed for inter-lid pressure monitoring.
· Monitoring of the heat removal function during storage
  - Periodic measurement of temperature shall be performed at representative points on the surface of the center part of a metal cask body.
· Usage of metal gaskets
  - Double-lid structure shall be adopted, and containment monitoring shall be conducted.
  - Negative pressure shall be maintained during normal storage and at abnormal events.
  - Positive pressure shall be applied between the double lid, and pressure walls for preventing the release of radioactive material shall be arranged.
· Design requirements
  - Metal gaskets used for storage shall maintain the containment performance which can withstand the vibration and external force generated during normal transport before storage and under general test conditions (excluding free drop) even during storage.
  - At the time of transport after storage, a metal gasket of a second lid shall be replaced with a new gasket or a rubber O ring, or a tertiary lid using the rubber O ring shall be used.
· Design standards: Reference leakage rate > Leakage rate of a metal gasket

4.2.5 Heat removal design

· Design requirements
  - Structure for removing decay heat so as to allow members having the basic safety functions and structural strength and spent fuel to maintain integrity shall be required.
· Design standards
  - Spent fuel cladding shall be set at equal to or under the temperature of maintaining integrity during handling and

normal transport, under general test conditions, during normal storage, and at abnormal events and accidents.

- The lowest temperature among the following shall be set as restrictive temperature of spent fuel cladding: the initial temperature of allowing accumulative creep to be 1 %; the temperature of generating the recovery of irradiation hardening equivalent to cladding strength as design standards; and the temperature of not allowing deterioration in mechanical property due to hydride reorientation to occur.

4.2.6 Structural strength design

· Design requirements

- Metal casks shall be designed not to damage spent fuel cladding during normal transport, under general test conditions, during storage, and at abnormal events and accidents.

· Design standards

- Stress generated in spent fuel cladding shall be within the elastic region during normal transport, under general test conditions, during storage, and at abnormal events and accidents.

5. Inspection on metal casks

5.1 Inspection steps and inspection items

· Inspection during a storage period

- In spent fuel, its heat power and radiation intensity gradually decrease as the storage period elapses. The safety design in consideration of this ageing is performed. The basis of the inspection is the confirmation of maintenance of the basic safety functions by both the appearance inspection and the confirmation of records of surface temperature monitoring and inter-double lid pressure monitoring.

- Evaluation of the failure probability during the storage period based on FMEA

The evaluation by "Failure Mode and Effect Analysis (FMEA)" is effective in the setting of items, contents, and frequencies of periodic inspection of various installations and components, etc. The examples that the items and frequencies of the inspection to be periodically performed on the metal casks were evaluated by comparing failure occurrence between in the transport containers and in the metal casks by applying FMEA are provided.

· Pre-shipment inspection after storage

- It is an inspection of confirming that the packages conform to the transport regulations, and it can be mostly conducted in the same way as pre-shipment inspection in the power plants. As to subcriticality inspection and spent fuel inspection of the pre-shipment after storage inspection, the lids of the transport containers are usually opened and the integrity of the baskets and the spent fuel is visually confirmed in the power plants. However, the visual inspection is not conducted in the interim storage facilities which do not have fuel repacking equipment. Also in this case, the integrity can be confirmed by the above-mentioned holistic approach proposed by the former Nuclear and Industrial Safety Agency. (See "Long-Term integrity of Metal Dry Casks and Spent Fuel in Spent Fuel Interim Storage Facilities Using Metal Dry Casks" (Nuclear Fuel Cycle Safety Subcommittee, Interim Storage WG and Transport WG, June 25, 2009))

The standards have not been endorsed by the government yet as of May 2013. However, in the technical evaluation by the former Nuclear and Industrial Safety Agency (http://www.nsr.go.jp/archive/nisa/shingikai/106/5/2/003/3-1.pdf), additional requirements were required for "Primary lid seal, secondary lid seal and through hole seal of each lid", "Influence on cladding strength due to hydrogen absorption", and "Recovery of irradiation hardening", etc., so that attention is needed to be paid.

2.1.5 Structural codes (Japan Society of Mechanical Engineers)

The Japan Society of Mechanical Engineers established "Codes for Spent fuel Storage Facilities: Rules on Transport/Storage Packagings for Spent Nuclear Fuel" on the basis of concepts of the safety design and inspection of the metal casks, which were established by the Atomic Energy Society of Japan. Until now, the codes were issued in 2001 and revised in 2007. The 2013 revision is in progress now.

(1) Constitution of codes

The constitution of the codes is shown in Fig. 2.7. The interpretation is attached to each section.

Chapter I General provisions
- Section 1 Scope of application
- Section 2 Definitions

Chapter II Mechanical test and nondestructive test
- Section 1 Mechanical test
- Section 2 Nondestructive test

Chapter III Metal cask
- Section 1 Scope of application
- Section 2 Definitions
- Section 3 Material (shielding container, basket, trunnion, intermediate shell)
- Section 4 Design (same as above)
- Section 5 Manufacture
- Section 6 Inspection

Chapter III Metal cask (MANDATORY APPENDIX)
- APPENDIX 3-1 to 3-6

Appended tables and appended figures
- 1. Appended tables
  - 1.1 Material used for metal casks
  - 1.2 Design stress intensity, allowable tensile stress, design yield point, etc.
- 2. Appended figures

Interpretation

Fig. 2.7 Constitution of "Codes for Spent Fuel Storage Facilities: Rules on Transport/Storage Packagings for Spent Nuclear Fuel"

(2) Scope of application

The roles prescribed by the codes are applied to the metal casks used for the storage in the spent fuel storage facilities and the transport to the outside of the facilities, among the equipment of the facilities. The following items (i) to (iv) related to the structural requirements that are required for securing the basic safety functions (containment, shielding, criticality prevention, and heat removal) defined in "Standards for Safety Design and Inspection of Metal Casks for Spent Fuel Interim Storage Facilities" by the Atomic Energy Society of Japan (hereinafter referred to as "AESJ standards") are prescribed:

(i) material, (ii) design, (iii) manufacture (including molding), (iv) inspection

Meanwhile, the codes are applicable to the metal casks used in the spent fuel storage facilities included in the nuclear facilities. As the examples of design concepts of the metal casks, there are the following types: a forged cask of steel-water shield type, a forged cask of steel-resin shield type, a multilayer cask of lead shield type, and a ductile cast iron- polyethylene shield type.

(3) Basic concepts

1) States to be evaluated and evaluation standards

The evaluation by uniform standards should not be performed on various states of transport and storage. The evaluation standards should be set according to the proneness of occurrence of the states and the degree of structural integrity to be maintained. The classification of the states to be evaluated is shown in Table 2.1

In the evaluation standards for stress generated in each member of the metal casks, strict stress limits were set for the states relatively easy to occur, and the limits corresponding to failure limits were set for the rarely-occurring states. Moderate allowance is included due to the uncertainty of evaluation methods and modeling and the variation in material strength.

Specifically, in the states relatively easy to occur , such as Design Event I, Design Event II, and General Test Condition, etc., initiation stress was limited in the range of allowing elastic limits (yield points) to have such allowance that plastic deformation and progressive deformation are not generated, because it is required that the components can be continuously used. In the states that rarely occur like Design Event IV and Specific Test Condition but on that strict load conditions are imposed, the initiation stress was limited in the range of allowing ultimate strength (tensile strength) to have such allowance that the components are not damaged.

Table 2.1 Classification of states to be evaluated in structural codes of metal casks

Stage of work	Classification	State
Storage and handling	Design Event I	Normal states of storage and handling
	Design Event II	Events caused by a single failure or a single operation
	Design Event IV	Severe events assumed for confirming safety
Transport	Normal Transport	Normal states of transport
	General Test Condition	Minor events, such as a single failure, which are assumed in the transport regulations
	Special Test Condition	Severe events which are assumed for confirming safety and which are assumed in the transport regulations

2) Safety functions of each component and corresponding classification

In the rules on the metal casks, it was considered that the components correspond to the prescribed classification of nuclear equipment as follows, and the evaluation standards were set.

a. Containment vessel- Containment boundaries are made for containing the spent fuel. The vessel corresponds to the class 4 container in Notice No.501 (similar to ASME Code Section III, Division 1) as a storage container, and to the class 1 container as a transport container. However, even at Design Event IV of the storage or under Specific Test Condition of the transport, the generated stress was limited to yield points or under so as not to make the plastic deformation generated in containment seal parts and containment lid closure bolts.

b. Basket- It directly holds the spent fuel assemblies, has the criticality prevention function, and corresponds to the core support structure in Notice No. 501. However, even at Design Event IV of the storage or under Specific Test Condition of the transport, amount of deformation is calculated in the case of generating the plastic deformation, and used for criticality prevention evaluation.

c. Trunnion- It is used for suspending the metal cask and fixing it to a transport frame, and corresponds to the class 1 support structure in Notice No.501.

d. Intermediate shell- In the case of the multilayer metal cask, it is a cylindrical structure reinforcing/supporting the containment vessel as a structure strength member of a shell part, and corresponds to the class 1 support structure in Notice No.501.

(4) Design method

Various loads are applied to the metal casks at various events such as General Test Condition and Special Test Condition prescribed in the transport regulations. "Design by Formula" means that the shapes of the members and the standardized formulas corresponding to the state where these loads are applied are set so that structural design is performed with required board thickness, etc. However, it is difficult to adopt the "Design by Formula", and the adaption limits the degree of freedom of design. Thus, "Design by Analysis" that the stress generated in each member is calculated so as to be compared with permissible levels is adopted, so that rational design is performed. The "Design by Analysis" is such a method that every possible fracture mode is considered, the integrity of a structure is evaluated in detail by the analysis on each fracture mode, and the conformance to the standards is confirmed. This method is the same as that applied to the class 1 container, the core support structure, and the class 1 support structure in Notice No.501.

2.1.6 Building substructure guidelines (Japan Electric Association)

The Subcommittee on Seismic Design of the Japan Electric Association had examined the technical details of "Research Report on Substructure Technique for Nuclear Facilities" (complied in 1999) since 2001, and released the results of design of pile foundations by issuing "Technical Guide for Substructure Design of Dry Cask Storage Buildings, JEAG 4616-2003" (hereinafter referred to as "JEAG 4616"). In association with the revision of "Regulatory Guide for Reviewing Seismic Design of Nuclear Power Reactor Facilities" in 2006 (decided by the Nuclear Safety Commission on September, 19, 2006), the technical code ("Technical Code for Seismic Design of Building Foundation of Spent Nuclear Fuel Interim Storage Using Dry Casks", JEAC 4616) was compiled in 2009 by adding the reflections of the revision and the design of improved ground.

The constitution of the guidelines is shown in Fig. 2.8. The interpretation is attached to each section and annex.

Chapter I General provisions	Chapter II Design of pile foundation
Section 1 Scope of application Section 2 Definitions and abbreviations of terms Section 3 Classification of importance of interim storage buildings in aseismic design Section 4 Load used for substructure design and load combination Section 5 Evaluation of seismic motion Section 6 Response evaluation and liquefaction evaluation of subsurface ground Annex 4.1, 6.1	Section 1 Basic matters Section 2 Design of pile foundation Annex 2.1-2.8 Chapter III Design of improved ground Section 1 Basic matters Section 2 Design of improved ground Annex 2.1-2.4

Fig. 2.8 Constitution of "Technical Code for Seismic Design of Building Foundation of Spent Nuclear Fuel Interim Storage Using Dry Casks"

Foundation types as objects in the guidelines are a pile foundation in the case of supporting interim storage buildings on supporting soil with piles and a spread foundation in the case of supporting the interim storage buildings on the supporting soil via the improved ground by a soil improvement method using cementitious solidification materials. The guidelines are applied to the design of the piles of the pile foundation and the design of the improved ground of the spread foundation. The design of the spread foundation in the case of supporting the interim storage buildings on the supporting soil is performed by the same method for the buildings and structures of the power reactor facilities on the basis of JEAC 4616.

The design of building superstructures and foundation slab structures is performed by the same method for the buildings and structures of the power reactor facilities on the basis of JEAC 4616.

Meanwhile, the design of a foundation slab of the pile foundation is based on JEAC 4616, and is needed to be performed in consideration of reaction force applied from pile heads.

The improved ground is basically the ground and not in a category of the structures. However, according to the guidelines, it is designed as well as the structures. Here, the improved ground that is obtained not by replacing the original ground with concrete by a replacement method, etc. but by the soil improvement method using the cementitious solidification materials is set as an object. The intended soil improvement method is described in "Chapter III Design of improved ground, Section 1 Basic matters".

2.2 Concrete cask storage

The safety review guidelines for concrete cask storage have not legislated yet. Here, the outline of the technical requirements, etc. by the former Nuclear and Industrial Safety Agency is described as follows.

2.2.1 Technical requirements (former Nuclear and Industrial Safety Agency)

The former Nuclear and Industrial Safety Agency compiled "Technical Examination Report on Spent Fuel Storage Facilities (Interim Storage Facilities) Using Concrete Casks" in June 2004. The basic matters (technical requirements) considered to be technically important, and the matters to be technically considered in safety review (matters to be considered), etc. were compiled in this report, for the sake of the safety review of the "spent fuel storage facilities" defined by Article 43, 4 "Act on the Regulation of Nuclear Source Material, Nuclear Fuel Material and Reactors (Act No. 166 of 1957)".

In the compilation, "Technical Examination Report on Spent Fuel Storage Facilities (Interim Storage Facilities)" (Agency of Natural Resources and Energy, December 2000), "Guidelines of Safety Review for Spent Fuel Interim Storage Facilities Using Metallic Dry Casks" (Decided by Nuclear Safety Commission on October 3, 2002), and other guidelines established by the Nuclear Safety Commission were used as references. The constitution of the technical requirements is shown in Fig. 2.9.

I. Roles and scope of the technical requirements II. Definitions of terms III. Site conditions (Requirements 1-3) IV. Basic safety functions (Requirements 4-7) V. Radiation control and environmental safety (Requirements 8-10) VI. Other safety measures (Requirements 11-22)	References 1. Matters to be considered (with respect to Requirements 2-8, 11, 12, 18) 2. Reflections of latest technical expertise, etc. regarding spent fuel storage facilities 3. Referential figures

Fig. 2.9 Constitution of "Technical Requirements Concerning Spent Fuel Storage Facilities (Interim Storage Facilities) Using Concrete Casks"

The technical requirements can be read through the following address.

http://www.nsr.go.jp/archive/nisa/oshirase/2006/files/181011-4.pdf

The following is main contents of the requirements. The titles of items intimately related to this textbook are enclosed in boxes, and the notable contents are underlined.

III. Site conditions (Requirements 1-3)

The following deference exists as compared to the requirements for the metal casks.

· In "Requirement 2 Normal- period condition", the consideration for tritium passing through canisters and radioactive materials generated by activation of the air cooling the canisters is added.

IV. Basic safety functions (Requirements 4-7)

· In "Requirement 4 Containment function", the maintenance of negative pressure and the monitoring design that are required for the metal casks are not required. The reasons are: (i) the canisters have welded structure, so as to have a highly-reliable containment function; (ii) the maintenance of the containment function can be confirmed by performing periodic confirmation as SCC measures from the outside; and (iii) continuous monitoring is not performed also in the US. Instead, the proper multilayered welded structure is required for lid parts of the canisters. It is because the canisters are designed and manufactured in the same manner as the class 1 containers of the reactor facilities. One-sided partial penetration welding is adopted for the design of welds of the lids. Thus, the radiographic examination as a volumetric inspection cannot be conducted on the welds, so that a state of uranami (penetration) beads of welding cannot be inspected.

Meanwhile, the term "multi-layering" here means that the lid welds of the canisters have at least double structure and not that the whole lid is multi-layered.

· In the interpretation of Requirement 4, the use of high corrosion resistant material and the reduction of weld residual stress and salt adhesion environment are listed as measures to be taken if there is the possibility of stress corrosion cracking due to the adhesion of salt brought into the canisters by cooling air.

· For obtaining "proper welded structure", the certainty and reliability of workability should be previously and sufficiently confirmed by a mock-up test, etc. Moreover, the inspection of the lid welds should be conducted by a multilayered penetrant testing (multi-layered PT, which is conducted at intervals of not exceeding the allowable maximum defect size) and an ultrasonic testing (UT). The reasons is that, if a latent defect not appearing on a surface still exists even after the PT confirms the absence of a surface defect, the possibility that the latent defect might grow beyond the PT-conducted surface due to subsequent welding cannot be completely eliminated.

· In "Requirement 5 Shielding function", it is required that the containment design of a concrete storage container should be performed in consideration of the streaming of neutrons and gamma rays from air inlets and outlets, the impairment of containment ability in association with aged deterioration of concrete, and the impairment of containment ability due to detachable corrosion products on a steel liner arranged on an inner surface of the concrete storage container.

VI. Other safety measures (Requirements 13-21)

· "Requirement 15 Consideration for cooling failure such as the closing of cooling air channels and the like" has contents related to "Requirements 7 Heat removal function" of the metal cask storage. The difference is that the closing of cooling air channels of the concrete storage containers can be detected by measuring the temperature of the air inlets and outlets when it occurs, and also that the heat removal design that the basic safety functions are not impaired even if heat remains in canister repacking apparatuses for a long time is required.

· "Requirement 18 Considerations for earthquakes" has approximately the same contents as "Requirement 17" of the metal casks storage. However, in the case of the metal cask storage, the earthquake resistance class is not required for the casks, and instead the design of maintaining the basic safety functions relative to seismic design force is required on the basis of the requirements of structural strength relative to spent fuel packages in the guidelines of the former Nuclear Safety Commission. In this regard, the concrete casks are not the packages, so as to be handled differently from the metal casks.

2.2.2 Safety design standards (Atomic Energy Society of Japan)

The Atomic Energy Society of Japan (AESJ) considered that the concrete casks would become a mainstream of a storage system as well as the metal casks on the basis of the global trend, so as to legislate "Standard for Safety Design and Inspection of Concrete Casks and Canister Transfer Machines for Spent Fuel Interim Storage Facility : 2007".

The constitution of the standards is shown in Fig. 2.10.

1. Scope of application 2. Reference standards 3. Definitions 4. Safety design of concrete casks 4.1 Basic requirements 4.1.1 Consideration for double use for transport and storage 4.1.2 Consideration for ageing 4.2 Method of safety design 4.2.1 Basic specifications 4.2.2 Containment design 4.2.3 Shielding design	4.2.4 Criticality prevention design 4.2.5 Heat removal design 4.2.6 Structural strength design 5. Safety design of canister repacking apparatuses 6. Inspection of concrete casks 6.1 Canister 6.2 Concrete storage container 7. Inspection of canister repacking apparatuses Annex 1-9 Interpretation 1-6

Fig. 2.10 Constitution of "Standard for Safety Design and Inspection of Concrete Casks and Canister Transfer Machines for Spent Fuel Interim Storage Facility : 2007"

The following is main contents of the standards. The titles of items intimately related to this book are enclosed in boxes, and the notable contents are underlined. In what follows, the numbers of sections and divisions correspond to the numbers of the AESJ standards.

4.1.2 Consideration for ageing

· Concrete casks shall be designed to reduce or prevent the influence of aged deterioration on their components. Also, they shall be designed to reduce or prevent the influence of aged deterioration on stored spent fuel assemblies by maintaining inert environment inside canisters and limiting a temperature rise.

4.2 Method of safety design

· Containment structure of canisters- All connections of containment boundaries of canisters shall have welded structure. Lid parts of the canister shall have double lid structure constituted by a primary lid, and a secondary lid or a sealing ring.

· Heat removal structure of concrete casks- Concrete casks shall have structure of removing heat by installing inlets, outlets, and channels of cooling air.

· Monitoring- Concrete casks shall be designed to be able to confirm the maintenance of a heat removal function by monitoring the outlet temperature and a temperature deference between the inlet and outlet in them.

4.2.6 Structural strength design

· Design requirement item

Concrete casks shall be designed not to damage spent fuel cladding under general test conditions, during normal storage, and at abnormal events.

· Design criteria

Initiation stress shall not exceed yield stress under general test conditions, during normal storage, and at abnormal events.

6.1.2 Inspection procedure of canisters

c) Exterior inspection 3) Inspection during storage period - The appearance and installation state of canisters shall be confirmed by remote viewing. The canisters to be inspected shall be selected in consideration of the combination of types of the canisters and concrete storage containers, the specification of contents, and the storage period, in terms of external corrosion due to sea salt particles. (It is interpreted as meaning that not all canisters, but the canisters possible to be rusting shall be inspected as inspection objects, and that the canisters which have small possibility of rusting and which are included in other canisters can be excluded from the inspection objects.)

d) Method of and standards for welding inspection- Welded joints related to structural strength shall be inspected according to the concrete cask structure standards.

e) Airtight leakage inspection- It shall be periodically confirmed that canisters maintain the containment function on the basis of appearance inspection records of c) 3).

o) Inspection of the contents 3.3) Inspection during storage period - The following records shall be confirmed: inspection of the contents records of inspection before transport from power plants and inspection before storage, and canister appearance inspection records and heat removal inspection records during a storage period.

p) Contents inspection 2) Airtight leakage inspection - In pre-shipment inspection after storage, vacuum drawing shall be performed on the inside of a canister for transporting cask and a rate of leakage of helium from a canister shall be measured by a helium leak inspection device, in such a state that the canister is in the canister transporting cask.

p) Contents inspection 4) Stored fuel inspection - In pre-shipment inspection after storage, the following records shall be confirmed: stored fuel inspection records of inspection before transport from power plants and inspection during a storage period, and canister appearance inspection records and airtight leakage inspection records of the pre-shipment inspection after storage.

6.2.2 Inspection procedure of concrete storage container

c) Appearance inspection 3) Inspection before storage and inspection during storage period - They shall be conducted by remote viewing of an inner surface of a concrete storage container in addition to by visual confirmation of an outside surface of the container. An object to be inspected by remote viewing shall be the concrete storage container storing a canister that is selected by 6.1.2, c) 3) Inspection during storage period.

e) Shielding performance inspection 3) Inspection during storage period - In a representative concrete cask, gamma-ray and neutron dose rates on the surface of a concrete storage container shall be measured by a survey meter in such a state that the contribution of the dose rate from other concrete casks may be ignored or corrected.

f) Heat removal function inspection 2) Inspection before storage and inspection during storage period- Temperature monitoring record of the temperatures of air inlets and outlets of concrete casks shall be confirmed.

2.2.3 Structural codes (Japan Society of Mechanical Engineers)

The Japan Society of Mechanical Engineers (JSME) established "Codes for Construction of Spent Nuclear Fuel Storage Facilities-Rules on Concrete Casks, Canister Transfer Machines, and Canister Transport Casks for Spent Nuclear Fuel-(JSME S FB1-2003) " in December, 2003 ahead of the Atomic Energy Society of Japan.

(1) Constitution of codes

The constitution of the codes is shown in Fig. 2.11. The interpretation is attached to each section.

(2) Chapter III Canister, Section 3 Material (Containment container)

If it is assumed that the concrete cask storage facilities are located near the sea, the canisters are exposed to salt air environment. The measures are required because of the concern of stress corrosion cracking (SCC) due to three factors: stainless steel, weld residual stress, and salt. In the current JSME structural codes, SCC measures by using high corrosion resistant stainless steel are examined.

On the other hand, in terms of economic efficiency and scientific rationality, the measures of reducing the weld residual stress by using regular stainless steel such as SUS304L and SUS316L could be considered. The studies on the measures will hereinafter be mentioned in this book.

Chapter I General provisions
Section 1 Scope of application
Section 2 Definitions
Chapter II Mechanical test and nondestructive test
Section 1 Mechanical test
Section 2 Nondestructive test
ChapterIII Canister
Section 1 Scope of application
Section 2 Definitions
Section 3 Material (Containment container, basket, canister hanger)
Section 4 Design (same as above)
Section 5 Manufacture
Section 6 Inspection
MANDATORY APPENDIX
APPENDIX 3-1 to 3-2
Chapter IV Reinforced concrete storage container
Section 1 Scope of application
Section 2 Definitions
Section 3 Material (reinforced concrete part, steel part)
Section 4 Design (load combination and permissible level, reinforced concrete part, steel part)
Section 5 Manufacture (reinforced concrete part, steel part)
Section 6 Inspection (manufacturing inspection, inspection before storage, inspection during storage periods)

Chapter V Concrete-filled steel storage container
Section 1 Scope of application
Section 2 Definitions
Section 3 Material (steel part, infilled concrete part)
Section 4 Design (load combination and permissible level, steel part, infilled concrete part)
Section 5 Manufacture (steel part, infilled concrete part)
Section 6 Inspection (manufacturing inspection, inspection before storage, inspection during storage periods)
Concrete-filled steel storage container
(MANDATORY APPENDIX)
Appendix 5-1

Chapter VI Canister repacking apparatus
Chapter VII Canister transporting cask

Appended tables and appended figures
Appended table 1-1 to 1-3,1-9 to 1-12
(material to be used, design stress intensity, etc)
Appended figures 1-1 to 1-10, 1-15 to 1-18, 1-25 to 1-27)
(design fatigue diagram, external pressure chart)

Fig. 2.11 Constitution of "Codes for Construction of Spent Nuclear Fuel Storage Facilities-Rules on Concrete Casks, Canister Transfer Machines, and Canister Transport Casks for Spent Nuclear Fuel-(JSME S FB1-2003) "

(3) Chapter III Canister, Section 5 Manufacture

1) Nondestructive test on lid weld (Table CCN-2300-1)

The lid welding containment after loading spent fuel into the canisters within pools of the nuclear power plants is performed by the one-sided partial penetration welding. In the structural codes for the concrete casks by JSME, the classification of such a joint was newly established and prescribed as E. In the nondestructive test on the category E weld joints after welding, the penetrant test is conducted on the root layer, the place of one-half of welded joint thickness, and the final surface when the number of welded layers is four or more. As an alternative test, the ultrasonic testing and the penetrant test after the completion of welding are conducted.

This is inconsistent with the technical requirements of the former Nuclear and Industrial Safety Agency, which is "the inspection of lid welds shall be conducted by a multilayered penetrant testing (multi-layered PT, which is conducted at intervals of not exceeding the allowable maximum defect size) and an ultrasonic testing (UT)". In this book, the scientifically rational proposals and the research results contributing the evidence to them will be described hereinafter.

2) Mechanical test on lid welds

As to the category E weld joints, it is prescribed that mechanical test plates should be manufactured by performing welding on the same classification execution point by an execution method under the same condition as the category E weld joints before the welding of the canister.

Originally in the rules on welding, the mechanical tests after welding are required for both-sided complete fusion penetration butt welding, but not required for the one-sided partial penetration welding. In this book, the scientifically rational proposals and the research results contributing the evidence to them will be described hereinafter.

2.3 Overseas regulations and standards

It would be important to know the overseas regulations and standards, in the mood of globalization of the Japanese regulations and standards.

2.3.1 IAEA

The documents related to the safety standards of spent fuel storage published by the nuclear safety department of IAEA include the followings.

(1) Predisposal Management of Radioactive Waste General Safety Requirements Part 5, Series No. GSR Part 5, May 19, 2009.

This document stipulates general requirements on confinement, retrievability, monitoring, inspection, consideration of burdens to the future generations at long term storage, etc.

(2) Specific Safety Guide No. SSG-15, Storage of Spent Nuclear Fuel, 2012

The contents are as follows.

Introduction, Protection of Human Health and the Environment, Roles and Responsibilities, Management System, Safety Case and Safety Assessment, General Safety Considerations for Storage of Spent Nuclear Fuel (Design, Commissioning, Operation, Decommissioning)

Appendix I Specific Safety Considerations for Wet or Dry Storage of Spent Nuclear Fuel

Annex I Short Term and Long Term Storage

Annex II Operational and Safety Considerations for Wet and Dry Spent Fuel Storage Facilities

Annexes III to VII Operating Procedures, Site Conditions, Processes and Events, Posturated Initiating Events

After the Fukushima accidents, addenda to SSG-15 were recommended as follows.

1. The operator should ensure that monitoring equipment for key safety functions is available and operating after Beyond Design Basis Accident.
   Addendum to that effect should be made at para. 6.61 (Instrumentation and control).
2. Para 6.98 (Operational aspects) should be amplified to the extent that combinations of initiating events are considered.

(3) Guidance for preparation of a safety case for a dual purpose cask containing spent fuel (Draft)

The contents are as follows.

Part 1: General Principles and Technical Information

Version and contents list of the DPCSC, Administrative information, Specification of contents, Specification of the DPC, Storage and transport conditions, General design considerations and acceptance criteria, Ageing considerations, Compliance with regulatory requirements, Operation, Maintenance plan, Emergency plan, Management systems, and Decommissioning.

Part 2: Specific Technical Assessment

Common provisions for all technical analyses in part 2 of the safety case, Structural analyses, Thermal analyses, Activity release analyses, External dose rate analyses, Criticality safety analysis.

Annex Example for the Holistic Approach of DPCSC for an Operational Scenario (of Japan)

2.3.2 USA

(1) 10CFR72 Licensing requirements for the independent storage of spent nuclear fuel, high-level radioactive waste, and reactor-related greater than class C.

(2) NUREG

The NUREGs are reports or brochures on regulatory decisions, results of research, results of incident investigations, and other technical and administrative information published for licensee and related people.

* Publications Prepared by NRC Staff NUREG-(nnnn)

* Publications Prepared by NRC Contractors NUREG/CR-(nnnn)

Major NUREGs include the followings.

NUREG-1536 Standard Review Plan for Spent Fuel Dry Storage Systems at a General License Facility - Final Report

NUREG-1567 Standard Review Plan for Spent Fuel Dry Storage Facilities (Commercial independent spent fuel storage installations may be co-located with a reactor or may be away from a reactor site.)

NUREG-1927 Standard Review Plan for Renewal of Spent Fuel Dry Cask Storage System Licenses and Certificates of Compliance - Final Report

NUREG-CR 6745 Dry Cask Storage Characterization Project - Phase 1:?CASTOR V/21 Cask Opening and Examination (This report documents visual examination and testing conducted in 1999 and early 2000 at the Idaho National Engineering and Environmental Laboratory on CASTOR V/21 (PWR) spent fuel dry storage cask.)

(3) ISG (Interim Staff Guidance Used by the Spent Fuel Project Office)

Interim staff guidance documents are used by the NRC staff to provide guidance to the staff concerning issues not currently addressed in a standard review plan (SRP) or issues where clarification of SRP text is necessary.

As of November 2014, there are 26 ISGs posted on the NRC web page. Major ISGs include the followings.

ISG-1 Damaged Fuel, ISG-4 Cask Closure Weld Inspections, ISG-5 Confinement Evaluation, ISG-8 Burnup Credit in the Criticality Safety Analyses of PWR Spent Fuel in Transport and Storage Casks, ISG-11 Cladding Considerations for the Transportation and Storage of Spent Fuel, ISG-12 Buckling of Irradiated Fuel Under Bottom End Drop Conditions, ISG-15 Materials Evaluation, ISG-18 The Design and Testing of Lid Welds on Austenitic Stainless Steel Canisters as the Confinement Boundary for Spent Fuel Storage, ISG-23 Application of ASTM Standard Practice C1671-07 when performing technical reviews of spent fuel storage and transportation packaging licensing actions, ISG-25 Pressure Test and Helium Leakage Test of the Confinement Boundary for Spent Fuel Storage Canister.

(4) ASME Code

The ASME Code stipulate technological design procedures, etc. and have received internationally prestigious reputation. The Japanese notification 501 on technological criteria for structure of nuclear power generation facilities was based on the ASME Code Section III, Division 1, and used for design and construction of the nuclear power generation facilities in Japan. The ASME Code for transport and storage of spent nuclear fuel are developed and published after the Divison1 as follows.

ASME Code Section III, Division 3 Containments for Transportation and Storage of Spent Nuclear Fuel and High Level Radioactive Material and Waste (2013). The contents are as shown in Fig. 2.12.

As compared with the structural code of cask in JSME, the ASME Code is substantial in terms of quality assurance and certification system. The ASME Code is thereby authoritative because they certifies cask designs, etc.by their own responsibility.

By the way, the materials criteria of ductile cast iron cask specified in WB-2331.4, WC-2332.2, etc. were proposed by CRIEPI, Japan and approved. The strain based criteria specified in WB-3700 and WC-3700 was introduced form the edition of 2013, which is for an alternative design method in an external event with a limited energy to be loaded on cask, such as the 9 m drop test condition, etc. The criteria have not yet been incorporated in the JSME Code.

Subsection WA General Requirements	Subsection WB Class TC Transportation Containments
1000 Scope of Division 3	Subsection WC Class SC Storage Containments
2000 Design Basis for Containments	1000 Introduction
3000 Responsibilities and Duties	2000 Materials
4000 Quality Assurance	3000 Design
5000 Authorized Inspection	4000 Fabrication
7000 Reference Standards	5000 Examination
8000 Certificates of Authorization, Nameplates, Certification Mark, and Data Reports	6000 Testing
9000 Glossary	7000 Nameplates, Stamping with Certification Mark, and Reports

Fig. 2.12 Structure of ASME Code Section III, Division 3 Containments for Transportation and Storage of Spent Nuclear Fuel and High Level Radioactive Material and Waste (2013)

2.3.3 Germany

(1) Guidelines for dry cask storage of spent fuel and heat-generating waste (2013.10.6 revision)

Main characteristics are as follows.

1) Major applicable inventories: Spent fuel assemblies from light water reactors using uranium oxide or uranium-plutonium oxide as nuclear fuel, and canisters with vitrified radioactive waste from reprocessing (article 1.1).

2) Storage period: Since a decision on the related repository concept and its realization has not been made yet, the actual time needed for storage cannot be stated. The period of 40 years, on which previous storage licenses are based, may be referred to as an appropriate time scale. Should it appear that this period will not be sufficient, appropriate additional analyses are to be performed (article 1.1).

3) Protection goals: The fundamental protection goals derived from it are confinement of radioactive material, stable decay heat removal, maintenance of subcriticality, and avoidance of unnecessary radiation exposure (article 1.2).

4) In case of repair of the primary lid, the cask can be transferred into the hot cell of a nuclear facility which may be adjacent to the interim storage facility, or it may be transported to it on public roads on the basis of a transport

permit. Alternatively, an additional lid is to be provided, sealed by means of welding instead of the metal seal, which is placed on the intact secondary lid barrier, thus restoring the two-barrier concept in the interim storage facility (article 2.2).

5) Aircraft crash, blast wave and ingress of toxic substances are, in general, beyond design basis accidents. The reduction of damage caused by aircraft crash and blast wave may either be achieved by the casks or by a combination of cask and storage facility/building (article 9.2.2).

References

1) The 18th Study Team Concerning New Regulation Standards for Nuclear Fuel Facilities, etc.: Material 1-5 Framework Plan for New Regulation Standards for Spent Fuel Storage Facilities (Revised), September 2, 2013.
2) The 18th Study Team Concerning New Regulation Standards for Nuclear Fuel Facilities, etc. : Material 2-5 Comments on Framework Plan for New Regulation Standards for Spent Fuel Storage Facilities and Responses to Them (partially amended), September 2, 2013.

CHAPTER 3 METAL CASK STORAGE

3.1 Outline of metal cask storage

Table 3.1-1 shows the history of technological development of spent fuel storage in the world[1]. The spent fuel storage began with the pool storage at and away from reactors, and the current major dry storages such as metal cask and concrete cask are following to date.

The dual purpose metal cask storage has potential issue of compliance with the future transport regulations after storage of several decades. If the transport regulations would not change significantly, the issue would be negligible. On the other hand, the concrete cask storage method would not have such issue of compliance with the future transport regulations, because the spent fuel would be transported in separate transport casks.

Table 3.1-1 History of technological development of spent fuel storage methods

Options	Year				
	1950–1960	1970	1980	1990	2000 ~
Wet	Most of the AR and AFR pools				
Dry		Vault (1971 Wylfa, 2000 Paks)			
		Concrete Silo (1977 Whiteshell)			
			Metal Cask (1986 Surry, 1992 Gorleben)		
			Concrete Module (1986 HB Robinson)		
				Concrete Cask (1993 Palisades)	

3.1.1 Background and needs

The metal cask was originally developed for transport cask for radioactive material, which withstands for transport accidental test conditions such as mechanical impact and fire accident and maintains safety functions of sub-criticality, confinement, heat removal, and shielding. When the metal cask is used for storage, it is normally fulfilled with helium gas for heat removal and preventing oxidation. It further has a double lid structure using metal gaskets to give durability and monitoring function to the confinement performance. By the way, "cask" is a small wooden barrel used for storing wine, etc. (See Fig. 3.1-1). Metal cask storage has merits of modularity that allows expansion of storage capacity as needed and of low investment risk that requires relatively low and constant investment.

Fig. 3.1-1 Casks storing wine

3.1.2 Design concept

Metal cask is a container to transport and store spent fuel. It consists of main body, kids, basket, etc. and is tie down on the floor as needed. Fig. 3.1-2 shows and example of the design concept. The metal cask has four safety functions, i.e. 1) confinement function, 2) shielding function, 3) criticality preventing function, 4) heat removal function.

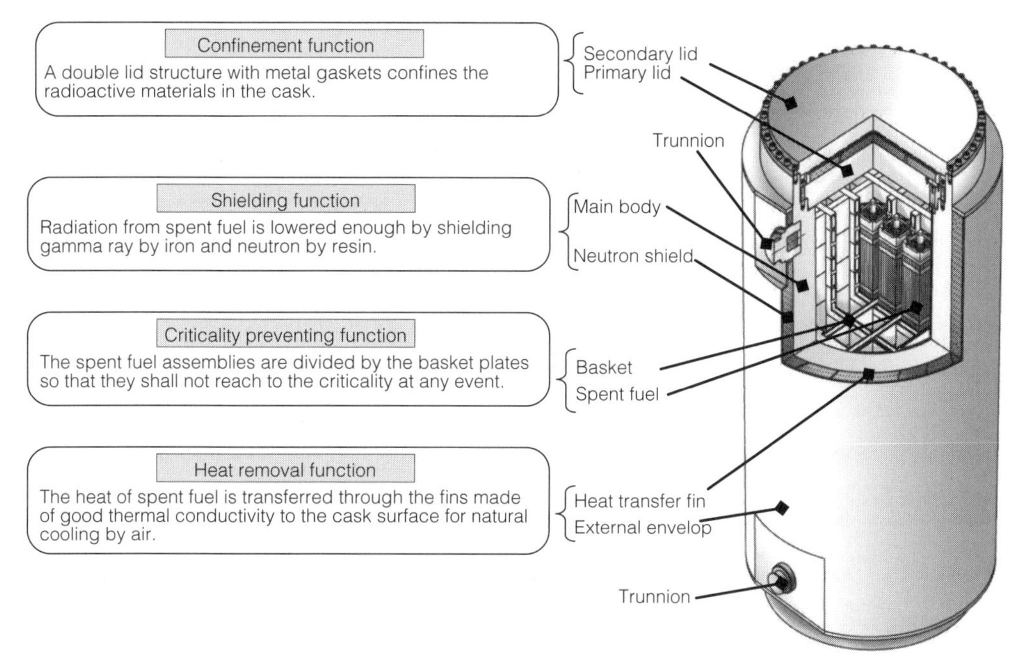

Confinement function
A double lid structure with metal gaskets confines the radioactive materials in the cask.
Shielding function
Radiation from spent fuel is lowered enough by shielding gamma ray by iron and neutron by resin.
Criticality preventing function
The spent fuel assemblies are divided by the basket plates so that they shall not reach to the criticality at any event.
Heat removal function
The heat of spent fuel is transferred through the fins made of good thermal conductivity to the cask surface for natural cooling by air.
Secondary lid
Primary lid
Trunnion
Main body
Neutron shield
Basket
Spent fuel
Heat transfer fin
External envelop
Trunnion

Fig. 3.1-2 Example of metal cask and lid structure for confinement (Cask diameter: 2.5 m, Height: 5 m, Total weight with fuel: 120 t) (26 PWR or 69 BWR)

In Japan, by the year 2013, two metal cask storage facilities were installed within the premises of nuclear power plants. One is at the Fukushima Daiichi Nuclear Power Plant of Tokyo Electric Power Company being operated from 1995 using nine storage-only metal casks. This facility was originally used for storing transport casks in horizontal attitude prior to sea transportation. These casks were attacked by the Tsunami at Great East Japan Earthquake on March 11, 2011. The casks maintained the safety functions and moved to the temporary storage facility (see Fig. 3.1-3) on the same premises after precise inspections[2]. In the temporary storage facility, each cask is covered with a self-shielded concrete module. The modules were placed on the improved soil

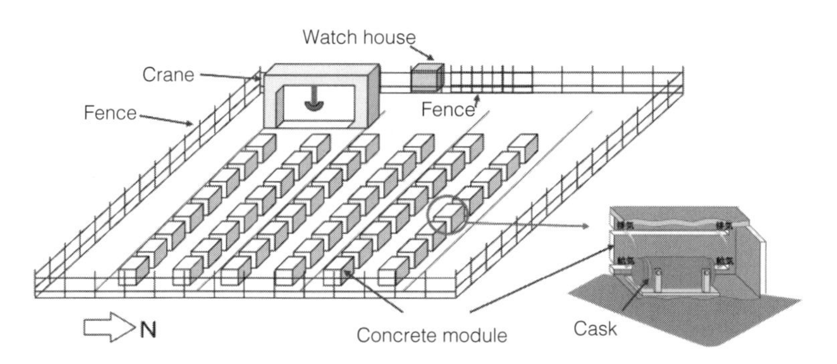

Fig. 3.1-3 Outline of temporary dry cask storage facility (Provided by Tokyo Electric Power Co, Inc.)

foundation. In the modules, the casks were tied down on the support frames with bolts. The casks are handled by a crane with a double lifting device. The current number of the casks are 50 and will be increased by 15. Among those casks, 20 casks are for storage only and the rest are for transport and storage.

The other storage facility is on the premises of Tokai Daini Nuclear Power Plant of Japan Atomic Power Company and is being operated using storage-only casks since 2001 (see Fig. 3.1-4). The casks are standing upright and tied down on the floor. The total capacity is 24 casks in the facility[3]. In the future, a new plan has been announced to store metal casks for transport and storage in the premises of Hamaoka Nuclear Power Plant of Chubu Electric Power Company. They plan to store 400 tU[4].

On the other hand, Recyclable Fuel Storage Company (RFS) that was established by Tokyo Electric Power Company and Japan Atomic Power Company is constructing an interim storage facility outside the premises of the nuclear power plant, at Mutsu city of Aomori prefecture[5]. The storage building is built on a foundation supported by piles, 130 m in length, 60 m in width, 30 m in height, and stores 3,000 tU of spent fuel by 288 metal casks for transport and storage. They are planning to store 5,000 tU in total finally for 50 years (see Fig. 3.1-5, -6).

In overseas countries, cask storage are implemented in USA, Germany, Spain, etc. In USA, the Yucca Mountain project to dispose spent fuel was cancelled. After the Blue Ribbon Committee's study, Department of Energy announced a plan to operate a pilot plant for interim storage facility by 2021, a large scale interim storage facility by 2025, and a disposal facility by 2048[6]. By that time, spent fuel is stored in the premises of power plants. In USA,

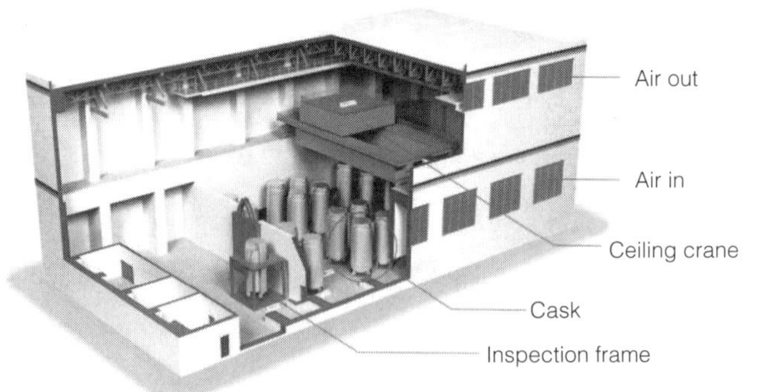

Fig. 3.1-4 Design concept of cask storage facility of Japan Atomic Power Company and the storage-only casks in the Facility (Provided by Japan Atomic Power Company)

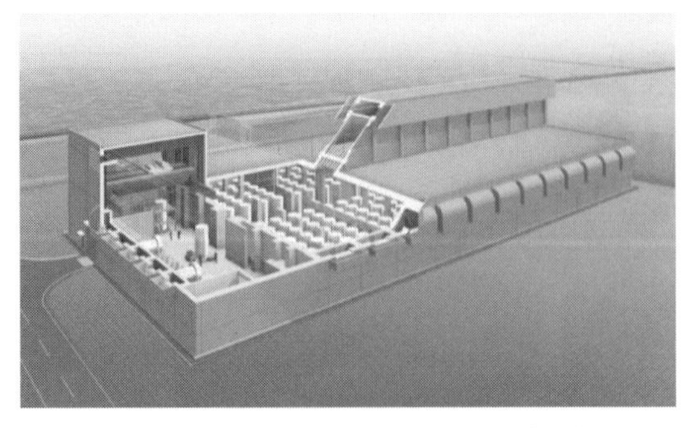

Fig. 3.1-5 Image of interim storage facility (Provided by RFS)

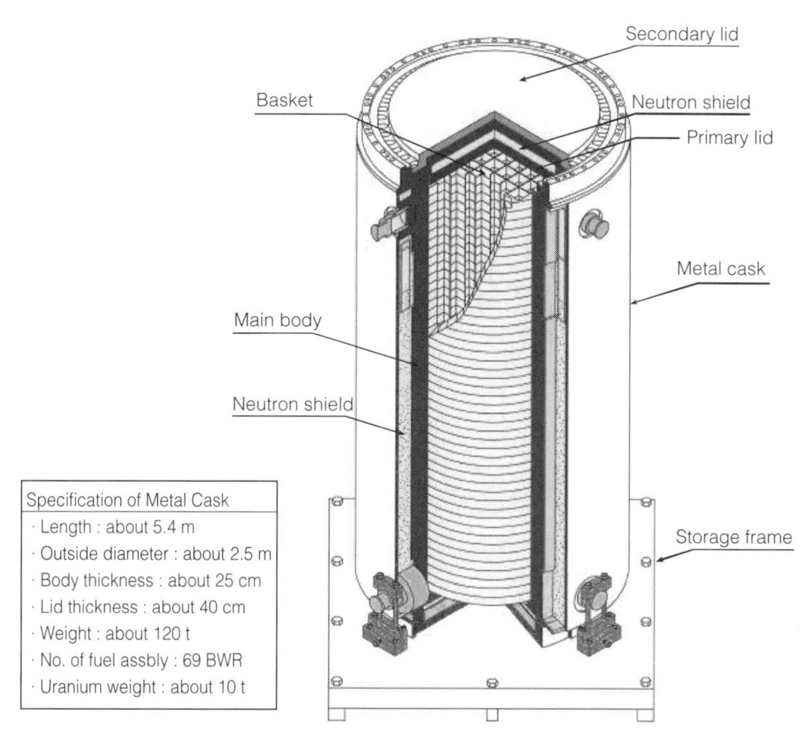

Fig. 3.1-6 Example of concept of transport and storage cask to be stored in the interim storage facility (Copyright: RFS)

the premises of nuclear power plant are wide and they do not need storage building for reducing the radiation level low at the site boundary. They store casks vertically on a concrete pad outdoors (Fig. 3.1-7). In Germany, interim storage facilities were built and commenced operation (Fig. 3.1-8) in Gorleben and Ahaus, but after 2005 transportation of spent fuel was prohibited by law and spent fuel is now being stored in the premises of the nuclear power plants[7]. On the other hand, reprocessed high level wastes are being stored in Gorleben.

Fig. 3.1-7 Metal cask storage in USA (Prairie Island NPP)(Copyright TN)

Fig. 3.1-8 Interim storage facility at Gorleben (Copyright GNS)

3.1.3 Economy

Economic comparison of metal cask storage and pool storage including sensitivity analyses with respect to storage duration, etc. was carried out. This study was carried out about 15 years ago as a reference for the selection of storage method before commercialization of away from reactor storage of spent fuel in Japan[8].

(1) Index of economy

Unit storage cost was employed as an index of economy, which is a cost required to store unit spent fuel, \$/kgU. The unit storage cost was defined as a levelized cost based on discounted cash flow for income and expenditure that has been used in CRIEPI's economic study for spent fuel storage[9]. The levelized cost is calculated as follows.

The total expenditure [$\Sigma C_t/(1+i)^t$] for the expense (Ct) for the construction, operation, etc. is equal to the total income [$\Sigma Q_t/(1+i)^t$] that was levelized value at a reference year, T, (a constant storage unit cost (C) x amount of spent fuel received by the storage facility (Q_t), as defined by the following equation.

$$C = \sum_{t=0}^{N} C_t / (1+i)^t \Big/ \sum_{t=0}^{N} Q_t / (1+i)^t$$

where,

C : Unit storage cost represented by real price [\$/kgU]

Ct : Expenditure in the year t (represented by real price) [\$]

Qt : Amount of spent fuel received by the storage facility in the year t [kgU]

i : Real discount rate [1/year]

t : Year of the expenditure in the assumed storage scenario

(2) Preconditions for evaluation

1) Storage facilities : Pool storage facility and metal cask storage facility (Figs. 3.1-9 and 3.1-10).

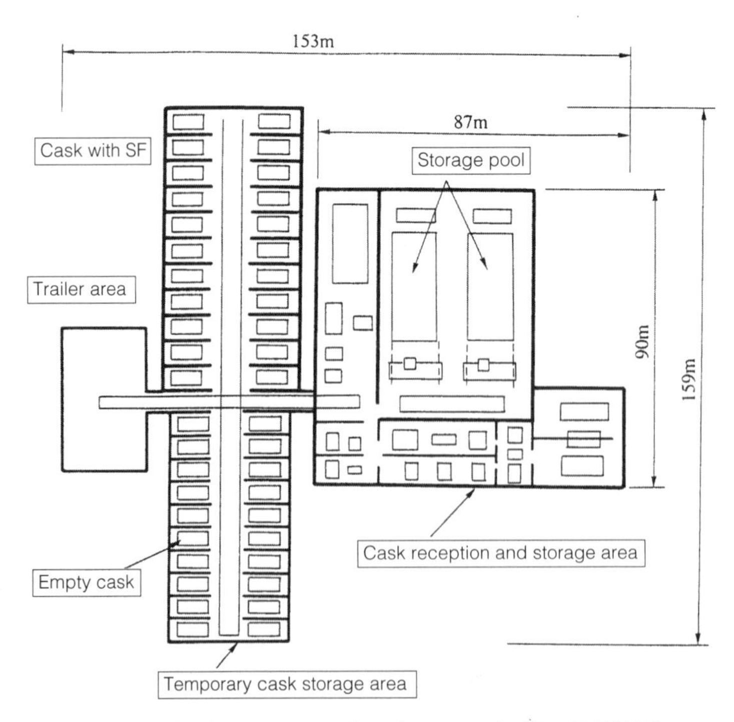

Fig. 3.1-9 Design concept of pool storage facility (5,000 tU)

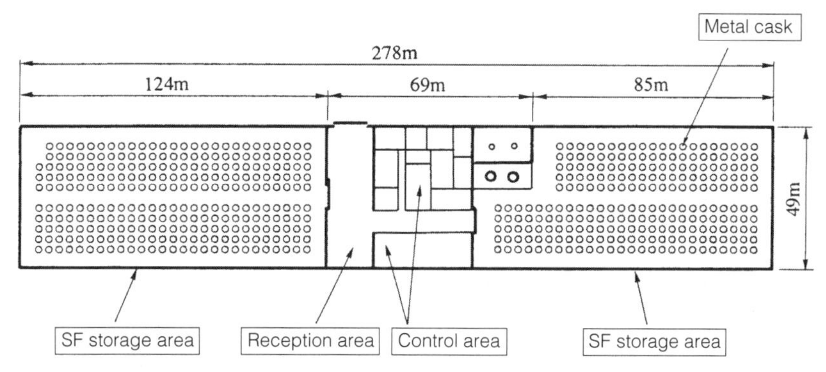

Fig. 3.1-10 Design concept of cask storage facility (5,000 tU)

2) Fuel conditions: Burn-up of the spent fuel was 40 GWd/t.

Ratio of storage of BWR fuel and PWR fuel was equal to the ratio of the power generation, 55:45.

3) Amount of storage and duration: 3000 tU/5000 tU/10000 tU and 40 years.

4) Discount rate : 5 %/year

The discount rate is defined as a rate to determine required price taking account of performance of business, price rise, money rate, etc. For instance, when \$ 10 K is expected to incur in 10 years, \$ 6.1 K should be prepared assuming the discount rate is 5 %.

$\$10\ K/(1+0.05)^{10} \fallingdotseq \$6.1\ K$

(3) Results of evaluation

1) Influence of storage method and amount of storage on economy

It was found that when the storage capacity is small, the unit storage cost by the pool method is more expensive than the metal cask storage method. As the storage amount increased, the unit storage cost for the metal storage method approached to ¥ 30,000/kgU (Fig. 3.1-11). This is because the storage

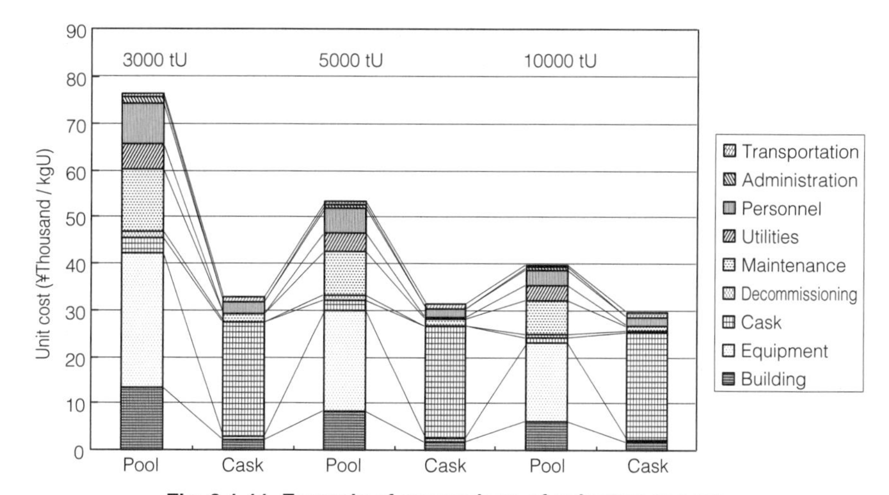

Fig. 3.1-11 Example of comparison of unit storage cost

expense due to the increase in the storage capacity does not linearly increase for the case of the pool storage method (merit of the storage amount). In the case of cask storage method, the increase of the storage amount does not contribute to the merit of the storage amount, but the storage unit cost is almost constant because of the structure of the cost.

2) Influence of the metal cask cost on the economy

Metal cask cost constitutes about 80 % of the unit storage cost. The cost down of the metal cask cost is significantly effective (Fig. 3.1-12).

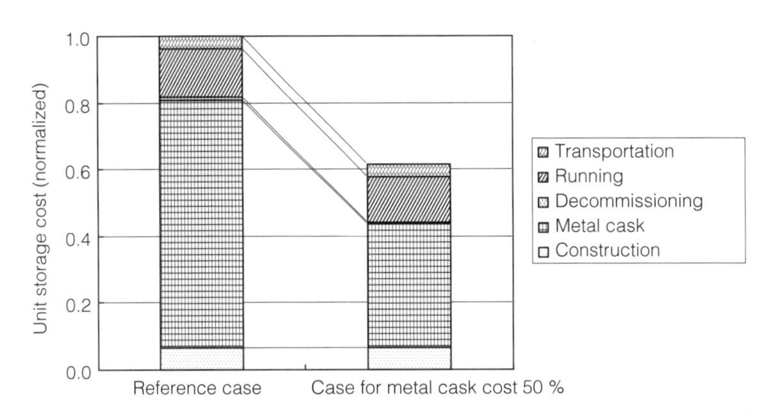

Fig. 3.1-12 Influence of metal cask cost on the unit storage cost

3) Influence to the power generation cost

We did trial calculation of storage cost [¥/kWh] that is the unit storage cost divided by the amount of electricity generated by the nuclear fuel. The storage cost was ¥ 0.09/kWh by the metal cask storage and ¥ 0.15/kWh by the pool storage for 5-year cooled spent fuel (Table 3.1-2). The difference seems to be a little large, but the total power generation cost was ¥ 5.9/ kWh according to the report by the Nuclear Energy Subcommittee of the Advisory Committee for Natural Resources and Energy of the Japanese government as of December 1999, of which 1.5 % is storage cost by the metal cask and 2.5 % is storage cost by the pool storage.

Table 3.1-2 Storage cost (¥/kWh)

Storage capacity	3,000 MTU		5,000 MTU		10,000 MTU	
Cooling time	5 years	15 years	5 years	15 years	5 years	15 years
Metal cask storage	0.095	0.058	0.091	0.056	0.085	0.052
Pool storage	0.214	0.132	0.150	0.092	0.112	0.069

4) Sensitivity analysis (Influence of the discount rate on the storage unit cost)

In this evaluation, the discount rate was assumed to be 5 % in the calculation. However, the 5 % may be criticized to be high for the spent fuel storage business that is for public benefit business. Then, we studied the cases of the discount rate being 0 % and 2 %. In addition, assuming that the spent fuel storage business may be similar to the general storage business by civilian industry, we studied the cases of the discount rate being 8 % and 10 %. As the result, the influence of the change of the discount rate on the storage unit cost was relatively small (Fig. 3.1-13).

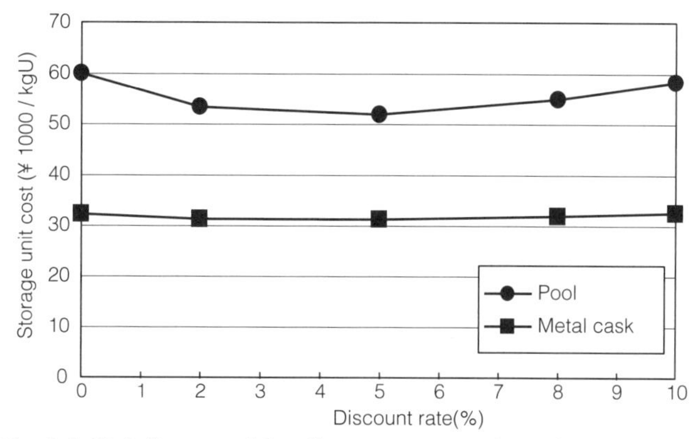

Fig. 3.1-13 Influence of the discount rate on the unit storage cost

References

1) IAEA: "Costing of Spent Nuclear Fuel Storage", No. NF-T-3.5 (2009)

2) Tokyo Electric Power Company: "Implementation of Construction of Temporary Dry Cask Storage Facility", http://www.meti.go.jp/earthquake/nuclear/pdf/120625/120625_02cc.pdf

3) Japan Atomic power Company: homepage, http://www.japc.co.jp/project/cycle/drycask01.html

4) Chubu Electric Power Company: "Change in Construction Plan of Dry Storage Facility for Spent Fuel at Hamaoka Nuclear Power Plant", Press release on July 31, 2014

5) Recyclable Fuel Storage: Homepage, http://www.rfsco.co.jp/about/about_1.html

6) Department of Energy of USA: "Strategy for the Management and Disposal of Used Nuclear Fuel and High-Level Radioactive Waste", January 2013.

7) Federal Republic of Germany: "Joint Convention on the Safety of Spent Fuel Management and on the Safety of Radioactive Waste Management, Report for the Fourth Review Meeting in May 2012".

8) C. Ito, K. Nagano, and T. Saegusa: "Economical Evaluation on Spent Fuel Storage Technology away from Reactor", CRIEPI Report U99047, May 2000.

9) K. Yamaji, K. Nagano, and T. Saegusa: "Comparative Economic Evaluation of Spent Fuel Storage Technology", CRIEPI Report L87001 (1987).

3.2 Heat removal performance

3.2.1 Heat removal performance of cask[1]

(1) Basic points to be taken account for heat removal performance evaluation in normal conditions

Metal cask for dry storage of spent fuel consists of cask body (body and bottom plate), lid, basket, and spent fuel. The cask body and lid usually have a multi-layer structure in order to ensure shielding performance against gamma and neutron rays.

There is a gas layer between the basket and spent fuel. Therefore, not only the heat conduction of the filled gas (typically, inactive helium gas of high heat conductivity) but also the heat transfer due to the convection of the gas and the mutual radiation of components (between basket and cladding tube of the fuel or between the tubes) need to be taken into consideration for the evaluation of the heat removal performance.

1) Heat propagation path of storage cask, etc.

Strictly speaking, the heat propagation path in radial direction of the storage cask depends on the structure of each cask, but an overview is given below.

(i) The heat produced from the cladding surface in the central part of the spent fuel propagates to the outer claddings by heat conduction, convection and mutual radiation, and to the inner surface of fuel storage tubes that make up the basket. Most of the heat is then transferred to the basket constructional material holding the fuel storage tubes.

(ii) The heat transferred to the basket constructional material next to the inner surface of the cask internal cavity, outer surface of the fuel storage tubes, or to the outer surface of the basket external panel (if any) propagates to the inner surface of the cask internal cavity by heat conduction, convection, and mutual radiation.

(iii) The heat transferred to the inner surface of the cask internal cavity propagates to the outer surface of the cask body mostly by the heat conduction of the body.

(iv) The heat transferred to the outer surface of the cask body finally propagates to the surrounding air by the heat convection and mutual radiation.

Therefore, for appropriate evaluation of the temperature of the fuel cladding tubes or at the seal boundary, which are critical parts in the thermal property evaluation of spent fuels, it is necessary to evaluate the following in addition to the thermal physical quantities of the materials of the cask components (thermal conductivity (with the temperature dependence taken into account), radiation factor, and density (only in case of an accident)).

(i) Atmospheric temperature around the cask. (Contribution of the outer air temperature increase around the cask placed in the central area of a storage building, which is caused by the heat release from the cask)

(ii) Major wind velocity around the cask (Determination of the heat conductivity of the outer surface of the cask)

(iii) Amount of heat released vertically from the cask (Heat would be released not only from the lid but also from the bottom plate to concrete floor panel if the cask is placed vertically.)

For strict evaluation of the thermal property of spent fuels based on the thermal analysis, it is necessary to evaluate not only heat conduction of the cask body's solid materials but also a heat flow effect of the filling gas in the cask and that of the natural convection of the air around the cask in the storage building. It is therefore needed to partly combine the heat conduction analysis and the thermohydraulic analysis, which would be highly difficult analysis.

2) Idea of heat conduction analysis model

As in the above, the heat conduction analysis based on a three-dimensional detail model would be adequate for the structural complexity of the dry storage cask. However, to improve the current analysis level, analysis based on a two-dimensional model (including combination with an axially symmetric model or cross sectional model) would be reasonable for the following reasons.

(i) For detailed analysis using a three-dimensional analysis model with the finite element method, precise simulation of the cask shape can be performed if the number of the element is increased. However the analysis is not practical since it is difficult to acquire thermal properties suitable for the number of the elements and evaluate three dimensional data of thermal resistance between constructional materials (connected by bolts or others).

(ii) If the number of the elements is increased, calculation of view factor necessary for the mutual radiation becomes extremely complicated.

(iii) If the number of the elements is increased, convergence of solution[2)] needs to be ensured.

(iv) If the division number is made smaller for the area having small temperature gradient, taking the aspect ratio of the elements into account would be a problem to the analysis precision.

(v) Namely, in detailed analysis with the three-dimensional model, it is difficult to evaluate the thermal properties of small parts of the cask components where the element division and dimensions become a problem.

(vi) Therefore, it is necessary to develop an analysis model that would be logically consistent with macroscopic property evaluation of these parts, where the area, volume, and other factors are taken account of, by introducing an idea of "equivalent property" with focus on the high heat conductivity of metal materials.

3) Heat transfer analysis based on measurement result of storage cask

As mentioned above, high precision analysis would be difficult in consideration of the analytical difficulty due to the structural complexity of the storage cask. However, since the measurement data of heat transfer tests conducted by Central Research Institute of Electric Power Industry with a full-scale storage cask and the following temperature analysis results are in good agreement with sufficient practical precision, we consider that making analysis with the above two-dimensional model would be rational in principle.

The procedure of the heat transfer analysis is shown below.

Storage casks in Japan are mostly placed in the vertical position in storage buildings. Therefore for high-precision analytic evaluation of the temperature of spent fuel cladding tubes, it is crucial to evaluate the temperature and wind speed around the cask appropriately and analytically, which could be used as boundary conditions of the heat transfer analysis.

In addition to these conditions, dimensions of the cask constructional materials, amount of heat from spent fuels, and thermal property specific to the cask should be studied by conducting a test in advance. With these data, the heat transfer analysis could be performed.

(2) Heat removal test in normal conditions

The method and evaluation result example of a heat removal test (hereafter called heat transfer test) for testing thermal property of the cask in normal storage conditions are given below.

1) Test method

For the heat transfer test, a full-size test cask that could store 21 PWR (Pressurized Water Reactor) type spent fuels was used. Also, heat transfer test equipment that could simulate the temperature and wind velocity conditions in the central area of the storage building was used. The test was conducted with parameters such as cask posture (vertical/ horizontal positions) and filling gas type (helium, nitrogen, and vacuum). (See Fig. 3.2.1-1.)

In the present test, for simulating decay heat (23 kW) of the spent fuel assembly, the cask that stored a spent fuel assembly model (a single unit simulating the shape and heat value for one PWR assembly) and heating elements (twenty elements simulating the heat capacity and heat value) was left in a heat transfer test hood with the wind velocity and temperature kept constant. The temperature increasing process until the cask reached a steady state and the temperature of various cask components and fuel cladding tubes in the steady state (which was reached in about 10 days) were measured. The pressure inside the cask cavity and the wind velocity in the heat transfer test hood were also measured.

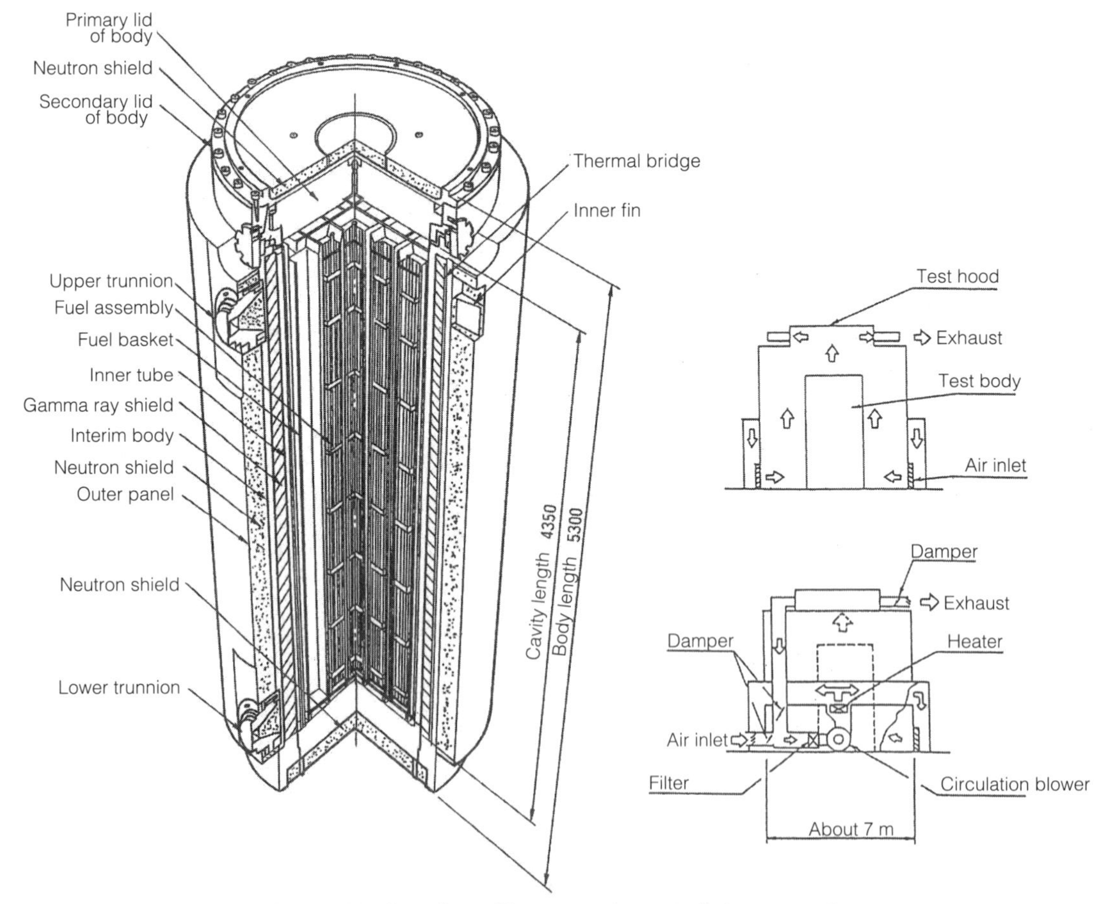

Fig. 3.2.1-1 Overview of heat transfer test of storage cask

As a supplementary evaluation method to the thermal performance test, a sealing test was conducted to check the sealing property before and after the test in each test case.

The present test cases are shown in Table 3.2.1-1.

Table 3.2.1-1 Filling gas in cask and cask posture

Gas type* / Cask posture	Helium	Nitrogen	Vacuum*** (Reduced pressure)
Vertical	○**(Case1)	○(Case2)	○(Case3)
Horizontal	○ (Case5)	○(Case4)	—

* Filling gas inside cask (Three kinds of filling gases were selected to separately evaluate the effect of major heat transfer inside the cask: Heat conduction, convection-based heat conduction, and radiation. The filling gas designated in the design is helium.)

** A reflector was installed to simulate a thermal effect that the target cask mounted in the center of the hood receives from the surroundings. It was used as a parameter for the heat transfer tests whether the reflector was used or not. Here both cases were tested. (In other heat transfer test cases, the reflector was used for the cask in the vertical position but not for the cask in the horizontal position.)

*** In the actual heat transfer test, we could not obtain an ideal vacuum state but a reduced pressure state.

2) Test results

The following is the knowledge obtained from the heat transfer test results.

(i) Days to reach steady state

It was found that the cask reached a stationary state in about 10 days after the start of the heat transfer test in any case.

(ii) Maximum temperature of fuel cladding tubes

The maximum temperature of the fuel cladding tubes, which is one of the important thermal evaluation items for the dry storage condition, was 282 °C, 316 °C, and 338 °C with helium gas (He), nitrogen gas (N_2), and vacuum condition respectively when the cask was placed in the vertical position.

Here, "vacuum condition" was not really vacuum since the average pressure inside the cavity was 15 Torr (1 atm=760 Torr) due to the occurrence of degassing.

(iii) Temperature distribution along axis in cask placed in the vertical position

In the heat transfer test with nitrogen gas filling the cavity of the cask in the vertical position, the maximum temperature was reached near the center of the axis (slightly in the upper side, i.e. lid side) of the fuel cladding tubes due to buoyancy (heat flow effect) of the nitrogen gas. (See Fig. 3.2.1-2.)

(iv) Temperature of cask components

The temperature of each component of the cask used for dry storage (fuel cladding, lead, and lid seals) was well below the design standard in every test case. (See Table 3.2.1-2.)

From the above test results, it is shown that the thermal integrity of the storage cask can actually be analyzed and the dry storage of the spent fuel cask can be made from a viewpoint of the thermal integrity (heat removal capability) even with the existing technology.

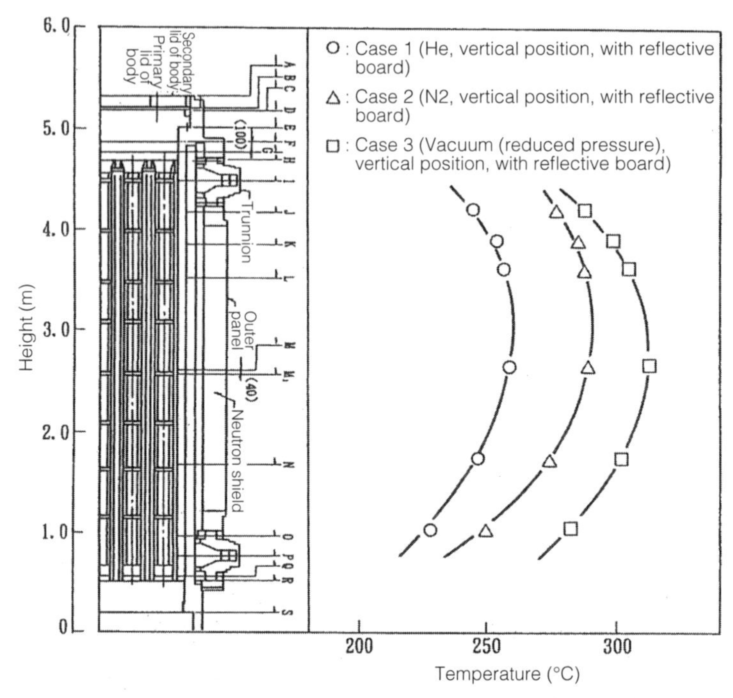

Fig. 3.2.1-2 Influence of filling gas type on temperature distribution along axis (Fuel assembly model with central heater)

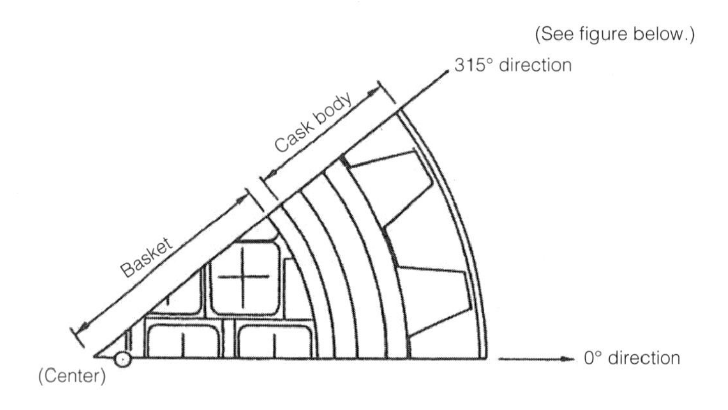

Table 3.2.1-2 Maximum temperature of evaluated component (in normal storage condition)

(°C)

Evaluated component	Case 1[*1]		Case 2[*2]	Case 3[*3]	Case 4[*4]	Case 5[*5]	Design criteria for cask
	w. reflector	w/o. reflector					
Cladding tube	282	274	316	338	324	265	380
Lead	154	143	151	153	150	149	327(melting point)
Primary lid seal	118	113	123	113	115	108	180
Secondary lid seal	94	91	97	91	96	92	180

Note) *1: He, vertical, *2: N_2, vertical, with reflector, *3: Vacuum (reduced pressure), vertical, with reflector
*4: N_2, horizontal, without reflector, *5: He, horizontal, without reflector

(3) Heat removal analysis in normal conditions[3),4)]

Based on the knowledge obtained from the heat transfer tests made for the cask in the normal storage condition, we propose the following heat transfer analysis method for high precision acquisition of the temperature of fuel cladding tubes and others.

1) Analysis method

(i) Overview of analysis

In this analysis method, ABAQUS code* was employed, the code of which is an improved version of the finite element method which has a good record of thermal evaluation of casks. The following points were taken account of based on the knowledge obtained from the heat transfer tests of the heat transfer test cask in the normal storage condition.

(a) Heat loss from the end parts of the cask in both the vertical and horizontal positions was taken into consideration.

(b) Heat flow effects inside the cask cavity were taken into consideration by incorporating a flow analysis code as user's subroutine.

The heat transfer analysis flow for a cask in the vertical position is shown in Fig. 3.2.1-3.

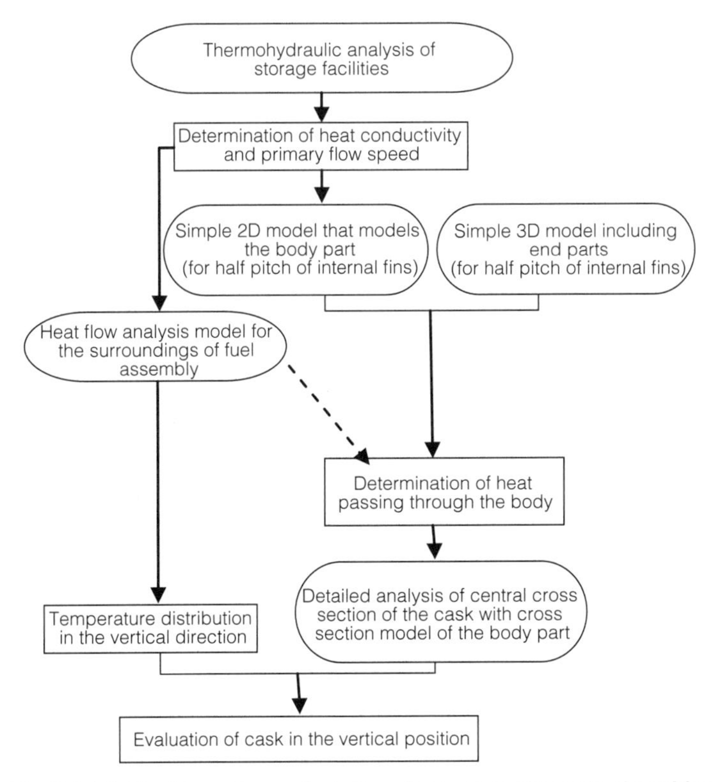

Fig. 3.2.1-3 Heat transfer analysis flow for cask in the vertical position

(ii) Thermohydraulic analysis for determining atmospheric conditions around the cask

The maximum temperature of the fuel cladding tubes depends on the atmospheric temperature or air flow. It is therefore necessary to obtain atmospheric conditions of these parameters inside the heat transfer test hood.

In the analysis, the primary flow speed and temperature of the air around the cask were calculated by using the thermohydraulic analysis code SOLA** based on the difference method. In addition, the heat transfer coefficient of the cask surface was derived from thus obtained primary flow speed and temperature of the air.

(iii) Analysis for determining amount of heat passing through the cask body

In the analysis, the amount of heat passing through the cask body in the radial direction was obtained by using a simple two-dimensional model that assumed insulation of the end parts of the cask and a simple three-dimensional model that allowed heat release from the end parts and by comparing the temperature gradient in the cask body between the two models. The above-mentioned ABAQUS code improved by the finite element method was used for the analysis.

(iv) Temperature analysis of cask body with cross section model

Based on the atmospheric conditions in the heat transfer test hood obtained in (ii) and on the heat passing through the cask body obtained in (iii), maximum temperature of the fuel cladding tubes and others on the central cross section of the cask body was calculated. For the heat transfer analysis, a cross section model shown in Fig. 3.2.1-4 was used.

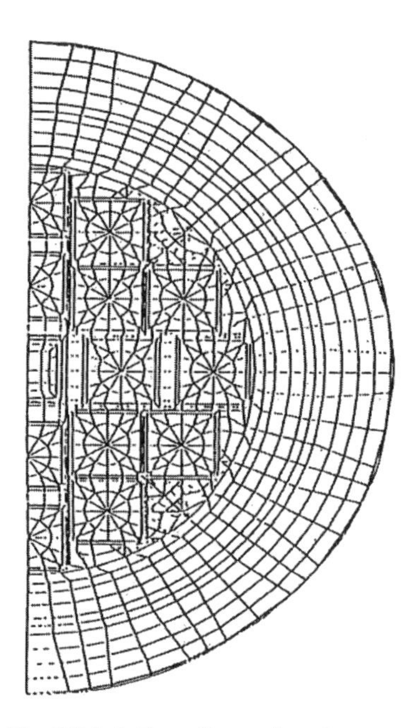

Fig. 3.2.1-4 Two-dimensional cross section of body part (divided to elements)

The analysis code used was the same as the one in (iii). From the knowledge obtained from the heat transfer tests, it was found that the distance between the inner surface of the cavity and the outer surface of the basket largely affected the temperature distribution of the cask body and stored objects. Therefore we took account of not only the thermal gap between various components of the cask but also a contact effect between the inner surface of the cavity and the basket in the lower part of the cask and the deformation of the basket due to thermal expansion or due to the weight (about 20 tons) of the fuel assembly model especially when the cask was placed in the horizontal position.

(v) Temperature analysis of lid seals

In the analysis, an axially-symmetric model was used to obtain the temperature of the cask body, stored objects such as fuel cladding tubes, and lid seals at the end parts of the cask. An improved ABAQUS code (which was developed by partially changing an improved SOLA code for taking account of a convection effect of the air and used as subroutine) was used to take into account of the convection around the fuel cladding tubes in the cavity.

2) Analysis result

(i) Temperature and flow speed in test hood

Table 3.2.1-3 compares the analysis result and the test result. From these results, one can see that both results are in good agreement and the validity of the analysis model and method is verified.

Table 3.2.1-3 Thermohydraulic analysis of heat flow in test hood

Item		Central cross section position along cask axis	
		0°	180° direction
Wind velocity (cm/sec)	Analysis	10.5	40.0
	Measured value	6.5	25.0
Atmospheric temperature (°C)	Analysis	47.0	47.2
	Measured value	46.8	46.8
Thermal conductivity (kcal/m²hr°C) (Note)		2.11	2.65

Note) 1 kcal/m²hr°C=1.163 W/m²K

(ii) Temperature of cask body

Table 3.2.1-4 shows the analysis result of the temperature of various components and fuel cladding tubes of the cask in the vertical position. Fig. 3.2.1-5 shows a typical example of the temperature distribution in the radial direction of the cask. One can see from these results that the analysis result and the measurement result are in good agreement. The validity of the analysis model and method is thus verified.

Table 3.2.1-4 Detailed analysis results of cask body based on cross sectional model (Comparison between analysis and measurement results with cask in the vertical position)

Test and analysis cases / Evaluation point	Helium gas		Nitrogen gas		Reduced pressure	
	Analysis result	Test result	Analysis result	Test result	Analysis result	Test result
Outer surface of cask	98.0~123.9	98.8~119.7	98.0~124.2	97.1~118.3	98.1~123.8	97.5~119.3
Outer surface of intermediate body	136.5~145.0	139.1~147.0	136.4~145.6	136.4~144.8	136.7~145.0	138.4~138.5
Inner surface of intermediate body	140.2~147.7	142.7~148.3	140.1~148.4	139.6~146.1	140.5~147.7	140.6~147.7
Inner surface of lead	150.6~154.3	145.6~153.7	150.6~155.5	143.0~151.4	150.8~154.2	145.0~152.9
Inner surface of internal cavity	153.9~159.7	152.0~158.8	153.9~161.6	149.7~157.3	153.8~159.3	151.2~158.3
Outer surface of basket	161.5~179.4	165.5~190.8	166.4~192.4	170.7~207.3	167.4~186.4	177.5~217.3
Max temperature of fuel assembly model	284.9	282.3	319.2	315.7	331.6	335.5
Primary lid seal	119.2	116.5~117.9	123.0	121.1~122.9	115.6	111.1~113.3
Secondary lid seal	98.5	92.4~94.0	101.2	94.3~96.6	96.0	88.2~91.0

(iii) Temperature of lid seals

Table 3.2.1-4 also shows the temperature of the lid seals of the cask in the vertical position, which is a thermally-severe condition. Fig. 3.2.1-6 shows a typical example of the temperature distribution along the axis of the cask. One can see from these results that the analysis result and the measurement result are in good agreement. The validity of the analysis model and method is thus verified.

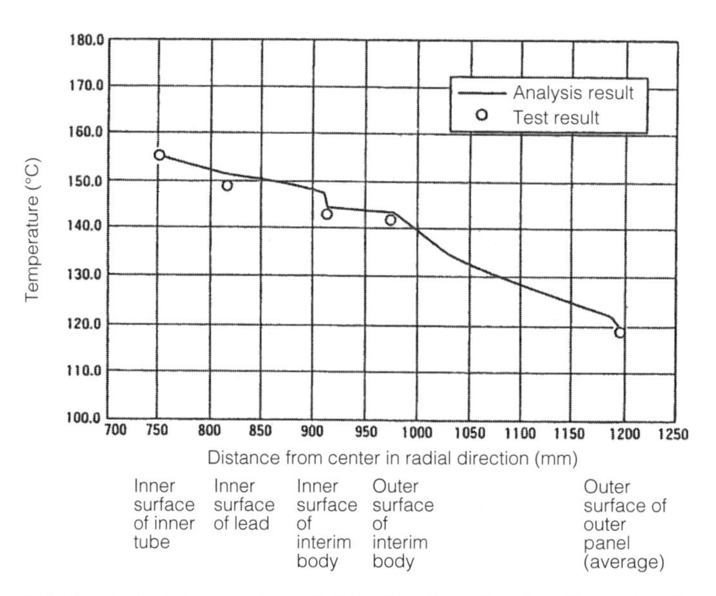

Fig. 3.2.1-5 Typical temperature distribution in radial direction of cask on two-dimensional cross section

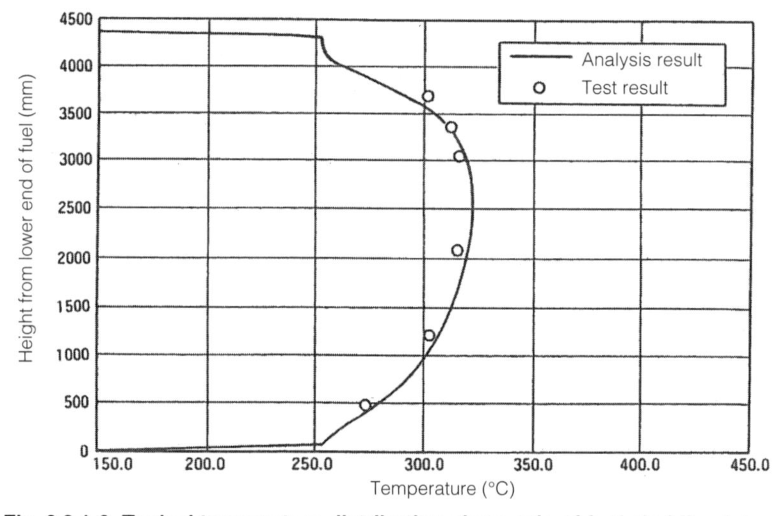

Fig. 3.2.1-6 Typical temperature distribution along axis of fuel cladding tubes (for cask in the vertical position for normal storage: Nitrogen gas used)

ABAQUS*: This is developed by adding an analysis function of heat flow phenomena around the fuel cladding tubes and an analysis function based on the equivalent properties to a versatile code of the finite element method.

SOLA**: This is developed by adding a simple heat transfer analysis function of solids to an open code of heat flow analysis based on the difference method. It can be used for the analysis of heat flow around the cask in the storage building.

References

1) Central Research Institute of Electric Power Industry: "Fiscal year 1990 Report on "Verification test for spent fuel storage technology at nuclear power plants" (commissioned by Ministry of International Trade and Industry), March, 1991.

2) Yoshitaka Watanabe: "Numerical Calculation and Error - How is floating point calculation reliable? - (July 31, 2013)":
<http://yebisu.cc.kyushu-u.ac.jp/~watanabe/RESERCH/MANUSCRIPT/KOHO/FLOATING/float.pdf>.

3) Ed. by Karlsson and Sorenser Inc.: "ABAQUS ver.4.6 User's Manual", 1988.

4) L.d. Clourman, C.W. Hin & N.C.Romero: "SOLA-ICE : A Numerical Solution Algorithm for Transient Compressible Flows," LA6236 UC-34 Los Alamos Scientific Laboratory of the University of California, U.S.A, 1976.

3.2.2 Heat removal performance of cask storage building

(1) Test method with heat removal model

Devices which use a natural circulation phenomenon and do not used active functions from a view point of passive safety are utilized for heat removal at nuclear-related facilities. The natural circulation phenomenon occurs around heated objects. When the phenomenon is applied to the design of nuclear devices, the information about temperature and flow rates under natural circulation is necessary in order to evaluate the design, and analysis is generally used for an examination method. The temperature and flow rates under natural circulation are determined based on balance between buoyancy and flow residence, so that the flow rate is small and unstable. Thus, it is important to obtain experimental data for verification of an analysis code. However, it is difficult to perform tests by using a full scale model because of high cost, so that a scale model is commonly used. In this study, a spent nuclear fuel storage facility using natural circulation for heat removal is an object. We performed tests using the scale model, and proposed an experimental method using similarity laws for natural circulation[1].

1) Similarity laws

A flow pattern in the storage facility having metal casks is shown in Fig. 3.2.2(1)-1. There are two kinds of flows. One is the main flow which passes between an inlet and an outlet. The other is a local vertical flow which occurs around a cask surface. Thus, different similarity laws should be used for each flow.

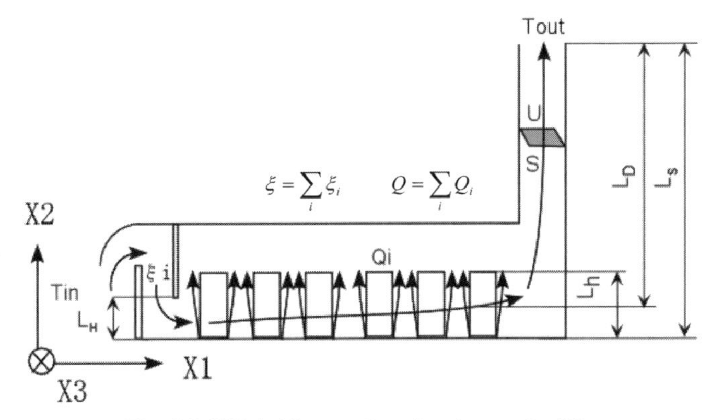

Fig. 3.2.2(1)-1 Flow pattern in storage facility

a. Similarity law for flow in whole storage facility

The whole flow rate and temperature difference between the inlet and outlet are determined under natural convection condition where buoyancy force is balanced with flow residence. When the length which contributes to the buoyancy force can be defined as L_D (L_D=L_s-L_h/2), the relation between the buoyancy force and flow residence is expressed by the following equation.

$$g\beta\rho\Delta TL_D = \frac{1}{2}\xi\rho U^2 \quad \text{..... (1)}$$

Here, ξ is a pressure loss coefficient. ΔT is defined as Tout-Tin.

When the equation (1) is divided by ρU^2,

$$\frac{g\beta\Delta TL_D}{U^2} = \frac{1}{2}\xi \quad \text{..... (2)}$$

The left hand side of the equation (2) expresses Ri number. This fact suggests that when the test in which the pressure loss coefficient coincides with an actual facility is conducted, we can perform the test in which Ri number coincides with the actual facility.

Heat generated from metal casks is conveyed by air in the facility, heat balance is expressed by the following equation (3).

$$Q = \rho cU\Delta TS \quad \text{..... (3)}$$

Here, Q is the total amount of heat Qi from each cask. S is a cross section area of a flow duct.

The relations between ΔT (4) and U (5) are established under different test conditions. These relations are obtained from Eq. (1) and (3). However, these relations are formed only in the case that the flow is turbulent and is unaffected by viscosity.

$$(\Delta T_1/\Delta T_2) = \left(\frac{k_1}{k_2}\right)^{1/3} \times \left(\frac{\xi_1}{\xi_2}\right)^{1/3} \times \left(\frac{\beta_1\rho_1^2c_1^2}{\beta_2\rho_2^2c_2^2}\right)^{-1/3} \times (L_1/L_2)^{-5/3} \times (Q_1/Q_2)^{2/3} \quad \text{..... (4)}$$

$$(U_1/U_2) = \left(\frac{k_1}{k_2}\right)^{-1/3} \times \left(\frac{\xi_1}{\xi_2}\right)^{-1/3} \times \left(\frac{\beta_1\rho_2c_2}{\beta_2\rho_1c_1}\right)^{1/3} \times (L_1/L_2)^{-1/3} \times (Q_1/Q_2)^{1/3} \quad \text{..... (5)}$$

When Ri numbers are matched between different tests, ξ_1=ξ_2 holds in Eq. (4) and Eq. (5). k means the distortion of models. When there is not the distortion, k_1=k_2 holds.

b. Similarity law for flow around heated object

A rising flow of natural convection is generated around a heated object as shown in Fig. 3.2.2(1)-2. Here, the shape of the heated object is a cylinder. Qi means a heat rate, and q means a heat flux.

The thickness (δy) of a boundary layer is expressed approximately by Squire[2].

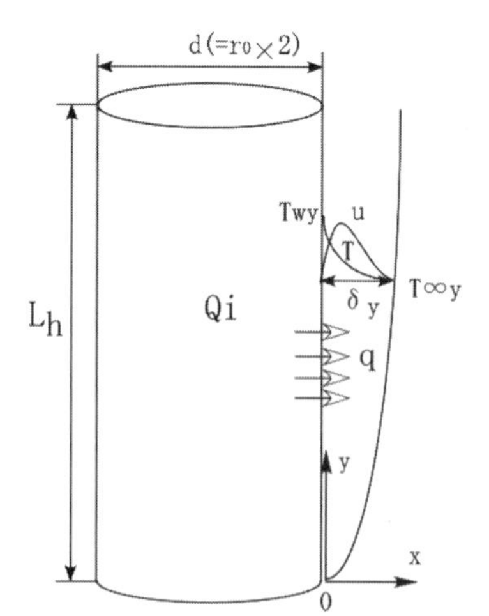

Fig. 3.2.2(1)-2 Boundary layer around heated object

$$\delta_y = 3.93\,\mathrm{Pr}^{-1/2}(\mathrm{Pr}+\frac{20}{21})^{1/4}\left[\frac{g\beta(T_{wy}-T_{\infty y})}{\nu^2 y}\right]^{-1/4} \quad \ldots\ldots (6)$$

Here, temperature distribution in the boundary layer is shown as follows.

$$\frac{T(x,y)-T_{y\infty}}{T_{wy}-T_{y\infty}} = (1-\frac{x}{\delta_y})^2 \quad \ldots\ldots (7)$$

As shown in Eq. (6), the thickness of the boundary layer is functions of Pr number and also of a distance in a vertical direction. Thus, the thickness of the boundary layer depends on the kind of fluid and the scale of a test model.

c. Application of similarity laws

In the facility having metal casks, there are the two kinds of flows: One is the main flow in the facility, and the other is the local flow around each metal cask surface. When we perform a heat removal test by using the scale model, a thermal hydraulic phenomenon can be simulated under conditions of enough turbulence and the same flow residence between the actual facility and the scale model. On the other hand, the thickness of the boundary layer around the heated object is the functions of Pr number and of the distance in the vertical direction. Thus, the temperature distribution in the boundary layer does not depend on the main flow.

2) Test device and test method

a. Test device

A test device is shown in Fig. 3.2.2(1)-3. A metal cask in the facility was simulated by using a heated object in an acrylic tank. The object was made from chloride vinyl cylinder which was wound with a Nichrome wire electrical heater and covered by a thin cupper plate.

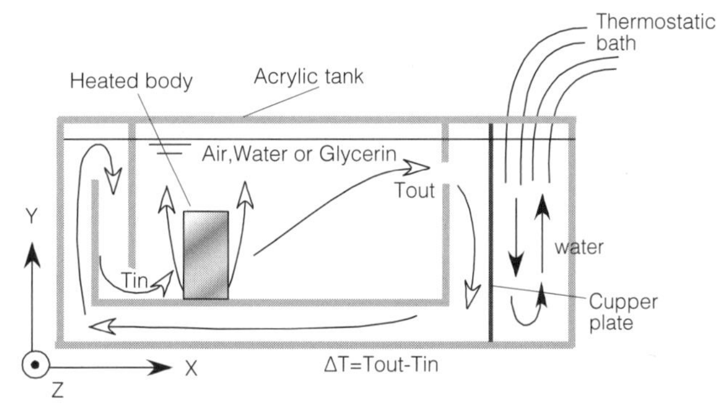

Fig. 3.2.2(1)-3 Test device for heat removal

Fluid is heated on a cylinder surface, rises and flows out from an outlet of the tank, and is cooled by the cold copper plate and goes down. The cold copper plate keeps 10 °C by cold water. The fluid flows through a duct at the tank bottom, and enters from an inlet. Thus, natural circulation occurs.

We used three scale models for the experiment. A scale ratio of Model I : Model II : Model III was 1:1/2:1/2.6. When the height of an actual cask is 5.5 m, the ratio of Model I : Model II : Model III is 1/22:1/42:1/56.

b. Test method

Natural circulation tests were performed using three different scale models and three different working fluids (air, water, glycerin), while changing a heat rate (1.6~200 W). Test conditions are shown in Table 3.2.2(1)-1. We measured vertical temperature distribution at the center of the heated object in the tank.

Table 3.2.2(1)-1 Test conditions for heat removal

	Air		Water		Glycerin	
	Heat rate (W)	Case name	Heat rate (W)	Case name	Heat rate (W)	Case name
Model I	40	AL40	200	WL200	—	—
	20	AL20	100	WL100	—	—
	10	AL10	50	WL50	—	—
Model II	7.9	AM8	107.3	WM107	—	—
	5.5	AM6	53.1	WM53	—	—
	2.6	AM3	26.0	WM26	—	—
Model III	4.1	AS4	26.4	WS26	28.2	GS28
	2.8	AS3	17.6	WS18	17.6	GS18
	1.6	AS2	8.7	WS9	8.7	GS9

3) Test results

ΔTai means a difference between Tmax and Tmin, and Tmai means an average of Tmax and Tmin. To obtain Gr* number and Ra* number, Tmai was used for characteristic temperature of physical properties. When the heating rate, the height and the diameter of an actual cask are 20 kw, 5.5 m and 2.5 m respectively, Ra* is 1.4×10^{15} .

An adaptable range of the similarity law for natural circulation was investigated by comparing non-dimensional temperature distributions.

a. Air test

The non-dimensional temperature distributions were obtained using Eq. (8).

$$T^{*} = \frac{T - T_{min}}{T_{max} - T_{min}} \quad \text{..... (8)}$$

The non-dimensional temperature distributions are shown in Fig. 3.2.2(1)-4.

All the figures are similar regardless of different test cases.

In the range of $3.1\times10^{7}<Ra^{*}<3.8\times10^{9}$, it was found that the similarity law can be used.

b. Water test

The non-dimensional temperature distributions are obtained by the same method used in the air test a.

In the range of $9.7\times10^{8}<Ra^{*}<2.4\times10^{11}$, it was found that the similarity law can be used for the water test.

c. Glycerin Test

The non-dimensional temperature distribution are obtained by the same method used in the air test a.

In the range of $1.4\times10^{7}<Ra^{*}<8.7\times10^{7}$, it was found that the similarity law can be used for the glycerin test.

d. Consideration of difference of fluids

In order to investigate the effect of working fluids, the tests using air, water and glycerin were performed under conditions of the same test model (Model III) and the almost same Ra* number. The non-dimensional temperature distributions are shown in Fig. 3.2.2(1)-5. The boundary layers of natural convection are described in these figures. The thickness of the boundary layer was calculated by Eq. (6). Temperature gradient of the layer in the case of air is shaper compared to the case of water and glycerin. The shapes of thermal stratification in the cases of water and glycerin are similar.

Normalized temperature difference (ΔTai) is shown in Table 3.2.2(1)-2.

In the cases in which the same model was used, the temperature distributions were normalized by the heat rate

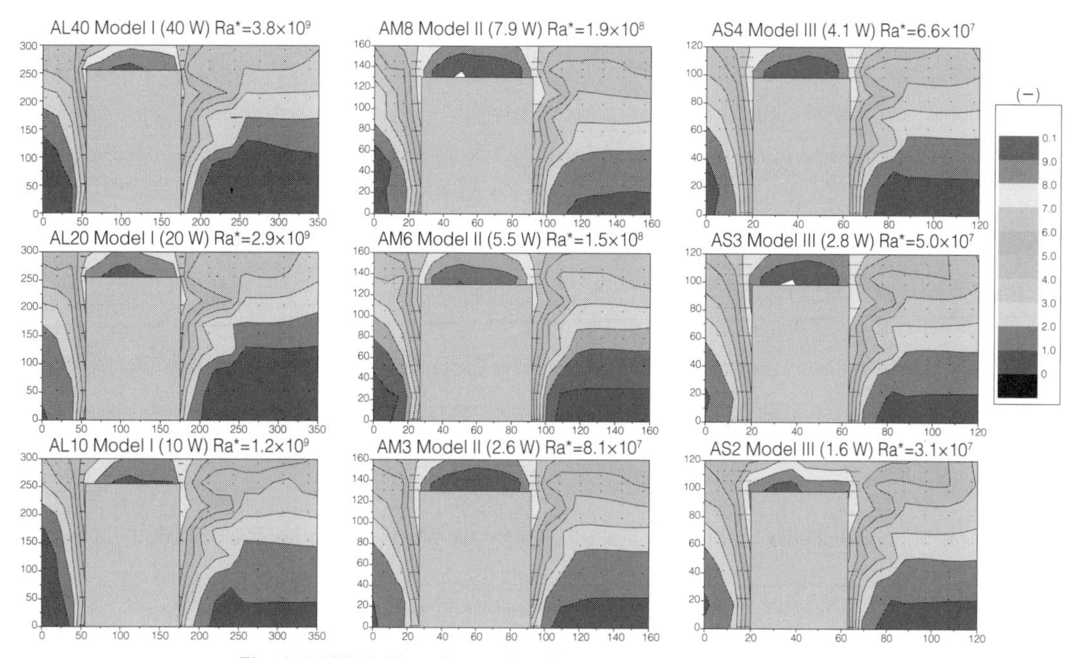

Fig. 3.2.2(1)-4 Non-dimensional temperature distribution (Air)

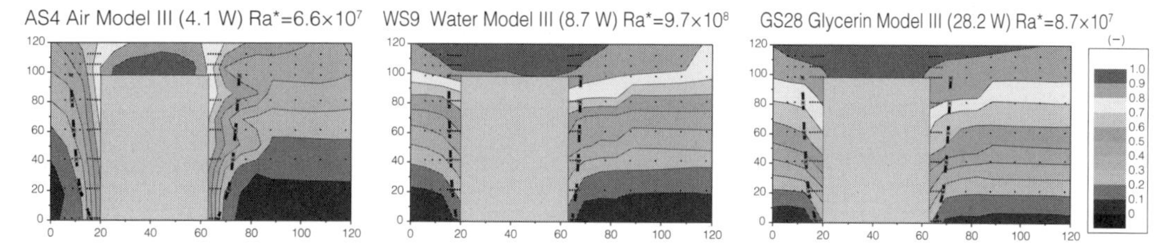

Fig. 3.2.2(1)-5 Non-dimensional temperature distributions and boundary layer (Air, Water, Glycerin)

Table 3.2.2(1)-2 Normalized temperature difference (Air, Water, Glycerin)

Case name	ΔTai(°C)	$\Delta Tai/(Q^{2/3}(\beta\rho^2c^2)^{-1/3})$	ξ	$\Delta Tai/(Q^{2/3}(\beta\rho^2c^2)^{-1/3}\xi^{1/3})$
AS4 (Air)	41.9	254	9	122
WS9 (Water)	4.8	1620	7822	82
GS28 (Glycerin)	36.7	5926	694451	67

(Q) and physical properties ($\beta\rho^2c^2$). In the results, the values of $\Delta Tai/(Q^{2/3}(\beta\rho^2c^2)^{-1/3})$ were markedly different. This result means that the pressure loss coefficients were different when the tests using the different fluids were performed. It is difficult to make Ri numbers same in different fluid tests even if the same model is used. In other words, it is difficult to match the pressure loss coefficients (ξ) by using different fluids.

Thus, ξ was obtained using characteristic velocity, and ΔTai was normalized by $(Q^{2/3}(\beta\rho^2c^2)^{-1/3}\xi^{1/3})$.

The differences between the values of $\Delta Tai/(Q^{2/3}(\beta\rho^2c^2)^{-1/3}\xi^{1/3})$were small. The reason of the difference between the non-dimensional temperature distributions could be caused by the different Ri numbers. Thus, the natural circulation test by air is difficult to be simulated by water or glycerin.

4) Consideration of test method taking account of similarity law

A flowchart of a test method for simulating the thermal-hydraulics phenomenon in the metal cask storage facility is shown in Fig. 3.2.2(1)-6, which takes account of the results of the natural circulation tests.

As design conditions, the shape and size of the facility, inlet air temperature, and the size and heat rates (Qp) of the actual casks are set. Next, the size of a model and the kind of fluid are determined. The heat rate (Qm) is preferably set in the range where accurate temperature data can be obtained.

As mentioned before, the two kinds of the similarity laws for the flows exist in the cask storage facility. One is for the main flow in the facility, and the other is for the flow around the heated object. For the similarity law of the main flow, air should be selected as a working fluid in order to match Ri numbers between the actual facility and the test model.

The shape of the model should be same as that of the facility. When the model is too small compared to the

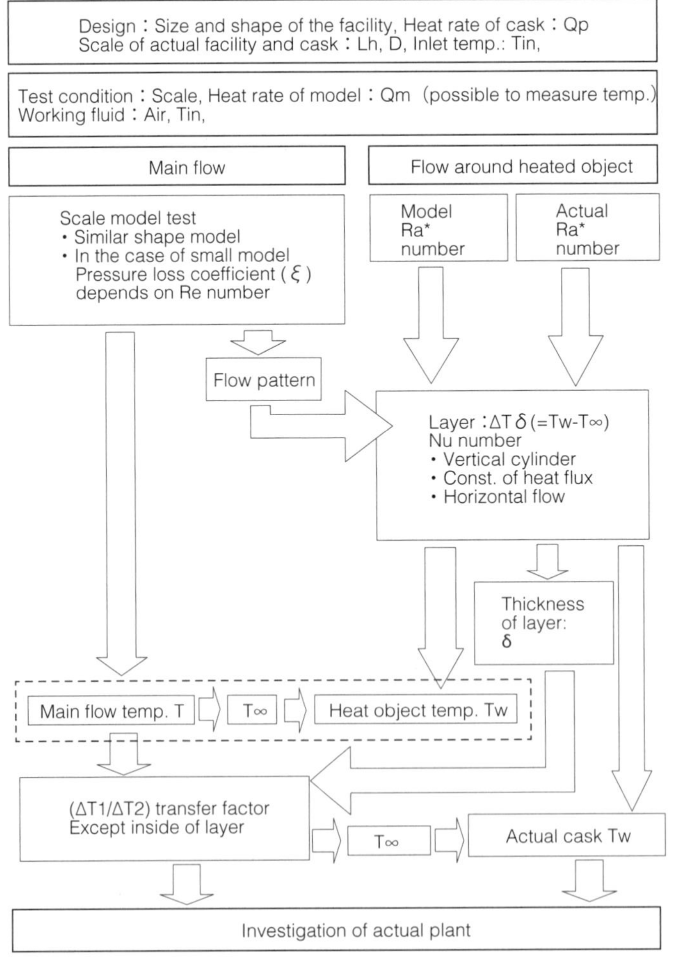

Fig. 3.2.2(1)-6 Flowchart of test method for heat removal

facility, the pressure loss at each part of flow paths depends on Re number. Thus, the pressure loss coefficient should be investigated by using the flow residence handbook[3].

The temperature of the facility can be calculated by using the test data and Eq.(4). On the other hand, as for the flow around the heated object, Ra* numbers of the model and the actual cask should be calculated using their heat rate and scale. Next, the temperature difference ($\Delta T\delta$) between the boundary layer (δ) and the object surface can be calculated by a heat transfer correlation equation (Nu number) for a vertical cylinder under a constant heat flux. When a horizontal flow colliding with the heated object exists, the heat transfer correlation equation (Nu number) modified by the effect of the horizontal flow should be used.

Temperature data obtained by the experiment can be utilized for verification data of analysis codes. When we obtain the temperature of the actual plant from the data obtained by the experiment, we should take care of the two different similarity laws.

5) Summary

We performed the experiments using 1/22, 1/42 and 1/56 scale models of an actual cask whose height is 5.5 m, under the condition that the heat rate range of the models were 1.6~200 W and working fluids are air, water and glycerin. The results were obtained as follows.

a. Two kinds of flows exist in the cask storage facility. One is the main flow and the other is a local flow around a heated object. There is a boundary layer around the heated object, so that a temperature distribution in the boundary layer should be treated differently from one of the main flow. Thus, two different similarity laws should be used.

b. The non-dimensional temperature distributions in the air test model are similar in the range of $3.1\times10^{7}<Ra^{*}<3.8\times10^{9}$.

c. It is difficult to simulate an air model test by using other fluids. For the similarity law for the main flow, air should be selected as a working fluid to match Ri numbers between an actual facility and a test model.

References

1) H. Takeda, (2007), "Proposal of the Heat Removal Test Method by Scaled Model for Metal Cask Storage Facilities (N06032)", CRIEPI report, (in Japanese)

2) S. Goldstein, ed., (1938), "Modern Developments in Fluid Dynamics, Vol.2", Oxford Press.

3) Flow Resistance : A Design Guide for Engineers, Hemisphere Publishing Co.

Nomenclature

Re = Rynolds number = UL_D/ν

Ri(h) = Richardson number = $\beta g\Delta Th/U^2$

Pe = Peclet number = ν/α

Pr = Prandtl number = UL_D/α

Gr* = Modified Grashof number = $\beta g q_c z^4/\lambda\nu^2$

Ra^* = Modified Rayleigh number = Gr^* Pr

U = Air velocity, m/s

L = Representative length, m

h = Cask height, m

L_D = Draft height, m

L_S = Stack height, m

H= Inlet height, m

c = Specific heat capacity, kJ/kg/°C

ΔT= Temperature difference between outlet and inlet, °C

ΔTai= Temperature difference between Tmax and Tmin , °C

g = Gravitational acceleration, m/s^2

Nu(z) = Local Nuselt number, = $h'z/\lambda$

Q = Total heating rate, kW

q = Heat flux per volume, kW/m^3

Tin = Inlet air temperature, °C

Tout = Outlet air temperature, °C

Tmoi= Mean temperature between Tin and Tout, °C

Tmax= Maximum temperature, °C

Tmin= Minimum temperature, °C

Tmai= Mean temperature between Tmax and Tmin , °C

z = Distance from the bottom to the top of the cask, m

Greek letters

α = Thermal diffusivity coefficient, m^2/s

β = Thermal expansion coefficient, 1/°C

λ = Thermal conductivity, kW/m/°C

ν = Kinetic viscosity, m^2/s

ρ = Density, kg/m^3

ξ = Friction coefficient, (-)

(2) Heat removal performance test of high-ceiling building

In the heat removal design of a cask storage building, it is particularly important to understand natural cooling phenomena reflecting mutual relations such as cask position. In the present heat removal performance test of the cask storage, the heat removal from the cask to the building is studied by investigating the flow status inside the building and the temperature and flow speed distributions[1)].

1) Test equipment

The target building was an actual storage building of the size 41.4 m × 238 m × height 30 m (full dimensions including storage area) where a total of 420 casks can be placed in 10 rows × 20 columns and 10 rows × 22 columns. We used a 1/5-scale model for the test by taking account of the building's symmetry. (See Fig. 3.2.2 (2)-1 to -3.)

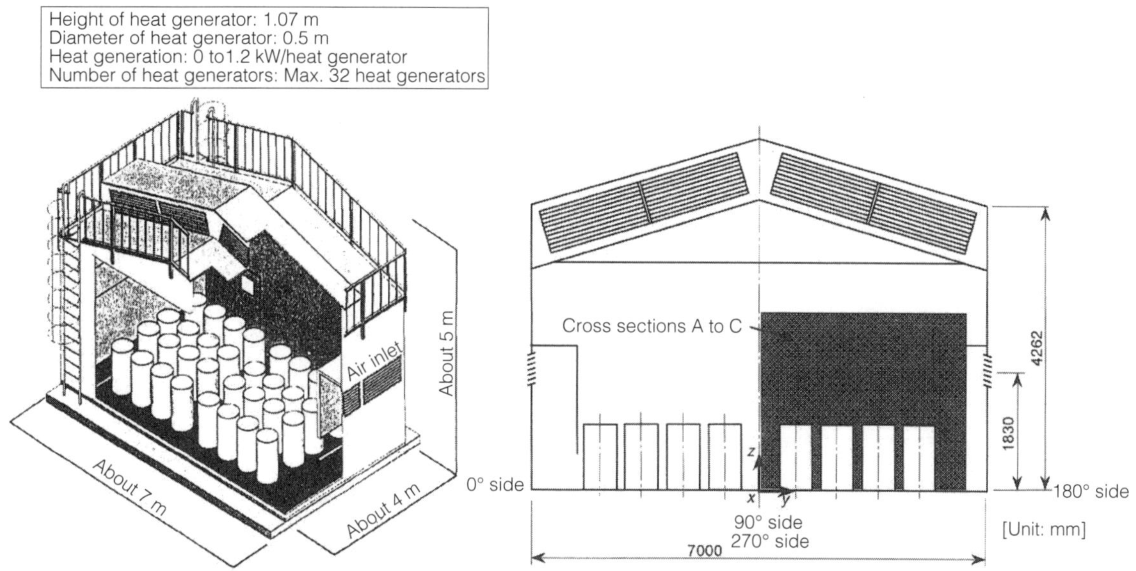

Fig. 3.2.2(2)-1 Overview of cask heat removal test equipment (1/5-scale model)

Fig. 3.2.2(2)-2 Elevation view of cask heat removal test equipment

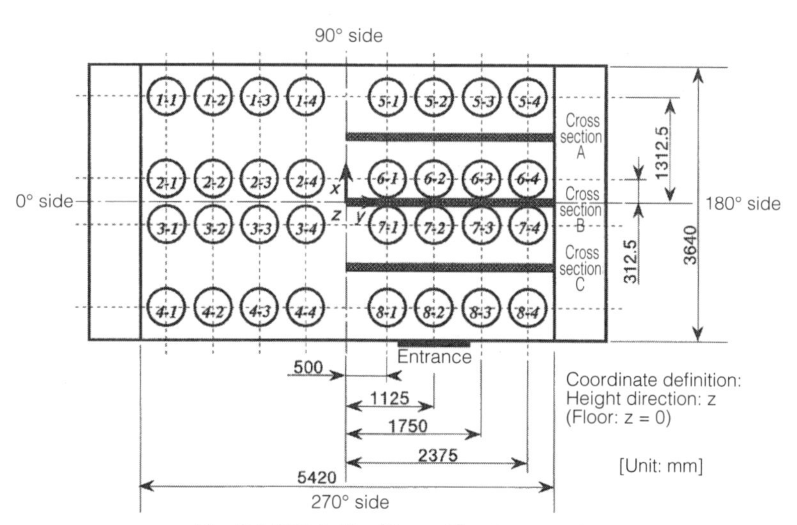

Fig. 3.2.2(2)-3 Position of heat generators

2) Idea of similarity law

The natural convection phenomena of the cask storage system can be described by the equation of continuity (mass conservation law), equation of motion (Navier-Stokes Equation: momentum conservation law), and energy equation (energy conservation law). In order to find the heat flow property of the cask storage system in the model test, it is necessary to set the *Re* number, *Eu* number, *Pe* number, and *Ri* number to be suitable for the actual building. When one of the *Ri* number and *Eu* number matches the actual building data, the other also matches the data.

3) Test method

In the test, air was used as fluid as in actual facilities. If the temperature dependence of the air flow properties is taken into account, the temperature increase of the air in the test should be almost equivalent to that in actual facilities. Also, the air flow speed of natural convection should be as large as possible so that it could be easily measured. In the test, we set the scale ratio of the model to be 1/5 and the heat values to be 0.5 kW/heat generator (n_Q = 1/40) and 1 kW/heat generator (n_Q = 1/20) under constant heat flux condition. Here n_Q is the reduction ratio.

4) Test result and discussion

The air flow condition in the building was observed to find the position and size of the stagnation area and high temperature area.

(i) Flow situation around heat generators

Fig. 3.2.2(2)-4 shows an elevation view of the air flow. The two dimensional flow view was obtained by a visualization technique with laser beam and smoke.

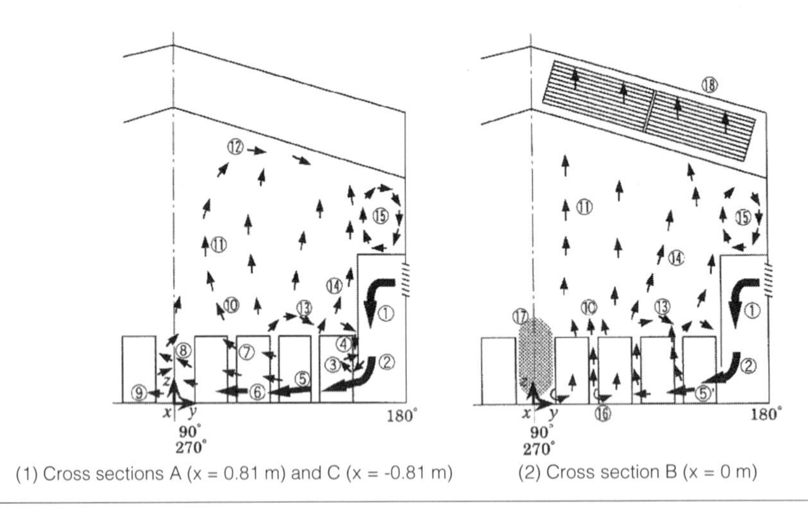

(1) Cross sections A (x = 0.81 m) and C (x = -0.81 m)　(2) Cross section B (x = 0 m)

(i) Downward flow from air-in louver
(ii) Fast obliquely-downward flow to storage unit
(iii) Vortex around wall of storage unit
(iv) Downward flow along duct wall (on storage unit side)
(v) Fast flow along floor (cross sections A and C)
(v′) Fast flow along floor (cross sections B)
(vi) Deaccelerating flow along floor
(vii) 2-dimensional diffusion
(viii) Collision and rise of flows from 0° and 180° sides
(ix) Intermittent flow from 180° to 0° sides
(x) Upward flow tilted to the center
(xi) Upward flow to the top
(xii) Flow along ceiling to 180° side
(xiii) Horizontal flow to 180° side
(xiv) Upward flow tilted to 180° side
(xv) Stagnation and circulated flow at pocket above air-in louver
(xvi) Flow in from cross sections A and C
(xvii) Flow in x direction and vortex
(xviii) Exhaust from model storage building

Fig. 3.2.2(2)-4 Air flow condition in whole storage area (yz cross sectional flow diagram)

(ii) Temperature distribution in storage area

The spatial temperature distribution on cross sections A and B are shown as difference from the incoming air temperature (external air temperature) in Fig. 3.2.2(2)-5. The heat generation condition was set to be 0.53 kW/heat generator (16.8 kW in whole facilities). The air flow speed near the floor ($z \leq 1.1$ m) around the heat generators on cross section A was high and the temperature in this region on cross section A was lower than that on cross section B. In the upper space ($z > 1.1$ m) above the heat generators, on the other hand, the temperature at $z = 1.1$ m near the top of the heat generators on cross section A was higher on the air inlet side than on the center side. However the air became mixed and the temperature became uniform as the measurement point went up toward the upper side (air outlet). At $z = 3.0$ m, the temperature was almost uniform and the maximum temperature difference on the xy cross section was about 1 °C.

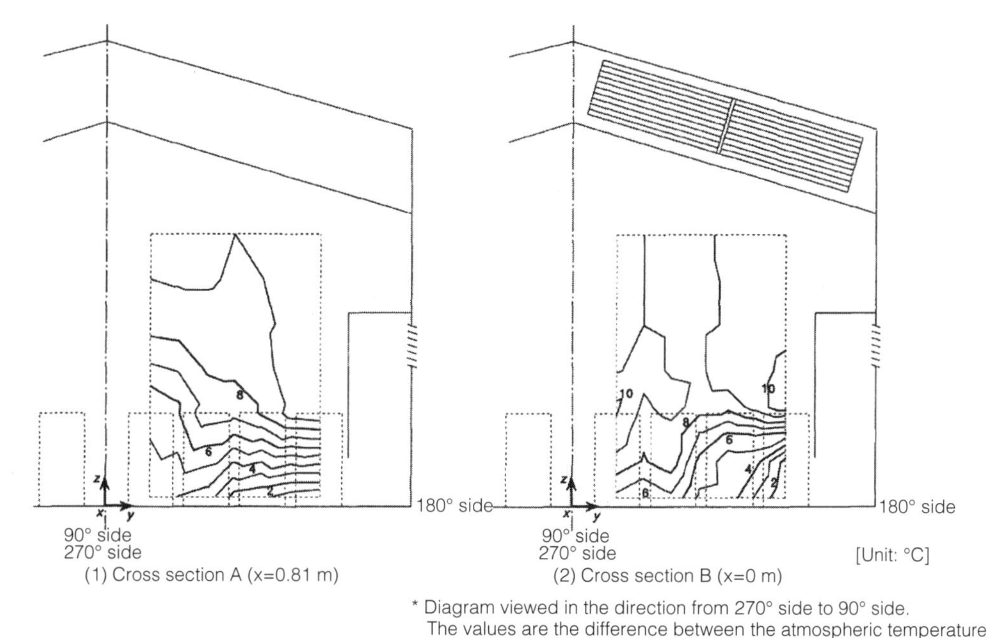

Fig. 3.2.2(2)-5 The yz cross sectional temperature distribution in storage building (heat value: 0.53 kW/heat generator)

(iii) Temperature distribution of heat generators

The heat generation condition was set to be 0.53 kW/heat generator (16.9 kW in whole facilities). Generally, the surface temperature increases with the height of the measurement point, but the temperature near the top was relatively low. The temperature decrease near the top could be due to air circulation above the heat generators (back flow from the center of the storage area to the air inlet).

The air flow condition in the storage area shows that the air flew more in cross sections A and C where the distance between the heat generators was large than in cross section B where the distance was small. The surface temperature distribution in the heat generator circumference direction reflected this flow condition. The heat removal of the heat generators was found be affected by not only the upward flow due to the buoyance near the surface but also the horizontal flow from the air inlet to the center of the storage area.

(iv) Influence of test parameters

(a) Influence of heat value

For studying the influence of the heat value to the air temperature in the storage area, a test was conducted under the condition of 1.04 kW/heat generator (33.2 kW in whole facilities) and the result was compared with that obtained with 0.53 kW/heat generator. It was found that the temperature profile in the storage area was almost the same irrespective of the heat value. Since the temperature distribution is controlled mostly by the air flow speed distribution, it could be considered that the flow condition in the storage area was also not affected by the heat value.

(b) Heat removal performance in the initial placement of casks

Casks are placed after the operation of the facilities starts. Even when the heat value in the whole facility is small or when the casks are placed in only a part of the storage area, cooling air has to be sufficiently supplied to the stored casks and spent fuels have to be appropriately cooled.

We studied a change of the heat removal performance by changing the number of the heat generators in the storage area to simulate the initial placement of casks in actual facilities. We performed tests by choosing 24, 16, or 8 heat generators out of the 32 heat generators in Fig. 3.2.2(2)-3 to study the effect of an asymmetric layout of the heat generators in the storage area and the flow condition and temperature distribution in the chosen cases. Table 3.2.2(2)-1 shows the relation between the air inflow volume and the temperature increase in the three cases where 24, 16, and 8 heat generators were placed in the equipment. In every case, there was no significant difference in the air inflow volume between the 0° side and the 180° side. Namely the volumes of the

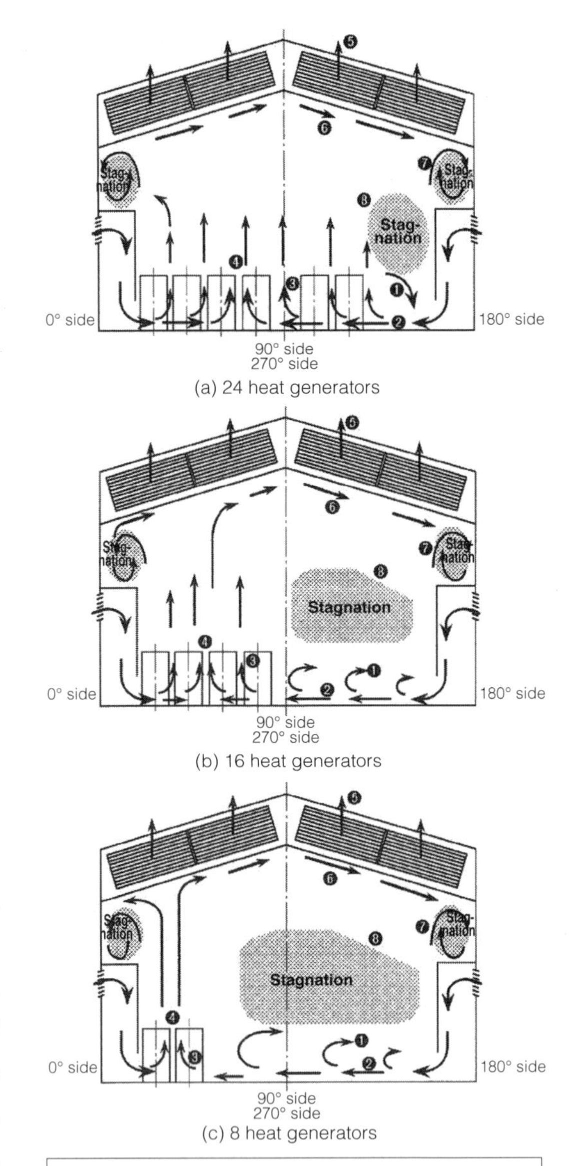

(i) Inverse flow to air inlet
(ii) Flow along floor
(iii) Upward flow near model heat generator
(iv) Collision and rise of air from both air inlet
(v) Exhaust from model storage building
(vi) Flow along ceiling from 0° side to 180° side
(vii) Stagnation and circulation at pocket above air intake louver
(viii) Stagnation at a place where no model heat generator is placed

Fig. 3.2.2(2)-6 Flow condition in storage area with initially-placed casks

air inflow from the both air inlets were balanced even when the heat generator layout was asymmetric. Next we observed the flow condition in the storage area by using smoke for visualization of the air flow. The result is shown in Fig. 3.2.2(2)-6.

It was confirmed that, even in the initial placement of casks, individual draft forces were produced depending on the heat values of the heat generators and a necessary amount of the air for the heat removal could be supplied appropriately. Namely a necessary amount of the cooling air could be supplied by its own draft forces and hence temperature increase of any heat generator due to unbalanced heat generation in the storage area would not be observed. However if there are only a small number of heat generators in the storage area, the air inflow volume would be small. In this case there would be a small air-spraying effect to the casks in comparison to the case where all the casks are placed in the storage area, and the upward flow near the cask surface would contribute more to the heat removal effect.

Table 3.2.2(2)-1 Air inflow volume and temperature increase with initially-placed casks

Number of heat generators	Power meter read Q_0: kW	Air inflow volume: m^3/s			Temperature increase: °C
		0° side	180° side	Total	
24	12.5	0.65	0.73	1.38	7.1
16	8.3	0.71	0.62	1.33	5.5
8	4.1	0.51	0.44	0.95	3.6

5) Extrapolation to actual facilities

The air flow speed distribution and temperature distribution in actual facilities were examined by using the above test results and the similarity law. Since the heat value was assumed to be about 20 kW per cask in actual facilities, the total heat value of 32 casks in actual facilities (where the casks are place in the standard positions shown in Fig. 3.2.2(2)-3) would be about 640 kW. Therefore we set Qp = 640 kW and Lm/Lp = 1/5. So, if the standard heat value in the test is Qm = 16.3 kW, the air volume flowing into the facilities is 86.04 m^3/s and the air temperature increase is 7.0 °C (difference in mixed mean temperature).

For example, if we assume the inlet air temperature to be 38 °C as designed, the exhaust air temperature is 45 °C. Since the air temperature and the building concrete temperature is almost the same, the concrete temperature is way below the allowable value 65 °C (allowable temperature of other parts in stationary state) and hence the long-term robustness of the building would be ensured at least from a viewpoint of the temperature.

Also the air flow distribution and temperature distribution in actual facilities were evaluated. The evaluation was made with Qm = 16.8 kW (0.53 kW/heat generator) in the test and Qp = 640 kW (20 kW/cask) in actual facilities and with Lm/Lp = 1/5. The temperature distribution evaluation result for cross sections A to C in Fig. 3.2.2(2)-3 is shown in Fig. 3.2.2(2)-7. The maximum flow speed in the storage area was 2.0 m/s near the air inlet floor. We can assume that the maximum temperature would be observed near the cask surface. In our study of the three cross sections, the highest temperature, i.e. the inlet temperature plus 8.3 °C, was found near the top of the cask closest to the air inlet on cross section B. This is sufficiently lower than the allowable temperature.

6) Summary

In this section we performed the heat removal performance tests of dry storage casks by using a 1/5-scale model of whole facilities in order to study the air flow conditions inside the storage area and the temperature and flow speed distributions of the air and examined the heat transfer from the model cask to the air. We also investigated the influence of the parameters such as heat value and the number of heat generators on the heat removal performance of the facilities. In addition, the flow speed distribution and temperature distributions expected in actual facilities were evaluated based on a similarity law. As a result, the heat removal safety of the cask storage was verified and the air flow was revealed in detail.

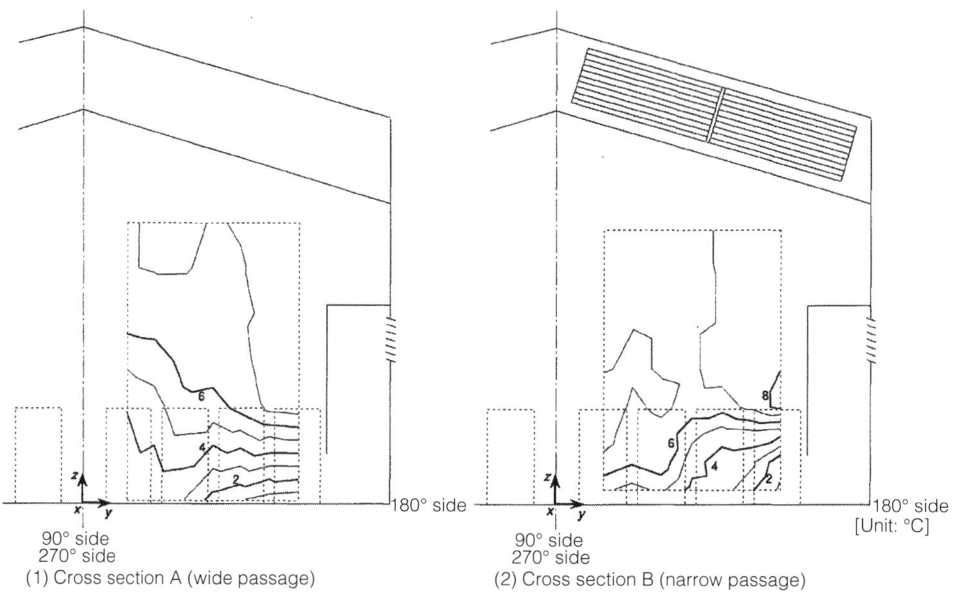

Fig. 3.2.2(2)-7 The yz cross section in cask storage extrapolated from the test results (heat generation: 20 kW/cask)

[Symbols]

x : Length in x direction (m)

y : Length in y direction (m)

z : Length in z direction (m)

L : Typical length (m)

Q : Heat value (kW)

Re : Reynolds number ($= dU / \nu$)

Eu : Euler number ($= \Delta P / (\rho U^2) = \xi / 2$)

Ri : Richardson number ($= g\beta\Delta TL / U^2$)

Pe : Peclet number ($= Pr\ Re$)

$Gr^*(z)$: Locally modified Grashof number in z direction ($= Gr(z)Nu(z) = g\beta qw\, z^4 / (\lambda\nu^2)$)

$Nun(z)$: Local Nusselt number in z direction with natural convection ($= h(z)z / \lambda$)

$Nu(z, \phi)$: Local Nusselt number in z and ϕ directions

$C(Re, \phi)$, $C(Re)$: Coefficients representing an effect of flow spraying to the heat generators (horizontal component of the flow)

Subscripts

y : y direction

z : z direction

m : Test equipment

p : Actual equipment

ϕ : Circumferential direction

Superscripts

* : Dimension-less value

Reference

1) K. Sakamoto, T. Koga, Y. Hattori, M. Wataru, Y. Gomi and E. Kashiwagi : "Development of Evaluation Method for Heat Removal Design of Dry Storage Facilities (Part 3) - Heat Removal Test on a Cask Storage System -", CRIEPI Report U98003, 1998.

(3) Heat removal performance test of stack type cask storage facility

A spent fuel storage facility with a high ceiling as shown in section 3.2.2(2) has been used. In order to reduce costs and a period of construction, a new-type cask storage facility for nuclear spent fuel is proposed. Ceiling height of a storage area is made low and a stack is set up on the opposite side of an air inlet. Compared with the conventional storage facilities, this stack-type facility has a characteristic cooling flow path in the storage area.

Influence of the ceiling height and the stack height on a thermal-hydraulic phenomenon was examined by the experimental survey using a 1/5 scale model. Then, actual phenomenon could be predicted by applying a similarity law to the experiment results[1),2)].

1) Test device

The 1/5 scale model of the facility was used. Twenty-four cask models were placed upright in the experimental facility. A bird's-eye view of a heat removal test device is shown in Fig. 3.2.2(3)-1.

Vertical and horizontal views of the test device are shown in Fig. 3.2.2(3)-2.

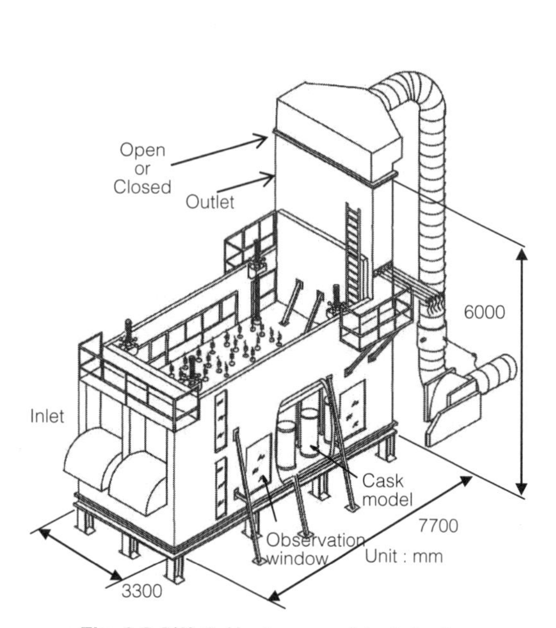

Fig. 3.2.2(3)-1 Heat removal test device

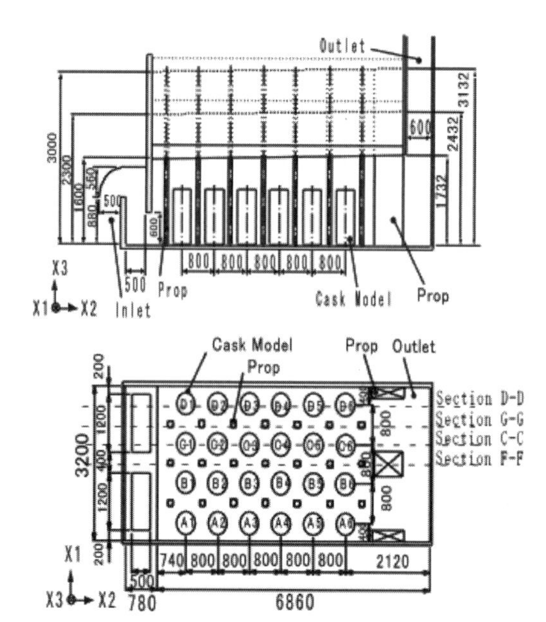

Fig. 3.2.2(3)-2 Vertical and horizontal views of heat removal test device

2) Similarity law

The thermal-hydraulic phenomenon in the facility is governed by a continuity equation, momentum equation, and energy equation.

These basic equations can be transformed to non-dimensional forms. The following five non-dimensional numbers are used as independent parameters in the transformation:

$Re=UL/\nu$, $Ri=\beta g\Delta TL/U^2$, $Eu=\Delta P/\rho U^2=\xi/2$, $Pe=UL/\alpha$ and $q^*=q/\rho cU\Delta TL^2$.

All the above non-dimensional numbers must be equated between an experiment and the actual phenomenon to complete similarity. But it is not possible in a simulation experiment using a scale model.

When Re number and Pe number are large enough, diffusion terms in the momentum equation and energy equation become negligibly small in comparison with the other terms. Under this condition, the other non-dimensional numbers, Ri number, Eu number and q*, have significant impact on the phenomenon. Thus, we must give first priority to these parameters in setting of similarity.

By conducting the following tests, the actual phenomenon in the facility can be predicted (See the reference[1] for detail).

3) Test results

a. Influence of the ceiling height

First, we performed experiments to investigate the influence of the ceiling height. In these experiments the ceiling height and Ri number were varied as parameters. The ceiling height (Hc) was set to 1.6 m (about 1.5 times the cask height), 2.3 m (about 2 times the cask height) and 3.0 m (about 3 times the cask height). The Ri number was set in the range of 2< Ri(h) <12.

Fig. 3.2.2(3)-3 shows temperature distributions inside the storage area.

When Hc was 1.6 m, a large area of the ceiling was heated up in comparison with the other ceiling height cases, so that it was found that the ceiling height influences the temperature of the ceiling.

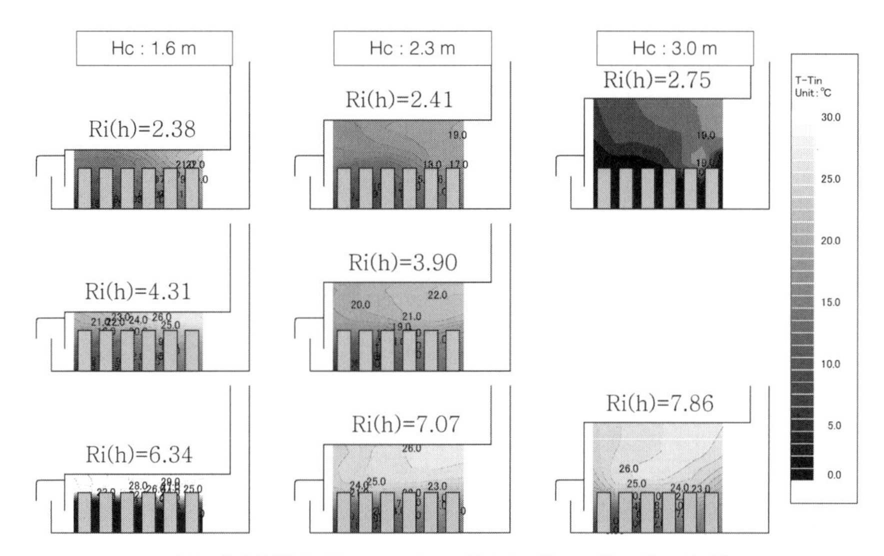

Fig. 3.2.2(3)-3 Temperature distributions (Section D-D)

Fig. 3.2.2(3)-4 is a diagrammatic sketch of a flow pattern in the storage area observed by a visualization test, under the condition of Hc is 2.3 m and Ri(h) is 7.07. A part of flows from the inlet collided with the first-row casks and rose up, but the main stream went along the floor towards the final-row casks. An upward flow was observed behind the 4th-row casks. The flow collided obliquely with the ceiling and divided into two flow branches. The main branch was inhaled into the stack and the other returned in an inlet direction along the ceiling. Some stagnation occured in the bottom behind the last row of casks.

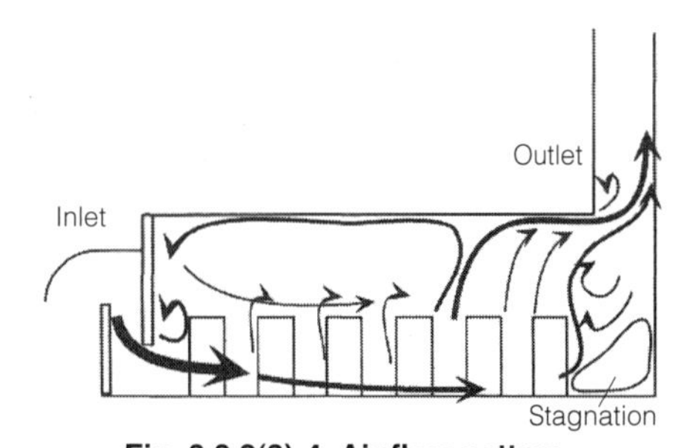

Fig. 3.2.2(3)-4 Air flow pattern (Hc=2.3 m, Ri(h)=7.07)

Fig. 3.2.2(3)-5 shows characteristics of heat transfer of the casks in three cases of ceiling heights. The Ri numbers are almost same in the all cases. C1 means the first-row casks and C6 is the final-row casks. In these graphic charts, the horizontal axes show the height of the cask from its bottom, and the vertical axes show local Nu number.

The Nu number changed along a circumferential axis in the first-row casks. On the other hand, Nu number did not change in the final-row casks. The heat transfer correlation (Vliet et al.)[3] is also shown in the graphic charts for comparison. From Fig. 3.2.2(3)-5, it was found that the ceiling height hardly influences the heat transfer of the casks.

We could conclude that the ceiling height should be decided in consideration of restriction of the ceiling temperature.

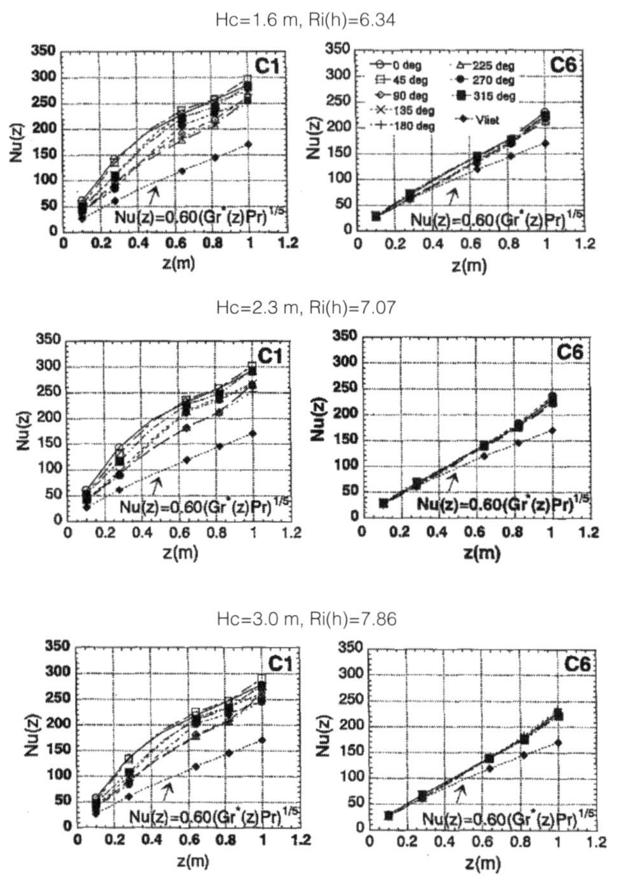

Fig. 3.2.2(3)-5 Heat transfer rate (each case of ceiling height)

b. Influence of stack height

Next, we performed experiments to investigate the influence of the stack height. Reference ceiling height was set to 2.3 m (about 2 times the cask height) based on the test results of ceiling height.

When the stack got higher, the higher flow rates could be obtained. We derived the relation between the heat removal rate and the stack height, and another relation between Ri (h) number and the stack height by one-dimensional thermal hydraulics analysis that is considered as the balance of buoyancy force and pressure loss of the facility. Pressure loss coefficients in a code of the analysis were originally given by a resistance handbook and modified by using natural convection test results.

Fig. 3.2.2(3)-6 shows the relation between the heat removal rate and the stack height. The higher stack brought

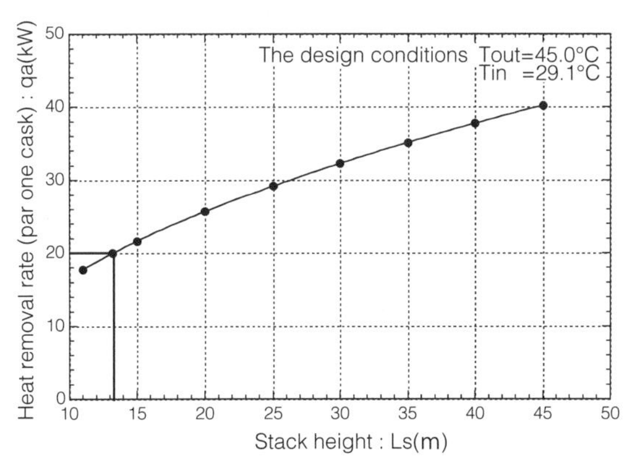

Fig. 3.2.2(3)-6 Relation between Ls and q_a

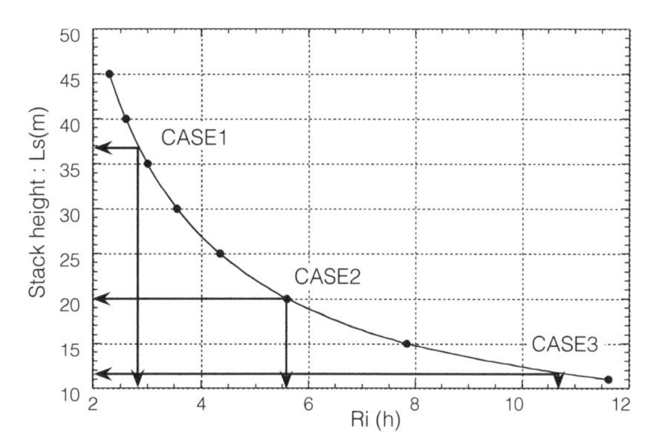

Fig. 3.2.2(3)-7 Relation between Ri(h) and Ls

about, the higher heat removal rate under the condition of constant temperature rise.

Air temperatures at the inlet and the outlet were set to 29.1 °C and 45.0 °C respectively as design conditions. It was clarified that if the stack in the actual facility is 13 m or higher, 20 kW per actual cask can be removed.

Fig. 3.2.2(3)-7 shows the relation between Ri(h) number and the stack height of the actual facility. We set CASE2 as a reference, in which Ls is 20 m. In CASE1, Ls is 37 m. In CASE3, Ls is 12 m.

Test conditions and the results are shown in Table 3.2.2(3)-1. "E" in the table means enthalpy rise between the inlet and outlet.

Table 3.2.2(3)-1 Test conditions for heat removal

	Hc (m)	E (kW)	Q (kW)	Flow rate (kg/min)	U (m/s)	ΔT (℃)	Ri (h)	Heat balance (E-Q) / Q*100 (%)	Tin (℃)
CASE1	2.3	20.68	24.07	65.5	0.500	18.4	2.78	-14.3	13.8
CASE2	2.3	20.05	24.17	50.2	0.395	23.8	5.60	-17.1	22.9
CASE3	2.3	19.04	24.13	39.5	0.314	28.8	10.7	-21.0	20.9

It was found that the higher Ri(h) number is, the higher the temperature in the storage area becomes. In this table, we can see that the smaller Ri(h) number brings about the stronger horizontal flow.

Fig. 3.2.2(3)-8 shows a relation between Ra* number and local Nu number. In this figure, the heat transfer correlations [3)] are also shown for comparison.

The test results cover the range of 10^8<Ra*<10^{13}. The heat transfer rates obtained in this test is higher than the values of the correlation proposed by Vliet et al. Especially, we can see this tendency in the first-row casks (C1). There are two kinds of flow in the storage area of the facility. One is an upward flow induced by buoyancy force on a surface of the cask and the other is a horizontal flow induced by stack effect. In the first-row casks, the horizontal flow collides directly with the surface of the cask, and the layer near the surface is disturbed. Thus, it seems that the heat removal of the cask might be promoted by the horizontal flow in comparison with natural convection by only upward flow.

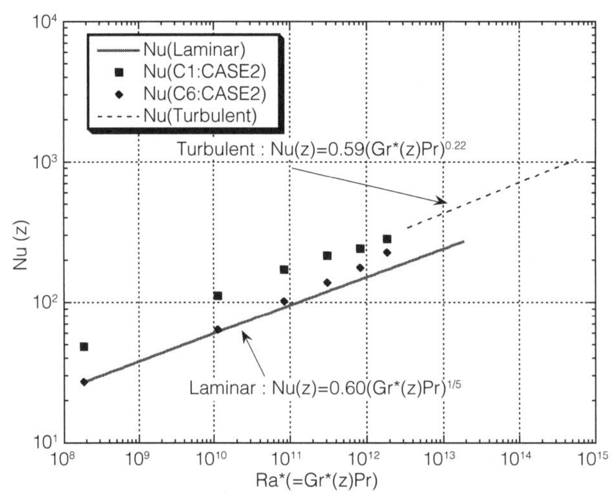

Fig. 3.2.2(3)-8 Relation between Ra* and Nu(z)

4) Prediction of phenomenon in actual facility

It was indicated that the experiments were performed in the range where the effects of the number of Re and Pe numbers were negligibly small.

We set Ls to 20 m, q_a to 20 kW per cask, and Tp-in to 29.1 °C as actual conditions. It was predicted that exhaust temperature might be 42.6 °C and the ceiling temperature might be 45.0 °C or lower. Mean velocity in the stack might be 0.668 m/s.

5) Summary

The following results were obtained in this study.

a. The ceiling height hardly influences the heat removal characteristics.

b. The ceiling temperature is seriously affected by the ceiling height. Thus, the ceiling height should be decided in consideration of restriction temperature of concrete and electrical parts.

c. There are two kinds of flows in the storage area of the facility. One is the upward flow induced by buoyant force on the surface of the cask, and the other is the horizontal flow induced by stack effect. These two flows contribute to cooling of the casks.

d. The stack height directly influences the heat removal characteristics.

References

1) H. Takeda, et al (2002), "Evaluation of Heat Removal Characteristic of Cask Storage Facility - Heat Removal Test for stack-type facility -" (in Japanese), AESJ-NUCE, Vol.8, No2, pp.145-154.

2) H. Takeda, et al, "Heat Removal Study for a New Type Cask Storage Facility", Proc. of ICONE9, April 8-12, 2001, Nice Acropolis, France

3) G.C. Vliet, and C.K. Liu, (1969), "An Experimental Study of Turbulent Natural Convection Boundary Layers;" J. Heat Transfer, Nov., Vol.91, pp.517-531.

Nomenclature

Eu : Euler number (-) : $\Delta p/\rho U^2=\xi/2$

c = specific heat capacity, kJ/kg/℃

ΔT= Temperature difference between outlet and inlet, ℃

ΔP= Pressure difference between outlet and inlet, Pa

g = gravitational acceleration, m/s^2

Gr*(z) = local modified Grashof number $=\beta g q_c z^4/\lambda \nu^2$

h = cask height, m

Hc = ceiling height, m

L = representative length, m

L_S = stack height, m

Nu(z) = Local Nuselt number, = h'z /λ

Q = Total heating rate, kW

q= heat flux per volume, kW/m^3

q_a = heat flux, kW

q_c = heat flux per area, kW/ m^2

Pe = Peclet number = ν/α

Pr = Prandtl number = $U L_D /\alpha$

Ra* = Modified Rayleigh number = Gr* Pr

Re = Rynolds number = $U L_D /\nu$

Ri(h) = Richardson number = $\beta g \Delta T h /U^2$

Tin = Inlet air temperature, ℃

Tout = Outlet air temperature, ℃

U = Air velocity, m/s

z = distance from the bottom to the top of the cask, m

Greek letters

α = thermal diffusivity coefficient, m^2/s

β= thermal expansion coefficient, 1/℃

λ= thermal conductivity, kW/m/℃

ν= kinetic viscosity, m^2/s

ρ= density, kg/m^3

ξ =friction coefficient, (-)

3.2.3 Heat removal effectiveness of tunnel type storage system

In Japan, nuclear power plants are often constructed in an area with hill and slope. A "tunnel storage system" is thought to be one of the effective methods for making use of this actual situation. One of the problems in designing this system is how to evaluate the effectiveness of heat removal from spent fuels by natural convection cooling. In the traditional systems for installing spent fuels into buildings on the ground surface, the airway network in whole systems is simple in many cases, so it is relatively easy to evaluate the efficiency of heat removal. By contrast, in the case of tunnel storage system, the complicated airway network is often adopted under the restriction of complex terrain of the power plant site. In such cases, sophisticated network analysis tools are needed for evaluating heat removal efficiency. In addition, seasonal change in temperature difference between inside and outside of tunnels influences in heat removal efficiency by natural convection. Furthermore, latent heat of condensation or vaporization on the rocks around airways also influences in heat removal efficiency. Summarizing the above, in order to evaluate heat removal efficiency from spent fuels of tunnel storage systems, it is needed to consider

a. sophisticated airway network analysis,

b. seasonal change in air temperature, and

c. latent heat of condensation or vaporization on the rocks around airways.

In order to evaluate heat removal efficiency of the designed tunnel storage systems in advance by considering three factors mentioned above, we have developed two analysis codes. The one is a "one-dimensional analysis code for calculation of airway network flow (1-D code)" and the other is a "three-dimensional analysis code for calculation of complicated thermal flow field of spent fuel depository (3-D code)". In this section, we present the outline and applied examples of these two analysis codes. These details are described in the liferature[1].

(1) One-dimensional analysis code for calculation of airway network flow (1-D code)

1) Outline

Our 1-D code is made based on an "underground ventilation" theory[2),3)]. Analyses for the ventilation of underground mines with complicated tunnels are carried out by applying this theory[4)]. In many cases of underground mines, the object of ventilation analyses is to secure required amount of fresh air for mine workers. In such cases, the main target of analyses is the natural ventilation under weak wind condition. In the present case to analysis the passive cooling of spent fuels, on the other hand, strong natural circulation and its accompanied strong wind occur caused by the large quantity of heat released from fuels. In such cases, an evaluation method is needed to calculate flow rate in the tunnel with high accuracy by considering air temperature at each location of the tunnel.

Our 1-D code is composed of two analysis codes. The one is a code for analyzing unsteady airflow in underground spaces (TRANCLIM) and the other is for analyzing airway network flow steadily (VENTCLIM)[5),6)]. Both of these two codes have good results to the analyses of ventilation in actual tunnel sites. By considering heat released from fuels additionally, these codes are applicable to the evaluation for the heat removal efficiency from spent fuels. By combining these two codes, our 1-D code enables us to analyze complicated airway network flow unsteadily with consideration of daily and seasonal changes in temperature outside the tunnel.

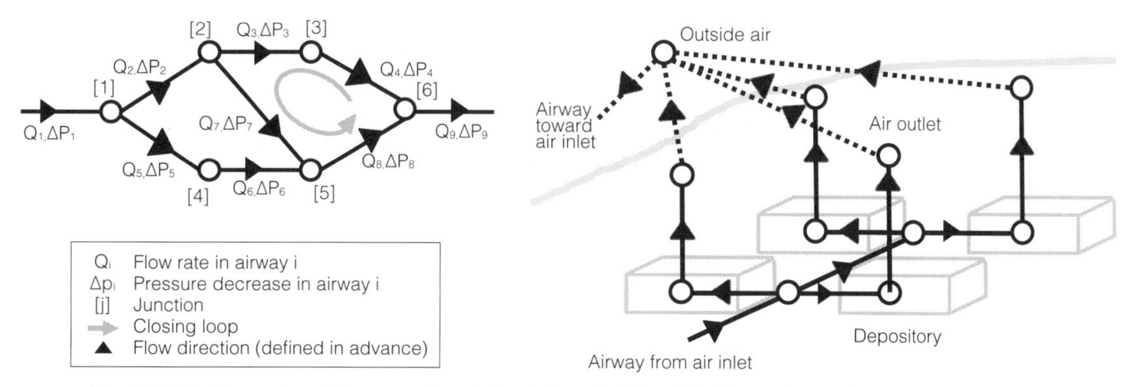

Fig. 3.2.3-1 Example of airway network for 1-D code (the right figure is a schematic image for applying the code for tunnel storage system)

In our 1-D code, flow rate in each airway is calculated under the limited condition of "mass-balancing" at all junctions and "pressure-balancing" in all closed loops of airways. A mass-balancing is the limited condition that the total flow rate of all airways linking to a junction is zero. At the junction [1] in Fig. 3.2.3-1, for example, the equation (1) must be consistent.

$$Q_1 - Q_2 - Q_5 = 0 \quad ..... (1)$$

In the equation (1), inflows to the junction [1] are marked positive (+) and outflows from the junction [1] are negative (-). A pressure-balancing is the limited condition that the total pressure change through all closed loops of airways is zero. Through the gray closed loop in Fig. 3.2.3-1, for example, the equation (2) must be consistent.

$$-\Delta P_4 - \Delta P_3 + \Delta P_7 + \Delta P_8 = 0 \quad ..... (2)$$

In the equation (2), the positive flow direction at each airway is defined in advance. If the direction of loop is same as positive direction of an airway, ΔPj is marked positive (+). And if not, ΔPj is marked negative (-). ΔPj in each airway is calculated by the equation (3).

$$\Delta P_j = R_j \times v_j^2 - P_N - P_F \quad ..... (3)$$

In the equation (3), turbulent air flow is assumed in all airways. vj is flow speed of the airway j, and Rj is specific resistance (degree of flow resistance). The variable R depends on the characteristics of airway, for example, wall material, configuration of cross section, curve of airway, and existing obstacles[7)]. P_N and P_F are increase in pressure between two ends of the airway caused by natural circulation and fan, respectively. P_N is caused by the difference in air temperature and height level between two ends of the airway due to the stack effect, and it also depends on thermal insulation performance of the airway wall. The flow rate in each airway Qj can be obtained by calculating simultaneous equations (1), (2) and (3) for all junctions, closing loops and airways.

Air temperature and wall surface temperature of each airway are calculated by the heat balance equation at the wall surface and the heat conduction equation in the wall element by using the boundary condition of air temperature outside the tunnel. In the heat balance equation, condensation and evaporation of water at the wall surface is also considered. In addition to the temperature, it enables us to calculate the air humidity in the airways by using the boundary condition of air humidity outside the tunnel. More details will be explained in the next subsection (2).

2) Application

We applied our 1-D code to an imaginary tunnel storage system and calculated the flow rates through one year. The boundary condition of air temperature outside the tunnel is the standard meteorological phenomena data (climatic normal value) at a specific site in Japan provided by Japan Meteorological Agency. A schematic image of the adopted imaginary tunnel storage system is shown in Fig. 3.2.3-2. This system is supposed to be located at a sloping land area. Open air inlets are set at the lowest height position of the tunnel. Spent fuel depositories are set at the same height position of air inlets. Outlets are set at high position just above the depositories in order to discharge the heated air due to the stack effect effectively. In each spent fuel depository, it is supported that the heat sources of 120kW are installed. In this system, all spent fuel depositories are connected with two inlets in order to secure required ventilation performance if one of these two inlets is closed by some accident for a limited time. Natural ventilation due to the stack effect is brought from the difference in height between the outlets and the spent fuel depositories.

Fig. 3.2.3-3 shows the calculated amount of heat discharged from this system through one year as one of the examples by our 1-D code. Although total amount of heat released from fuels is 960 kW (120 kW x 8 depositories), minimum and maximum discharged amount of heat from outlets is 706 kW (in July) and 879 kW (in January) respectively. The shortage of the discharged heat amount from released heat amount 960 kW is caused by heat transfer into the bedrock of the tunnel. Air temperature and flow rate in the tunnel also change seasonally. In winter compared to summer, air temperature in the tunnel is 2 °C higher (about 17 °C in winter and 15 °C in summer) and air flow rate is 5 % larger (figures are omitted). As shown in the above example, our 1-D code enables us to evaluate the validity of the ventilation performance of whole systems and the flow rate balance in each depository.

In the above example, the evaluation is done under the condition that the amount of spent fuels, that is, the released heat is supposed to be maximum amount of the targeted system. For the practical use of the storage system, it is assumed that spent fuels are sparsely arranged in the depository. So, the evaluation under such partial-load conditions may be also needed in order to detect the adequate distribution of fuels in plural depositories for effective ventilation. Furthermore, the validation under accidental situations that inlet or outlet of tunnel is closed partially and temporarily should be done if needed.

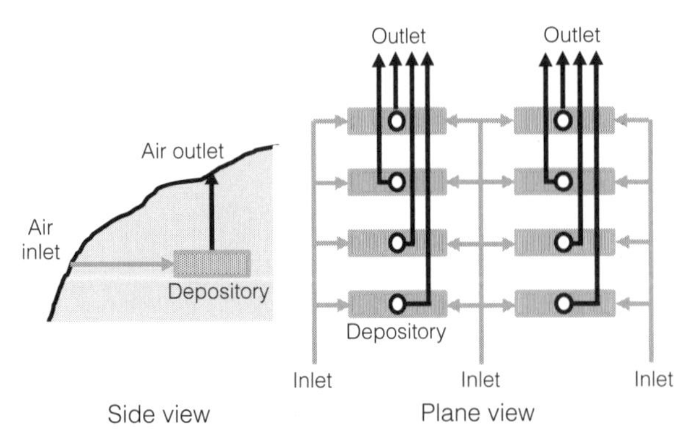

Fig. 3.2.3-2 Imaginary tunnel storage system for analysis by 1-D code

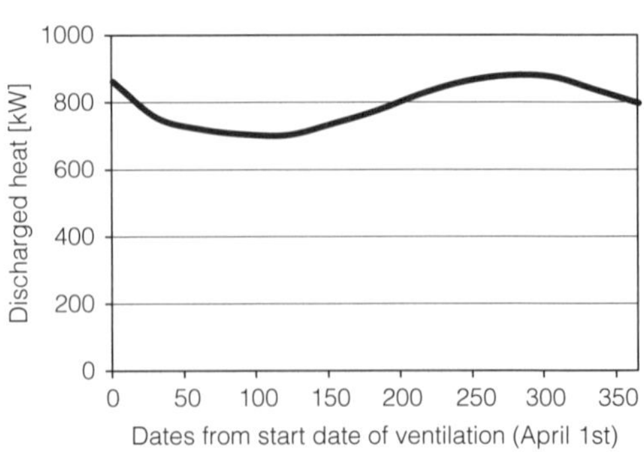

Fig. 3.2.3-3 Evaluated discharged heat by 1-D code (through one year)

(2) Three-dimensional analysis code for calculation of complicated thermal flow field of spent fuel depository (3-D code)

1) Outline

As mentioned in the previous subsection (1), we can evaluate the validity of the ventilation performance of whole storage systems by our 1-D code. But this code has a restriction that the evaluated flow rates and temperatures must be the spatially averaged values at the representative points in the storage system. So, it is difficult to evaluate local phenomena in each depository, for example, partial condensation, uneven distribution of heat, secondary flow, and so on by using this 1-D code. Our 3-D code is designed in order to evaluate such local phenomena in each depository by making use of the results calculated by our 1-D code as its boundary conditions. Our 3-D code is based on a universal thermal fluid analysis code PHOENICS[8]. The strong point of the PHOENICS is that users can insert their original equations (codes written in FORTRAN language) in addition to the precomposed basic equations such as equation of motion, mass conservation equation, and transport equation.

By making use of this function, we coupled our original equations to PHOENICS for modifying the accuracy in calculation of air temperature near the surface of fuel storage cask, calculating transportation of air humidity and calculating condensation and vaporization at wall and fuel surface. It enabled us to obtain more accurate results especially surface temperature of fuel storage casks.

In the equation for calculating condensation and vaporization at surface, the principal variables to be evaluated are air humidity f, attached water amount per unit wall surface W, and amount of vaporization (or condensation if the mark is negative) per unit wall surface and unit time ρ_s. Air humidity f is calculated by transport equation of air humidity (diffusion coefficient is calculated by a standard k-epsilon model). Amount of vaporization ρ_s is calculated by the equation (4).

$$\rho_s = \alpha \cdot (f_{wall} - f)/(C_p \cdot \Delta x) \quad \ldots\ldots (4)$$

In the equation (4), α is heat conductivity, S is surface area, C_p is specific heat of air, and dx is thickness of computational grids in air just neighbor to the wall surface. The parameter f_{wall} is air humidity at the dew point of wall temperature. If $f_{wall} < f$, condensation occurs to wall surface. If $f_{wall} > f$ and attached water amount W is not zero, evaporation occurs from wall surface. Change in W per unit time ΔW is calculated by the equation (5).

$$\Delta W = -\rho_s \cdot \Delta x \cdot \Delta t \quad \ldots\ldots (5)$$

2) Application

We applied our 3-D code to a thermal fluid analysis in an imaginary spent fuel depository. As shown in Fig. 3.2.3-4, the analysis area is a depository with a metal cask on the floor at the central part of area, an air inlet at the lowest height of the depository and an air outlet at the highest height of the depository. In this case, only the right half of this figure is calculated by using a symmetry boundary condition. The boundary conditions of air temperature, air humidity and air flow rate at air inlet and outlet are set up by using the calculation results of our 1-D code mentioned in the previous subsection (1).

Calculation results of attached water amount at surfaces, surface temperature and air humidity in the symmetry plane is shown in Fig. 3.2.3-5. These results are under mostly steady-state (only attached water amount at surface is

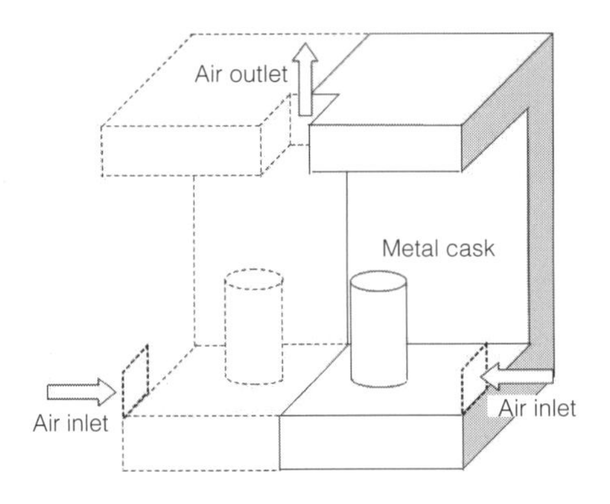

Fig. 3.2.3-4 Analysis area by 3-D code

not steady) by using the fixed boundary condition of air inlet and outlet.

In this case, the air temperature of outlet is 44 degrees. This value is 17 degrees higher than that of inlet (27 degrees). Air humidity near air outlet is less than that of air inlet. This is caused by condensation of water, that is, decrease in air humidity due to the difference in wall and air temperature (the former is lower). In addition to the areas with low wall temperature, the amount of attached water is also large near air outlet. This is caused by the increase in temperature difference between air and wall near air outlet due to the rising in air temperature by released heat from spent fuels. In reference to such results obtained by our 3-D model, we can judge, for example, whether water drain system should be prepared or not in the design stage.

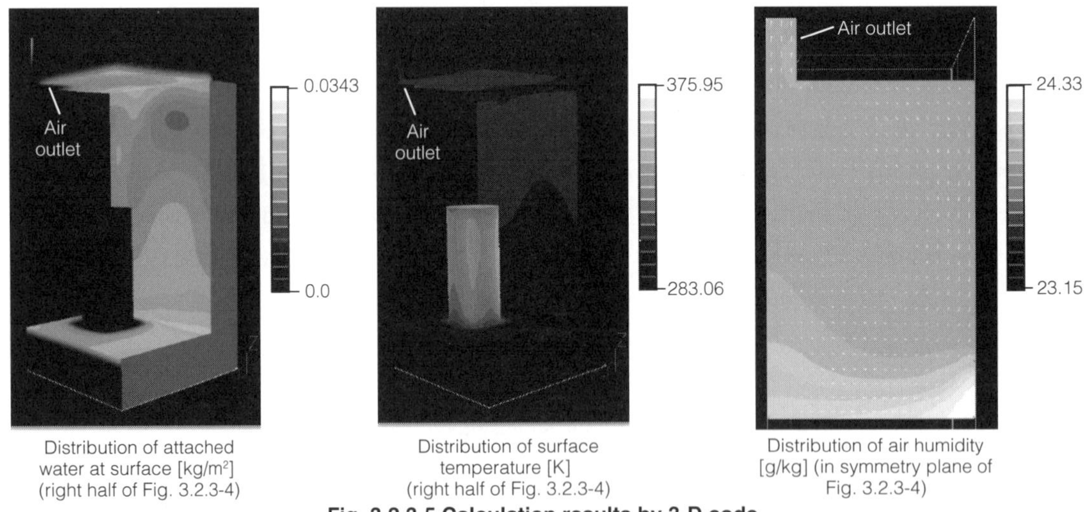

Distribution of attached water at surface [kg/m²] (right half of Fig. 3.2.3-4)

Distribution of surface temperature [K] (right half of Fig. 3.2.3-4)

Distribution of air humidity [g/kg] (in symmetry plane of Fig. 3.2.3-4)

Fig. 3.2.3-5 Calculation results by 3-D code

Although the case mentioned above is an imaginary one, we have also been applying our 3-D code for the cases under the same environment of the wind tunnel test[9),10)] and validate the results of the code by comparing with measured surface temperature of metal canisters. Consequently, it is found that the distribution of surface temperature of canisters well coincide with its measured value under the appropriate condition of turbulent model and simulation grid size especially near the surface of canisters[1].

As stated above, by using the "one-dimensional analysis code for calculation of airway network flow" and the "three-dimensional analysis code for calculation of complicated thermal flow field of spent fuel depository" in turn, we can evaluate amount of heat removed from a tunnel with its seasonal changes, surface temperature of spent fuels and walls and so on, in advance.

References

1) T. Koga and H. Tamura : Development of computer codes for thermal-hydraulics in tunnel storage facility of spent nuclear fuel, CRIEPI Report, N07014, 2007 (in Japanese)
2) Y. Hiramatsu : "Ventilation", Uchidaroukakuho- shinsya, 1974 (in Japanese)
3) K. Amano and Y. Mizuta : Prediction calculation of climate in road tunnel, Journal of JSCE, 387/II-8, pp. 219 - 228, 1987 (in Japanese)
4) Y. Hiramatsu et al, "Cooling in development workings in very hot ground", 2nd Mine Ventilation Congress
5) K. Amano : "Development of Ventilation Design System in Tunnel by Personal Computer" Journal of MMIJ, 102, pp.753-761, 1986 (in Japanese)
6) Y. Mizuta and V. S. Vutukuri, "(Prof. K. Amano Memorial Programs) Computer program for predicting ventilation and climatic conditions in mines and tunnels", Yamaguchi University, Ube JAPAN, 1990
7) I. E. Idel chik : The Handbook of Hydraulic Resistance, 3rd Edition, 1994
8) PHOENICS Over View, CHAM Technical Report: TR 001
9) B. Duret, JC. Bonnard, T. Chataing, S. Bournaud and D. Colmont : Experimental results on mixed-convection around a vertical heating cylinder cooled by a cross-flow air-circulation, Turbulent, Heat and Mass Transfer 5, pp.523-526, 2006
10) S. Benhamadouche, S. Bournaud, Ph. Clement, B. Duret, and Y. Lecocq : Large eddy simulation of mixed convection around a vertical heated cylinder cooled by a cross-flow air circulation, Conference on Modelling Fluid Flow (CMFF'06), The 13th International Conference on Fluid Flow Technologies, 2006

3.3 Containment performance

In order to evaluate long-term containment performance of metal gaskets, we conducted tests using scale models and full-scale models[1),2)].

3.3.1 Accelerated test using scale model for long-term containment of metal gasket

Containment system of the metal cask consists of primary and secondary lids. The containment function is secured by inserting gaskets between the cask body and the lids, and then bolting them together. The metal gaskets are used for long-term durability at high temperature. Therefore, it is very important to clarify the influence of the stress relaxation of the gaskets on the spring-back force and containment performance of the metal gaskets with time. In order to find temperature and time dependency on the containment performance of metal gasket, confinement tests using small flange models were carried out. The flange model consisted of two flanges with single metal gasket and the flanges are bolted together. The flange was made of stainless steel. Four types of gasket (ID: ϕ 176 mm, diameter of gasket cross section: ϕ 5.5 mm) shown in Fig. 3.3-1 were used. The flange models were put in the furnace and heated for maximum 10000 h. The maximum heating temperature was 300 °C.

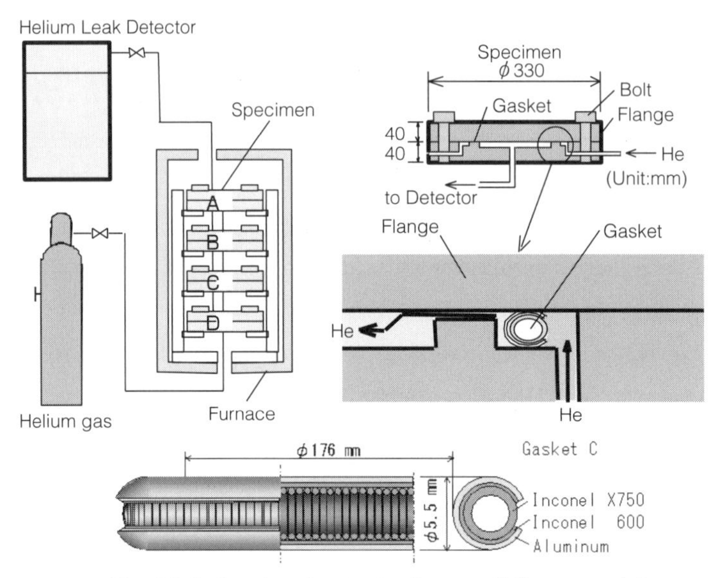

Fig. 3.3-1 Acceleration test using small flange models

We assume that gasket deformation is related to Larson-Miller Parameter (LMP) and then leakage from the gasket is related to LMP, which is given by the Arrhenius equation.

$$LMP = T\times(C+\log_{10}(t)) \quad ..... (1)$$

In this equation, T is absolute temperature (K) and t is time (h). C is a constant value. In this study, we assume C is 14. When LMP is used, the estimate for long-term at lower temperature can be predicted using the result for short time at higher temperature.

Fig. 3.3-3 shows the relation between LMP and plastic deformation rate of the gasket as shown in Fig. 3.3-1. Plastic deformation rate (Dp) is defined as follows.

$$D_p = \frac{d_0 - d_2}{d_0 - d_1} \times 100(\%) \quad ..... (2)$$

Fig. 3.3-2 shows d_0, d_1, d_2. Fig. 3.3-3 shows that LMP and plastic deformation rate has a good relation. Fig. 3.3-4 shows the relation between LMP and leak rate. When LMP exceeds 8050, leak rate increases. As a result, we assume the threshold of leakage occurrence from the metal gasket is 8050.

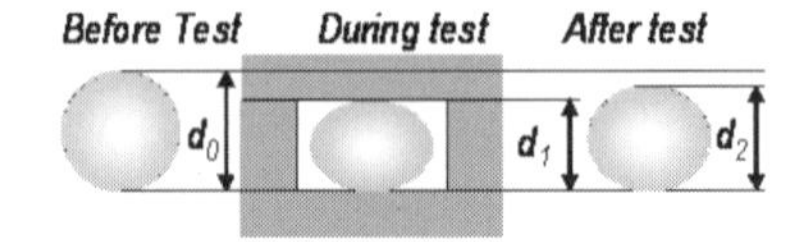

Fig. 3.3-2 Deformation of metal gasket

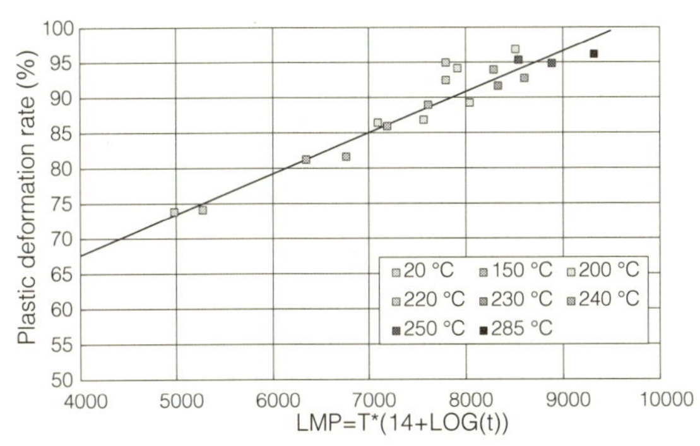

Fig. 3.3-3 Relation between LMP and plastic deformation rate of metal gasket

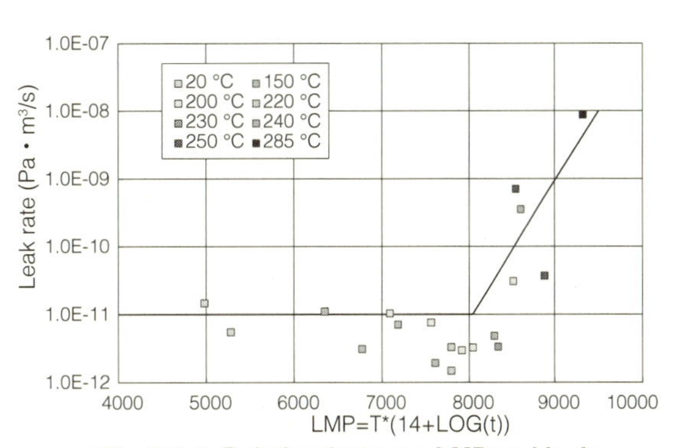

Fig. 3.3-4 Relation between LMP and leak rate of metal gasket

3.3.2 Long-term containment performance test and evaluation of metal gasket using full-scale model

We demonstrated long-term containment of the metal gaskets and aimed to establish a method to evaluate containment performance by analysis.

(1) Long-term containment test of metal gaskets

Using two models of full-scale metal cask lid structure, we conducted long-term containment performance test from October 1990 to January 2010 (Fig. 3.3.2-1).

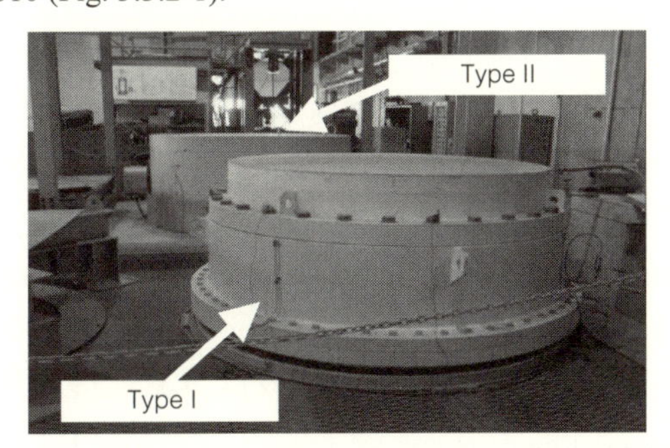

Fig. 3.3.2-1 Full-scale models of metal cask lids

1) Leak rate of the metal gaskets of the two lid models were maintained at constant values for about 19 years and 4 months, demonstrating long-term containment (Fig. 3.3.2-2). Calculating LMP using the measured temperature, LMP values of Type-I and Type-II are 7942 and 7781. Namely, containment performance of the gasket covered by aluminum is assured until LMP = 7942 and that of gasket covered by silver is assured until LMP = 7781. These kind of demonstrative tests are rarely performed in the world. The LMP values obtained in these tests have been employed in the safety examination for licensing.
2) Fig. 3.3.2-3 shows the relation between temperature of the intersection points and time, which is the relation between initial temperature of the second lid gasket and assured period of sealing performance. Assuming that the storage period is 60 years, sealing performance is assured for the gasket covered by aluminum, if the initial gasket temperature is 134 °C or less.

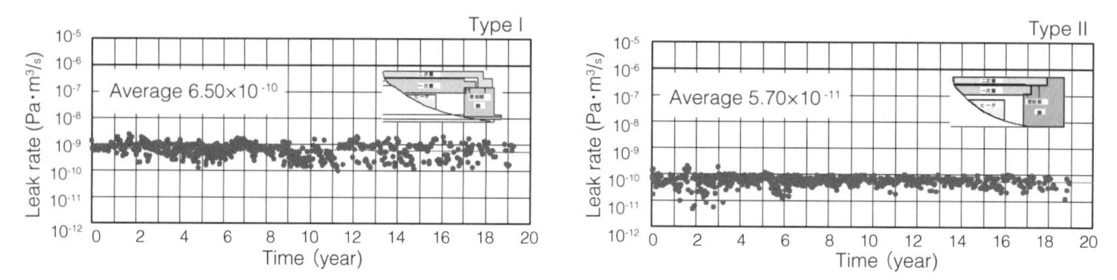

Fig. 3.3.2-2 Measurements of leak rate of metal gaskets

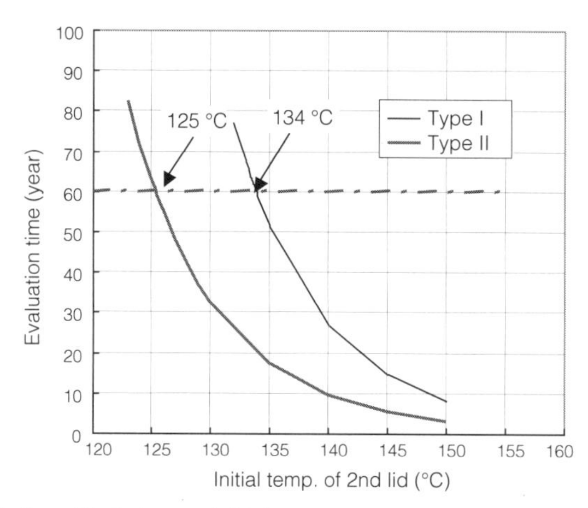

Fig. 3.3.2-3 Relationship between initial temperature and evaluation time of metal gaskets

(2) Observation of inside the lid after long-term containment tests

We opened the lids after the long-term containment tests and found status of deterioration of metal gaskets, etc.

1) The metal gasket was adhered to the cask body. After unscrewed the bolts, the lid was forced to open by a crane and the gasket deformed (Fig. 3.3.2-4)

2) There was observed no degradation by corrosion affecting the integrity, but were oily dirt materials that were possibly derived from seizure-proof agent for bolts were observed near the metal gasket and the underside of the secondary lid (Fig. 3.3.2-4).

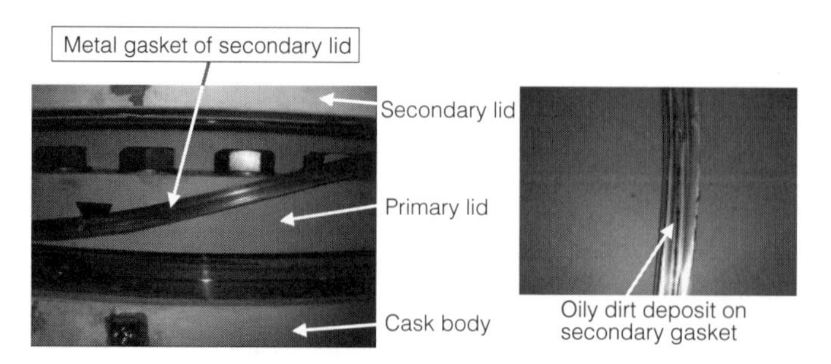

Fig. 3.3.2-4 Gaskets of the secondary lid during opening the lid of type I

3) Axial force was measured using the bolt with strain gauges. Leak rate was measured using the helium leak detector. Fig. 3.3.2-5 shows the relation between the cycles of unscrewing and the axial force of the bolt in case of second lid of Type II. Fig. 3.3.2-6 shows the relation between the cycles of unscrewing and the displacement of the lid in case of secondary lid of type II. During the unscrewing process of the thirteen's cycle, the leak rate increased rapidly from 10^{-10} to 10^{-7} Pa · m^3/s. Using the relation between the axial force and the displacement, the degradation of the metal gasket can be evaluated. These data are utilized for verification of stress relaxation analysis of metal gaskets.

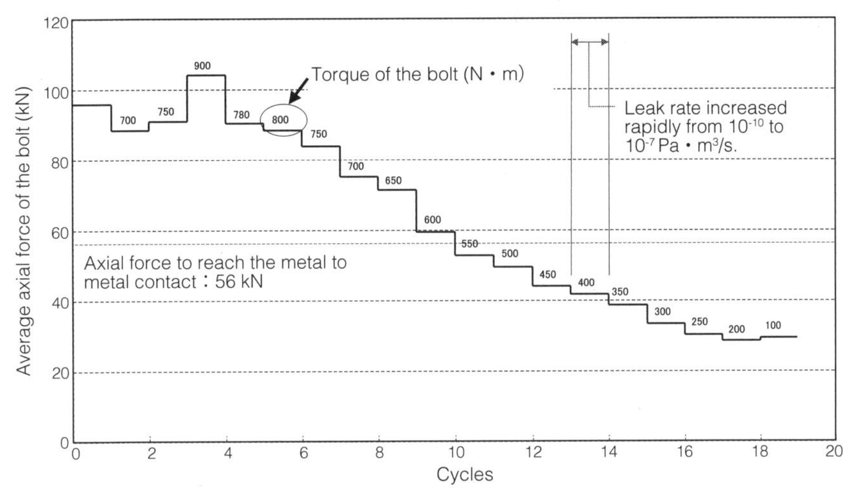

Fig. 3.3.2-5 Change of axial force of bolts with the cycles of unscrewing (Secondary lid of Type II cask)

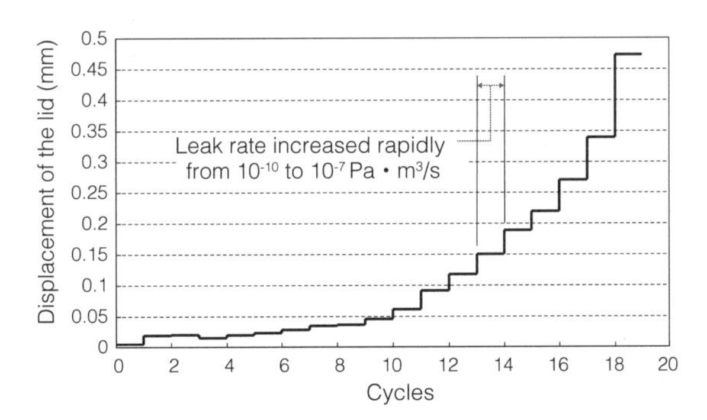

Fig. 3.3.2-6 Relation between the cycles of unscrewing and the displacement of the secondary lid of type II

(3) Influence of lid fixing conditions on containment performance of metal gaskets

In order to find conditions of contact of the lid and seal surface, we measured relationship between the number of bolt tightening and the lid displacement. In addition, in order to investigate the influence of the lid fixing and temperature conditions, we measured the axial bolt force change with heating the lid model (Fig. 3.3.2-7).

1) We found that bolt tightening by 5 to 10 times are necessary to close the lid completely (until the lid contacts the seal surface completely) (Fig. 3.3.2-8).

Fig. 3.3.2-7 Bolting full-scale cask lid model with measuring displacement, torque, etc.

The primary lid is fixed by 40 bolts and the secondary lid 32 bolts. The lid was fixed controlling the bolt torque with a torque wrench, and the lid displacement, bolt force, etc. were measured.

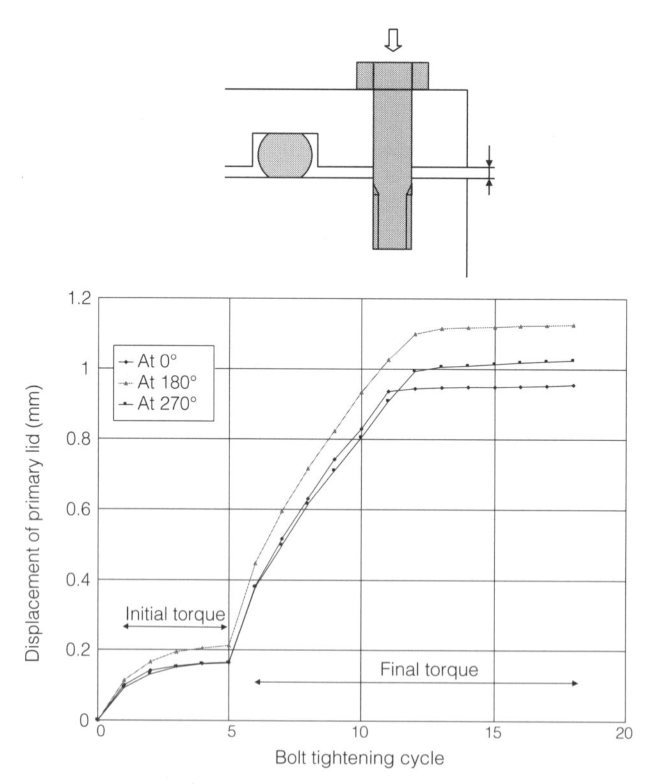

Fig. 3.3.2-8 Bolt tightening cycle and displacement of primary lid

Cycle means number of tightening all bolts per circle of the primary lid. Bolts were tightened with two steps of torque. After 12 to 13 cycles, the lid displacement reached to a saturated value, which means completion of lid closure.

2) When the temperature was increased from room temperature to 140 °C and 7 to 9 days elapsed after fixing the lid to contact completely with the seal surface, the bolt axial force decreased by 10 to 20 % from the initial force. We confirmed that this decrease of the bolt force can be reproduced by thermal stress analysis. On the other hand, when the lid were not completely contacted with the sealing surface, the axial force dropped significantly, i.e. 24 bolts out of 32 bolts showed the axial force drop by 50 % (Fig. 3.3.2-9).

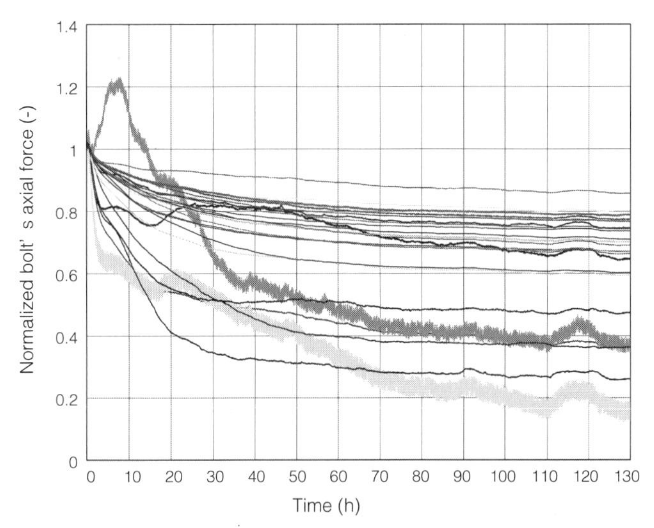

Fig. 3.3.2-9 Change of bolt's axial force after tightening bolts of the secondary lid for three times in circle and heating to about 140 °C.

The vertical axis shows normalized bolt' s axial force. Measurements of 24 out of 32 bolts were conducted. Six bolts showed drop of the force by 50 %.

From these results, we confirmed that it is important to fix the lid until the lid and the sealing surface contact completely in order to secure the long-term containment.

(4) Stress relaxation analysis to evaluate containment performance of metal gasket

We established analytical method to evaluate containment performance of metal gasket based on creep analysis of gasket material.

1) We newly developed creep constitution formula based on compressive creep tests. The analysis was verified by comparing with the experimental results of the 1/10 scale model.

2) We made creep analysis for the full-scale metal cask used for the long-term containment tests. We found that residual spring-back force of the metal gasket exceeding the threshold value for leak remains for the storage of 60 years with an initial temperature of 139 °C (Fig. 3.3.2-10).

From these results, we confirmed the long-term reliability of the metal gasket.

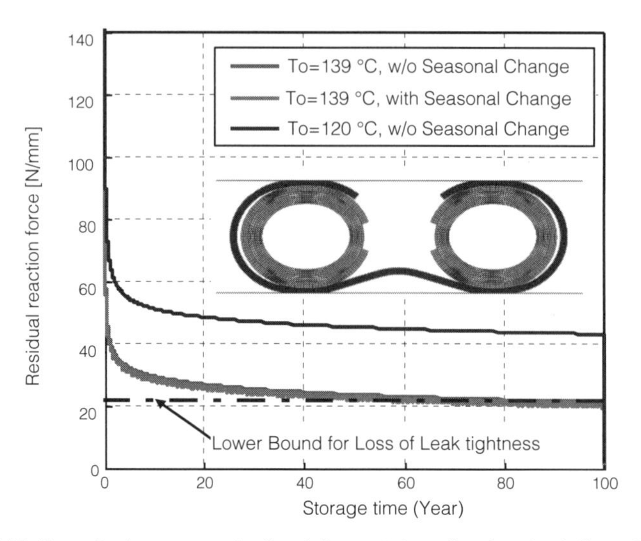

Fig. 3.3.2-10 Result of creep analysis of the metal gasket for the full-scale metal cask

It was confirmed that the remaining spring-back force of the gasket with an initial temperature of 139 °C will be exceeding the threshold value for leakage (22 N/mm) and that the containment performance will be maintained for 60 years.

References

1) O. Katoh and C. Ito: "Long-term sealability of gaskets for spent-fuel storage casks", CRIEPI Report U92009, 1992.

2) Masumi Wataru, Koji Shirai, Toshiari Saegusa, Chihiro Ito: "Long-term Containment Performance test of Metal Cask for Spent Nuclear Fuel Storage", Proc. PSAM 11 / ESREL 2012 - June 25-29, 2012, Helsinki, FINLAND

3.4 Criticality prevention performance

Criticality prevention performance (subcriticality performance) of storage metal cask is mainly supported by the inner basket. The inner basket is a very important support structure in the cask, in order to separate each spent fuel and to keep subcriticality of the cask system, and to maintain the integrity of fuel assemblies in cask by removing the decay heat. The inner basket structure has an effect on the number of fuel assemblies per cask (namely storage density) as well as on the subcriticality. Also, it has an effect on structural integrity in the drop accident, on heat removability and partially on the shielding performance.

3.4.1 Basket structure materials for subcriticality[1)]

(1) Preliminary conditions for consideration

We assumed the cask stores PWR spent fuel as shown in Table 1.3 of section 1.1, which has the maximum values both of the radiation strength and heat generation. Cooling time of the fuels is assumed 5 years for PWR high burnup spent fuel (max. 55 GWd/tU) and 10 years for PWR MOX spent fuel (max. 48 GWd/tU). In comparison with the low burnup PWR U spent fuel (max. 48 GWd/tU), the high burnup spent fuel in this section has about 1.2 times both the heat generation and radiation strength of the U spent fuels. The MOX spent fuel in this section has about 3 times the heat generation and about 18 times the radiation strength of the U spent fuels. By this reason, evaluation of neutron shielding performance considerating the cask body structure is necessary.

(2) Problems in case of using conventional basket materials and countermeasures

Stainless steels and aluminum (Al) alloys are used as the typical inner basket structural materials, by adding about 1 wt% boron (B) into these base metals in order to give higher subcriticality, or by combining structural materials with the neutron absorption materials as the simple substance[2)-5)].

In this section, several problems are discussed in case of the inner baskets materials that improve the subcriticality and heat removability by combining stainless steels and aluminum alloys, apply to the storage casks for high burnup and MOX spent fuels.

As the results, in case of high burnup spent fuel, improvement of subcriticality is necessary, because initial U concentration of this fuel is high and the requirement of effective multiplication factor ($k_{eff} + 3\sigma$) < 0.95 could not be satisfied for the same number of fuel assemblies per cask for the low burnup PWR U spent fuel.

In case of the MOX spent fuel, improvement of heat removability and shielding properties are necessary, because both the neutron discharge rate and heat generation are very high.

In this study for storage of the high burnup and the MOX spent fuel, inner basket structural materials containing over 1 wt% B were considered.

Also, enriched boron instead of natural boron may be used, if the addition of high content of boron was not enough for the neutron absorption requirement.

(3) Borated stainless steel

Borated stainless steel is used for various purposes, such as shielding material of nuclear reactor and control rod, and many papers for borated materials up to 2 wt% B have been published[6)-9)].

The type 304 borated stainless steels containing 0.2~2.25 wt% B was standardized by ASTM(American Society for Testing and Materials) in 1989[10)].

In Japan the use of borated stainless steels for the basket material was stipulated in Appendix 3-4 of JSME S FA1-2007 [11)] that is a structural code of metal cask.

Upper solubility limit of B in solid state of austenite stainless steel is about 100 ppm. This value is very small and most of the B precipitate dispersedly in the base metal as $[(Fe \cdot Cr)_2B]$ that is hard and brittle[7),12)]. The properties give large effect on the hot workability and the ductility of the materials.

In this study, four test materials A, B, C, and D, as shown in Table 3.4.1-1 were manufactured by a melting method that can make basket structural material with 4 m in length, and several data were obtained by material tests. Detailed procedure and data are found in the literature[13)].

Tensile test results are shown in Fig. 3.4.1-1. The elongation (EL) and reduction of area (RA) for the parameter of ductility decreased by the addition of B. Standard requirement of type 304L stainless steel for the base metal is that both EL and RA should be over 40 %. On the other hands, EL of the borated stainless steel with 1.1 wt% B was about

Table 3.4.1-1 Specifications and requirements of test materials of borated stainless steel

Use	Requirement / Test Materials	Test materials	Boron content (wt.%)	Thickness $\times 10^{-3}$ (m)	Size (m) Width×Length
Basket material for MOX spent fuel	Requirement of basket material Note 1)	–	1.0 (natural boron)	20	1.2×4.0 (approximately, maximum)
	Borated stainless steel with 1.1 wt.% natural boron	A	1.14	17	1.0×1.0
		B	1.08	17	1.0×1.0
Basket material for high burnup spent fuel	Requirements of basket material Note 1)	–	1.3 (enriched boron)	20, 30	1.2×4.0 (approximately, maximum)
	Borated stainless steel with 1.4 wt.% natural boron	C	1.42	17	1.0×1.0
		D	1.43	17	1.0×1.0

Note 1) According to the result of 3.4.1 (6).

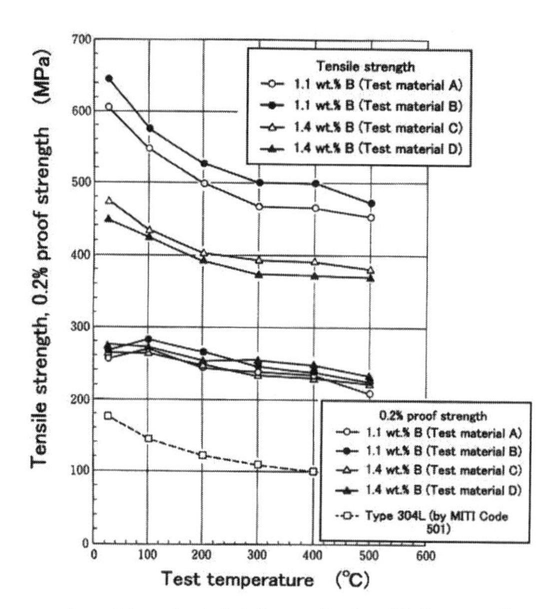

Fig. 3.4.1-1 Tensile properties of borated stainless steels with temperature (transverse direction)

20 % at room temperature, and both EL and RA of the borated stainless steel with 1.4 wt%. B were much lower values. These decreased phenomena of the ductility are affected by the precipitate phase of the Boride remarkably.

Nowadays, it has been reported that one of the borated stainless steels B-SUS304 P-1 does not use weld joint[14]. However, generally borated stainless steel for basket structural materials is fabricated by welding. Weldability of the borated stainless steel is same as that of type 304L base metal.

In case of the borated stainless steel, strength increases with B content, but its ductility decreases remarkably with B, as compared with the type 304L base metal. The boron content of the borated stainless steel used for basket structure should be better under 1.3 wt% B in order to keep its ductility. Additional problem of the borated stainless steel is lower thermal conductivity as compared with borated aluminum alloy. This problem should be compensated by design of basket structure.

(4) Borated aluminum alloy

Borated aluminum alloy has been developed as a basket structural materials, because it has high thermal conductivity and is corrosion resistant in spite of low mechanical strength[15].

There are many fabricating procedures for basket using borated aluminum alloy, e.g. molding method, plate assembly method, the square pipe method, powder metallurgy method, etc. In this study, application of square pipe method is discussed.

In recent years, development of aluminum alloy's square pipe that dispersed B_4C powder up to 9 mass% by using powder metallurgy method has been reported[16]. But, in this chapter borated aluminum alloy manufactured by the traditional melting method was selected for study.

Aluminum alloys A5356 in the non-heat-treated Al-Mg group and A6351 in the worked and heat-treated Al-Mg-Si group are well known for the base metal of borated aluminum alloy. In this chapter A6061 alloy that shows excellent strength properties was selected as the base metal for adding B to it. The A6061 alloy that belongs to A1-Mg-Si group is strengthened by artificial age hardening treatment after solid solution treatment, whereas its ductility is reduced. The A6061 alloy is recognized that mechanical strength gradually deteriorates during long-term ageing[17),18], and it should be necessary to confirm in advance that the deteriorated strength for the inner basket structural material during storage periods is within a safety range.

Several material data were obtained with two heats of the borated aluminum alloys, namely test material A and B, as shown in Table 3.4.1-2 [13].

Table 3.4.1-2 Specifications and requirements of test materials of borated aluminum alloys

Requirement / Test materials	Boron content (wt.%)	Thickness ×10^{-3} (m)	Allowable stress Note 2)
Requirement of basket material for high burnup and MOX spent fuel Note 1)	1.0 (^{10}B:60 at.%)	10	for high burnup spent fuel: 146 MPa (14.9 kgf/mm^2) for MOX spent fuel: 244 MPa (24.9 kgf/mm^2)
Test material A	0.7~0.8 (^{10}B: 100 at.%)	12	–
Test material B	0.8~0.9 (^{10}B: 100 at.%)	12	–

Note 1) According to the result of 3.4.1 (6).
Note 2) (Primary membrane stress + Primary bend stress).

In the borated aluminum alloy, gross weight of ^{10}B should be more than 0.6 wt% in order to satisfy subcriticality. Namely, enriched B, of which content is 60 at% ^{10}B, needs to be more than 1.0 wt%. Both A and B material were assumed to contain enriched B, of which content is 100 at% ^{10}B, so that these materials contain ^{10}B more than 0.6 wt%. In the material tests, ^{10}B and ^{11}B were considered to have the same mechanical properties and natural B was used in place of enriched B.

Chemical compositions of the test materials are shown in Table 3.4.1-3. Temperature dependence of the 0.2 % Proof stress (PS) is shown in Fig. 3.4.1-2. The 0.2 % PS values of the borated aluminum alloys are almost equal to that of A6061 alloy for the base metal. The tensile strength decreases with increasing temperature.

Creep rupture test result is shown in Fig. 3.4.1-3. The horizontal axis of this figure shows Larson-Miller Parameter (LMP) (LMP = T · (20 + log tr), T : temperature (K), tr : time(hr)) of which formula is based on the thermally activated process of the metal.

Table 3.4.1-3 Chemical composition of borated aluminum alloys

Test materials	Chemical compositions (wt.%)								
	B	Si	Fe	Cu	Mn	Mg	Cr	Zn	Ti
A−(T)	0.7	0.71	0.46	0.29	<0.005	1.11	0.15	<0.005	0.004
A−(B)	0.8	0.74	0.46	0.29	<0.005	1.14	0.16	<0.005	0.004
B−(T)	0.9	0.71	0.45	0.29	<0.005	1.17	0.17	<0.005	0.005
B−(B)	0.8	0.73	0.46	0.29	<0.005	1.14	0.14	<0.005	0.004
Reference	equal to ≧0.6	0.40~ 0.80	≦0.70	0.15~ 0.40	≦0.15	0.80~ 1.20	0.04~ 0.35	≦0.25	≦0.15

Note) Si is due to gravimetric analysis and other elements are by ICP analysis.
Reference specification except of boron is due to JIS 6061.

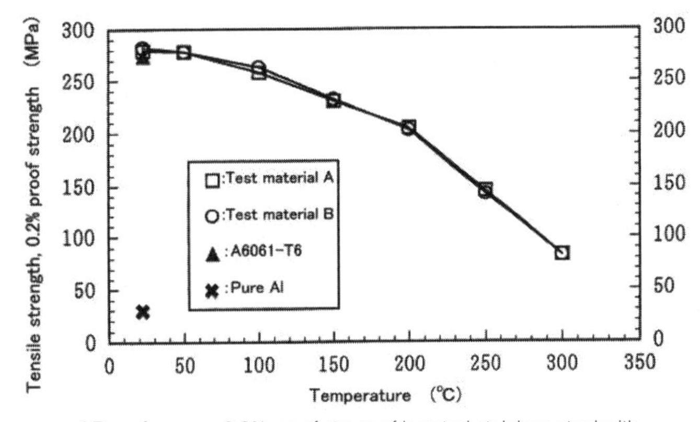

(For reference : 0.2% proof stress of borated stainless steel with 1.1wt%B shows 256 MPa at room temp. and 238 MPa at 300 °C.)

Fig. 3.4.1-2 Proof stress (0.2 %) of borated aluminum alloys with temperature (transverse direction)

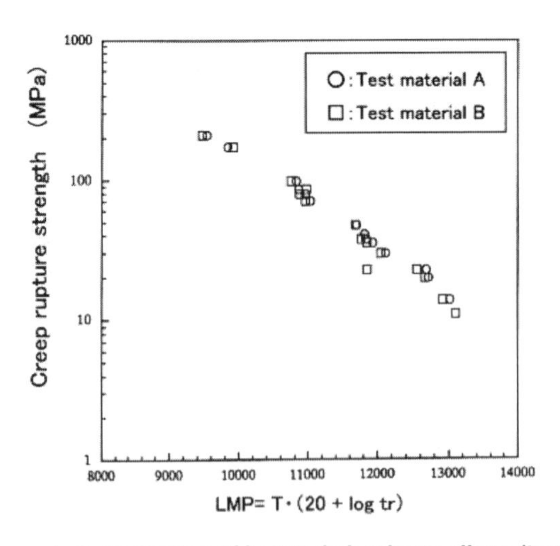

Fig. 3.4.1-3 Creep rupture properties of borated aluminum alloys (transverse direction)

In order to examine the change of the mechanical properties of the borated aluminum alloy with time during storage, the temperature history of the inner basket was given to the specimens before tensile tests and Charpy impact tests.

Table 3.4.1-4 shows temperature history of the inner basket calculated by analysis.

Table 3.4.1-4 Temperature cycle of inner basket of borated aluminum alloys (by analysis)

The time elapsed (hr)	Temperature of basket (K)	The time elapsed (hr)	Temperature of basket (K)
0	545 (272 ℃)	6000	522 (249 ℃)
1000	541 (268 ℃)	7000	519 (246 ℃)
2000	536 (263 ℃)	8000	516 (243 ℃)
3000	532 (259 ℃)	9000	513 (240 ℃)
4000	529 (256 ℃)	10000	510 (237 ℃)
5000	525 (252 ℃)		

Fig. 3.4.1-4 shows the tensile test results of the material A at room temperature that was given the temperature history of the actual inner basket before the test. All the fractography after the tensile test showed ductile fracture in spite of the length of time history.

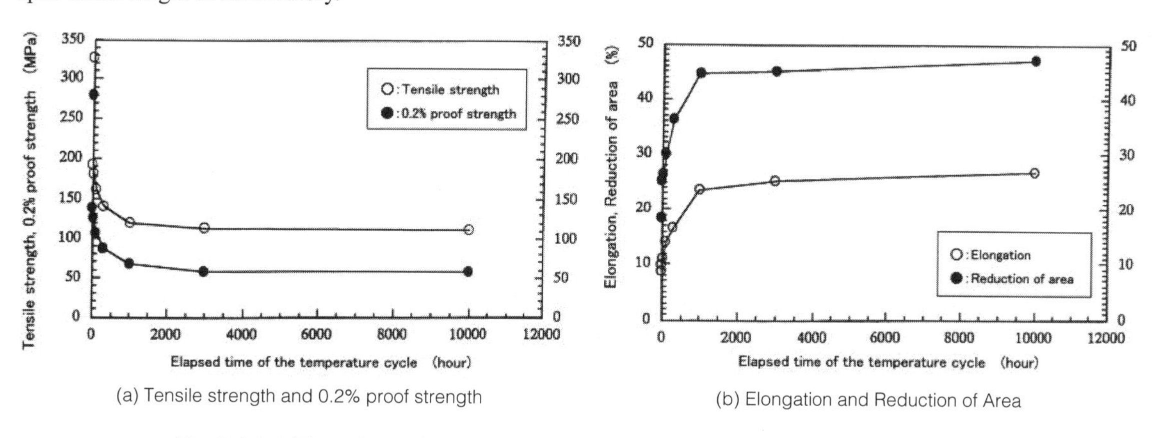

(a) Tensile strength and 0.2% proof strength

(b) Elongation and Reduction of Area

Fig. 3.4.1-4 Time dependency of tensile properties of borated aluminum alloys (Test material A in transverse direction. Tensile test results at room temp. after giving the thermal history.)

The absorption energy by Charpy V-notch impact test, Elongation (EL) and Reduction of Area (RA) increased after receiving the thermal history.

The borated aluminum alloy used in this section is A6061 alloy that is an age hardening type, and shows recovery phenomena after receiving the thermal history. Namely, the Mg_2Si precipitates grow up and become coarse by receiving the thermal history, and then effect of strengthen mechanism become small. Drop of the strength by receiving the thermal history seems to be stop, or saturate in 10,000 hours.

The deterioration of strength by ageing effect during storage periods can be estimated by using Larson-Miller Parameter (LMP) [19]. The deterioration of strength by ageing should be considered in structural design of the inner basket. The creep rupture strength should be also considered to determine the Allowable Tensile Design Stress (S) [11].

(5) Borated three layered clad metal

The three layered clad metal is composed of two layers of borated stainless steel and a copper layer in the middle (see Fig. 3.4.1-6). The borated stainless steel is superior in mechanical strength and neutron absorption performance. The middle layer of copper (Cu) compensates the low thermal conductivity of the borated stainless steel. The interface between the three layered clad metal keep good condition for thermal contact by metallic binding all over the binding area.

The corrosion resistant borated stainless steel covers the outside of the borated three layered clad metal as a whole.

With above consideration, several material data of the borated three layered clad metal were obtained using A and B test materials as shown in Table 3.4.1-5[13].

Table 3.4.1-5 Specifications and requirements of test materials of borated three layered clad metal

Requirement / Test materials	Boron content (wt.%)	Thickness $\times 10^{-3}$ (m)	Thickness ratio of each layer B-SUS/Cu/B-SUS
Requirement of basket material for MOX spent fuel Note 1)	1.0 (natural boron)	30	10/10/10
		20	7/6/7
Requirement of basket material for high burnup spent fuel Note 1)	1.0 (natural boron)	20	7/6/7
		Use together with 10 mm thickness of borated stainless steel.	−
Test material A Note 2), Note 4)	1.14	20	7/6/7
Test material B Note 3), Note 4)	1.42	20	7/6/7

Note 1) According to the result of 3.4.1 (6).
Note 2) Borated stainless steel (B-SUS) for cladding is taken from test material B of Table 3.4.1-1.
Note 3) B-SUS material for cladding is taken from test material D of Table 3.4.1-1.
Note 4) Copper (Cu) material for cladding is used JIS H3100 C1020P, which is oxygen-free Copper (O.F.C.).

Fig. 3.4.1-5 shows tensile test results of the borated three layered clad metal. Fig. 3.4.1-6 shows an example of the three layered specimens after tensile test.

In this figure, L direction means Longitudinal direction (namely, rolling direction) and C direction means Transverse direction (namely, crossing to the rolling direction).

In addition, several test results were conducted and reported, for example, bending test, Charpy V-notch impact test, bending test of fillet weld joints, etc.[13]

(6) Example of inner basket structure for subcriticality performance[1]

1) Effect of ^{10}B content to effective multiplication factor (keff)

We assumed that high burnup spent fuel is stored in the inner basket structure shown in Fig. 3.4.1-7 and Fig. 3.4.1-8. Criticality analysis was conducted using Monte Carlo computer analysis code KENO—V.a[20),21)] in order to find

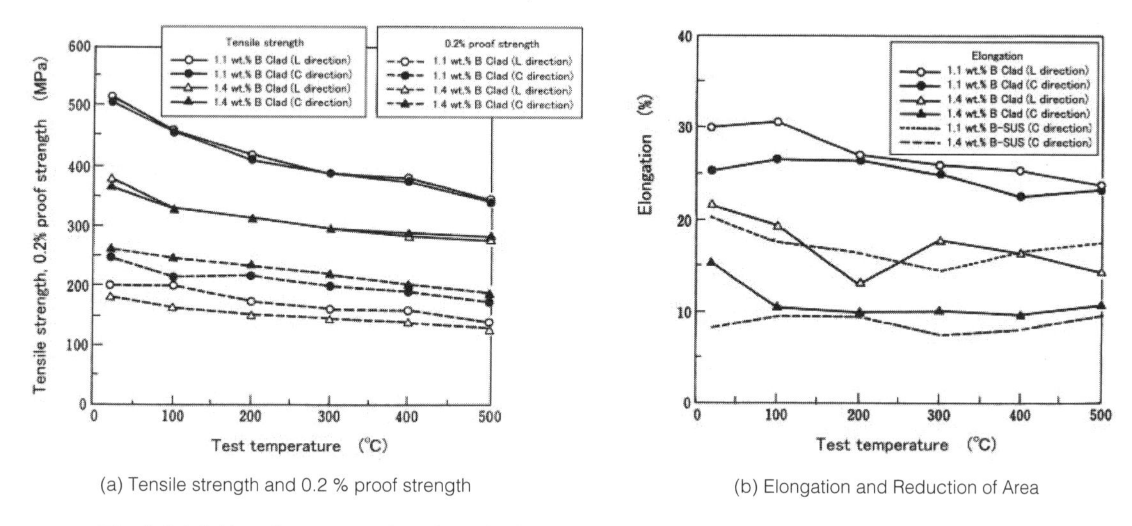

(a) Tensile strength and 0.2 % proof strength

(b) Elongation and Reduction of Area

Fig. 3.4.1-5 Tensile test results of total thickness specimen of borated three layered clad metal

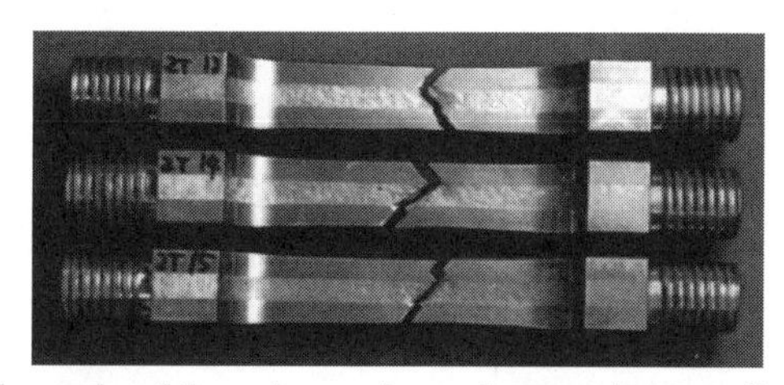

Fig. 3.4.1-6 Example of three layered specimens after tensile test at 400 °C

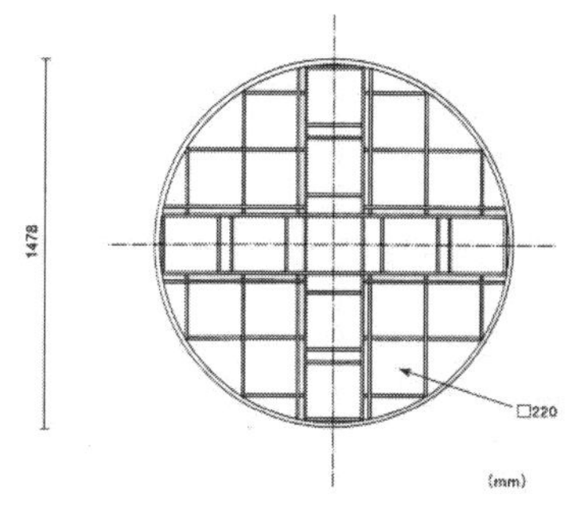

Fig. 3.4.1-7 Example of inner basket structure made of borated stainless steel for 21 PWR spent fuel assemblies with 35 GWd/tU[1)]

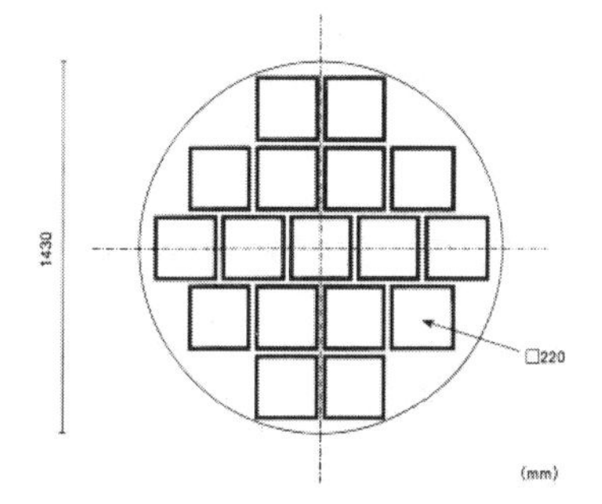

Fig. 3.4.1-8 Example of inner basket structure made of borated aluminum alloy for 17 PWR high burnup spent fuel assemblies[1)]

required ^{10}B content for subcriticality.

Result of the analysis is shown in Fig. 3.4.1-9 showing effect of ^{10}B content to effective multiplication factor (keff + 3σ). Requirement of subcriticality is assumed that (keff + 3σ) value is under 0.95. From this figure we can estimate that required ^{10}B content to keep subcriticality should be over 0.9 wt% in case of the borated stainless steel, and over 0.6 wt% in case of the borated aluminum alloy.

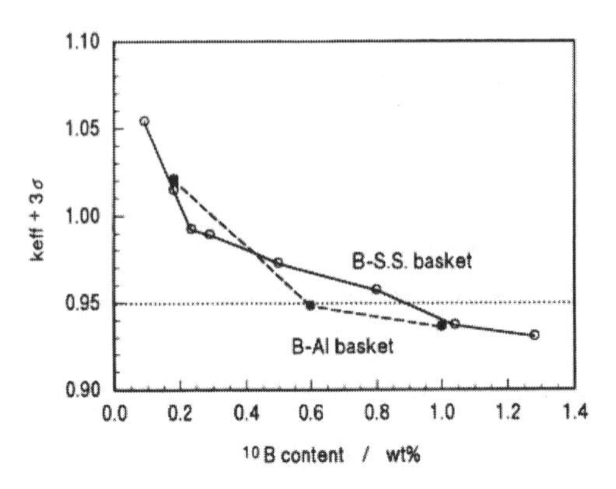

Fig. 3.4.1-9 Effect of ^{10}B content of basket material to effective multiplication factor (keff + 3σ)[1)]

2) Effect of water gap

Providing appropriate water gaps between lattice plates of the inner basket and to give neutron moderation effect by water is very effective method to enhance subcriticality. Fig. 3.4.1-10 shows a simplified model of the system for the effect by water gap.

The result of the critical analysis with a parameter of thickness of water gap is shown in Fig. 3.4.1-11 for the case of basket made of borated aluminum alloy with uniform water gap.

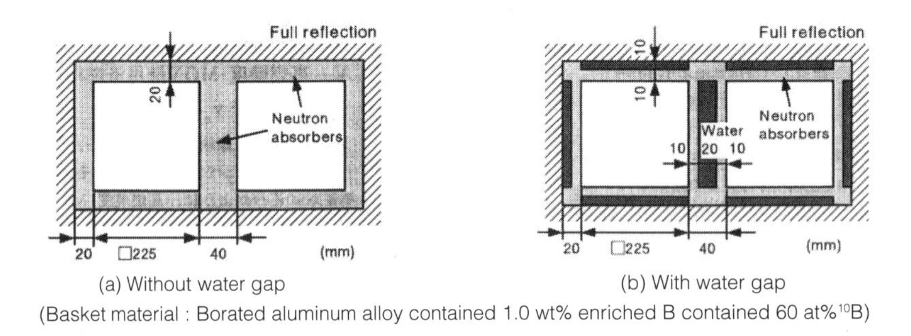

Fig. 3.4.1-10 Design model of the basket system with and without water gap[1)]

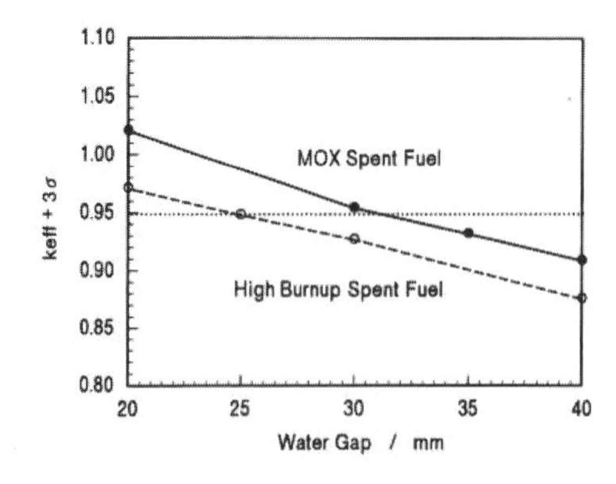

Fig. 3.4.1-11 Effect of thickness of water gap to effective multiplication factor (keff + 3σ)[1)]

The effective multiplication factor (keff + 3σ) becomes small and the performance of subcriticality improves, with increase in thickness of water gap.

3) Example of structure design of inner basket using advanced neutron absorption structural materials

Fig. 3.4.1-12 shows an example of structure design of inner basket using borated stainless steel. In this case, the number of water gap was reduced and the thickness of all basket structural materials increased up to 20 mm in order to improve heat conductivity. By decreasing the storage density from 21 assemblies to 12 assemblies per cask, the heat removal requirement could be satisfied.

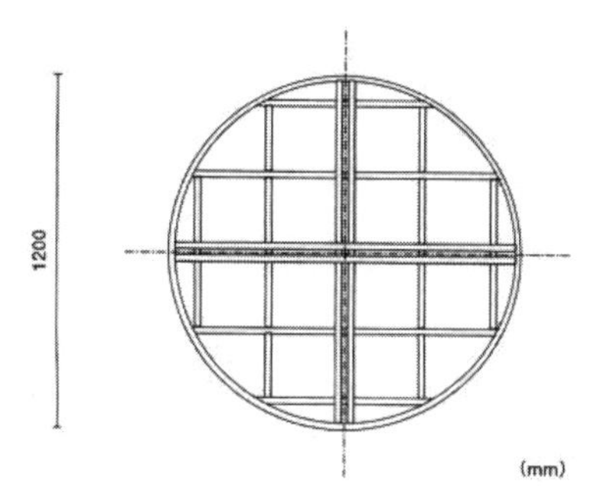

Fig. 3.4.1-12 Example of inner basket structure made of stainless steel for 12 PWR high burnup spent fuel assemblies[1)]

Fig. 3.4.1-8 shows an example of structure design of inner basket using borated aluminum alloy. Borated aluminum alloy shows high performance of heat removal, because thermal conductivity of borated aluminum alloy is 10 times higher than that of borated stainless steel. However, boron solubility per unit volume of borated aluminum alloy is smaller than that of borated stainless steel, namely density of borated aluminum alloy is about one-third of that of stainless steel. More water gaps are needed. By using enriched boron, 17 assemblies can be stored per cask.

Fig. 3.4.1-13 shows an example of structure design of inner basket using borated three layered clad metal. In this design, borated three layered clad metal of 20 mm in thickness with thickness ratio of 7/6/7 is used at the center position of the basket.

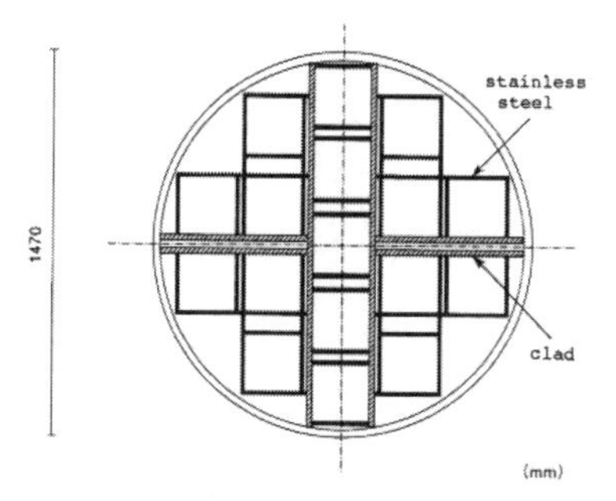

Fig. 3.4.1-13 Example of inner basket structure made of borated three layered clad metal for 17 PWR high burnup spent fuel assemblies[1)]

In this example, sufficient margin for heat removal was obtained and the water gap was effectively allocated for prevention of criticality. As the result, using 1.0 wt% of natural boron without enriched boron, the storage density of 17 assemblies per cask , instead of 12 assemblies per cask in the case of borated stainless steel, became possible.

References

1) T. Ajima, A. Kosaki, Y. Inohara, and H. Yokoyama, "Application of Advanced Neutron Absorbing Materials to Structural Material of Inner Basket of Storage Casks for High Burnup and MOX Spent Fuels", J. At. Energy Soc. Japan, Vol.39, No.2, pp.156-165 (1997).

2) J. L. Crosthwate, and J. A. Chada, Proc. 1989 Joint Int. Waste Management Conf., (1989).

3) G. Sert, Proc. PATRAM' 83, Vol.1, 763, (1983).

4) H. Yokoyama, et al., Proc. PATRAM' 83, Vol.1, 771, (1983).

5) R. Diersch, et al., Proc. PATRAM'92, Vol.1, 264, (1992).

6) E. Miyoshi, et.al, SUMITOMO KINZOKU, Technical Report of Sumitomo Metal Industries, Ltd., 12 [2], 415, (1960).

7) S. Yamamoto, et.al, The Thermal and Nuclear Power, 41 [9], 1149, (1990).

8) J. J. Stephens, et al., Proc. PATRAM' 92, Vol.3, 1477, (1992).

9) J. J. Lomburdo, WAPD-Sep-Fe-192, (1995).

10) ASTM Standards, ASTM A887-89, Vol.01.03, (1989).

11) JSME, JSME S FA1-2007, (2007).

12) T. Matsuda, et.al, SUMITOMO KINZOKU, Technical Report of Sumitomo Metal Industries, Ltd., 47 [4], 71, (1995).

13) Central Research Institute of Electric Power Industry (CRIEPI), "Test Report of Verification Tests on the Storage Technology of Spent Fuel from Nuclear Power Plant", each annual report of 1992-1996 (Japanese language only), entrusted by MITI, JNES Nuclear Library.

14) S. Doumori, S. Kawauchi, and T. Hiranuma, "Mechanical Properties of B-SUS304P-1 Used for Basket of Transport and Storage Cask", Trans. of JSME (Series A), 76, 772, pp.308-310, (2010).

15) V. Roland, et al., Proc. PATRAM' 89, Vol.1, 379, (1989).

16) T. Matsuoka, D. Ishiko, and T. Maeguchi, "Mechanical Properties of BC-A6N01SS-T1 for Basket Material of Transport and Storage Cask", Trans. of JSME (Series A), 78, 786(2012-2), pp.105-107, (2012).

17) H. Tsuji, and K. Miya, Nucl. Eng. Des., 155, 527, (1995).

18) H. Tsuji, and K. Miya, Nucl. Eng. Des., 155, 547, (1995).

19) J. Shimojo, and H. Akamatsu, "Mechanical Properties of 1%B-A6061-T6/T651 and A6061-T6/T651 Used for Basket of Transport and Storage Cask", Trans. of JSME (Series A), 76, 771, pp.141-143, (2010).

20) SCALE-4, A modular code system for performing standardized computer analyses for licensing evaluation, Vol.2 : Functional modules (Part 1, 2).

21) L. M. Petrie, and N. F. Landers, KENO V.a, An improved Monte Carlo criticality program with super grouping, Sec. F11, NUREG/CR-0200, Vol.2, (1984).

3.4.2 Uniformity of boron concentration in metal cask basket

Boron in a form of born carbide, etc. will be added to the basket materials for cask or canister in order to absorb neutron and prevent criticality. The boron particles should be small enough in the powder diameter and distribute uniformly so that they avoid neutron streaming that affects criticality evaluation.

For evaluation of neutron absorption performance in the basket materials, a minimum value of boron contents (weight % or atom number density) is given and confirmed by the following inspection and quality assurance.

- Size, weight: Measured by size and weight inspection
- Boron contents: Measured by the added weight, mass spectrometry (ICP atomic emission spectrometry) or neutron transmission examination

This section introduces results of investigation and inspection on uniformity of the boron distribution in the basket material to be used in the metal cask for the interim storage facility in Mutsu city of Aomori prefecture in Japan.

(1) Specimens of basket material for inspection of boron addition

Inspected basket materials were 1) Boron-added aluminum alloy and 2) Boron-added stainless steel. We took seven specimens of 20 cm x 30 cm from the respective materials. The boron-added aluminum alloy had three-layered clad structure with a thickness of 2.5 mm and the boron (B_4C) content was about 40 wt.%. On the other hand, the boron-added stainless steel had a thickness of 11 mm and the boron content (B) was about 1.0 wt.%. The boron-added aluminum alloy may not be used as the structural material, but the boron-added stainless steel may be used as the structural material.

(2) Inspection method of uniformity of boron distribution

We grasped the overall characteristic by sampling multiple specimens from each basket material. In order to inspect the uniformity of the material the following analyses were made.

- Mass spectrometry: Measurement of boron content
- Observation of micrograph : Observation of local boron distribution by SEM
- Measurement of boron powder radius distribution: Measurement of the particle diameter of boron powder before mixing into the specimen by optical diffraction method.

In addition, we investigated the uniformity of the boron distribution in the whole specimen by neutron transmission method (neutron radiography) and imaging plate.

(3) Result

Fig. 3.4.2-1 shows an example of SEM image of B_4C particles in the boron-added aluminum alloy. We investigated and examined the boron-added stainless steel similarly. Average B_4C particle size obtained from SEM observation in the horizontal direction on specimen was about 28.5 μm for the aluminum alloy and about 10.5 μm for the stainless steel.

Fig. 3.4.2-2 shows a result of inspection by neutron transmission method using the imaging plate. From the result of the neutron radiography, we confirmed that there was no significant heterogeneous neutron transmission

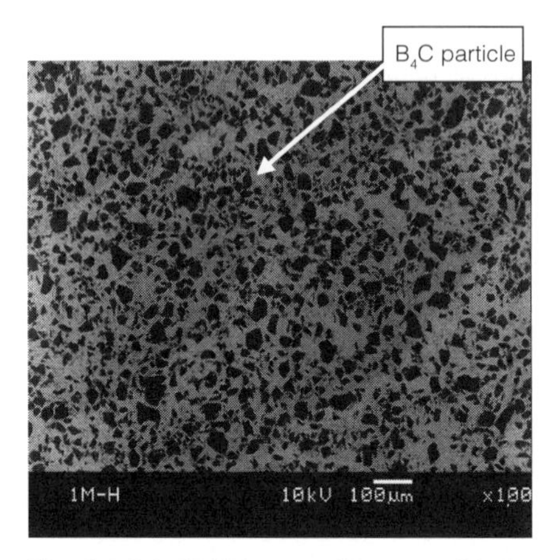

Fig. 3.4.2-1 SEM image of boron-added aluminum alloy (horizontal cross section)[1]

Fig. 3.4.2-2 Inspection of uniformity of boron distribution by neutron radiography[1]

distribution in the specimen. On top of that, we quantified the amount of neutron transmission by treatment of the light intensity distribution. With respect to the aluminum alloy, about 99 % of the light distribution was within 5 % difference from the average value, and for the stainless steel about 94 % was. We obtained the similar results of SEM observation and neutron radiography from the other 6 specimens of the boron-added aluminum alloy and boron-added stainless steel.

From the uniformity observed by the SEM and the neutron transmission distribution of the aluminum alloy and the stainless steel, we concluded that the B_4C or B particle size was small enough, that the distribution was uniform, and that there was no systematic change due to sampling location from the plates.

Reference

1) A. Sasahara, T. Saegusa, M. Wataru, and T. Sano: "Verification of uniformity of boron concentration in metal cask basket", presented at autumn meeting of Atomic Energy Society of Japan at Hokkaido University, 2010.

3.5 Structural integrity

3.5.1 Cask materials

The JSME Code specifies forged steel, stainless steel, and ductile cast iron (DCI) for cask main body (containment) materials[1]. Because the ductile cast iron is not as ductile as the stainless steel, DCI cask requires evaluation for brittle fracture against mechanical impact at low temperature. Therefore, DCI cask is needed for fracture toughness data, inspection of casting flaw, and brittle fracture evaluation method for its commercial service. The followings are the related research results. Further details are found in the literature[2].

(1) Fracture toughness of ductile cast iron (DCI)

The DCI casks are designed to withstand accidental test conditions in transport at the lowest service temperature of -40 °C. Namely, the DCI casks shall not brittle fracture against the drop impact. The CRIEPI developed reference fracture toughness values of DCI as shown in Fig. 3.5.1-1[3]. This figure was obtained from fracture toughness data obtained by six to eight inch-CT specimens taken from four full-scale DCI casks and eight short length DCI casks with full-scale diameter.

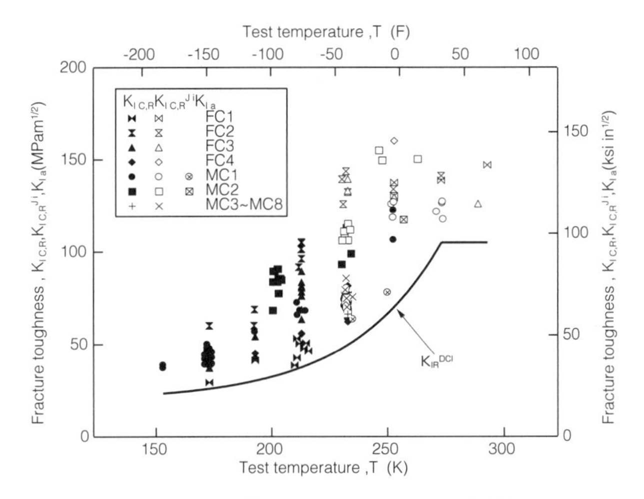

Fig. 3.5.1-1 Fracture toughness of DCI

$K_{IC,R}$: Dynamic fracture toughness
$K_{IC,R}^{J1}$: Value converted from J-integral
K_{Ia} : Crack arrest toughness
FC1~FC4 : Material data sampled from full-scale DCI casks
MC1~MC8 : Material data sampled from short length DCI casks with full-scale diameter

The curve is expressed by the following equation.

For temperature at 0 °C or below,

$$K_{IR} = 20.0 + 1.34\,e^{0.0261(T_C+160)}\ \mathrm{MPa\text{-}m^{0.5}} \quad \ldots\ldots (1)$$

For temperature exceeding 0 °C,

$$K_{IR} = 108\ \mathrm{MPa\text{-}m^{0.5}} \quad \ldots\ldots (2)$$

where K_{IR} : reference fracture toughness ($\mathrm{MPa\text{-}m^{0.5}}$)

T_C : the service temperature (°C).

Based on the fracture mechanics theory, crack propagation can be judged by comparison of stress intensity factor K loaded to a hypothetical crack in the DCI cask with the fracture toughness of the material. Using this theory, the JSME code stipulates that crack does not propagate when the following inequality is satisfied.

$$K < K_{IR} \quad \ldots\ldots (3)$$

Namely, when the material's temperature is 0 °C or below,

$$K < 20.0 + 1.34\,e^{0.0261(T_C+160)}\ \mathrm{MPa\text{-}m^{0.5}} \quad \ldots\ldots (4)$$

When the material's temperature exceeds 0 °C,

$K < 108$ MPa-m$^{0.5}$ ..... (5)

where K: stress intensity factor (MPa-m$^{0.5}$)

On the other hand, calculation of the stress intensity factor is not practically easy. It would be beneficial to evaluate the crack propagation possibility based on fracture toughness value of the material. Namely, the threshold value at -40℃ was stipulated as follows.

$$(\text{average})K_{IC,R} - 3\sigma_{SD} \geq 50 \text{ MPa-m}^{0.5} \quad ..... (6)$$

where $(\text{average})K_{IC,R}$: average of dynamic fracture toughness (MPa-m$^{0.5}$)

σ_{SD} : standard deviation (MPa-m$^{0.5}$)

These average and standard deviation can be calculated assuming Weibull distribution of the tested results.

(2) Applicability of fracture mechanics to DCI

Fracture tests of reduced scale cylindrical models made of ductile cast iron were conducted in order to demonstrate applicability of fracture mechanics to DCI [4].

1) Reduced scale model

Nine reduced scale models were cast. Fig. 3.5.1-2 shows shape and dimensions of the cylindrical models. Six of them were used as cast (DCI A). Three of them (DCI B) were subjected to heat treatment to obtain lower toughness level than K_{IR}^{DCI}. Then, artificial semi-elliptical surface flaw was machined by arc-machining at the mid of outer surface of each model.

2) Dynamic fracture toughness of the reduced scale model

Two inch thick compact tension (2TCT) specimens were extracted from the prolongation of the nine reduced scale models. Results of the fracture toughness test was summarized and statistical analyzes were performed for the data set obtained at 233 K for each material. The stress intensity factor speed, dK/dt, was 3.0 x 10^4 MPa-m$^{0.5}$/s and more. This value was obtained conservatively assuming the strain rate, dε/dt, be 10/s that was generated at the 9 m drop test[5),6)] from the following equation[7)].

$$dK/dt = (1.8\,\rho\,\sigma_y\,E/K_I)\,(d\varepsilon/dt) \quad ..... (7)$$

The J-integral value was obtained as Jc value at the point of crevice fracture initiation and converted to the dynamic fracture toughness, $K_{IC,R}$ through the method of ASTM E813-81. Table 3.5.1-1 shows the results after statistic treatment.

Table 3.5.1-1 Results of fracture toughness tests at 233 K (-40 °C)

DCI	Weibull parameter		$K_{IC,R,ave}$*	σ_{SD}*	$K_{IC,R,ave}-3\sigma_{SD}$*
	α	β*			
A	6.8	126.9	118.5	20.5	57
B	13.1	58.6	56.3	5.2	40.7

*unit : MPa-m$^{0.5}$

3) Fracture tests of reduced scale model

Fracture tests of the reduced scale models of Fig. 3.5.1-2 were performed by three points bending with loading span 1000 mm. Table 3.5.1-2 shows the results. The stress intensity speed, dK/dt, was 500 MPa-m$^{0.5}$/s.

Fig. 3.5.1-3 shows an example of fracture surface after the test. The stress intensity factor, K_{IF}, was calculated by the following equation.

$$K_{IF} = (M_m \sigma_m + M_b \sigma_b)(\pi a/Q)^{0.5} \quad ..... (8)$$

where a: flaw depth, M_m,M_b: correction factor for membrane stress and bending stress, σ_m,σ_b: membrane stress and bending stress at fracture, Q: flaw shape factor

4) Comparison of stress intensity factor and fracture toughness

Tables 3.5.1-1 and 3.5.1-2 demonstrate the applicability of the fracture mechanics showing that the values of $K_{IF,ave}$ and $K_{IC,R,ave}$ are approximately equal to each other for the different materials, DCI A and DCI B, with different fracture toughness properties. On the other hand, the values for $K_{IF,ave}$-$3\sigma_{SD}$ and $K_{IC,R,ave}$-$3\sigma_{SD}$ are not well correlated. Because, these values include margins for design stress values and could not be used to discuss the applicability of the fracture mechanics.

Table 3.5.1-2 Fracture test results of the reduced scale models (-40 °C)

DCI	Weibull parameter		$K_{IF,ave}$*	σ_{SD}*	$K_{IF,ave}$-$3\sigma_{SD}$*
	α	β*			
A	11.6	118.7	113.6	11.9	77.9
B	8.1	62.2	58.6	8.6	32.8

*unit : MPa-m$^{0.5}$

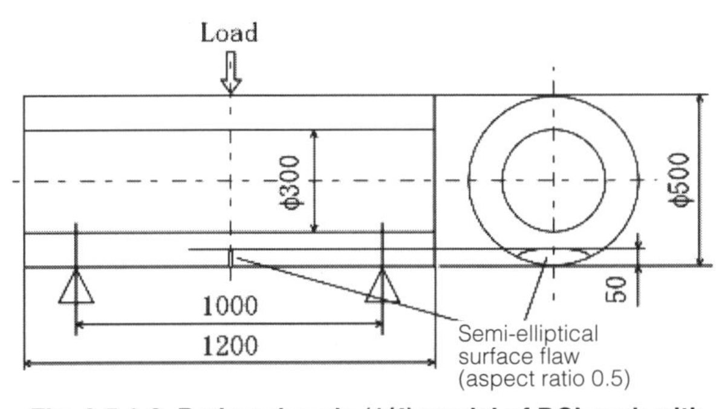

Fig. 3.5.1-2 Reduced scale (1/4) model of DCI cask with artificial flaw for fracture test (mm)

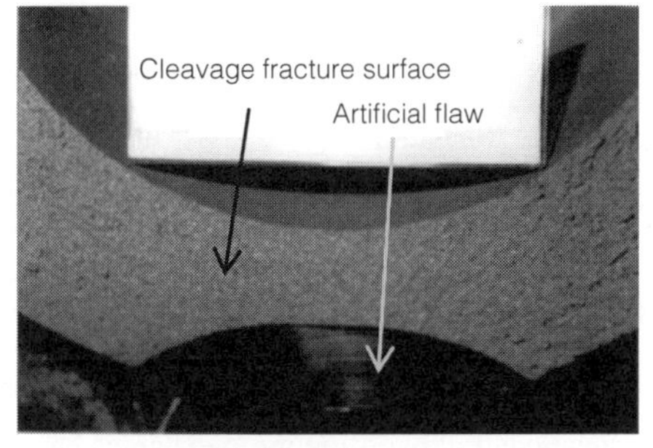

Fig. 3.5.1-3 Example of fracture surface of the reduced model (DCI A)

(3) Drop test of full-scale DCI cask

Demonstration tests were performed to verify if a full-scale DCI cask with the above-mentioned fracture toughness withstands the mechanical drop test impact[8),9)] (Fig. 3.5.1-4). The DCI cask cooled to -40 ℃ was dropped horizontally on to the unyielding target. An artificial flaw (half-ellipsoidal shape with 83.5 mm in depth and 510 mm in length) was introduced by electron discharge method on the cask surface where the maximum stress would be generated.

Fig. 3.5.1-4 The 9 m drop test of full-scale DCI cask at -40 °C

1) Fracture toughness of the DCI cask

Table 3.5.1-3 shows the results of dynamic fracture toughness tests of the specimens taken from the prolongation of the DCI cask attached at the casting. Assuming Weibull distribution, the average values and the standard deviation were calculated using median-rank method that is used to obtain accumulated distribution probability, as follows.

Weibull distribution parameter α=12.3, β=74.8 MPa-m$^{0.5}$

$$(\text{average})K_{IC,R} = 71.7 \text{ MPa-m}^{0.5} \quad \ldots\ldots (9)$$

$$\sigma_{SD} = 7.09 \text{ MPa-m}^{0.5} \quad \ldots\ldots (10)$$

$$(\text{average})K_{IC,R} - 3\sigma_{SD} = 50.4 \text{ MPa-m}^{0.5} \quad \ldots\ldots (11)$$

The fracture toughness of this material is almost equal to the threshold value (50 MPa-m$^{0.5}$) given by the equation (6), and would be representative property of DCI cask.

Table 3.5.1-3 Dynamic fracture toughness of the DCI cask materials (MPa-m$^{0.5}$)

Material	Fracture toughness	Average	Standard deviation	Average -3x Standard deviation
DCI	81.5			
JIS G	71.6	71.7	7.09	50.4
5504	68.8			
	70.1			

2) Stress intensity factor

The stress intensity factor at the drop test was 38.2 MPa-m$^{0.5}$. (This was calculated using the above-mentioned equation (8) with input of stress σ_m : 60 MPa, σ_b : 16 MPa that were converted from the measured strain values generated at the tip of the artificial flaw at the drop test.

3) Margin between drop test result and fracture

After the drop test, no crack propagation was observed using an optical micro scope at the tip of the artificial flaw.

As mentioned above, fracture mechanics can be applied to DCI cask. Namely, DCI cask will fracture when the stress intensity factor becomes equal to the fracture toughness of the material. The margin of the DCI cask material with the average fracture toughness for the fracture at the 9 m drop test was calculated as 71.7/38.2=1.88.

On the other hand, the margin of the DCI cask material with the average fracture toughness - 3 x standard deviation ($_{(\text{average})}K_{IC,R} - 3\sigma_{SD}$ = 50.4 MPa-m$^{0.5}$) for the fracture at the 9 m drop test was calculated as 50.4/38.2=1.32.

If the equation (4) is used for a practical design, a margin of 1.32 was demonstrated. On the other hand, the IAEA guideline[10] recommends 1.4.

(4) Detectable size of flaw by ultrasonic test

Round robin tests (where a same sample is tested by multiple organizations) for ultrasonic test were performed to define detection limit for DCI[11),12)]. As the results, it was found that one can detect a flat bottom defect as large as 6 mm by vertical beam method and 3 mm by angle beam method.

In a practical code[13] for inspection by UT, UT quality level is defined in accordance with thickness of the cast product. Because the surface flaw would be the most severe location for the DCI cask drop test condition, quality level 1 regardless of the overall thickness of the cast product is applied for the volume of the castings within 1 inch of the surface and is 38 mm or less at maximum.

References

1) Japan Society of Mechanical Engineers (JSME): Codes for Spent Fuel Storage Facilities: Rules on Transport/ Storage Packagings for Spent Nuclear Fuel" (2007 to be revised in 2014) JSME S FA1-2007.

2) T. Saegusa, and T. Arai, : "ASME Codification of Ductile Cast Iron Cask for Transport and Storage of Spent Nuclear Fuel", CRIEPI Report N11027 (2012).

3) T. Arai, T. Saegusa, G. Yagawa, N. Urabe, and R. Nickell, : "Determination of Lower-Bound Fracture Toughness for Heavy-Section Ductile Cast Iron (DCI) and Estimation by Small Specimen Tests", ASTM STP No.1207, p.355 (1995).

4) T. Arai, T. Saegusa, and N. Urabe, : "Fracture Toughness of Ductile Cast Iron and Applicability of Fracture Mechanics to DCI Casks", Proc. PVP 2004, 2004 ASME/JSME Pressure Vessels and Piping Division Conference, July 25-29, 2004, San Diego, CA ,USA.

5) C. Ito, Y. Kato, and K. Shirai, : "Establishment of Cask-Storage Technology for Spent Fuel -Evaluation of Cask Integrity at Postulated Mishandling Drop-", CRIEPI Report U92035(1992)

6) Y. Kato, S. Hattori, C. Ito, et al, "Storage Cask Drop Test on Reinforced Concrete Slab", Proc. PATRAM 1992, Sept. 13-18, 1992, Yokohama, p.1443.

7) ASTM Standard E813-87, "Test Method for Plain-Strain Fracture Toughness".

8) K. Shirai, C. Ito, T. Arai and T. Saegusa : "Integrity of Cast Iron Cask against Free Drop Test-Verification of Brittle Failure Design Criterion.", RAMTRAN, Vol.4, No.1, pp.5-13 (1993).

9) K. Shirai, Y. Kato, C. Ito, K. Shimazaki, and T. Saegusa, : "Estimation of Integrity of Cast-Iron Cask to Free Drop Test Impact -Part (3) Verification of Brittle Failure Design Criterion" , CRIEPI Report U90001 (1990).

10) IAEA, Guidelines for Safe Design of Shipping Packages against Brittle Fracture, IAEA TECDOC-717 (Vienna: IAEA).

11) H. Imaeda, "Estimation of the Critical Size of Detectable Flaws in Ductile Cast Iron Using Ultrasonic Testing", Nondestr. Test. Eval., 1994, Vol. 11, pp.43-62.

12) H. Imaeda, "Estimation of the Critical Size of Natural Flaws in Ductile Cast Iron Using Ultrasonic Testing", Nondestr. Test. Eval., 1994, Vol. 11, pp.341-348.

13) ASME Section III, Division 3 "Containments for Transportation and Storage of Spent Nuclear Fuel and High Level Radioactive Materials and Waste", 2011a.WB-2570"Examination and Repair of Cast Products".

3.5.2 Cask integrity under accidental drop during handling

We carried out hypothetical drop accident tests and analyses of casks during handling in a storage facility in order to investigate the integrity and containment performance of the casks using full-scale casks and a model of reinforced concrete floor[1),2)].

(1) Drop tests

1) Method

We drop-tested full-scale casks with different orientations from various heights on to a reinforced concrete floor simulating floor of storage facility. Figs. 3.5.2(1)-1 and 3.5.2(1)-2 show the test casks without mechanical impact limiters. The test casks are made of ductile cast iron and equipped with metal gaskets for containment. Orientation of the casks may be vertical and oblique for the accidental drop during handling in the facility. The cask may be dropped horizontally when the cask is lifted down from a trailer to the facility floor. Therefore the drop tests were carried out for those three cask orientations. The design concept of the storage facility was as shown in Fig. 3.5.2(1)-3. The normal handling height was 1.5 m, the maximum lifting height in the facility was 7.5 m, and the limiting height for finding the design margin was 17.0 m. Helium leak tests were carried out before and after each drop test in order to check the containment performance of the casks. Table 3.5.2(1)-1 shows all the test conditions.

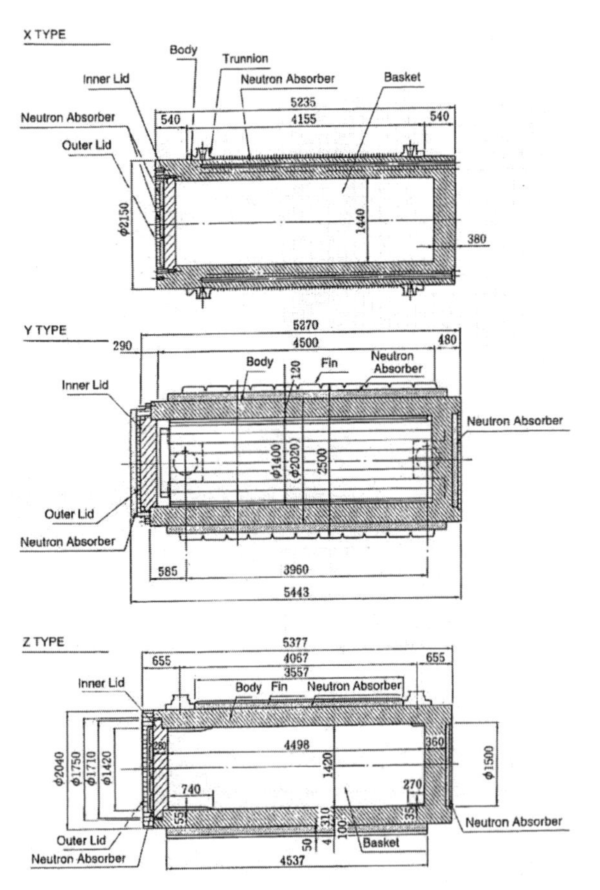

Fig. 3.5.2(1)-1 Full-scale casks for drop tests

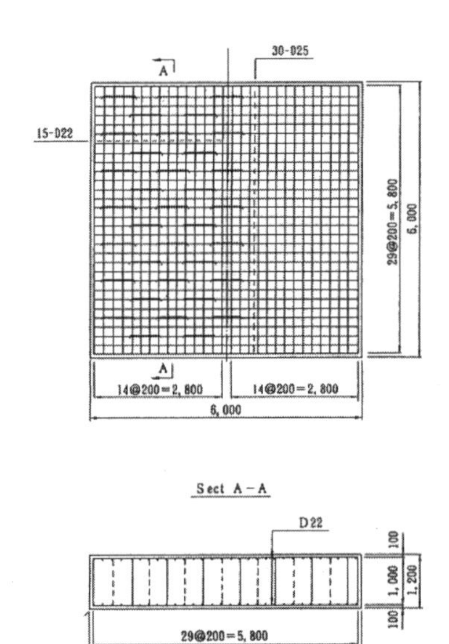

Fig. 3.5.2(1)-2 Reinforced concrete floor

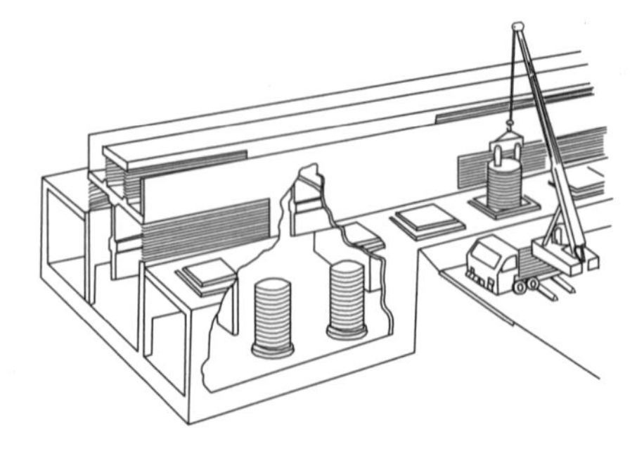

Fig. 3.5.2(1)-3 Design concept of storage facility

Table 3.5.2(1)-1 Conditions for cask drop tests without impact limiters (The gray columns are the conditions tested.)

Drop Height (m)		Orientation	Cask	Note
Normal Lifting Height	1.5	Vertical	X	Verify the package integrity for free drop of the normal operating height
		Horizontal	Z	
		Oblique	Y	Drop test was not conducted, because the strain generation at the oblique drop was sufficiently small as compared with that at the vertical and horizontal drops by pre-drop test analysis.
Maximum Lifting Height In Storage Building	7.5	Vertical	X	Verify the package integrity for free drop from the maximum lifting height in the designed storage building.
		Oblique	Y	
		Horizontal	Z	Drop test was not conducted, because the package was not lifted up to this height with the horizontal orientation.
Verification of the Package Margin	17	Vertical	X	Verify a margin in the integrity of packages against a drop accident.
	5	Horizontal	Y	
	17	Oblique	Z	

2) Test results

Fig. 3.5.2(1)-4 show the tests. The maximum acceleration and strain are shown in Fig. 3.5.2(1)-5. The acceleration and the strain increased with the drop height approaching to a saturated values after a certain drop height. The values for the cask with the oblique orientation showed less values than those for the other cask orientaions. This would be because the cask with the oblique orientation penetrated into the concrete floor and the drop energy was partially absorbed by the fracture of the concrete as compared with the other drop tests.

The strain generated in the bolts securing the lid of the cask dropped from the normal handling height was within the elastic region and maintained the cask integrity that showed no abnormal leak test results.

Fig. 3.5.2(1)-4 Casks drop tests

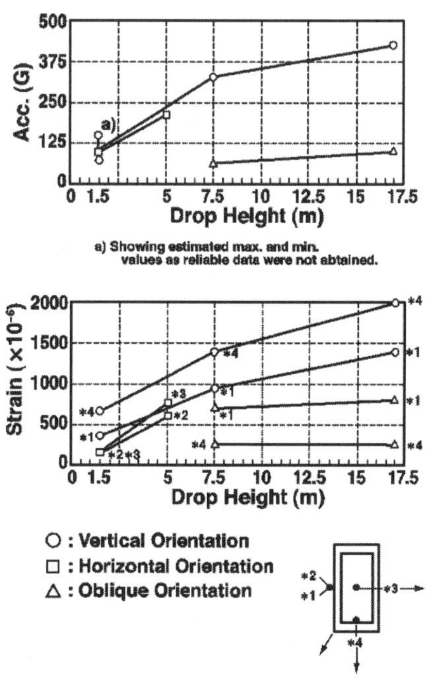

Fig. 3.5.2(1)-5 Drop test results

The strains generated in the bolts securing the lid of the casks dropped from the maximum lifting height and the limiting height for the drop test were also within the elastic region. Although the containment performance of the inner lid showed somewhat deteriorating, but that of the outer lid showed no abnormality in all cases. Table 3.5.2(1)-2 shows the all results of the leak tests.

Table 3.5.2(1)-2 Results of leak test before and after the drop tests

Orientation	Drop height (m)	Lids	Leak rate (atm · cc/sec)	
			Before drop test	After drop test
Vertical	1.5	Primary	$< 1.6\times10^{-9}$	$< 4.8\times10^{-6}$
		Secondary	$< 2.8\times10^{-8}$	$< 6.8\times10^{-8}$
	7.5	Primary	$< 5.8\times10^{-8}$	> Measurable limit
		Secondary	$< 1.3\times10^{-9}$	$< 5.3\times10^{-9}$
	17.0	Primary	$< 2.3\times10^{-9}$	> Measurable limit
		Secondary	$< 1.9\times10^{-9}$	$< 2.9\times10^{-9}$
Horizontal	1.5	Primary	5.7×10^{-9}	2.8×10^{-9}
		Secondary	$< 2.1\times10^{-10}$	$< 4.4\times10^{-10}$
	5.0	Primary	$< 3.2\times10^{-10}$	> Measurable limit
		Secondary	$< 5.6\times10^{-10}$	2.2×10^{-6}
Oblique	7.5	Primary	$< 1.5\times10^{-10}$	$< 2.9\times10^{-10}$
		Secondary	$< 3.8\times10^{-10}$	$< 2.1\times10^{-10}$
	17.0	Primary	$< 4.3\times10^{-10}$	$< 7.5\times10^{-10}$
		Secondary	$< 3.2\times10^{-10}$	$< 6.0\times10^{-10}$

(2) Analyses of drop tests

1) Outline of the analytical code

Mechanical impact analysis code DYNA-3D was used for the present analyses. Constitutional equation for the concrete was developed based on results of tri-axial material tests by CRIEPI. The model has the following characteristics.

a. Fracture condition for compressive fracture employs tri-axial stress conditions.

b. Fracture condition including tri-axial cracks was considered for tensile fracture.

c. Strain rate dependence was considered for both the compressive and tensile fracture strength.

2) Analytical model

The analyses were carried out for the drop tests with the vertical and horizontal orientations.

Fig. 3.5.2(2)-1 shows the analytical model of the cask and the reinforced concrete floor. The computational models were axial symmetry for the vertical drop test and half-plane symmetry for the horizontal drop test. The computation starts from the moment for the cask to touch the concrete floor with an input of the free drop speed in accordance with each drop height to each element.

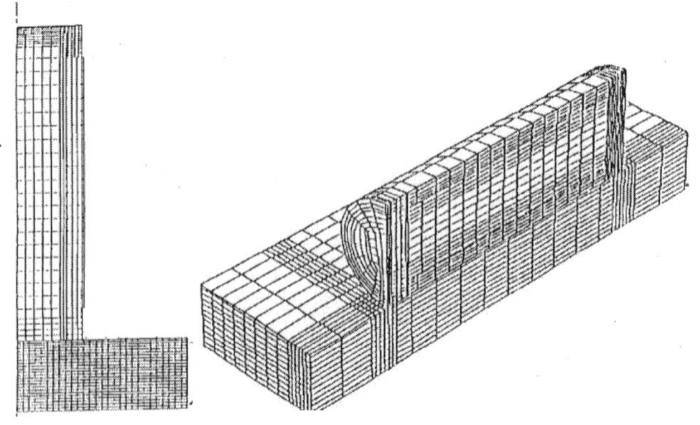

Fig. 3.5.2(2)-1 Analytical models of cask and reinforced concrete floor

3) Analytical result

Cask integrity was evaluated with respect to the following items and compared with the test results.

a. In order to maintain the containment performance of the cask, no plastic strain shall be generated at the lid fixture area of the cask body.

b. In order to secure cask integrity, no significant stress shall be generated at the part related to the cask integrity.

(a) Plastic strain at the lid fixture area

Fig. 3.5.2(2)-2 shows the analytical results on the plastic strain at the lid fixture area for the drop tests from the heights of 1.5 m, 7.5 m, and 17.0 m. No plastic strain was generated at the secondary lid fixture area according to the analyses, and the leak tests showed no leakage as shown in Table 3.5.2(2)-1. No plastic strain was generated at the primary lid fixture area for the drop test from the height of 1.5 m according to the analysis, and the leak test showed no leakage, either. On the other hand, plastic strain was predicted at the primary lid fixture area for the drop tests from the heights of 7.5 m and 17.0 m, and the leak test results showed some leakage.

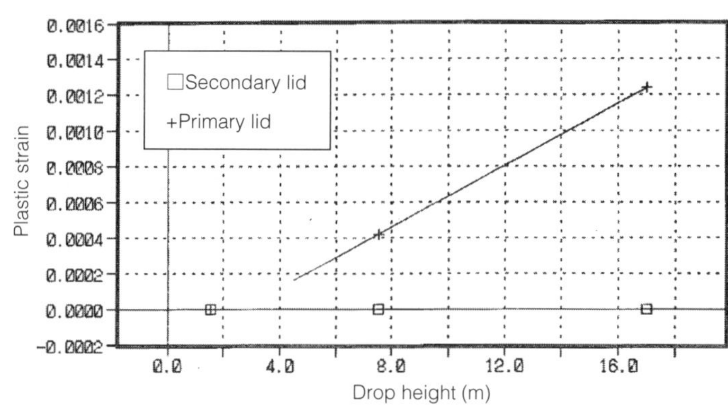

Fig. 3.5.2(2)-2 Analytical results of plastic strain at the lids

Table 3.5.2(2)-1 Results of leak tests (Vertical drop tests during handling)

Lid	Drop height (m)		
	1.5	7.5	17
Primary lid	No leak	Leak	Leak
Secondary lid	No leak	No leak	No leak

(b) Stress at the relevant location of the cask

Figs. 3.5.2(2)-3 and 3.5.2(2)-4 show the maximum strain generated at the cask body in the vertical and horizontal drop tests comparing the tests and the analyses. In either case, the analytical results showed conservative values as compared with the test results, which confirmed the applicability of the analytical method. In the near future, the concrete model may have to be improved to take into account of the strain rate effect on the concrete material property and non-linear behavior of the concrete with respect to shear deformation.

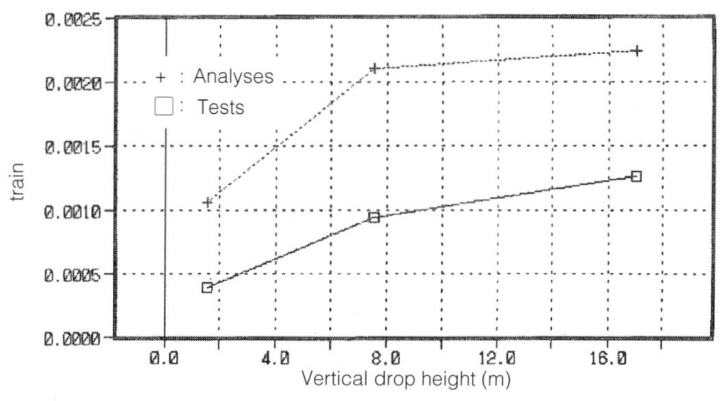

Fig. 3.5.2(2)-3 Comparison between analyses and tests on strain

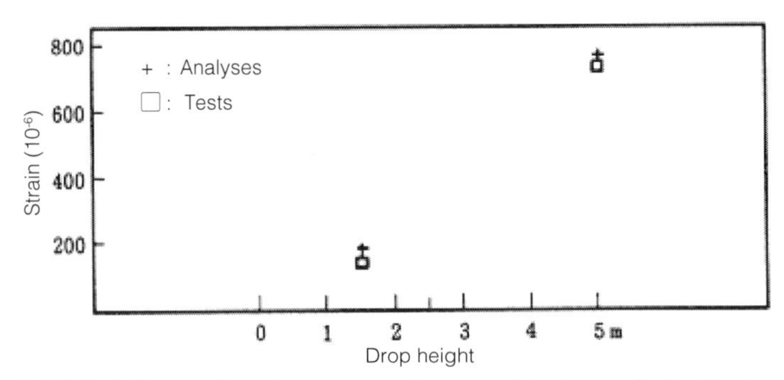

Fig. 3.5.2(2)-4 Comparison between analyses and tests on strain -Horizontal drop-

(3) Evaluation method of cask integrity at drop accident

1) Containment performance

As mentioned in the above section (2) 3) (a), there is a relationship between the presence of the plastic strain in the vicinity of the lid seal O ring and the actual containment performance. Although more data should be accumulated, we propose to evaluate the containment performance by the existence of plastic strain in the vicinity of the lid seal O ring.

2) Ductile fracture

For the cask integrity evaluation at the drop accident, fracture model of the concrete is important. Although there are some models, nothing is well determined. Designers should make benchmark tests before use the model in the analysis. We propose the analytical method described in the above section (2). For the evaluation of the cask body, the conventional method used for transport cask should be used.

3) Brittle fracture

We propose to use the method described in the section 3.5.1 after obtaining the stress generated in the cask using the method in the above section (2).

References

1) C. Itoh, Y. Katoh, K. Shirai, S. Hattori, O. Katoh, and Y. Ozaki: "Establishment of Cask Storage Method for Storing Spent fuel - Evaluation on Integrity of Cask at Postulated Mishandling Drop -", CRIEPI Report U92035,1992.12

2) Y. Kato, S. Shirai, C. Ito, et al., "Storage Cask Drop Test on Reinforced Concrete Slab", Proc. PATRAM '92, Sept.13-18, 1992, Yokohama, Japan, 1443.

(4) Instantaneous leakage evaluation of metal cask during drop test

As drop accidents of a metal cask, two cases are assumed: i.e. one is an accident during transportation, and the other is one during handling in a spent fuel storage facility.

Impact limiters are installed on a metal cask during transportation. On the other hand, they are not installed during storage. A lot of tests and analysis have been performed for evaluation of drop tests of metal casks [1),2)] before now.

However, no quantitative measurement has ever been performed for any instantaneous leakage from metal gaskets during the drop tests, which is caused by loosening of bolts of containment parts and lateral sliding of lids of the casks.

In order to determine a source term for radiation exposure dose assessment, it is necessary to obtain a fundamental data of the instantaneous leakage.

In this study, we performed leakage tests using scale models of a lid and a full scale metal cask without the impact limiters and simulated the drop accidents in the storage facility, with the aim of measuring and evaluating any instantaneous leakage at the impact[3)].

1) Leakage test of lid sliding in scale model of cask lid

Prior to the drop tests of the full scale metal cask, a series of leakage tests were conducted using a scale model (1/11) of the cask lid (Fig. 3.5.2(4)-1). These tests were performed to establish a measurement method and to obtain the relationship between the amount of lateral sliding of the lid and helium leak rate. The model consists of three flanges bolted together, and helium gas is installed in a groove of one of outer flanges.

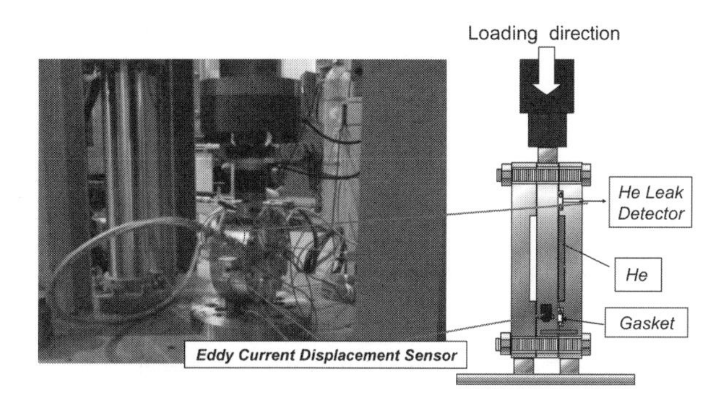

Fig. 3.5.2(4)-1 Scale model of a cask lid structure

Containment is maintained by metal gaskets. A sliding load and relative displacement were applied by pushing the middle flange with a loading test equipment.

Fig. 3.5.2(4)-2 shows the test results. The solid lines show the leak rates continuously measured as a function of sliding displacement of the lid at a loading speed of 0.01 mm/s. The measurement was repeated three times and double-digit variation of data was seen.

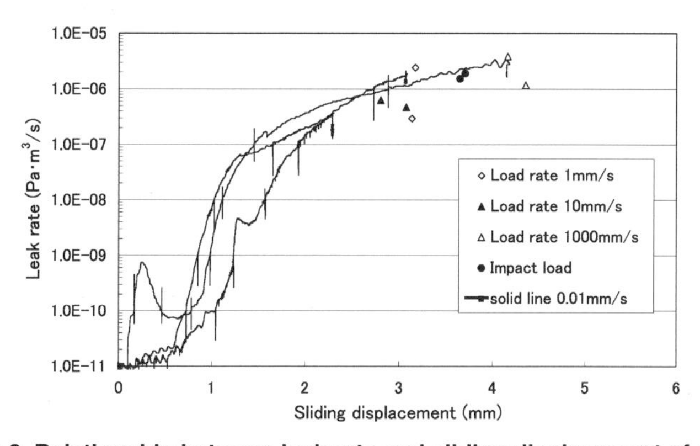

Fig. 3.5.2(4)-2 Relationship between leak rate and sliding displacement of scale model

Fig. 3.5.2(4)-2 also shows plotted measurement of the maximum leak rates of the maximum sliding displacement at a loading rate between 1 mm/s to 1000 mm/s. At the loading rate, the continuous measurement was not possible because the response of the helium detector was 0.2 s. Furthermore, an impact load was applied to the scale model, which was equivalent to that for a free drop from a height of 1 m. From the results, it was found that the relationship between the maximum sliding displacement and leak rate does not depend on the loading rate significantly.

2) Drop test of full scale metal cask

The drop tests of a full scale metal cask without the impact limiters were carried out by simulating drop accidents during handling in the storage facility. The floor into which the cask runs was designed to simulate a reinforced concrete floor in the facility. The first test was a horizontal drop from a height of 1 m (Fig. 3.5.2(4)-3). The second test was to simulate a rotational impact from a height of 1 m and with a lower trunnion of the cask as an axis.

Fig. 3.5.2(4)-3 Overall view of the horizontal drop test

The main measurement items were the sliding and lid opening displacements of the primary lid and the secondary lid, leak rates, and the pressure of helium in a space between the primary lid and the secondary lid. A lid structure of the cask and measurement positions of leak rate are shown in Fig. 3.5.2(4)-4. The double type metal gaskets are installed on the bottom of each of the primary lid and the secondary lid, and the containment is maintained by the metal gaskets.

Instantaneous leak rates were quantitatively measured at both the primary and secondary lids by the helium leak detectors. In this test, helium of 4 atm (gauge pressure) was filled in the space between the lids. On the other hand, eddy current displacement sensors (accuracy of ±0.01 mm) were used for displacement measurement of the lids.

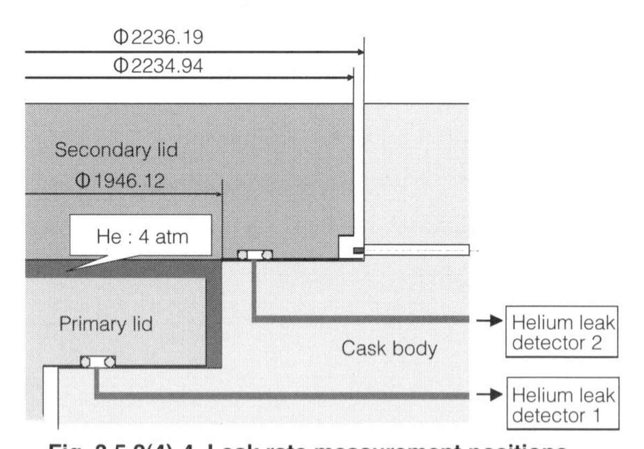

Fig. 3.5.2(4)-4 Leak rate measurement positions

a. Horizontal drop test

Fig. 3.5.2(4)-5 shows test conditions. The cask was dropped horizontally from a height of 1 m.

In this test, the front trunnion attacked the concrete floor, directly.

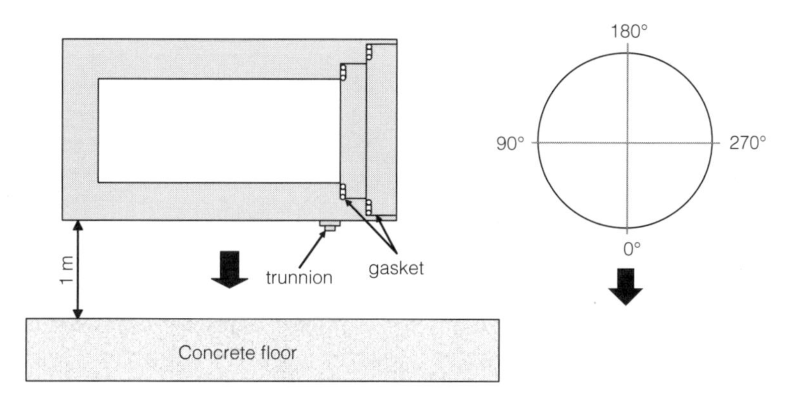

Fig. 3.5.2(4)-5 Horizontal drop test conditions

Table 3.5.2(4)-1 shows the summarized results of this test.

The amount of penetration to the concrete floor of the trunnion was about 100 mm and the average acceleration of the cask body center was about 50 G.

The maximum sliding displacements were about 0.4 mm and 0.3 mm in the primary lid and the secondary lid, respectively. They were observed at a 0 °direction (the drop direction).

The tendency of the sliding displacement was that the lids moved toward the shell in 0 ° direction, they did not move at 180 ° direction, and that they moved apart from the shell in 90 ° and 270 ° directions. Thus, it was considered that the cask body was transformed into an elliptical shape.

On the other hand, no significant opening displacement was observed in both of the primary lid and the secondary lid. Moreover, no decrease of pressure between the lids was observed right after drop impact.

Table 3.5.2(4)-1 Results of horizontal drop test

Acceleration	Main body	50 G
	Lid	16 G
Primary lid	Sliding	0.4 mm
	Lid opening	No significant change
Secondary lid	Sliding	0.3 mm
	Lid opening	No significant change
	Axial stress of bolt	No significant change
Maximum leak rate	Primary lid	2.38×10^{-10} Pa · m^3/s
	Secondary lid	2.85×10^{-9} Pa · m^3/s
Leak rate after 6 hours	Primary lid	1.52×10^{-11} Pa · m^3/s
	Secondary lid	7.90×10^{-12} Pa · m^3/s
Pressure between lids		No significant change

Fig. 3.5.2(4)-6 shows a time history of the leak rate of the primary lid. And Fig. 3.5.2(4)-7 shows a time history of the leak rate of the secondary lid.

Values of leak rate of the primary lid showed single-digit increase right after the impact, and at 10 minutes later, it decreased to the initial level. However, the leak rate increased again by a single-digit about 25 minutes later and restored to the initial level. After that, such a phenomenon was not observed within 6 hours. Thus, the leak rate seemed to be restored to the initial level completely.

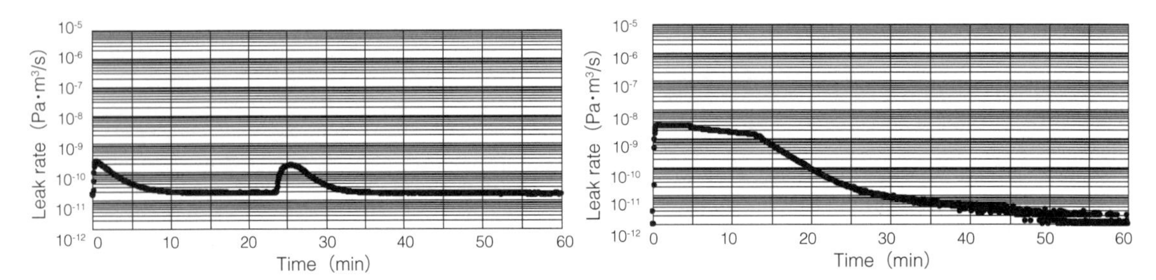

Fig. 3.5.2(4)-6 Time history of leak rate of the primary lid (Horizontal drop test)

Fig. 3.5.2(4)-7 Time history of leak rate of the secondary lid (Horizontal drop test)

On the other hand, the leak rate values of the secondary lid rose by double digits right after the impact. The higher leak rate than the initial revel was maintained for about 1 hour. After that, the leak rate seemed to be restored to the initial level completely.

The amount of helium gas leakage was calculated by integrating the leak rate with time. The total amount of helium gas leakage from the primary and secondary lids was 1.99×10^{-6} Pa · m^3. This value is 9.61×10^{-9} % of the initially installed helium gas. Thus, it was found that the amount of leakage was insignificant.

b. Rotational impact test

Fig. 3.5.2(4)-8 shows test conditions. This test was to simulate the rotational impact from a height of 1 m and with the lower trunnion of the cask as an axis. In this test, both of the front trunnion and the cask corner strucked the concrete floor directly.

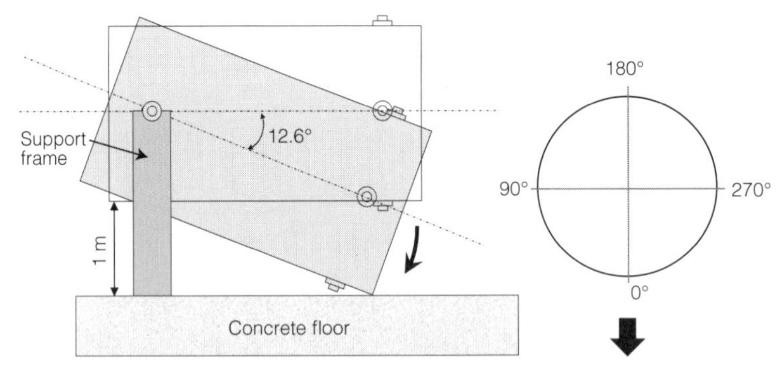

Fig. 3.5.2(4)-8 Rotational impact test conditions

Table 3.5.2(4)-2 shows the summarized results of this test.

The amount of penetration to the concrete floor of the trunnion was about 50 mm and the average acceleration of the primary lid center was about 48 G.

The maximum sliding displacements were about 0.6 mm and about 1.0 mm in the primary lid and the secondary lid, respectively.

Table 3.5.2(4)-2 Results of rotational impact test

Acceleration	Main body	16 G
	Lid	48 G
Primary lid	Sliding	0.6 mm
	Lid opening	0.11 mm
Secondary lid	Sliding	1 mm(0°), 0.6 mm(45°)
	Lid opening	No significant change
	Axial stress of bolt	Increase to 50 MPa
Maximum leak rate	Primary lid	3.86×10^{-9} Pa · m³/s
	Secondary lid	8.37×10^{-9} Pa · m³/s
Leak rate after 6 hours	Primary lid	4.91×10^{-10} Pa · m³/s
	Secondary lid	2.64×10^{-10} Pa · m³/s
Pressure between lids		Decrease of 0.006 MPa

It was considered that both 0 °direction and 180 °direction sides of the secondary lid contacted with the shell of the cask body.

The tendency of the sliding displacement was that lids moved toward the shell at 0 ° direction, and they moved apart from the shell in 90 ° and 270 ° directions.

The large lid opening displacement exceeding 0.1 mm was observed at 0 °direction of the primary lid. On the other hand, no significant opening displacement was observed in the secondary lid.

The axial stress of the secondary lid bolt increased to about 50 MPa at 0 °direction for 0.02 seconds during the drop. It was restored to the initial value after the impact.

The sliding displacement was larger than that of the horizontal drop test, and the lid opening displacement was observed. Therefore, the leak rate was bigger than that of the horizontal drop test.

Fig. 3.5.2(4)-9 shows time histories of the leak rate of the primary lid. And Fig. 3.5.2(4)-10 shows a time history of the leak rate of the secondary lid.

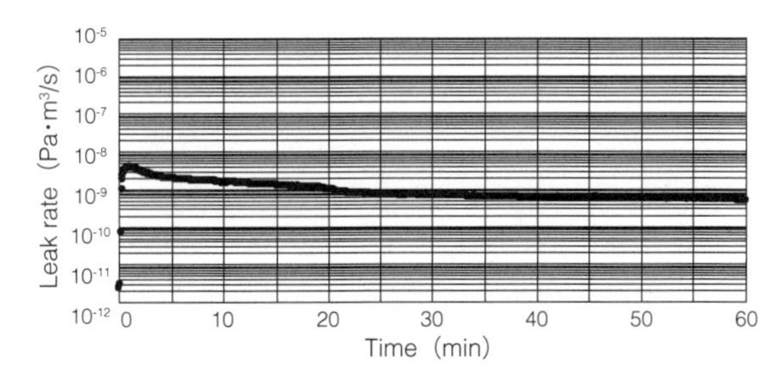

Fig. 3.5.2(4)-9 Time history of leak rate of the primary lid (Rotational impact test)

Fig. 3.5.2(4)-10 Time history of leak rate of the secondary lid (Rotational impact test)

The total amount of leakage from the primary and secondary lids was 1.74×10^{-5}Pa · m³, which is 8.45×10^{-8}% of the initially installed helium gas. This value was larger than in the horizontal drop test. Nevertheless, the amount of leakage was also insignificant.

The decrease of pressure between the lids was observed, which means that the helium gas leaked. The leakage seems to be from the helium filling port, not from the lid gaskets. Because the leak rate which was calculated based on pressure drop had much exceeded the detection range of the helium leak detector, in which case the detector must have malfunctioned.

3) Evaluation

Fig. 3.5.2(4)-11 shows the relationship between the maximum sliding displacements of the lids and leak rates in addition to the results that obtained by the scale models. Here, the result of the lid scale models was adjusted to the full scale metal cask.

It was found that the relationship between the maximum sliding displacements of the lids and the leak rates in the full scale metal cask can be evaluated by the results of the lid scale model.

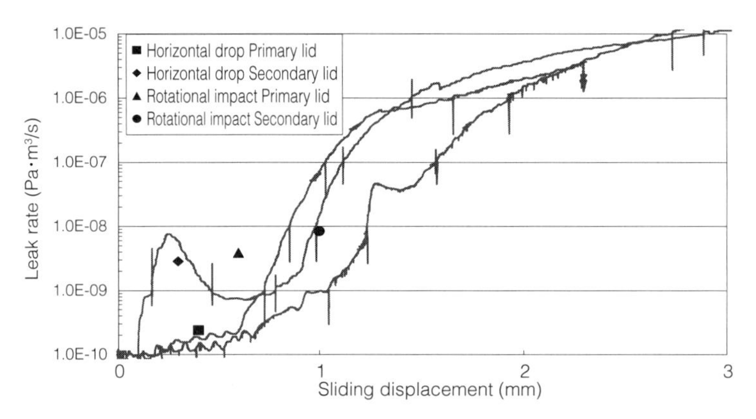

Fig. 3.5.2(4)-11 Relationship between leak rate and sliding displacement

4) Summary

The instantaneous leak rates were quantitatively measured during the drop tests of the full scale metal cask, which simulated the drop accidents in the storage facility.

From the results of a series of leakage tests using the scale model (1/11) of the lid, it was found that the relationship between the maximum sliding displacement and leak rate does not depend on the loading rate.

Two cases of drop test were performed using the full scale metal cask. The first test was a horizontal drop from a height of 1 m. The second tests was to simulate a rotational impact form a height of 1 m and with the lower trunnion of the cask as an axis.

Negligible helium leak was observed in both tests. In the rotational impact test, the amount of leakage was larger than that of the horizontal drop test. However, the amount of leakage was insignificant in both tests.

It was found that the relationship between the maximum sliding displacements of the lids and the leak rates in the full scale metal cask can be evaluated by the results of the lid scale model.

References

1) K. Shirai, et al., (1995),"Dynamic Interaction between Spent Fuel Storage Cask and Reinforced Concrete Slab

subjected to Impact Load", J. At. Energy Soc. Japan. Vol.37, No.5, pp.430-441.
2) T. Yokoyama, et al., (2004),"Integrity Assessment of Dual-Purpose Metal CASK after Long Term Interim Storage-Seal Performance under Transport Conditions", PATRAM2004, Berlin
3) H. Takeda, et al., (2006), "Instantaneous Leakage Evaluation of Metal Cask at Drop Impact", ICONE14, Miami, Florida, USA

(5) Evaluation of instantaneous leakage during the drop event

1) Evaluation for containment of cask lid

For evaluation of containment boundary which is consisted with metal gasket, it is suggested that both of sliding and opening displacement of cask lid should be evaluated separately, in the draft of "consideration for cask performance after long term storage" by JNES[1]. However, it also says that the containment performance can be evaluated by sliding displacement of metal gasket, when "stress generated on flange of lid, flange of body, and lid bolt" is lower than elasticity limit under instantaneous impact such as drop or tip-over event.

Relation between the sliding displacement and leak rate of metal gasket, which is determined based on some experimental results, has been studied by JNES. In this study, a regression line concerning relationship between sliding (total) displacement of the lid and estimated leak rate has been derived from a sliding experiment using a scale model of lid with a double metal gasket, whose hoop diameter and wire diameter are 250 mm and ϕ 10 mm respectively.

This metal gasket used for the experiment has been aged corresponding with 60 years storage.

If gasket that would be evaluated by this method differs from the gasket used for the experiment mentioned above, estimated leak rate should be compensated based on hoop diameter of the gasket[2]. Hoop diameter of primary lid's gasket used for the metal storage cask mentioned in Section 3.5.2 is ϕ 1738 mm, and hoop diameter of secondary lid's gasket is ϕ 2032 mm. Therefore, Fig. 3.5.2(5)-1 shows relationship between sliding displacement and estimated leak rate which is compensated for hoop diameter of these gaskets. In this section, leak rate of these gaskets would be estimated with using the regression line shown in the figure. Detailed method is described in the reference[3].

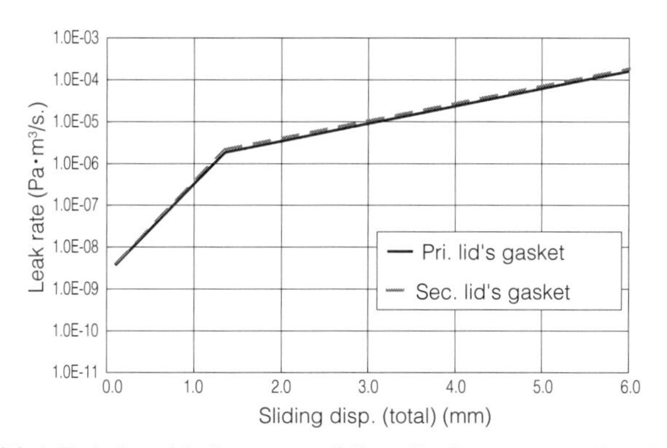

Fig. 3.5.2(5)-1 Relationship between sliding displacement and estimated leak rate (after ageing of 60 years storage)

2) Post analysis of the impact test using metal cask

To establish evaluation method of lid displacement under accidental event for metal cask, computer analysis for representing the drop tests using full scale metal cask (the horizontal drop test and the rotational impact test), which is mentioned in Section 3.5.2(4), has been performed.

The analysis has been performed for the horizontal drop test and the rotational impact test with using three dimensional dynamic analysis code, and purpose of these post analysis is evaluation of the lid displacement during these impact tests.

a. Analysis code

Three dimensional analysis code used for this analysis is the finite element analysis code LS-DYNA Ver.970 (for windows, dual precision). A user subroutine with non-liner constitutive equation[4] considering strain rate dependency and multi-axial fracture developed by CRIEPI has been applied to this LS-DYNA code.

b. Analysis model

This analysis model has been made as a 1/2 symmetric model considering the cask structure and the drop orientation.

c. Analysis condition

(a) Contact condition

a) Friction

On the contact surface between cask body (flange surface) and lids (primary and secondary lid), it is thought that behavior of sliding displacement is different from general metal contact surface, because a metal gasket is equipped on the contact surface. Therefore, friction coefficient on the surface is set to 0.6 in reference to "apparent friction coefficient on the flange surface" which is obtained from experimental results of "instantaneous leakage evaluate test using scale model of lid". The apparent friction coefficient has been measured in the experiment that static slide displacement has been applied to the scale model of a lid to which metal gasket is equipped.

Friction coefficients on the other surface have been set to 0.14, which is determined based on research in the past[5].

b) Tightening force by lid bolts

Tightening force of lid bolts has been set to equivalent force to tightening torque of 2400 N-m.

(b) Material model

a) Metallic material

Table 3.5.2(5)-1 shows material properties for metallic material used in this analysis.

b) Concrete material

Table 3.5.2(5)-2 shows material properties applied to concrete material model. For the concrete pad, material properties obtained from strength test result using specimen of sealed curing at job site has been applied, and the strength test has been performed just after the actual drop test. Moreover, for the light weight concrete, the value has been determined by strength test result using concrete core specimen obtained from the full scale cask for the drop test.

Table 3.5.2(5)-1 Material properties for metallic materials

Part	Density (kg/m³)	Young's modulus (MPa)	Poisson's ratio	Yield strength (MPa)	Hardening modulus	Parameter for strain rate*	
						P	C
Body and lid	7.85x10³	203460	0.3	205	2034	200	5
Lid bolt	7.85x10³	202000	0.3	890	2020	200	5
Outer shell	7.86x10³	203000	0.285	215	2030		
Trunnion	7.86x10³	195000	0.3	725	1950	200	5
Dummy weight	7.86x10³	203000	0.3	215	2030		
Reinforced bar	7.86x10³	206000	0.3	295	2060		

*Note) Coefficient for yield strength $1+\left(\frac{\dot{\varepsilon}}{C}\right)^{1/P}$ and $\dot{\varepsilon}$ is strain rate

Table 3.5.2(5)-2 Material properties for concrete materials

Part	Density (kg/m³)	Shear modulus (MPa)	Bulk modulus (MPa)	Compressive strength (MPa)	Tensile strength (MPa)
Concrete pad (Horizontal drop)	2.286×10³	11939	16177	36.92	2.31
Concrete pad (Rotational impact)	2.281×10³	11919	15706	35.93	2.39
Light weight concrete	1.543×10³	5876	8805	22.68	2.52

(c) Initial condition

As described in Section 3.5.2(4), the drop test has been performed for two cases, horizontal drop test onto trunnion side and rotational impact test that the rotational center axis is set at the center axis of rear trunnion.

a) Horizontal drop test

Since drop height of the horizontal drop test was 1 m, in this analysis, initial velocity 4430 mm/sec equivalent to 1m free drop has been set to the cask model.

The cask model has been dropped as the 0 degree side is lower side, same as the actual drop test.

b) Rotational impact test

Since height of the cask before rotation was 1 m, in this analysis, rotational angular velocity 1.09 rad/sec equivalent to 1 m rotational impact that rotational center axis is set at rear trunnion has been set to the cask model.

The cask model has been dropped as the 0 degree side is lower side, same as the actual drop test.

d. Comparison between drop test result and analytical result

(a) Horizontal drop test

Table 3.5.2(5)-3 shows comparison between the drop test result and the analytical result of the horizontal drop test.

Table 3.5.2(5)-3 Horizontal drop test result and analytical result of the horizontal drop

Item		Test result	Analytical result
Average deceleration on primary lid		97 m/s²	135 m/s²
Pri. Lid	Sliding displacement	Maximum 0.43 mm Total 0.68 mm	Maximum 0.44 mm Total 0.63 mm
	Opening displacement	No significant opening	None
Sec. Lid	Sliding displacement	Maximum 0.30 mm Total 0.42 mm	Maximum 0.20 mm Total 0.29 mm
	Opening displacement	No significant opening	Maximum About 0.02 mm

(b) Rotational impact test

Table 3.5.2(5)-4 shows comparison between the drop test result and the analytical result of the rotational impact test.

Table 3.5.2(5)-4 Rotational impact test result and analytical result of the rotational impact test

Item		Test result	Analytical result
Average deceleration on primary lid		340 m/s²	264 m/s²
Pri. Lid	Sliding displacement	Maximum 0.60 mm Total 1.02 mm	Maximum 0.70 mm Total 1.10 mm
	Opening displacement	About 0.11 mm	About 0.01 mm
Sec. Lid	Sliding displacement	Maximum 1.01 mm Total 1.89 mm	Maximum 0.91 mm Total 1.64 mm
	Opening displacement	About 0.01 mm	About 0.05 mm

(c) Summary of the analytical results

· Post analysis for the horizontal drop test and the rotational impact test have been performed to comparison with actual test results.

· In this analysis, initial tightening force of lid bolts and reaction force from metal gasket have been considered in detail. Friction coefficient on flange surface has been set to the value obtained from the scale model test result. Moreover, if necessary, initial lid position has been set equivalent to the test result.

· In horizontal drop test analysis, for maximum sliding displacement of lids, almost same value of the test result has been obtained.

· In rotational impact test analysis, there has been good agreement for sliding displacement of lids, with considering initial lid location of the rotational impact test.

As mentioned above, it is confirmed that sliding displacement of the lid during drop test can be evaluated by using this analytical method.

3) Analysis for tip-over events

Tip-over event from upright position or during tilting up operation to vertical position are more severe than the above mentioned drop tests condition. Analyses for these tip-over events have been performed and lid displacements during these events have been evaluated.

a. Analysis model for tip-over events

Same analysis code and analysis model used in the previous section has been applied for this analysis. Contact condition and material model has been also set to same as the previous section.

And material property for the concrete pad of rotational impact test has been selected from Table 3.5.2(5)-2 in the previous section.

b. Analysis condition for tip-over events

(a) Tip over from upright position

It is assumed for this analysis that the metal cask tips over on its side onto a concrete pad from upright position. As initial angular velocity corresponding to the tip-over event is set to 1.681 rad/sec whose rotational center is bottom edge of the cask.

(b) Tip over during tilting up operation

It is assumed for this analysis that the metal cask tips over on a transport frame when a lifting device comes out during tilting up operation of the cask.

As shown in Fig. 3.5.2(5)-2, cask position for this analysis has been set as lid of the metal cask is contact to a concrete pad, when the rotational center is center axis of rear trunnion.

Height of rear trunnion as a rotational center of tip-over event has been set to the same as the rotational impact test, which is described in Section 3.5.2(5)-2). Therefore, the inclined angle of the cask just before the impact is the same as the rotational impact test.

Angler velocity for just before the impact of the cask lid onto a concrete pad has been set to 2.57 rad/sec.

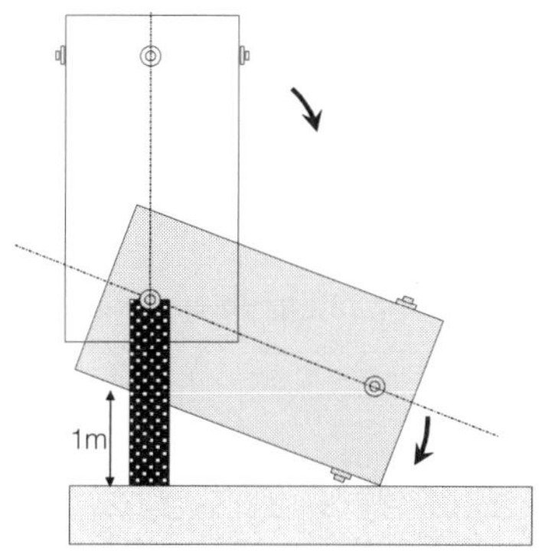

Fig. 3.5.2(5)-2 Tip over during tilting up operation

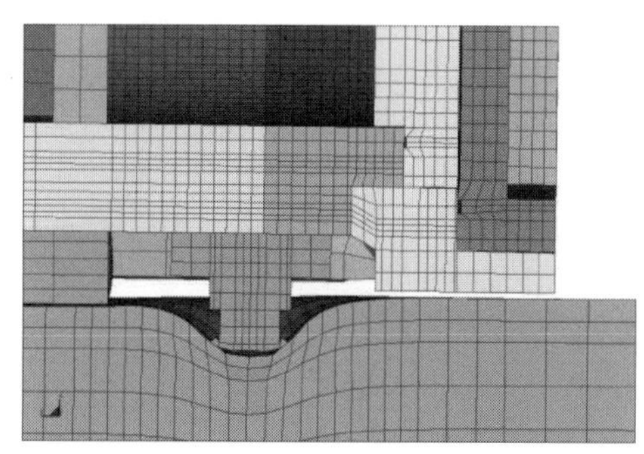

Fig. 3.5.2(5)-3 Deformation in the vicinity of lid at max. displacement time

c. Analytical result for tip-over events

(a) Tip over from upright position - trunion side is lower side -

Fig. 3.5.2(5)-3 shows deformation of vicinity of the lid at maximum displacement time of the cask, and it shows that there is no contact between side of the cask body at lid flange and a concrete pad.

Table 3.5.2(5)-5 shows analytical result of the tip-over event that trunnion side is lower side.

Table 3.5.2(5)-5 Analytical result of tip-over event (Trunnion side is lower side.)

Item		Result
Average deceleration on primary lid		323 m/s^2
Pri. Lid	Sliding displacement	Maximum 1.14 mm Total 1.59 mm
	Opening displacement	None
	Axial stress on lid bolts	No change
	Assumed leak rate	2.3×10^{-6} Pa m^3/s
Sec. Lid	Sliding displacement	Maximum 0.51 mm Total 0.57 mm
	Opening displacement	About 0.04 mm
	Axial stress on lid bolts	No change
	Assumed leak rate	4.3×10^{-8} Pa m^3/s

(b) Tip over from upright position - no trunion side is lower side -

Table 3.5.2(5)-6 shows analytical result of the tip-over event that no trunnion side is lower side.

Table 3.5.2(5)-6 Analytical result of tip-over event (No trunnion side is lower side.)

Item		Result
Average deceleration on primary lid		630 m/s^2
Pri. Lid	Sliding displacement	Maximum 1.12 mm Total 1.47 mm
	Opening displacement	None
	Axial stress on lid bolts	No change
	Assumed leak rate	2.1×10^{-6} Pa m^3/s
Sec. Lid	Sliding displacement	Maximum 0.62 mm Total 0.84 mm
	Opening displacement	About 0.05 mm
	Axial stress on lid bolts	No change
	Assumed leak rate	1.7×10^{-7} Pa m^3/s

(c) Tip over during tilting up operation

Table 3.5.2(5)-7 shows analytical result of the tip-over event during tilting up operation.

Table 3.5.2(5)-7 Analytical result of tip-over event during tilting up operation

Item		Result
Average deceleration on primary lid		619 m/s^2
Pri. Lid	Sliding displacement	Maximum 1.07 mm Total 1.63 mm
	Opening displacement	About 0.19 mm
	Axial stress on lid bolts	Max. about 380 MPa
	Assumed leak rate	2.4×10^{-6} Pa m^3/s
Sec. Lid	Sliding displacement	Maximum 1.07 mm Total 1.65 mm
	Opening displacement	About 0.27 mm
	Axial stress on lid bolts	Max. about 470 MPa
	Assumed leak rate	(Loss of containment)

d. Evaluation of containment during tip-over event

In standard of Atomic Energy Society of Japan "Standard for Safety Design and Inspection of Metal Casks for Spent Fuel Interim Storage Facility: 2010", it is described for definition of allowable leak rate under accident during storage that "allowable leak rate for accident is defined as that cavity pressure of a metal cask can be retained as negative pressure until necessary repair has been provided".

Thus, in this section, cavity pressure of the metal cask after tip-over event is calculated based on the assumed leak rate, which is shown in Table 3.5.2(5)-7.

It can be considered as a cause of increase of the cavity pressure, that pressurized enclosed gas between primary lid and secondary lid (interlid space) is leaked to cask cavity through a metal gasket on a primary lid. Under the tip-over event during tilting up operation, interlid pressure might be decreased to the almost same pressure to the ambient pressure due to leakage of interlid gas through secondary lid's metal gasket, because containment of secondary lid would be lost. In this calculation, it is assumed as safety side, that the initial interlid pressure (0.41 MPa for this metal cask) is kept. Moreover, cavity pressure of the metal cask would also be increased due to leakage of fissile gas from spent fuels and inflow gas from the interlid. This increase of cavity pressure would prevent the leak from interlid to cask cavity, therefore, it is assumed in this calculation as safety side, that initial cavity pressure (0.081 MPa for this metal cask) is kept.

As the calculation condition for the tip-over event during tilting up operation, if leak rate of primary lid (2.4×10^{-6} Pa m^3/s) is maintained for one week (6.05×10^5sec), total volume Q of inflow gas to cask cavity is calculated as follows.

$$Q = L\times\frac{Pu-Pd}{Ps}\times t$$
$$= 2.4\times10^{-6}\times\frac{4.1\times10^5-8.1\times10^4}{1.01\times10^5}\times6.05\times10^5 \quad \ldots\ldots (1)$$
$$\fallingdotseq 4.7\ Pa\cdot m^3$$

where:

L : Leak rate of primary lid ($2.4x10^{-6}$ Pa m^3/s)

Pu : Upstream pressure ($4.1x10^5$ Pa)

Pd : Downstream pressure ($8.1x10^4$ Pa)

Ps : Standard pressure ($1.01x10^5$ Pa)

t : Duration for evaluation ($6.05x10^5$ sec)

Moreover, since cavity volume of this cask is 6.3 m^3, increase of cavity pressure P' corresponding to the inflow gas is calculated as follows.

$$P' = 4.72/6.3 \fallingdotseq 0.75\ Pa \quad \ldots\ldots (2)$$

This increased pressure of cask cavity is less than 0.001 % of initial pressure (0.081 MPa).

The increased pressure of cask cavity per week after tip-over event during tilting up operation is negligible with less than 0.001 % of initial pressure. Therefore, it is concluded that damage of lids due to a tip-over accident would not affect in maintaining negative pressure of cask cavity, if it is short term.

4) Summary

As the evaluation of instantaneous leakage during the drop event, post analysis of drop tests of the cask and analysis for tip-over event have been performed and following knowledge can be obtained.

a. In the post analysis for the horizontal drop test, for maximum sliding displacement of lids, almost same value of the test result has been obtained.

b. In the post analysis for the rotational impact test, there has been good agreement for sliding displacement of lids, with considering initial lid location of the rotational impact test.

c. From above a. and b., it is confirmed that sliding displacement of lids during the drop tests can be evaluated by using the analytical method, which is used for the post analysis.

d. In the analysis for tip-over event from upright position, larger sliding displacement of lids than the drop test results have been observed. However, since there has been no opening displacement and plastic deformation of flange surface, it is considered that leak rate form lids can be evaluated based on the sliding displacements.

e. In the analysis for tip-over event during tilting up operation, containment of secondary lid would be lost.

However, since assumed leak rate of primary lid has been retained in low, the cavity pressure has been maintained as negative and inner gas of cavity would not be released to outside of the cask.

f. Maximum assumed leak rate under the tip-over condition has been 2.4×10^{-6} Pa · m^3/s. In this case, increase of cask cavity pressure per week is less than 0.001 % of initial pressure. Therefore, it is concluded that damage of lids due to a tip-over accident would not affect in maintaining negative pressure of cask cavity, if it is short term.

References

1) Japan Nuclear Energy Safety Organization: Fiscal Year Heisei 15 Recycled Fuel Resources Storage Technology Research etc. (Metal Cask Storage Technology Confirmatory Testing Report) Final Report, June 2004

2) Japan Nuclear Energy Safety Organization: Fiscal Year Heisei 16 Metal Cask Storage Facility Safety Assessment Report Final Report, January 2006

3) Central Research Institute of Electric Power Industry: Evaluation of Instantaneous Leakage from Metal Cask under Drop Accident -Evaluation of Containment during Handing Accident with Numerical Analysis-, CRIEPI Report N06005, December 2006

4) Central Research Institute of Electric Power Industry: Impact Analysis of Reinforced Concrete Structure by Finite Element Method -Description of Concrete Failure Model and Benchmark-, CRIEPI Report U93053, March 1994

5) Japan Nuclear Energy Safety Organization: Fiscal Year Heisei 14 Recycled Fuel Resources Storage Technology Research etc. (Metal Cask Storage Technology Confirmatory Testing Report), March 2003

3.5.3 Drop of heavy object onto storage cask by building collapse

Spent fuel storage cask is expected to maintain its integrity under the maximum class of earthquake (Ss class). Hence, the storage building may be constructed without the special classification of earthquake, provided that the cask would not receive any critical effect by the earthquake. In this case, drop of heavy object onto storage cask by earthquake may be considered.

We carried out drop tests of heavy objects onto full scale storage casks taking account of parameters of dropping objects (weight, structure, drop height, etc.) in order to demonstrate the integrity of the casks. In addition, the analytical method was also established. More detail information is found in the literature[1)-4)].

(1) Drop tests on to cask

1) Test method

A reinforced concrete slab was dropped simulating a ceiling of a storage building onto full-scale cask standing vertically. Fig. 3.5.3-1 shows the outline of the reinforced concrete slab.

The tested cask was the type X cask used for the drop test during the handling as described in the section 3.5.2. In this kind of tests, most of the drop energy is absorbed by the fracture of the concrete slab and thus stress generated in the casks will be small. However, because the slab directly hit the cask lid, the tests were carried out with special attention to the containment performance of the cask lids. Metallic seal (liner: silver, core material: Inconel) was used for the cask lid structure. The drop height was determined as follows.

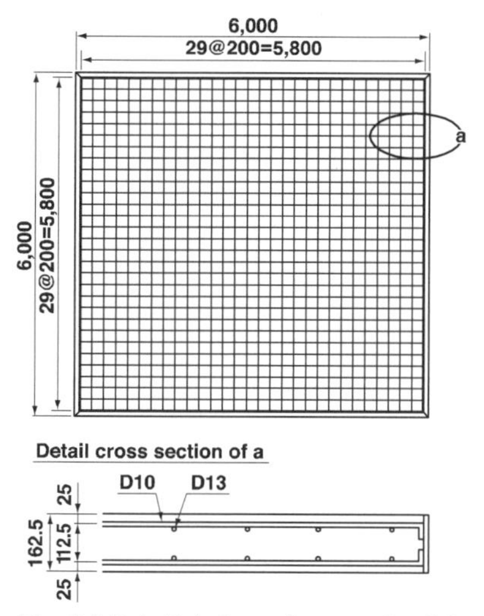

Fig. 3.5.3-1 Reinforced concrete slab

a. Height that would not damage the containment performance of the secondary lid (outer lid): 5 m (Preliminary analysis showed that no plastic strain would be generated in the lid structure by a drop form 5 m.)

b. Height of the ceiling of the building: 17.1 m

Before and after the drop tests, we carried out He leak tests in order to confirm the containment performance of the cask.

2) Test results

Fig. 3.5.3-2 shows the drop test. Fig. 3.5.3-3 shows the maximum strain generated at the drop tests.

At the drop test from the height of 5 m, the strain was within the elastic in the area of structure and lid bolts that would directly affect the structural integrity and containment. However, in the secondary lid area, approximately 1800 micro strain including approximately 300 micro residual strain (plastic) was generated.

Fig. 3.5.3-2 Drop tests of heavy object onto cask

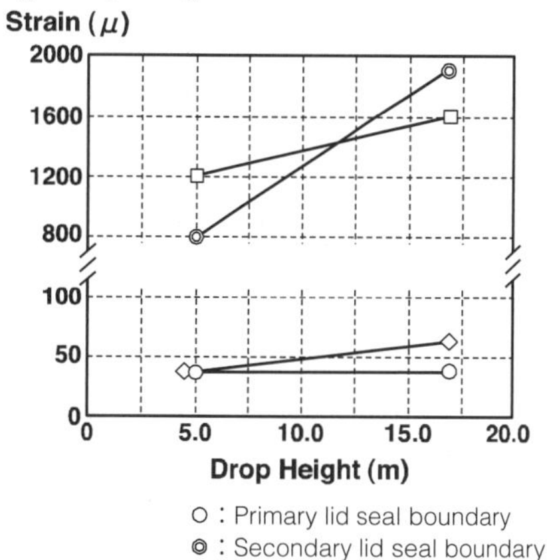

Fig. 3.5.3-3 Results of strain by drop tests

Table 3.5.3-1 shows that the containment performance was maintained before and after the drop tests for the heights of 5 m and 17.1 m.

Table 3.5.3-1 Results of containment measurement before and after the drop tests

	Drop height (m)	Tested part	Leak rate (atm・cc/sec)	
			Before drop test	After drop test
Drop object	5.0	Primary lid	< 9.0×10^{-8}	< 9.8×10^{-7}
		Secondary lid	< 2.1×10^{-8}	< 1.1×10^{-8}
Heavy weight	17.1	Primary lid	< 4.4×10^{-8}	< 4.2×10^{-6}
		Secondary lid	< 1.6×10^{-8}	5.1×10^{-8}

In conclusion, although the strain in the primary lid would be negligible and a large strain would be generated in the secondary lid, the casks would maintain the containment performance under the heavy object drop accident by the building collapse of the cask storage building.

(2) Mechanical impact analysis of storage cask

We carried out the mechanical impact analysis using the impact analysis code (DYNA-3D) that was used for the cask drop accident during handling. Fig. 3.5.3-4 shows the analytical model.

The characteristics of the analytical method are as follows.

a. The method applies compressive fracture condition considering tri-axial stress state.

b. It considers tensile fracture state including tri-axial direction cracks.

c. Strain rate dependence for the strength of compressive and tensile fracture.

Fig. 3.5.3-5 shows results of analysis and tests on the strain generated in the secondary lid. Although the maximum strain by the analysis showed larger value as compared with test result, the strain response is similar.

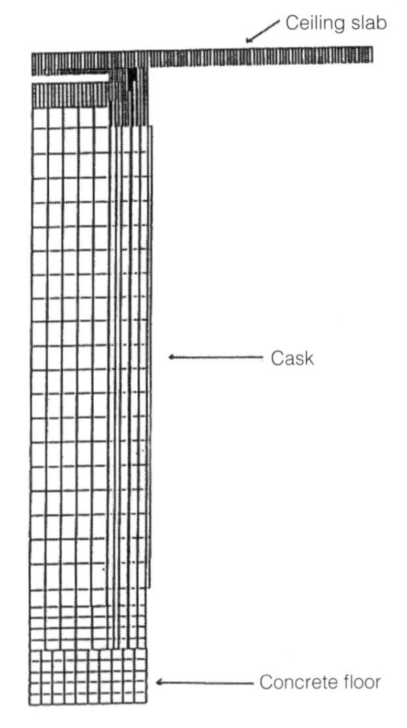

Fig. 3.5.3-4 Analytical model of cask and ceiling slab

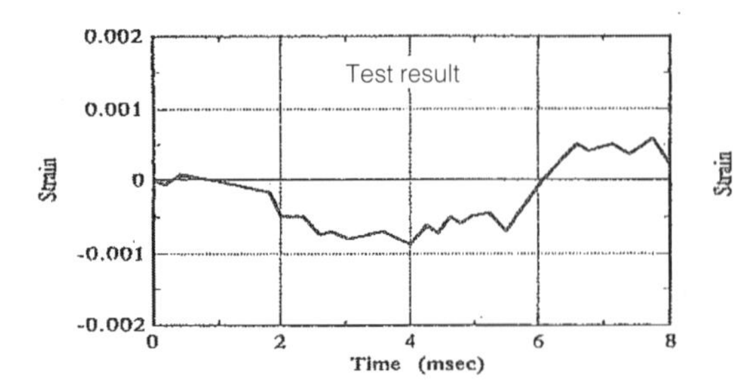

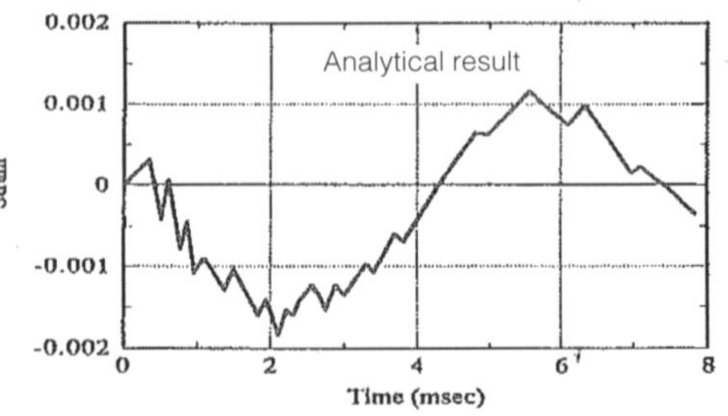

Fig. 3.5.3-5 Test result and analytical result on the strain generated in the secondary lid

Fig. 3.5.3-6 shows the results of analysis (plastic strain) for the 17.1 m drop test. It showed a plastic strain generated in the secondary lid, and coincides with the test result of Fig. 3.5.3-3. On top of that, there was no plastic strain generated in the seal area of the primary lid, and no abnormality in the containment performance. Then, the applicability of the containment evaluation method judging from the existence of the plastic strain was confirmed.

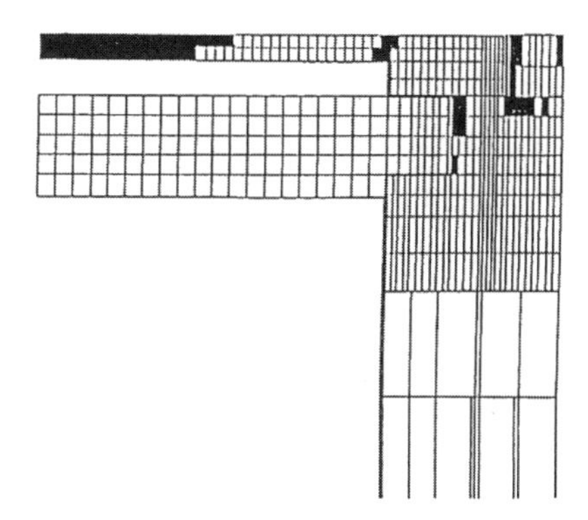

Fig. 3.5.3-6 Analytical result on plastic strain in the cask lid (Drop height : 17.1 m)

References

1) H. Ohnuma, S. Shiomi, Y. Maki, and T. Saegusa: "Safety Evaluation Dry-Cask Storage Facility for Spent Fuel during Earthquake", No.385009 (1985)
2) C. Ito and H. Ohnuma: "Dynamic Analysis for Evaluation of Integrity of Storage Cask against Impact load", CRIEPI Report U86046 (1987)
3) Y. Kato, S. Hattori, C. Ito, K. Shirai, S. Ozaki, O. Kato: "Drop Test of Reinforced Concrete Slab onto Storage Cask", Proc. PATRAM '92, Sept.13-18, 1992 Yokohama, Japan, 1443.
4) C. Ito, O. Kato, K. Shirai: "Establishment of Cask-Storage Technology for Spent Fuel - Evaluation of Cask Integrity against Dropping Heavy Weight -", CRIEPI Report U92036,1992.12

3.6 Seismic performance

3.6.1 Tip over of unfixed cask

Casks are stored vertically in order to enhance the storage density. It is important to evaluate their seismic stability of casks in Japan, where earthquakes are often experienced.

The seismic stability of the casks is essentially treated as dynamic vibration issue of cylinders. These tip over phenomena have been studied from the view point of estimation of seismic behavior of tomb stone[1], prevention of furniture or equipment from tip over by earthquake[2], etc.

There have been two kinds of analytical methods in the conventional studies, i.e. theoretical analysis method and numerical simulation method. The theoretical analysis method obtains a theoretical solution for sine wave under a limited condition of one dimensional sine wave earthquake by simplifying the basic equation[3]. Although this method is superior in understanding the basic characteristics of the phenomenon in a macroscopic way, it cannot analyze seismic responses for real earthquakes with irregular waves. The numerical simulation method is to numerically integrate nonlinear equation of motion directly in order to obtain seismic response successively[4]. In order to investigate the seismic stability of cask quantitatively, we may have to rely on the numerical simulation.

In this section, we considered the tip over phenomenon of the cask as a vibration phenomenon in the two dimensional problem of a rigid structure, and paid attention to Distinct Element Method (DEM) for the numerical simulation method[5]. The DEM was reportedly suitable for an analytical method for stability problem of bed rock, particularly fracture phenomenon such as tip over and slide down of slope[6]. We have made a numerical simulation for a vibration test using a small cylinder and found that the DEM is effective to investigate the tip over phenomenon[7].

We conducted cask tip over tests using a cask model and reinforced concrete floor simulating floor of storage facility on a shaking table in order to investigate response characteristics for cyclic input and seismic input. On top of that, we conducted tip over analysis by the DEM and found we can evaluate seismic stability of cask for tip over [8]. More detail is found in the literature[9].

(1) Tip over test

1) Cask model

Metallic containers that are about 2.5 m in diameter, about 5.5 m in height, and about 9800 N in weight are used for casks storing spent fuel. Considering a similarity law, the following conditions should be met.

a. Similarity ratio of acceleration should be 1, because equation of motion includes acceleration of gravity.

b. In order to take into account the condition of cask as rigid structure and repulsion property and friction property between cask and floor, similar materials should be used.

With these conditions, ratios of physical properties of the full-scale cask and the simulated model are given as follows.

$$\left.\begin{aligned} 1/N &= L_m/L_p : \text{ Scale ratio} \\ K &= E_m/E_p : \text{ Rigidity ratio of material} \\ M &= \rho_m/\rho_p : \text{ Density ratio of material} \end{aligned}\right\} \quad \text{..... (1)}$$

where, L : length, E : Young's ratio, ρ : density, subscript m : simulated model, p : full-scale cask

Considering similarity with the full-scale cask and performance limit of the shaking table, we determined the scale ratio as 1/3. As the result, the density ratio M of the material would become 3 and the density of the scale model cask ρ_m should have exceeded 20 g/cm^3 as the material condition. Such high density material does not exist. Then, we made the outer shell and bottom plate of the cask made of carbon steel, in which additional lead was fulfilled inside the model in order to make similarity of the apparent density. Table 3.6.1 shows similarity ratio applied for each parameter. The floor model simulating that of storage facility was reinforced concrete floor.

Table 3.6.1 Similarity law

Parameter	Notation	Dimension	Similarity ratio	
			General form	for N=3
Length	L	L	$L_m/L_p=1/N$	1/3
Weight	W	MLT^{-2}	$W_m/W_p=1/N^2$	1/9
Time	T	T	$T_m/T_p=1/N^{1/2}$	1/1.73
Velocity	V	LT^{-1}	$V_m/V_p=1/N^{1/2}$	1/1.73
Acceleration	A	LT^{-2}	$A_m/A_p=1$	1
Mass	m	M	$m_m/m_p=1/N^2$	1/9
Moment of inertia	I	ML^2	$I_m/I_p=1$	1
Surface force	P	$ML^{-1}T^{-2}$	$P_m/P_p=1$	1

(Note) Suffix p denotes the prototype, and suffix m denotes the model.

2) Friction test

We conducted friction tests in order to find dynamic characteristics between the cask bottom surface and the concrete floor surface.

The concrete floor test structure was fixed on the horizontal shaking table that is connected to a hydraulic actuator for shaking in the horizontal direction, and a block simulating the cask bottom surface was put on the concrete floor structure. The friction coefficient was measured from the ratio of vertical force to the horizontal force obtained from load cells attached in the vertical and horizontal directions after holding to a required pressure and vibration in the horizontal direction.

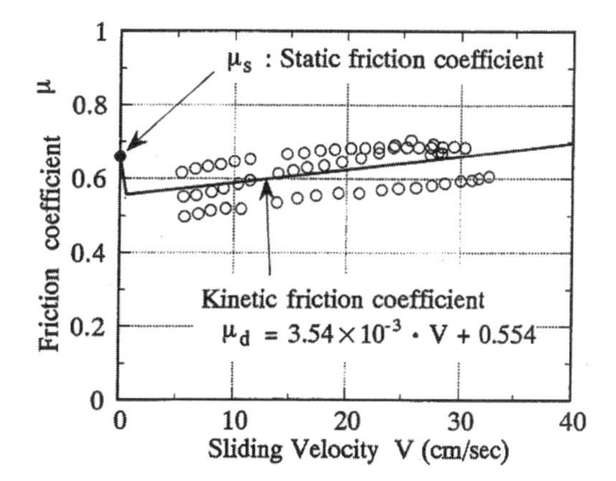

Fig. 3.6.1 Relationship of sliding velocity and kinetic friction coefficient

There was no significant dependency for the friction coefficient on the contact pressure. The dynamic friction coefficient tends to depend on sliding velocity. Fig. 3.6.1 shows a relationship between sliding velocity and the friction coefficient.

3) Free damping oscillation tests

We conducted free damping oscillation tests in order to study repulsion characteristics between the cask model and the floor model. In the test we measured the free damping oscillation waves of angular velocity of the model that was rocking free on the floor. The ratio of the angular velocity change (hereafter, angular velocity damping ratio, δ) before and after impact was defined as follows.

$$\dot{\theta} \rightarrow \delta \times \dot{\theta} \quad \text{at} \quad \theta=0 \quad (0<\delta<1) \quad ..... (2)$$

A measured value of δ was 0.963.

4) Vibration test

a. Test method

In order to grasp the tip over phenomenon of the cask, we conducted vibration tests using a cask model and floor model. The shaking table was electric oil pressure-type (Maximum movable load: 9800 N, Displacement stroke: ± 5 cm). Fig. 3.6.2 shows outline and measurement points of the shaking table. We measured rotation angle and angular velocity of the cask model by putting angle inclination meters and angular velocity meters on the top of the model that was placed on the shaking table with the floor model.

Fig. 3.6.2 Shaking table test apparatus

The kind of the vibration waves and directions were as follows.

(a) Vibration waves : Sine wave (2.0~6.0 Hz)

Seismic wave (El Centro, Hachinohe)

(b) Vibration direction : Horizontal

(c) Floor : Reinforced concrete

(d) Maximum acceleration : 100 to 700 cm/s^2

In the sine wave tests, we recorded the response wave shape of the steady state condition after the cask model was in a cyclic vibration mode with a high frequency because a stable cyclic rocking vibration is obtained at higher frequency [3].

In the seismic wave tests, El Centro wave (Imperial Valley Earthquake, California:1940, S00E) and Hachinohe wave (Tokachi-Oki earthquake : 1968, E-W) were employed. Fig. 3.6.3 shows the input wave. In the vibration test, we used a wave scaled to $\sqrt{1/3}$ for time that was converted based on the similarity law as shown in Table 3.6.1.

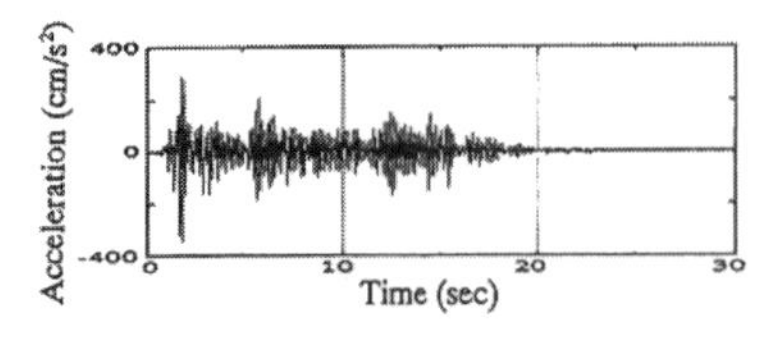

(a) El Centro wave (Imperial valley earthquake, California, 1940, S00E)

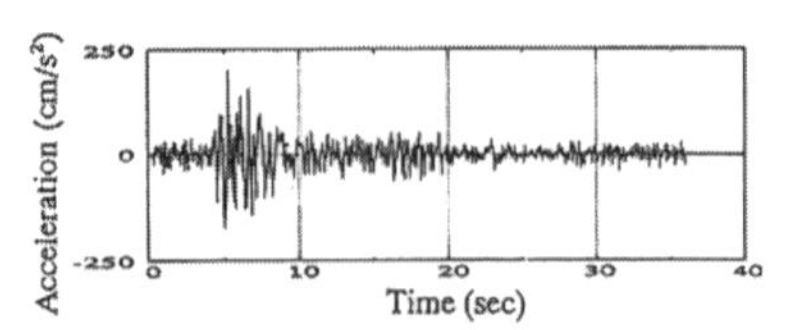

(b) Hachinohe wave (Tokachioki earthquake, 1968, E-W)

Fig. 3.6.3 Input waves for shaking table test

b. Sine wave vibration test

Fig. 3.6.4 shows a relationship between the input frequency and the maximum response angle and Lissajouss figures for the input of the sine wave displacement 1.0 cm to the cask model. The solid line is a response curve calculated form the periodic solution using the dimensions of the cask model and the measurements of angular velocity damping ratio.

As the vibration frequency gets small, the response angle of the cask model tends to be large. We found that the response to the cyclic input to the cask showed a good agreement with the periodic solution. Therefore it is considered that the tip over phenomenon of the cylindrical cask is treated as a two dimensional vibration problem.

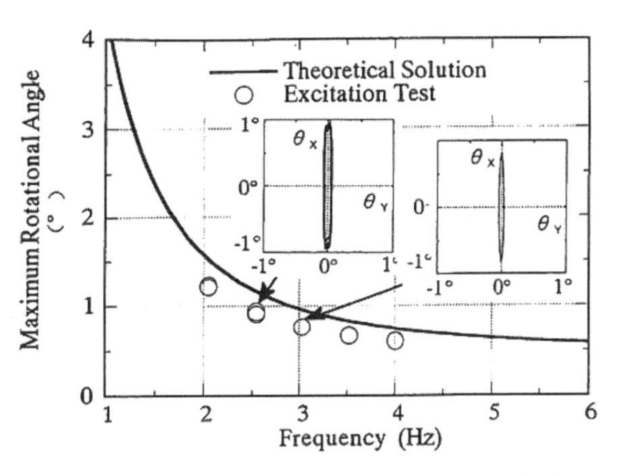

Fig. 3.6.4 Test results for sinusoidal wave excitation

c. Seismic wave vibration test

Fig. 3.6.5 shows a relationship between the maximum input frequency and the maximum response angle for the seismic wave. Rocking of the cask was rarely observed for the El Centro wave even if the input acceleration increased. On the other hand, rocking was observed for the Hachinohe wave when the input acceleration exceeded 300 cm/s^2, and the response angle became larger when the input acceleration exceeded 400 cm/s^2. This is because the rocking frequency becomes large as the response angle of the cask gets large. The response angle will be amplified for the wave with a relatively long-period component like the Hachinohe wave.

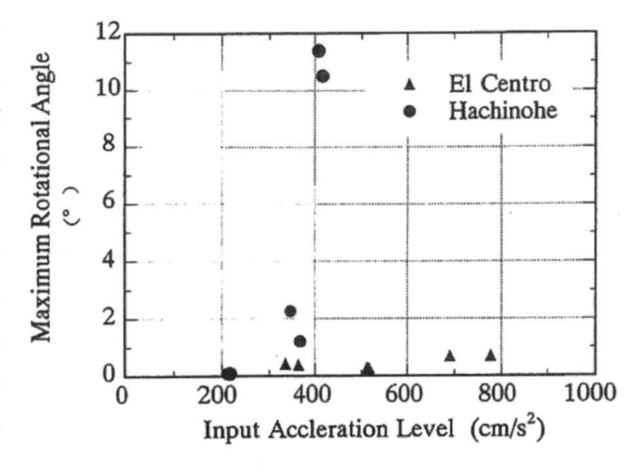

Fig. 3.6.5 Test results for natural earthquake wave excitation

The maximum response angle was about 11° for the stringent test condition for tip over (Input wave: Hachinohe, Maximum input acceleration: 400 cm/s^2 : corresponding to double the original acceleration), which was far below the critical angle for tip over 20°.

(2) Numerical analysis and evaluation

1) Characteristics of the analytical code

We used a two-dimensional DEM with polygon elements in order to evaluate the cask stability against tip over. The DEM is a numerical analysis code invented by Cundall[5)] and has been developed as one of the analytical code for the problem of rock bed stability[6].

The DEM can solve numerically a kinetic motion of the whole system state based on equations for the motion of center of gravity of each block that were represented for the cask model and the floor model with polygon shapes.

The DEM assumes that the cask model and the floor model contact only when one element's top contacts with the other element's side, or one element's side overlaps the other element's side. Discontinuity among elements, elastic deformation and friction characteristics due to contacts are expressed assuming spring-slider system among contacts, namely contact force among blocks without contacts are zero, thereby the separation state was expressed.

2) Analysis model

Fig. 3.6.6 shows the cask model for analysis and dimensions. Table 3.6.2 shows material data used in the analysis.

The analysis model is two-dimensional plane prism. The cask model and the floor model were polygon rigid models. Seismic waves (sinusoidal wave and natural earthquake wave) were input to the floor block. Specifications of the analysis model were assumed so that static tip over angle (ratio of radius of the cask bottom model to the height of center of gravity) of the cask model and the frequency parameter λ become equivalent. The floor model's density was determined so that the mass ratio of the shaking table used for the tip over tests to the cask becomes equivalent.

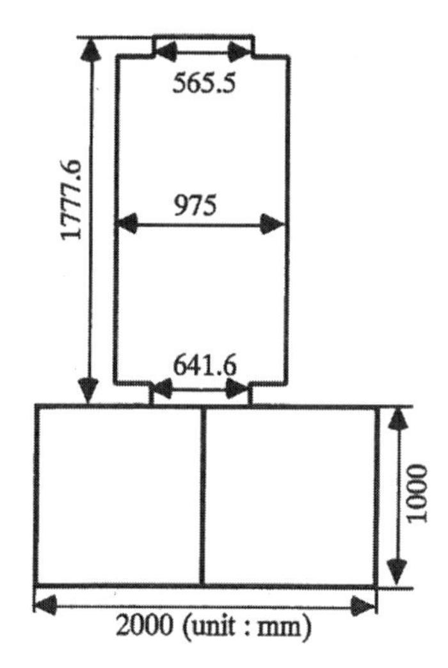

Fig. 3.6.6 DEM analysis model for model cask

Table 3.6.2 Spring constant

Mass density	8.743 g/cm^3
Spring constant (bar·cm)	
Normal direction K_n	9.51×10^6
Shear direction K_s	8.81×10^6
Degrading factor for time step	0.01
Rounding length	0.85 cm
Friction coefficient	see Fig. 3.6.1
Damping ratio	1.7 %
(Characteristic frequency)	(138 Hz)

3) Analysis for response to sine wave

We conducted rocking response analyses of the cask model under regular external force by inputting sine waves with constant displacement. The input of the sine waves were made so that required input displacement was given after the analytical model was rocking in the steady state, as in the vibration tests.

Fig 3.6.7 shows the maximum response angle with respect to frequency when the sine wave with a constant displacement amplitude (1.0 cm) was input to the analysis model, comparing the periodic solution, test results and analytical results. Because the analysis took into account of sliding and jumping, the analytical results showed a little smaller values than the periodic solution. The analysis showed good agreement with the test results, which demonstrated adequacy of the analytical method.

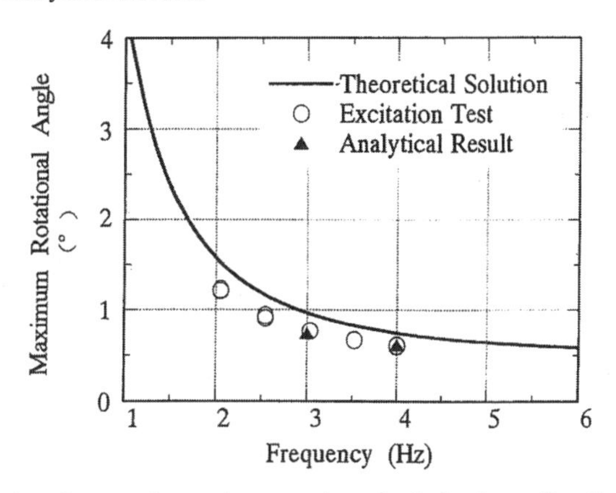

Fig. 3.6.7 Comparison of test and analysis for free vibration test

4) Analysis for response to seismic wave

Fig. 3.6.8 shows a relationship between the maximum response angle and the maximum input acceleration comparing the tests and the analysis for El Centro wave and Hachinohe wave. Fig. 3.6.9 shows time history of the response angle of the cask model comparing the test results and the analysis for Hachinohe wave (Maximum input acceleration 412 cm/s^2). The maximum response angle and the timing of rocking phenomenon agreed with the others.

From these results, the currently proposed method was found to be effective method to evaluate the cask stability against tip over.

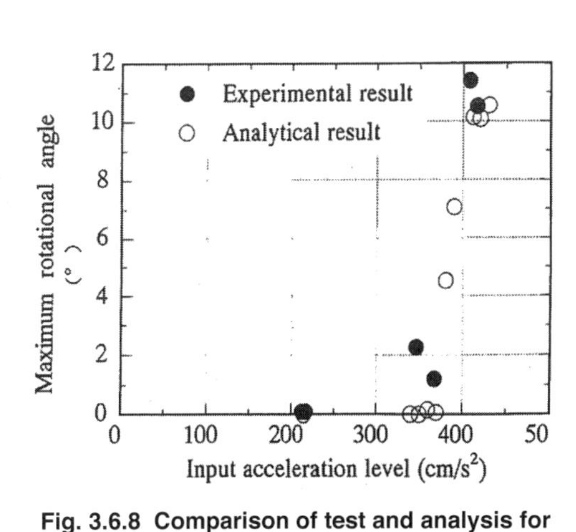

Fig. 3.6.8 Comparison of test and analysis for natural earthquake wave excitation

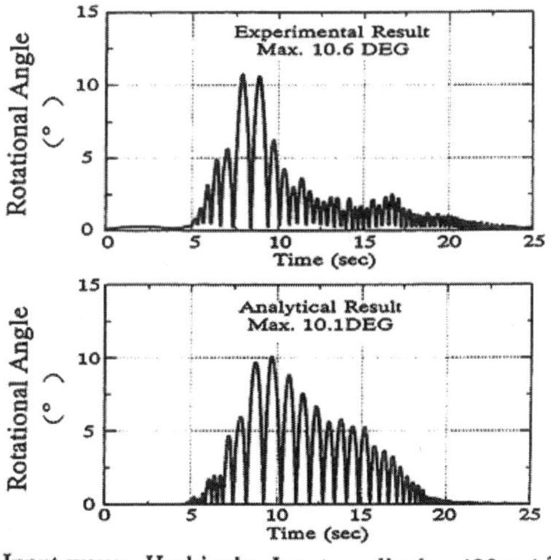

[Input wave : Hachinohe, Input amplitude : 400cm/s^2]

Fig. 3.6.9 Time history of rotational angle response for natural earthquake wave excitation

5) Tip over analysis of full-scale cask

We carried out the tip over analysis of the full scale cask by the proposed method in this section. Table 3.6.3 and Fig. 3.6.10

show the material properties and the analytical model used in the analysis. The floor was concrete and the input waves were El Centro and Hachinohe waves.

Table 3.6.3 Analysis parameters for prototype cask

Mass density	6.401 g/cm^3
Spring constant (bar·cm)	
Normal direction K_n	2.76×10^7
Shear direction K_s	2.55×10^7
Degrading factor for time step	0.01
Rounding length	1.47 cm
Friction coefficient	see Fig. 3.6.1
Damping ratio	2.0 %
(Characteristic frequency)	(138 Hz)

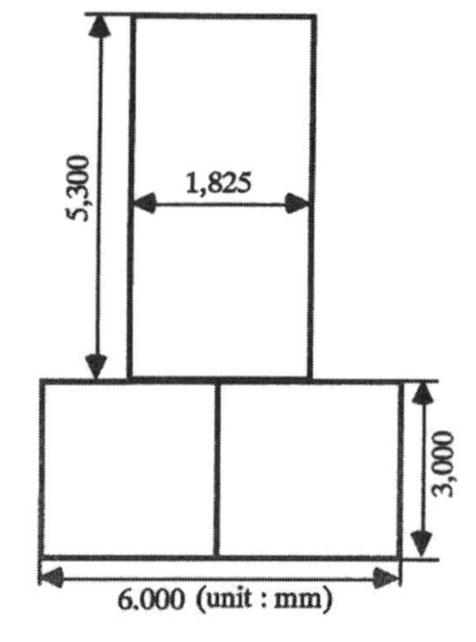

Fig. 3.6.10 DEM analysis model for prototype cask

Table 3.6.4 summarizes the tip over analysis conditions and the analytical results. The full-scale cask will not tip over for the input waves corresponding to double the original wave. Thereby, the full-scale cask has sufficient stability for tip over by the input waves chosen in this study.

Table 3.6.4 Evaluation results of stability of prototype cask subjected to seismic load

Input wave	Max. input acceleration (cm/s^2)	Max. response angle[†1]	Note
El Centro	341.7	0.11°	Original amplitude[†2]
	683.4	11.53°	Magnified amplitude[†3]
Hachinohe	203.3	0.01°	Original amplitude[†2]
	406.6	12.49°	Magnified amplitude[†3]

†1 Critical tipping-over angle of prototype cask=17.4°.
†2 Original amplitude was applied to demonstrate tipping-over stability.
†3 Magnified amplitude was applied to verify a margin for tipping-over.

(3) Summary

We carried out vibration tests with a cask model and numerical analyses on the cask stability against tip over by earthquake, and obtained following results.

1) For the sine wave vibration tests, the response and the transient behavior showed a good agreement with the periodic solution using the two-dimensional prism rocking model. Thereby, the cask response against the regular input can be evaluated by the periodic solution for the equations on motion of the two-dimensional prism.
2) For the natural earthquake wave tests, the cask response depended largely on the frequency component of the input wave and increased with the input wave having a long-term frequency.
3) We investigated numerical analyses using a two-dimensional DEM for the vibration tests of the cask model, and proposed a method for setting parameters (friction coefficient, spring constant, attenuation, time interval, etc.).

As a result of numerical simulation of a cask under natural earthquake, the simulation showed a good agreement with the test results with respect to the maximum response angle and rocking duration. Therefore it was found that cask stability against tip over at earthquake can be evaluated by the numerical analysis method utilizing the two-dimensional DEM that were proposed in this study.

3.6.2 Seismic stability of metal cask on storage frame in earthquake

An air pallet system (Fig. 3.6.11)[10] has an advantage in operability in a small space. When this air pallet system is used for a transfer system of a metal cask (hereinafter called as a "cask") in interim storage facilities, it is assumed that the cask attached to a storage frame (hereinafter called as a "frame") is vertically placed without being fixed to a concrete floor (hereinafter called as an "unfixedly grounded state"). A containment performance is impaired if the cask tips over and is damaged. Thus, the seismic safety evaluation of the cask attached to the frame in the unfixedly grounded state is essential.

In existing studies on tipping problems of rigid bodies, it is indicated that a start condition of rocking is determined based on the acceleration of seismic motions and a tipping condition is determined based on their velocity [11]. When displacement due to seismic motions is the same, input acceleration is relatively larger in seismic motions with short periods than in those with long periods and rocking is easy to generate. Thus, the existing studies [11],[12] were mostly conducted base on seismic motions with short periods, and findings about rocking by a seismic motion including a number of spectra with long-period components of 5 to 10 seconds (hereinafter called as a "long-period seismic motion") are few. However, when the maximum input acceleration is the same, the more the spectra with the long-period components are included, the higher the response velocity of the cask becomes and also the easier the cask is to tip over, so that the findings about the rocking by the long-period seismic motion were required.

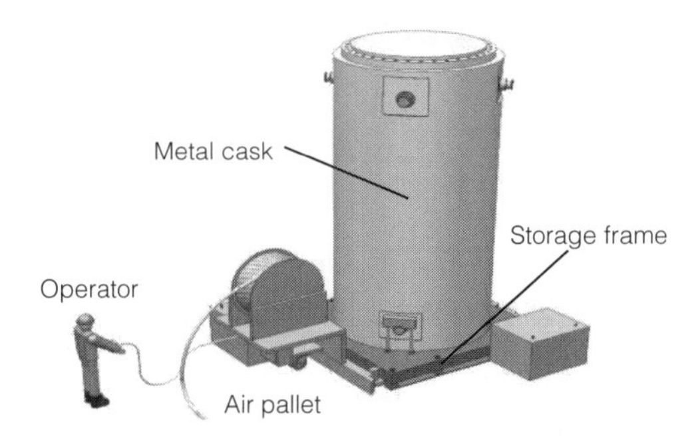

Fig. 3.6.11 Outline of air pallet system

(1) Seismic test

The following two series of seismic tests were conducted by using Type A and B of test scale models (Table 3.6.5).

1)Series I : Influence verification test of long-period components

To verify the influence of the long-period seismic motion, seismic tests (17 recorded seismic waves: NW01-NW17, and 6 simulated seismic motions: AW01-AW06) using the scale models (Type A and B) and input acceleration (300,

Table 3.6.5 Shape parameter of scale models and actual cask

	Unit	Actual cask	Type A scale model (similar model)	Type B scale model
Overall width of frame	m	3.400	1.360	1.000
Height of gravity center	m	2.867	1.160	1.210
Mass	t	138.6	12.465	11.913
Moment of inertia around gravity center	$t \cdot m^2$	588.9	6.795	6.007
Tipping angle	rad	0.535	0.530	0.392

600, 800, 980 gal) as parameters were conducted by using a single-axis large-sized shaking table. Also, in regard to the input wave of the simulated seismic motion (AW02) that had made Type B tip over, another test with a changed content rate of the long-period component was conducted by using Type A and its effect was evaluated.

2)Series II : Seismic performance test

To confirm if the cask attached to a frame in a grounded state would tip over and confirm and the influence of vertical input in the simulated seismic motions (AW01-AW04) and the recorded seismic waves (NW06, NW07, NW16, NW17), a seismic test was conducted by using a three-axis large-sized shaking table.

The evaluation of seismic performance was conducted by using Type A scale model under the condition that input levels of the simulated seismic motions were 1.0-fold, 1.3-fold, and 1.5-fold of the original wave. As to the recorded seismic waves, four waves having large response angles were selected from the result of Series I , and oscillation was applied while the maximum acceleration is 1000 gal, 1300 gal, and 1500 gal according to the original wave level (1030 gal) of AW01.

(2) Test result and applicability of tip over evaluation method

The following evaluation was performed to evaluate the influence of the seismic motions on the rocking response of the scale models.

a. Tip over evaluation based on determination indexes (tip over limit acceleration, and input energy velocity) for tip over of the rigid bodies in the existing studies is performed on the all test cases, so that its applicability is verified.
b. Tip over evaluation by a window energy spectrum method using the energy velocity that is input in a divided period of time is performed on the all test cases, so that its applicability is verified.
c. A seismic test based on the seismic motions having the various spectra is conducted to verify the influence of the long-period components, so that the tip over evaluation is performed. Also, the influence of long-period seismic waves is evaluated based on the relation between the dominate frequency of the seismic wave and a tip over angle of the cask by using AW02.
d. In the result of the seismic performance test on the original wave and the seismic motions exceeding it, the earthquake stability of the frame is evaluated by the window energy spectrum method. Refer to the reference[13] for details.

1) Evaluation by window energy spectrum method

Here, we describe the window energy spectrum method as an evaluation method of earthquake stability of the cask attached to the frame. This window energy spectrum V_{WES} focuses on a correlation between input energy effective in uplift of the cask and the energy input in a divided period of evaluation time (window width T_{window}), and it differs greatly from the existing energy spectrum of Akiyama, which targets at the gross energy of duration time of earthquake. Fig. 3.6.12 shows a conceptual diagram of the evaluation time that each of Akiyama's energy spectrum and the window energy spectrum targets at.

T_{window} is obtained by smoothing V_{WES} calculated every moment with the minimum response time T_0 when rocking is expected, while the maximum response time when the rocking generates is expressed as T_1.

This window width is uniquely obtained from a shape of a prismatic rigid body. For example, in the case of Type A, $T_0 = 0.05\sqrt{a} = 0.58s$ and $T_1 = 0.5\sqrt{a} = 5.8s$. In the case of Type B, $T_0 = 0.05\sqrt{a} = 0.57s$, and $T_1 = 0.5\sqrt{a} = 5.7s$. Here, a is the distance from a bottom rounded end to the gravity center of the rigid body.

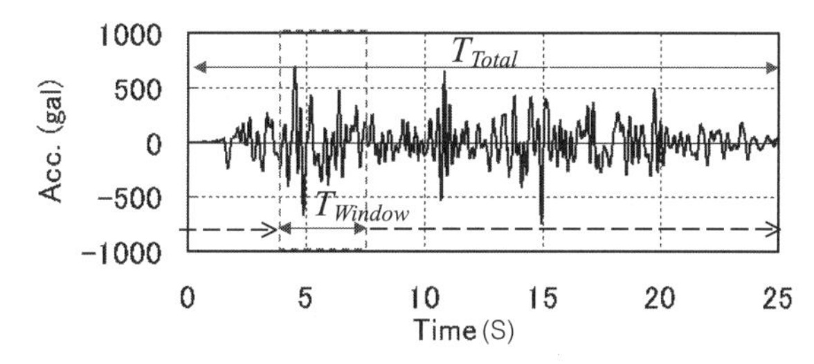

Fig. 3.6.12 Conceptual diagram of evaluation time T_{total} of existing energy spectrum and evaluation time T_{window} of window energy spectrum

Comparison of the maximum value of V_{WES} with V_{resp} is shown in Fig. 3.6.13 and Fig. 3.6.14. In either of Type A or B, there is a tendency that the proportion of energy contributing to the uplift decreases as V_{WES} increases and the values of V_{resp} reach the peak. In an input range where V_{WES} exceeds 100 kine, the response values are below the 45-degree line in the figures, and become input values on the safe side relative to V_{resp} even if V_{WES} is used for the input energy acceleration.

2) Influence of long-period components on rocking response

The result of evaluation of the seismic test of Series I , which is on the influence of the long-period components of the seismic motions on the rocking response, by the window energy spectrum method is shown as follows.

Fig. 3.6.15 shows the relation between an equivalent frequency F_e of the input wave, which was obtained by the seismic test, and the response energy velocity V_{resp}. When the input level of AW02 (Ss-d_ Free bed rock surface) is 800 gal, V_{resp} tends to increase as the long-period components increase. In the seismic test, the equivalent frequency F_e is slightly higher than a target value because the long-period components are slightly fewer than in the original waveform.

Furthermore, an estimated line was drawn based on the result of the response energy velocity of a corrected wave.

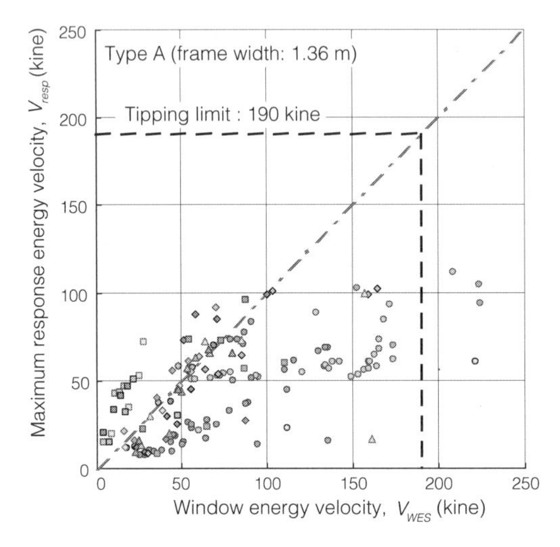

Fig. 3.6.13 Comparison of V_{WES} with V_{resp} obtained by tests (Type A scale model)

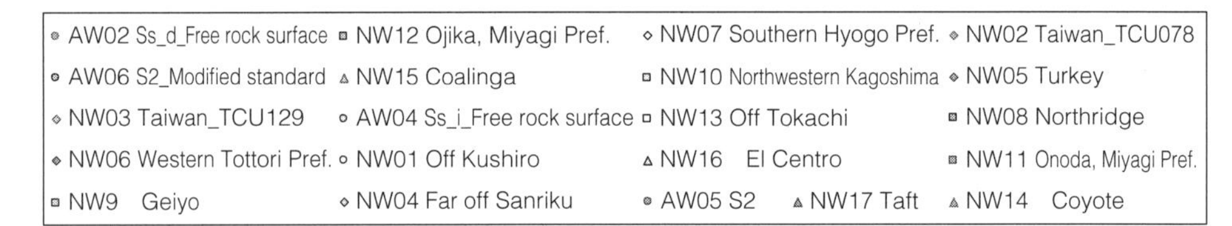

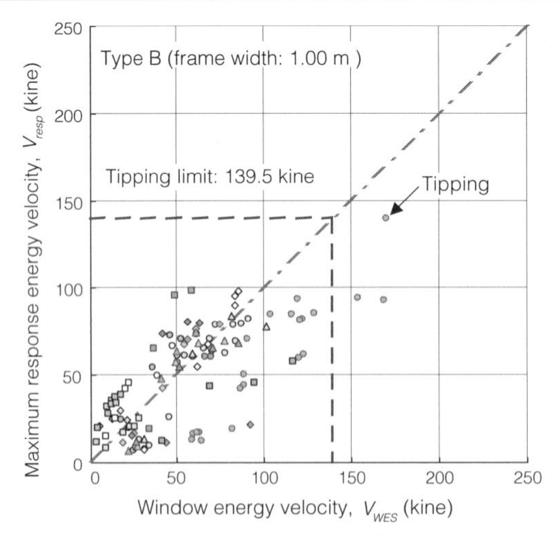

Fig. 3.6.14 Comparison of V_{WES} with V_{resp} obtained by tests (Type B scale model)

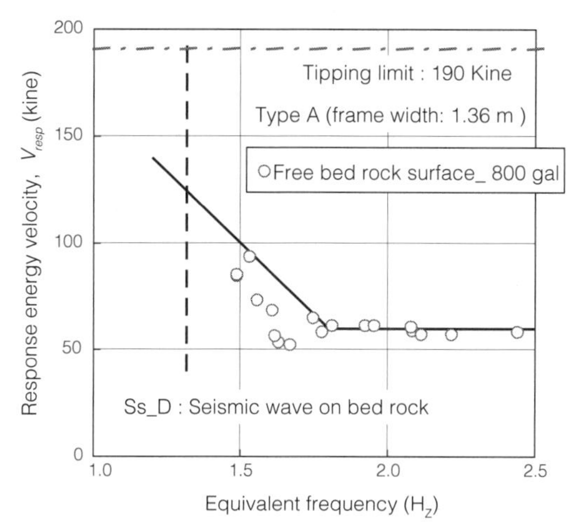

Fig. 3.6.15 Relation between equivalent frequency and response energy velocity

The estimated line has an inflection point at 1.8 Hz, and has a straight line containing the maximum value of the response energy velocity in the vicinity of 1.5 Hz. When extrapolation was performed with the estimated line up to the equivalent frequency (1.3 Hz) of the original wave, a tip over limit was not exceeded even in the equivalent frequency domain of the original wave.

In this study, it is assumed that the extrapolation is possible; however, its validity is required to be verified by analysis, etc.

3) Seismic performance test

Parameters of the seismic performance test of Series II are as follows.

· Types of seismic waves:

Simulated seismic motions (AW01, AW02, AW03, AW04)

Recorded seismic motions (NW06, NW07, NW16, NW17)

· Acceleration levels of the seismic waves: Original wave (or 1000 gal)

×1.0, 1.3, and 1.5

· Influence at the time of multi directional (horizontal and vertical) input

Fig. 3.6.16 shows the relation between the input energy velocity V_{WES} and the response energy velocity V_{resp} using an input direction as a parameter, taking AW01 (Ss-d_ Floor response) and NW07 (Southern Hyogo Pref.) for example.

As with the result of one-directional vibration, the response energy velocity V_{resp} tended to converge at about 100 kine in multi directional vibration. In horizontal two-directional vibration, sliding and saltation (the cask is uplifted, and displaced horizontally without friction) tended to occur easily, and the rocking response decreased compared to horizontal one-directional vibration. The result showed that when the input energy is excessive, the uplift due to rocking rarely occurs and the energy is converted into translational motions (sliding and saltation) of the cask.

Furthermore, when the vertical input was performed, the amplification of the rocking response was rarely shown

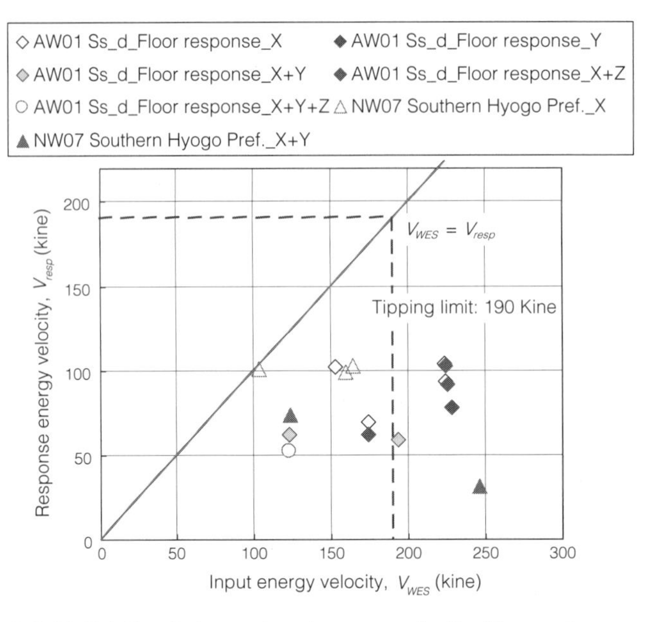

Fig. 3.6.16 Relation between input energy velocity V_{WES} and response energy velocity V_{resp} using input direction as parameter

compared to the rocking response in the case of inputting only horizontally. Thus, it seemed that the horizontal input is only necessary in the tip over evaluation.

(3)Summary

The main results obtained from this study are shown as follows.

1) As the result of the shaking tests using the scale models, it was verified that the dimensionless response angles of the scale models nearly agree with the theoretical solution of the prismatic rigid body using an angular velocity damping rate of about 0.9 and the evaluation by assuming that the cask attached on the frame is a prismatic rigid body is possible.

If the window width used for the calculation of the window energy spectrum V_{WES} is set to T_l and the time history of the energy spectrum is smoothed with T_0, the input energy contributing to the uplift of the scale models can be properly calculated in either of the scale models, Type A or B. Also, when the seismic motions proposed by Akiyama are considered as the repetition of a plurality of unit seismic motions and an effective tipping energy spectrum ${}_{ou}V_{Ef}$ obtained by dividing a tipping energy spectrum by the number of repeats of the unit seismic motion is compared with V_{WES}, ${}_{ou}V_{Ef}$ and V_{WES} have approximately-proportional relation and V_{WES} has a value obtained by decreasing ${}_{ou}V_{Ef}$ by 1/1.2.

2) As the result of the seismic tests conducted by preparing a plurality of seismic motions varying in the content rate of the long-period component, it was verified that the rocking response increases as the long-period components increase. Furthermore, it was verified that when the frequency domain where the long-period components are dominant is linearly extrapolated with the equivalent frequency, the occurrence of tip over due to a design seismic

motion (AW02 Ss-d_free bed rock surface response, 800 gal) can be avoided. (However, the validity of the extrapolation is required to be verified by analysis, etc.)

3) As the result of the seismic tests conducted by performing the horizontal one-directional vibration on the scale models, it was verified that rocking vibration is dominant in the dynamic behavior of the scale models and the response energy velocity V_{resp} is far below a tipping limit velocity of 190 kine at the level of the design seismic motion. Furthermore, when the input level is increased 1.3-fold, V_{resp} is about 100 kine and converges at a constant value (100 kine) due to the decrease of a restitution coefficient caused by penetration into a concrete floor slab of a frame leg part and the sliding and jumping at the time of excessive input. Thus, V_{WES} can be used for a conservative design target value.

4) As the result of the seismic test conducted by performing both the horizontal one-directional vibration and the horizontal one-directional and vertical one-directional vibration on the scale models, it was verified that there is no amplification due to the vertical input, so that the horizontal input is only necessary in the tipping evaluation. Also, in the horizontal two-directional vibration, the sliding phenomenon is dominant and the rocking response decreases relatively.

5) As the result of the seismic tests using the scale models that employ a similarity rule on the parameters (the gravity center and the restitution coefficient) affecting the rocking response, it was verified that the tip over does not occur. It seems that in generated rocking behavior and its response tipping angle, the behavior of the actual cask is simulated, and it can be estimated that also the actual cask does not tip over in response to each of the seismic motions used in the test.

References

1) T. Mochizuki and K. Kobayashi: “A study on acceleration of earthquake motion deduced from the movement of column”, Collection of Architectural Institute of Japan article reports, No.248, 1976.

2) Y. Ishiyama, et al.: “Vibration tests and theoretical study on tip over of furniture, etc. by earthquake”, 1978.

3) N. Ogawa: “Study on the overturning vibration of rigid structure”, Collection of Architectural Institute of Japan article reports, No.287, 1980.

4) A. Shibata: “Latest earthquake proofing structure analysis”, The latest architecture series 9, p.97 to 112, 1981.

5) P. A. CUNDALL : “A computer model for simulating progressive large scale movements in blocky rock system”, Proc. Symp. Int. Soc. of Rock Mechanics, Vol.1, No.II-8, 1971.

6) T. Ishida, et al.: “Programming for distinct element method and its experimental examination using block models”, CRIEPI Report 383014, 1983.

7) K. Shirai and H. Ryu: “Applicability of distinct element method to cask tip over”, Proc. the 21st JSCE Earthquake Engineering Symposium, 1991.

8) K. Shirai, et al.: “Establishment of cask-storage technology for spent fuel ? Evaluation of cask tipping over due to strong earthquake motion”, CRIEPI Report U92037, 1993.

9) K. Shirai, H. Ryu, C. Ito, and T. Saegusa : “Tip-Over Stability of Spent Fuel Storage Cask Subjected to Seismic Load”, J. Atomic Energy Society of Japan, Vol.36, No.11, p.1068 to 1078, 1994.

10) A. Kawamoto: "Plan and technical development of interim storage of spent fuel", Fiscal year 2005 meeting of Thermal Power Engineering Soc. of Japan, Tokyo International Forum, October 2005.
11) H. Akiyama, et al.: "Tip over prediction of rigid body using energy spectrum", Collection of Architectural Institute of Japan article reports, No.488, 1996.
12) H. Akiyama, et al.: "Relationship between energy spectra and velocity response spectra", Collection of Architectural Institute of Japan article reports, No.608, p.37~43, 2006.10.
13) S. Kawaguchi, K. Shirai, and K. Kanazawa: " Seismic stability of metal cask with storage frame for long period earthquake during transfer", Journal of JSCE A1 (structure and earthquake engineering), Vol.68,No.2, p.271~286, 2012.

3.7 Severe accidents

3.7.1 Cask burial test due to building collapse[1)]

(1) Thermal test simulating cask burial

The full-scale cask shown in section 3.2.1 was used for thermal tests simulating cask burial with debris due to building collapse by earthquake. The followings are the outline of the method and the evaluation results of the thermal test of cask burial.

1) Test method

Various conditions of cask burial were simulated by covering the cask with adiabatic materials (simulating thermal property of the thick concrete roof of the storage building) or controlling inflow ventilating air to the cask as shown in Table 3.7.1-1. The cask cavity gas is helium gas only as specified by the cask design condition. The cask posture was either horizontal or vertical.

In all the tests, cask leak tightness was measured before and after each test to confirm the cask integrity.

Table 3.7.1-1 Cases of heat transfer tests of cask buried by debris

Test Case \ Item	Cask Attitude	Cavity gas	Assumed burial state	Remarks
I	horizontal	He	a Debris fully covered the horizontal cask.	The whole cask surface was covered with adiabatic materials.
II	horizontal	He	b Debris covered the upper part of the horizontal cask.	The cask was surrounded with metal frames and metal wire-nets that were partially covered with sheets of heat-resisting boards as required. (The amount of inflow ventilating air was parametrically varied.)
III	vertical	He	c Debris covered the lower part of the vertical cask.	The cask was surrounded with thermal-reflecting boards simulating normal storage condition. Additional reflecting boards were placed to restrict inflow ventilating air to the lower part or the upper part of the cask. (The amount of inflow ventilating air was parametrically varied.)
IV	vertical	He	d Debris covered the upper part of the vertical cask.	

2) Test results

a. Case of whole cask burial (Case I)

This case hypothetically assumed that a cask was buried with debris of the storage building by earthquake. The condition was conservatively simulated by wrapping the cask with adiabatic materials. In the thermal test, the cask temperature gradually increased and the neutron shield material on the external surface of the cask melted and extruded from the melt-plugs as designed as well as the other parts in seven days after the test. The test was interrupted by shutdown the electric heater at this point, when the temperature of the cask components was below the design conditions. The final behavior of this case was evaluated by thermal analysis as described later.

The result is shown in Fig. 3.7.1-1. The temperature of the cask components would reach to the steady state condition in two months and the most critical temperature of the lid seal part would be approximately 245 °C. This means the cask leak tightness would be maintained if the cask were recovered in one month after the cask burial accident.

In addition, the steady state temperature of lead was approximately 240 °C, which is below the melting point, 327 °C, and showed enough safety margins.

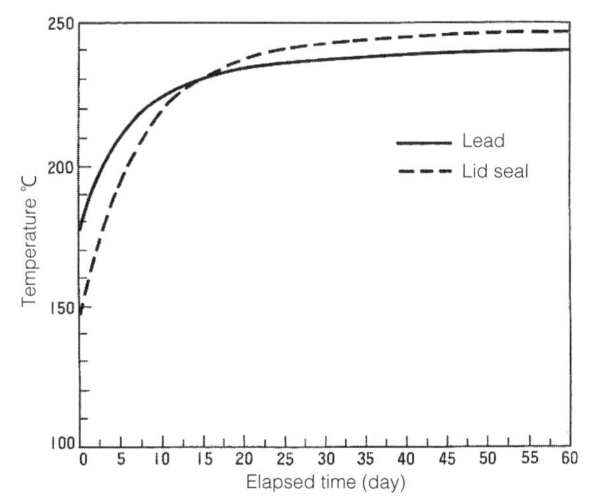

Fig. 3.7.1-1 Temperature change in lead and lid part (analysis) (assuming no melt and extrusion of the neutron shield material)

b. Case of partial cask burial with debris or wall (Cases II to IV)

Even if the cask were partially buried with debris, partial inflow of cooling air to the cask surrounding would result in a steady state condition in 10 days after the cask burial either in horizontal or vertical posture. Their maximum temperature values of the cask components were confirmed to be well below the design limits (as shown in Fig. 3.7.1-2).

During the cask burial tests, the cask integrity had been maintained. Based on the cask burial test results, a procedure of the thermal analysis to evaluate the temperature of the spent fuel claddings is described in the following.

(2) Thermal analysis simulating burial test

Principle and method of the thermal analysis are shown below.

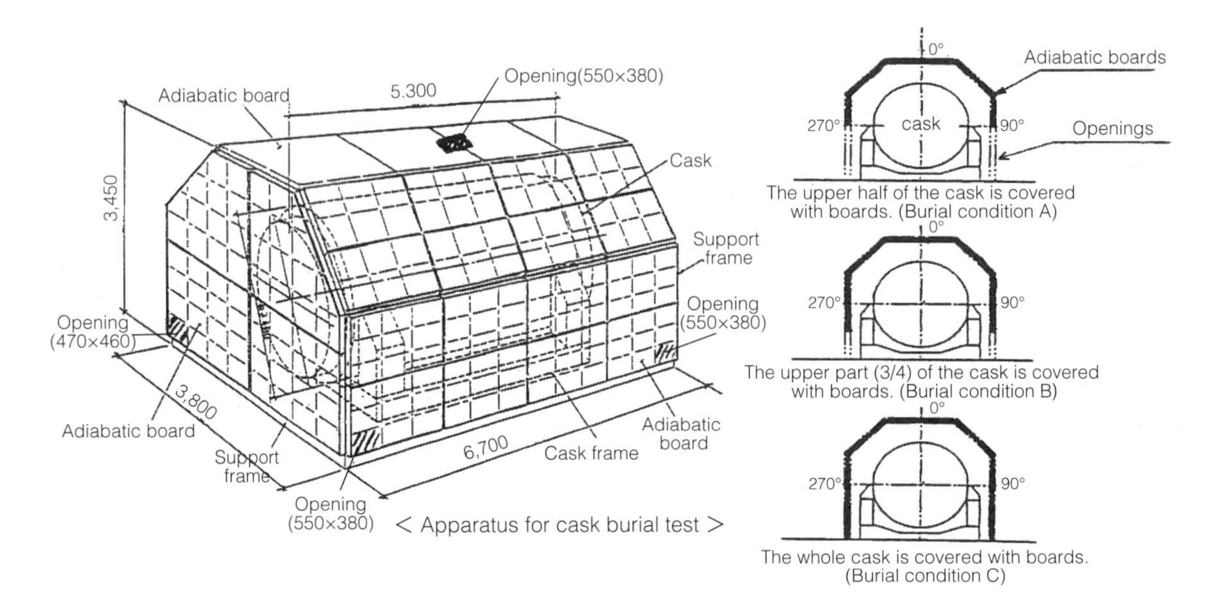

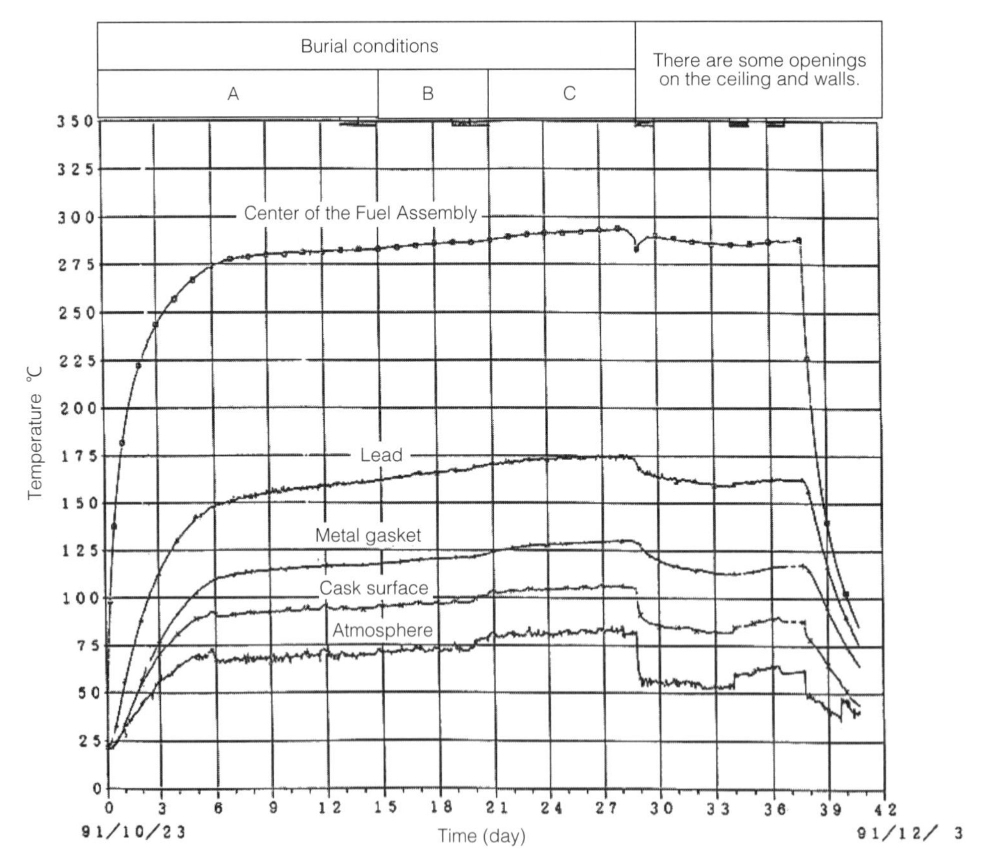

Fig. 3.7.1-2 Temperature change of the cask components in the burial tests (Case II)

1) General principle

The thermal analysis aims to obtain precise temperatures of the spent fuel cladding surface, the lid seal, etc. taking account of thermal hydraulics effect in the cask cavity (as shown in Fig. 3.7.1-3). Then, analytical method are basically available using a detail model of cross section of the central part of the cask in order to obtain the precise surface temperature of spent fuel claddings and the aforementioned method described in the section 3.2.1 for lid seal temperature.

The effect of cask burial is considered in the thermo-hydraulics analysis to obtain environmental temperature of the cask surroundings in a storage building (namely, the cask surrounding temperature in the apparatus simulating cask burial condition) and in the thermal analysis to obtain temperature profile of the cask body. Namely, in the above-mentioned analysis, the boundary conditions surrounding the cask should be changed in accordance with the burial conditions.

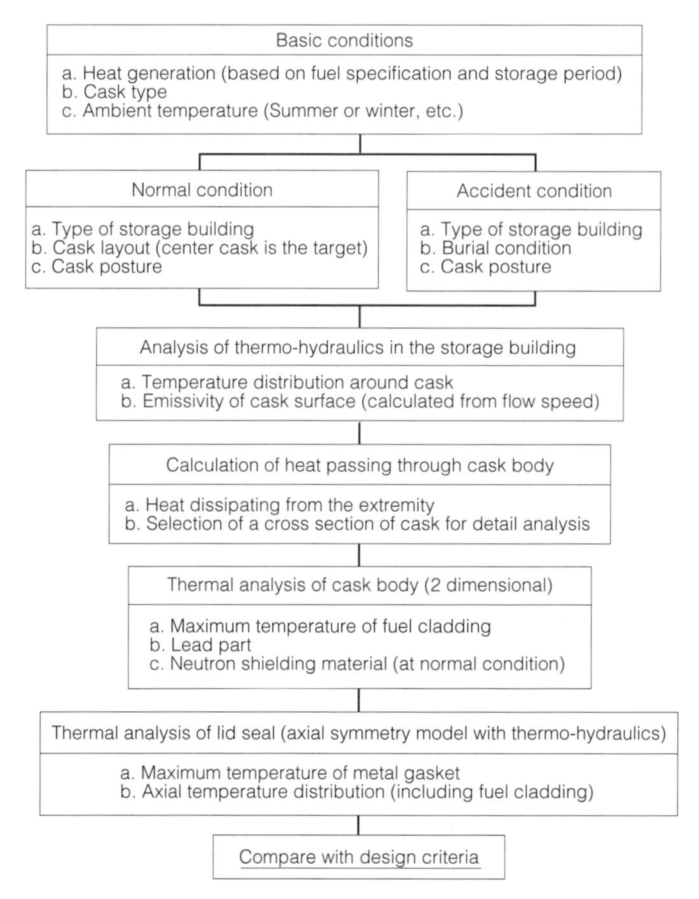

Fig. 3.7.1-3 Procedure of analysis to calculate temperature of fuel cladding, etc.

2) Method of evaluation

Thermal integrity of cask components including spent fuel claddings is based on the maximum allowable temperature (including allowable time) of the material (above-mentioned analytical results or actual measurements).

Location to be evaluated for the thermal integrity at the burial accident and their general temperature limits are shown below. In addition, because the allowable temperature of the material depends on the duration, the evaluation will be made by both the maximum temperature and its duration.

a. Spent fuel cladding: approximately 500 °C or a little more

b. Metal gasket* for seal: 250 °C (approximately one month for the present thermal test)

*Spring: Inconel, Inner liner: Inconel, Outer liner: Aluminum

c. Lead: 327 °C

3) Method of thermal analysis for burial accident

a. Method for cask as a whole buried by debris

a.1 Thermal analysis of cask body and lid seal structure

The analytical method described in the section 3.2.1 was modified in order to take into account the difference of the boundary condition of the cask surroundings. The cask body and the lid seal structure was modeled with axial symmetry to obtain the maximum temperature, the time (days), etc. to reach the steady state condition by unsteady state thermal analysis. In this case, the contents such as spent fuel claddings, basket, etc. are replaced by the equivalent property as a uniform heat generator.

a.2 Thermal analysis of spent fuel claddings

Because the analytical accuracy is not good enough if the aforementioned analytical method using the axially symmetric model is used for the maximum temperature of the spent fuel claddings that is essential for the thermal evaluation, a two dimensional thermal analysis (steady state analysis) was performed. Namely, the steady state temperature of the cask inner surface obtained from the aforementioned unsteady state thermal analysis was used as a boundary condition (fixed temperature) of a two-dimensional model in order to obtain the temperature distribution inside the cask cavity.

b. Method for cask partially buried by debris or wall

The same method as described in the section 3.2.1 is basically used.

4) Result of thermal analysis at the cask burial accident

a. Case for whole cask buried by debris

a.1 Temperature of cask body and lid seal structure

a.1.1 Comparison with the test result

Figs. 3.7.1-4 to 3.7.1-6 show temperature history results by both the unsteady state thermal analysis and the test for the internal surface of the cask, lead and lid seal structure for thermal integrity evaluation. From these results, the following conclusion was obtained.

a.1.1.1 It is possible to evaluate the temperature of lead and lid seal structure properly by using the axial symmetric model.

a.1.1.2 It is possible to apply the analytical method used for the thermal analysis for the normal storage condition to the temperature of the lid seal structure including the cask body.

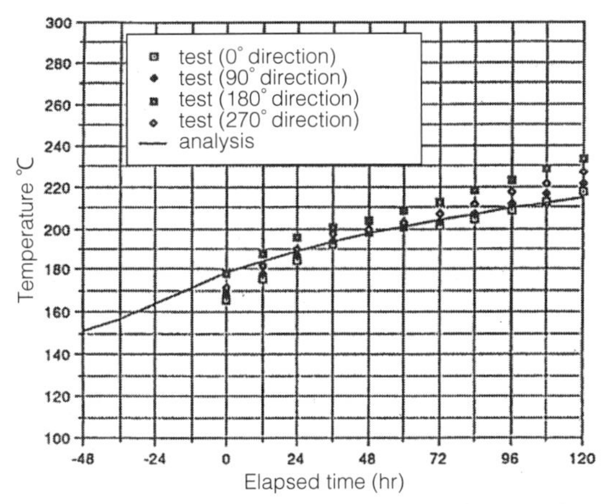

Fig. 3.7.1-4 Temperature history of the cask inner surface (Test and analysis (Case I))

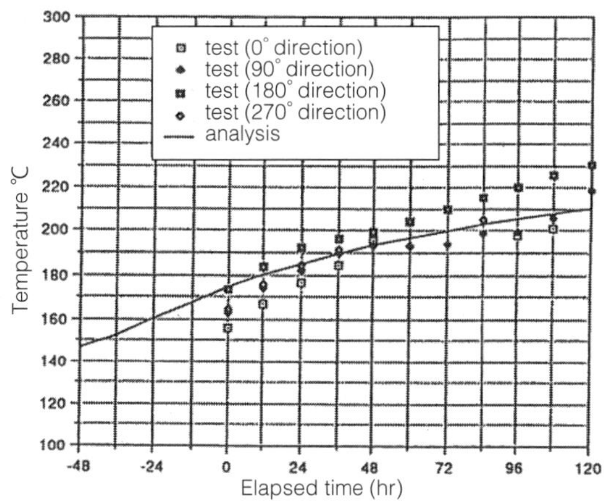

Fig. 3.7.1-5 Temperature history of the lead (Test and analysis (Case I))

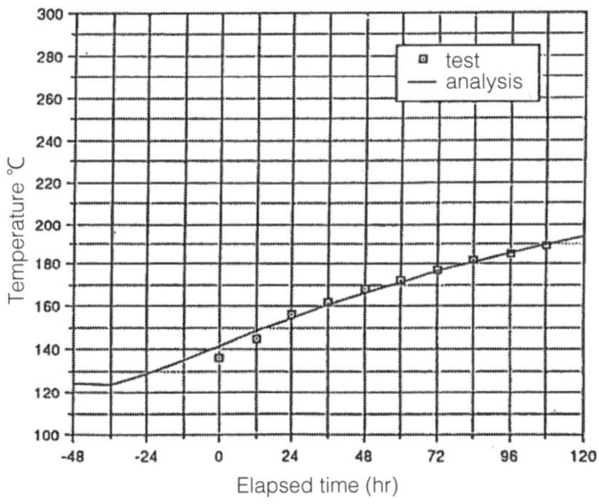

Fig. 3.7.1-6 Temperature history of the cask inner surface (Test and analysis (Case I))

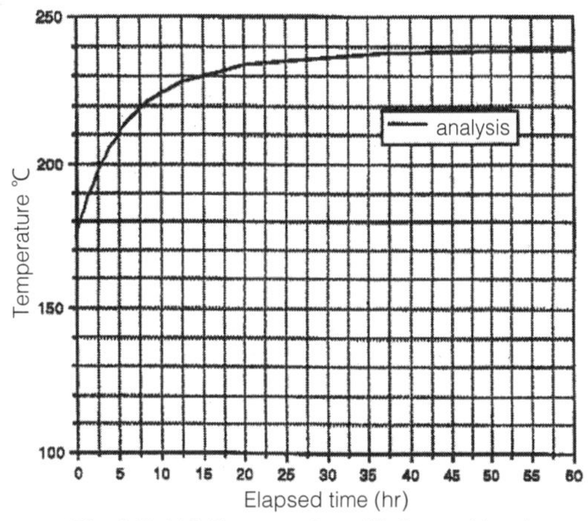

Fig. 3.7.1-7 Temperature history of lead (assuming no extrusion of molten neutron shielding material) (Case I supplemented)

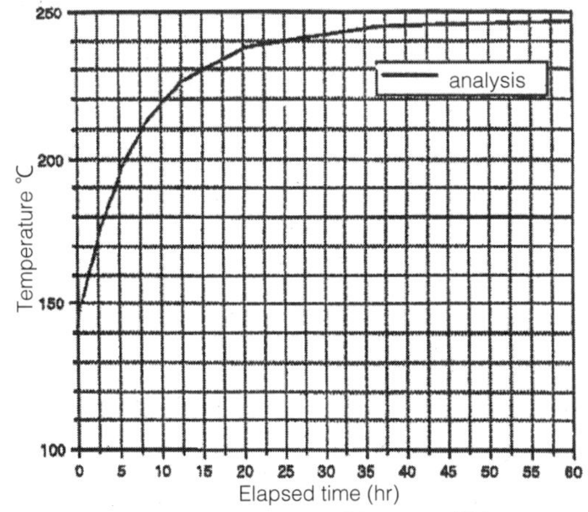

Fig. 3.7.1-8 Temperature history of lid structure (assuming no extrusion of molten neutron shielding material) (Case I supplemented)

Table 3.7.1-2 Maximum temperature of cask component (Assuming no extrusion of molten neutron shielding material)

Component	Analytical result (°C)
Cask surface	200.3
Intermediate shell surface	227.9
Inner surface of lead	239.2
Cask inner surface	243.4
Seal at primary lid	247.6
Fuel assembly*	423.4

*Use of two dimensional round slice model

a.1.2 Prediction of the maximum temperature, etc.

Using the aforementioned axial symmetrical model, unsteady state thermal analysis was carried out assuming the resin will not melt and extrude. Figs. 3.7.1-7 and 3.7.1-8 show the results. From these results the following conclusion was obtained.

a.1.2.1 The temperature of the cask components will reach to the steady state condition in about two months.

a.1.2.2 The maximum temperature of lead in the steady state condition is approximately 240 °C, which is below the allowable temperature at the burial accident.

a.1.2.3 The maximum temperature of the lid structure was a little less than 250 °C. If the recovery were made in one month, the thermal integrity would be maintained.

a.1.3 Maximum temperature of spent fuel cladding

Table 3.7.1-2 shows the maximum temperature of spent fuel claddings. The result showed that the maximum temperature of the spent fuel cladding was approximately 425 °C, which is below the allowable temperature limit at the burial accident.

b. Case for cask partially buried by debris or wall

b.1 Temperature and flow velocity in the test hood

Thermo-hydraulic analysis was carried out for the cask burial tests cases II to IV to obtain temperature and flow velocity around the cask (identification of thermal emissivity on the cask external surface). Furthermore, based on the result, temperature of the cask components was calculated.

b.2 Heat passing through the cask body (Heat dissipation from the cask extremity)

Heat dissipation from the cask extremity would be significant when the cask was buried in the severe accident. Such heat dissipation from the cask extremity was calculated a thermal analysis using an axial symmetric model for the cases II to IV. Using the results, heat passing through the cask body was calculated as shown below.

Case II (Debris covered the upper part of the horizontal cask.) : 85 %

Case III (Air inflow is restricted at the lower part of the vertical cask.) : 82 %

Case IV (Air inflow is restricted at the upper part of the vertical cask.) : 80 %

b.3 Temperature of the cask body

Temperature distribution in the cask body was obtained using the aforementioned results of thermos-hydraulic analysis and the heat passing through the cask body.

Table 3.7.1-3 shows the results revealing that the analytical results show approximately 20 °C higher than the test results. This is due to that the ambient temperature around cask calculated by the thermo-hydraulic analysis was higher than the test results by 20 °C. Then, it is considered that the analytical model to calculate temperature distribution of spent fuel claddings and the cask body was established. Furthermore, considering the temperature difference between the analytical result and the test result for the cask ambient temperature, the cask internal model for the normal condition of storage could be used.

b.4 Temperature of lid seal, etc.

Temperature of lid seal, etc. was obtained for the thermal tests simulating cask burial. The results are also shown in Table 3.7.1-3. The analytical results were slightly higher than the test results by the same reason as mentioned above. Nevertheless, this analytical model was shown to be able to evaluate the temperature of the lid seal conservatively.

Table 3.7.1-3 Comparison between analyses and tests (Cask burial cases II to IV)

Part	Test result with resin	Test result I losing resin (91/9/5-6)	Test result II losing resin (91/9/8-9)	Analytical result	Case II		Case III		Case IV	
					Test result (91/11/21)	Analytical result	Test result (91/12/25)	Analytical result	Test result (92/1/30)	Analytical result
External surface	83	75.8	79.2	83.9～108.2	105.9	90.7～120.1	109.7	108.6～131.3	125.2	125.5～147.4
	102	94.8	97.5		119.8		126.4		141.0	
Resin	106	121.1	122.7		148.0		142.4		—	
Intermediate shell (external/ internal)	131	148.1	147.5	144.3～157.8	155.2	153.6～167.8	170.2	168.1～181.2	—	184.6～197.4
	132	152.2	153.0	148.0～160.2	168.6	157.5～170.5	175.0	171.7～183.9	189.7	188.1～200.1
Lead (internal)	138	154.7	154.3		165.7		176.5		194.3	
Inner shell (internal)	144	160.5	159.6	157.9～168.2	174.0	171.2～186.3	181.1	184.4～199.9	198.2	200.7～215.9
	145	163.5	162.3		170.7		184.5		202.2	
Basket (external / heat conductance / cell)	163	176.1	175.6	160.9～176.8	195.5	177.7～212.6	201.5	191.2～224.9	195.3	207.4～240.8
	218	228.8	228.3		248.6		259.7		277.9	
	233	243.5	243.4		262.8		276.4		294.2	
Simulated fuel assembly	246	253.0	252.9		275.5		285.5		300.3	
	262	267.0	267.2	287.8	291.3		303.8		318.8	
	258	263.6	263.7		286.0		303.4		318.1	
	265	269.5	269.6		293.7	296.5	308.6	309.1	322.8	323.8
Atmosphere	46.5				43～81		81		103	

Note) Case II: Debris covered the upper part of the horizontal cask.
Case III: Debris covered the lower part of the vertical cask.
Case IV: Debris covered the upper part of the vertical cask.

5) Evaluation of analytical results

From the aforementioned results, it is considered that the thermal analytical model to obtain temperatures of spent fuel claddings, lid seal, etc. was established. Therefore, using this analytical method, temperature of the other cask including spent fuel claddings will be properly evaluated.

References

1) H. Yamakawa, Y. Gomi, S. Ozaki, and A. Kosaki: "Establishment of Cask Storage Method for Storing Spent Fuel -Evaluation on Heat Transmission Characteristics of Storage cask-", CRIEPI Report U92038, (January 1992), or Central Research Institute of Electric Power Industry: "Fiscal Year 1991 Report on Verification tests on Spent Fuel Storage Technologies at Reactor", for Ministry of Economy, Trade, and Industry of the Japanese Government, March 1992 (in Japanese), or H. Yamakawa, Y. Gomi, T. Saegusa, and C. Ito: "Thermal Tests of a Transport/ Storage Cask in Buried Conditions", Proc. PATRAM 1998, Vol. 2, page 675-682, Paris.

3.7.2 Mechanical impact of airplane engine to cask

After the terrorist attack in 2001 in USA, safety assessment for airplane crash to important nuclear facilities as severe accident has been actively conducted in USA and Germany[2].

In Japan, as to airplane crash, it depends on its probability whether it is considered in design. If it exceeds 10^{-7}/ facility/year, it shall be taken into account in the design as "credible external event".

In this section, a civilian airplane is assumed to crash on a storage facility as a severe accident of "credible external event". First of all, we considered an impact condition so that a highly rigid engine (hereafter, engine) penetrates the storage facility via local fracture of the facility and hits a storage cask. Then, we conducted tests of horizontal impact assuming a missile simulating the engine hit the cask in order to evaluate its containment performance of the cask against the impact. Furthermore, we simulated the horizontal impact test using the impact analysis code LS-DYNA in order to reproduce the impact behavior and the containment performance. More detail information is found in the literature[3),4)].

(1) Horizontal impact test to scale model cask

1) Containment structure of metal cask

Fig. 3.7.2(1)-1 shows overview of a metal cask. Metal gaskets (hereafter, gaskets) are installed between the cask body and the lids to assure containment. Pressure inside the cask cavity is maintained negative in order to prevent radioactive materials to leak from inside of the cask. Fig. 3.7.2(1)-2 shows the overview of the metal gasket. The metal gasket is subjected to a thermal loading due to the heat generated from the spent fuel in the cask. The spring back force of the metal gasket may be deteriorated due to creep deformation of aluminum jacket of the gasket and the containment performance may deteriorate. Therefore, the effect of stress relaxation due to the thermal loading to the gasket was taken into account.

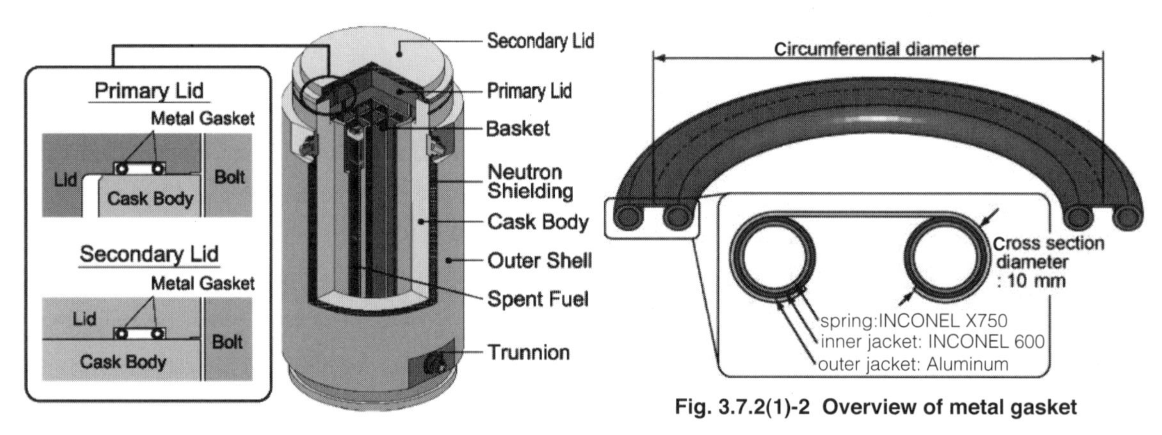

Fig. 3.7.2(1)-1 Overview of a metal cask

Fig. 3.7.2(1)-2 Overview of metal gasket

2) Assumed impact condition

The airplane was assumed Boeing 747 most often used in the assessment in USA and Germany. The missile was assumed a rigid engine of the airplane.

When a Boeing 747 crashes to a storage facility with a wall thickness of 85 cm with a speed of 90 m/s (same as the requirement of the type C package in the IAEA transport regulations) that exceeds the landing speed of 280 km/h (77 m/s), the engine is expected to penetrate the wall of the storage facility. This penetration was evaluated by the Degen formula that was used to evaluate penetration of a rigid missile into the concrete wall of the reprocessing facility at Rokkasho village in Japan. The residual speed after the penetration was calculated as 60 m/s using the formula in the "Guidelines for the Design and Assessment of Concrete Structures subjected to Impact" by UKAEA[5)].

Fig. 3.7.2(1)-3 shows time histories of force by aircraft engine crash with a speed of 60 m/s obtained through the impact analysis.

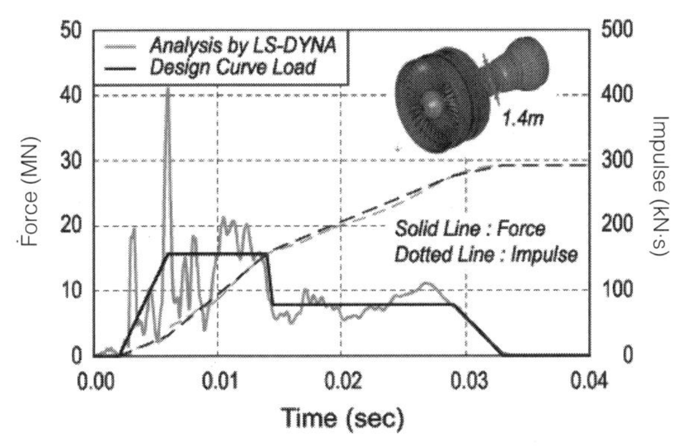

Fig. 3.7.2(1)-3 Time histories of force by aircraft engine crash with a speed of 60 m/s

3) Test conditions

a. Impact posture

The impact direction of the missile against the cask lid could be mainly horizontal or vertical. In this case, we conducted tests so that a missile impacts horizontally on the upper edge of the cask in the vicinity of the lid. The cask was a scaled model down to 2/5 (hereafter, scale cask).

b. High speed missile

The missile flies with a force by gunpowder and gives the load to the scale cask. The mass and the expected speed of the missile were determined using the engine specification and similarity law. The mass was 288 kg and the impact speed was 60 m/s. Fig. 3.7.2(1)-4 shows the converted force according to the similarity law. Shock absorber was installed inside the missile so that the loading duration (about 12 ms) of the converted force becomes equal to the impulse. The material of the shock absorber was incombustible and light weighted aluminum alloy that is foamed and has thin-walled cell structure (cylinder of ϕ 485×h283, density of 0.28 g/cm^3, and material property of Al-10Zn-0.3 Mg). Fig. 3.7.2(1)-5 shows the overview of the missile.

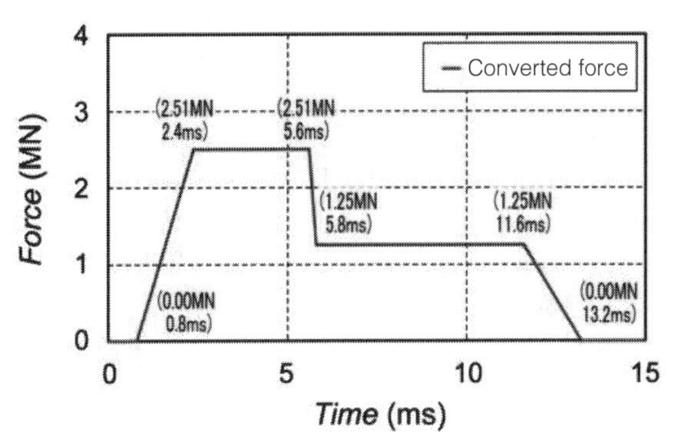

Fig. 3.7.2(1)-4 Converted force by missile

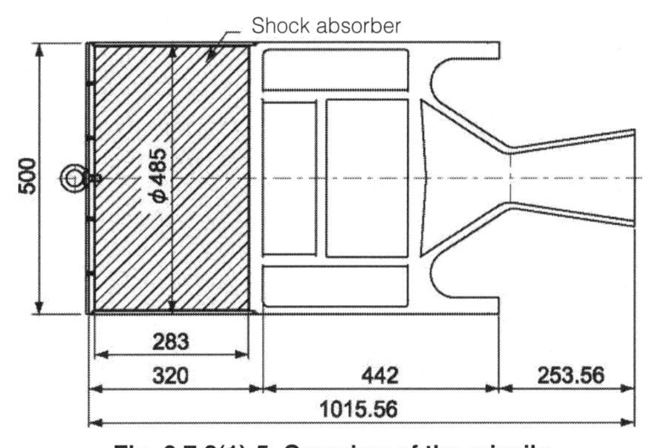

Fig. 3.7.2(1)-5 Overview of the missile

c. Scale cask

The scale cask is scaled down cask by 2/5 of a full size cask. The scale cask has a cylindrical shape with an outer diameter of about 0.9 m, height about 1.9 m, and mass about 4.5 tons. The containment performance is secured by metal gaskets installed between the cask body and the lid. Although the real cask has two lids, the scale cask for the test had only one lid assuming that the two lids will undergo the same force and behave similarly at the impact. The scale cask for the test will represent conservative behavior. In addition, the diameter of the metal gasket of the scale cask was about 6.1 mm in order to secure the reliability of the containment performance of the scale cask.

d. Apparatus of the horizontal impact test

The apparatus of the impact test consisted of launcher, launcher base, concrete base, protection cage that prevents the missile and the target from dispersion and protect measurement devises, etc., as shown in Fig. 3.7.2(1)-6.

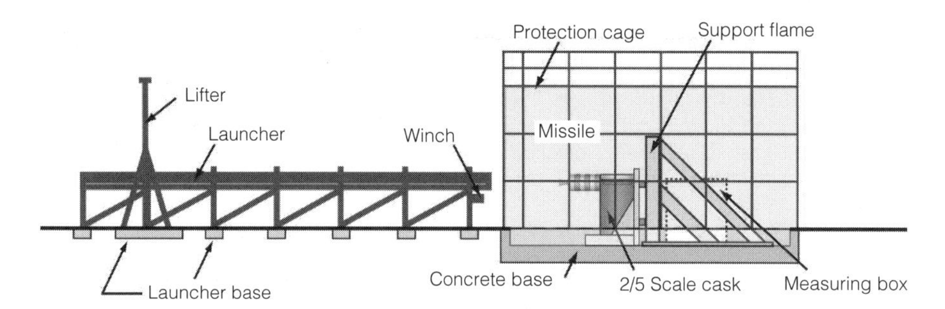

Fig. 3.7.2(1)-6 Outline of horizontal impact test

4) Horizontal missile impact test

a. Thermal load to the gasket

Thermal load was given to the scale cask in the vicinity of the lid with thermal insulator and electric heater. The heating condition to the metal gasket was 174.1 °C×32.5 hrs. This condition is expressed by LMP = 9618 using the following equation.

$LMP=T (C + \log t)$ ..... (1)

T : Absolute temperature (K) t : time (hr)

C : Constant (=20)

The temperature of the real cask and the lid decreases with time due to the decay heat of the spent fuel. Assuming the initial temperature of the cask being 120 °C, 115 °C, 105 °C, the thermal load condition (LMP = 9618) is equivalent to that the real cask undergoes ageing of 4 to 60 years. By the way, the impact test was conducted after the test body was cooled down to the ambient temperature.

b. Leak test

After the lid of the scale cask was closed, leak tests were conducted to secure that the scale cask satisfies the required containment performance. The leak tests were conducted after placing the cask on the frame, before and after the heating, and before the impact test. In any cases, we confirmed that the leak rate was less than 1.0×10^{-10} Pa $\cdot$ m^3/s.

5) Results of horizontal impact test

Although the axial stress of the lid bolts slightly increased, the increment was as small as about 7 % and significantly small as compared with the yield stress of the material (890 MPa). Thereby, we concluded that the generated stress was within the elastic region.

The maximum sliding displacement generated in this test was about 0.6 mm and accumulated sliding displacement that is discussed later was about 1.0 mm.

Helium leak rate at the gasket right after the impact increased rapidly from 8.2×10^{-11} Pa · m^3/s to 4.0×10^{-6} Pa · m^3/s. Then, the leak rate gradually decreased and converged to about 1.0×10^{-6} Pa · m^3/s at 20 hours after the impact test.

6) Leak rate evaluation of a full-scale cask

a. Method of evaluation of leak rate

With respect to the confinement function of casks used for interim storage, Atomic Energy Society of Japan shows "Method to evaluate leak rate of metal gasket at its post storage transportation" in its interpretation to "Standard for Safety Design and Inspection of Metal Casks for Spent Fuel Interim Storage Facility" (AESJ-SC-F002: 2010)[6]. It proposed that evaluation should be made each for opening displacement and for lateral sliding displacement. For instantaneous leakage at drop or tipping over event, the leakage can be evaluated by the lateral sliding if "the stresses generated at lid flange, cask body flange, and lid bolts retaining a metal gasket" are within the elastic region. Here, the opening displacement means displacement of lid movement in the vertical direction against the containment flange (see Fig. 3.7.2(1)-7).

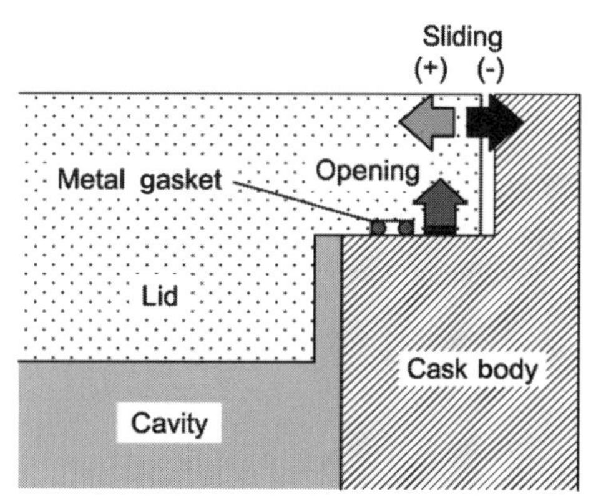

Fig. 3.7.2(1)-7 Sliding and opening displacement of cask lid

When one estimates the leak rate from accumulated lateral sliding, one should use a relationship between leak rate of the gasket with thermal load and sliding displacement. The relationship between leak rate of the gasket with thermal load and sliding displacement of a gasket with diameter of 10 mm was obtained by our research[7] on instantaneous leak test using a 1/10 scale model of a lid structure. Fig. 3.7.2(1)-8 shows a relationship between leak rate and sliding displacement that was calibrated for a full-scale cask studied in this section.

b. Estimation from accumulated sliding

We estimate leak rate of a full-scale cask from the relationship between leak rate and sliding displacement that was calibrated for a full-scale cask.

The accumulated sliding displacement of the scale cask by the impact test was about 1.0 mm, which is 2.5 mm for the full-scale cask. Leak rate corresponding to the sliding displacement of 2.5 mm is estimated as 7.9×10^{-6} Pa · m^3/s from the relationship shown in Fig. 3.7.2(1)-8. Assuming the diameter of the metal gasket of the full-scale cask be 10 mm, the leak rate of the full-scale cask is estimated as 3.5×10^{-5} Pa · m^3/s.

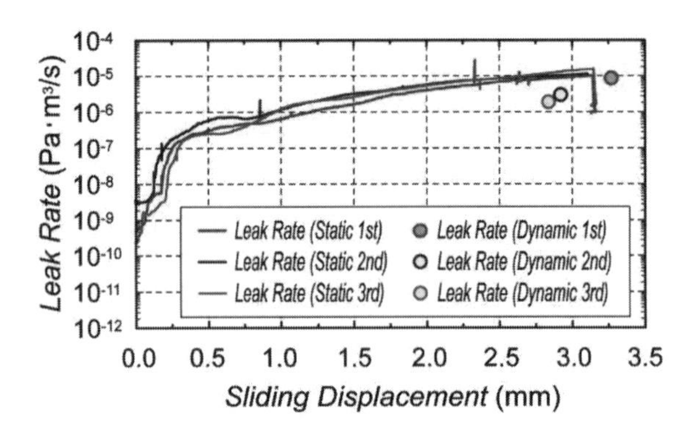

Fig. 3.7.2(1)-8 Leak rate and lateral sliding displacement

7) Analysis of impact response at the horizontal impact test

Since the scenario of the missile impact could be assumed in many ways including the impact angle to the cask, leak rate and response behavior should be evaluated efficiently using numerical analysis. Therefore, the horizontal impact test was simulated with respect to leak rate and response behavior by an analysis in order to confirm reproducibility of the numerical analysis. Three-dimensional analytical code was LS-DYNA Ver. 970 (for PC).

By the analysis of the horizontal impact test, we confirmed that no plastic zone exceeding 0.2 % was found in the lid flange, cask flange, and lid bolts retaining the metal gasket, and that they were within the elastic region. The analytical results were in good agreement with the test results as shown in Table 3.7.2(1)-1.

Table 3.7.2(1)-1 Comparison of analytical result and test result

Item	Analysis	Test
Sliding	Max. 0.64 mm	Max. 0.59 mm
	Accumulated 0.96 mm	Accumulated 1.05 mm
Opening	Max. 0.03 mm	Max. 0.02 mm
Bolt stress	Max. 253 MPa	Max. 259 MPa
Estimated leak rate of full-scale cask	7.3×10^{-6} Pa·m³/s	3.5×10^{-5} Pa·m³/s

(2) Vertical impact test to full-scale partial model cask

The containment structure of the cask and the impact conditions are the same as those in the previous section. More detail information is found in the literature[7].

1) Test condition

a. Impact posture

In the vertical impact test, we conducted tests so that a missile impact vertically on the center of the secondary lid of the cask lying horizontally. Fig. 3.7.2(2)-1 shows the outline of the vertical impact test.

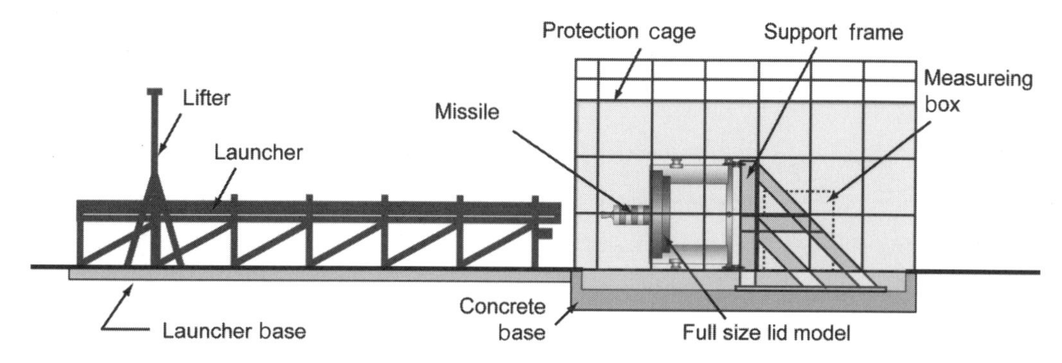

Fig. 3.7.2(2)-1 Outline of the vertical impact test

b. High speed missile

Conditions of the missile (missile structure, speed, etc.) were determined so that the impact damage is equivalent to that by the assumed force. The force was adjusted by the shock absorber attached to the missile similar to that used for the horizontal impact test in the previous section. Fig. 3.7.2(2)-2 shows the overview of the missile used for the vertical impact test. The absorber had a cylindrical shape with dimension of ϕ485 mm × h 141.5 mm. The mass was designed to be 298 kg and the impact speed was 65 m/s. The real product weighed 302.5 kg.

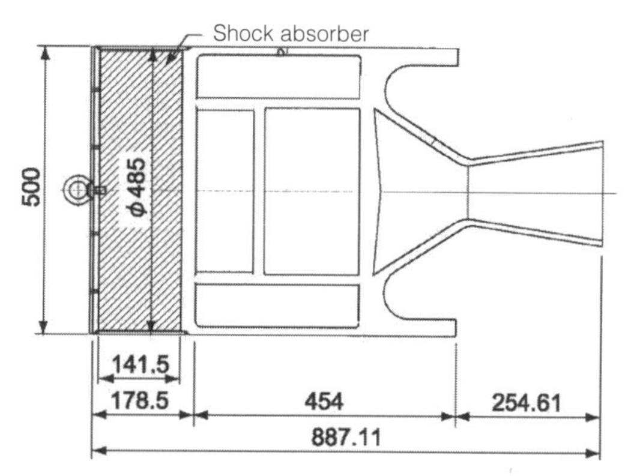

Fig. 3.7.2(2)-2 Overview of the missile

c. Lid model

The lid model was cut from a full-scale transport/storage cask that can install 69 BWR spent fuel assemblies and has double lid structure, and a bottom plate was welded to the cut surface. Fig. 3.7.2(2)-3 shows the test model cut from the full-scale cask. The lid model has the dimension of about 2.8 m in diameter, about 1.8 m in length, and weighs about 36 tons (Cask body 27 tons, primary lid 4 tons, secondary lid 5 tons). The lid structure consists of

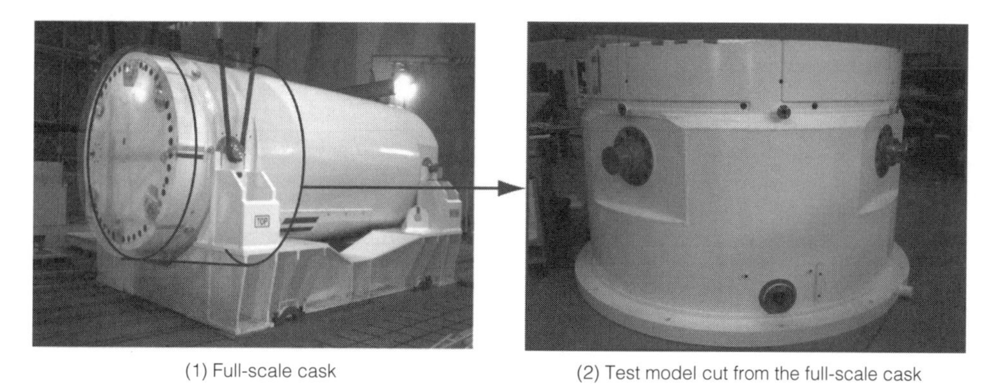

(1) Full-scale cask (2) Test model cut from the full-scale cask

Fig. 3.7.2(2)-3 Cask lid model

primary and secondary lids forming a space between the two lids. The containment is secured by a metal gasket with aluminum coating (Cross sectional diameter is 5.6 mm for the primary lid and 10 mm for the secondary lid).

2) Results of vertical impact test

Fig. 3.7.2(2)-4 shows the measurement results of the vertical impact test. The maximum acceleration generated at the cask body was about 50 G. The maximum acceleration of about 280 G was generated at the primary lid and a bending vibration of about 400 Hz was cyclically generated. Measurements of the secondary lid were not available due to failure of wires for measurement.

The axial stress of the secondary lid increased by about 100 MPa, which is within the elastic region and the lid bolts were sound. While, the axial stress of the bolts for the primary lid did not change before (average 311.0 MPa) and after (average 308.5 MPa) the impact test.

The leak rate of the primary lid gasket was less than 1.0×10^{-10} Pa · m^3/sec before the impact test and increased right after the impact up to 1.43×10^{-7} Pa · m^3/sec at maximum and gradually restored. On the other hand, the leak rate of the secondary lid gasket was less than 1.0×10^{-10} Pa · m^3/sec before the impact test and exceeded the detection limit of 1.0×10^{-3} Pa · m^3/sec right after the impact test and was not measured. Replacing the detector and measurement after three hours, the leak rate of the secondary lid gasket was 7.25×10^{-7} Pa · m^3/sec.

The pressure between the two lids, vibrated by the bending of the lids and decreased to 298 kPa (gauge pressure)

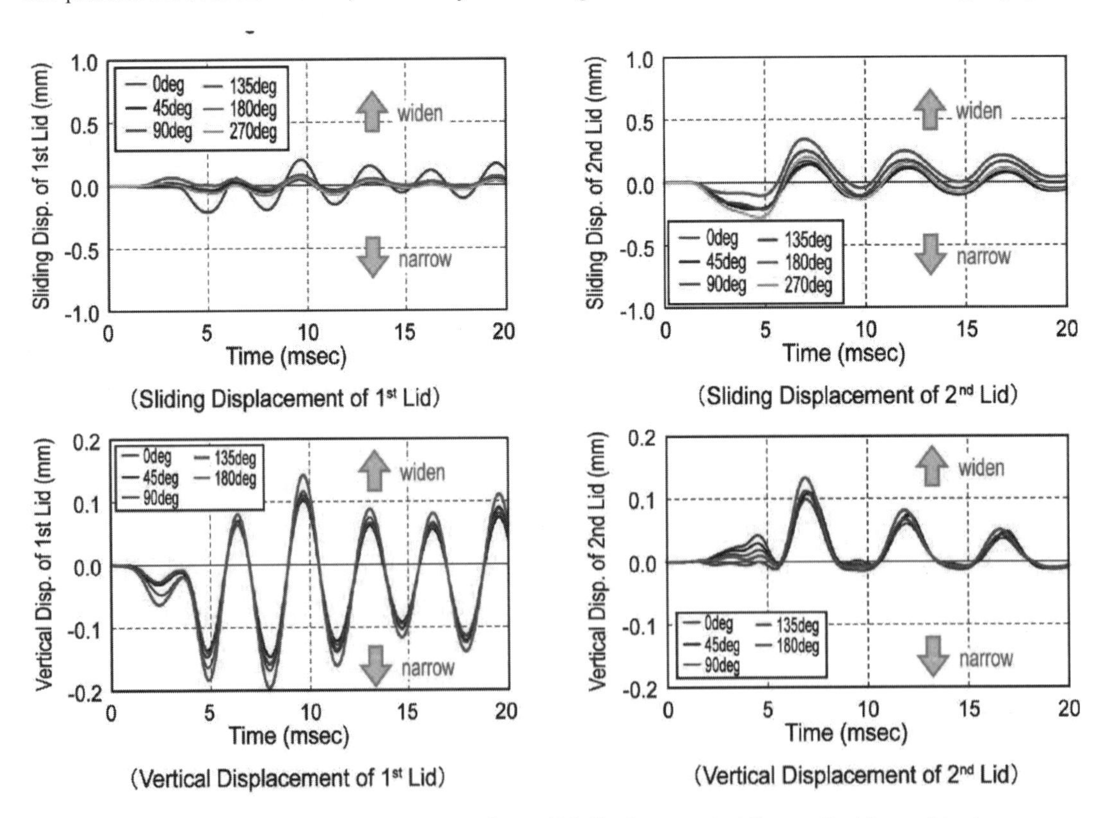

Fig. 3.7.2(2)-4 Measurement results on lid displacement at the vertical impact test

finally, which was 306 kPa before the impact test. The decrease of 8 kPa will be due to the leak of Helium gas.

Vertical displacement of the primary and the secondary lids showed similar response. They were about 0.13 mm for the secondary lid and about 0.2 mm for the primary lid. The cyclic displacement of the primary lid would have been accompanied by the cyclic bending vibration of the lid.

3) Analysis of impact response at the vertical impact test

The vertical impact test was simulated with respect to the response behavior of the cask by an analysis in order to confirm reproducibility of the numerical analysis and evaluate the behavior of the lid with the metal gasket.

Three-dimensional analytical code for the numerical analysis was LS-DYNA Ver. 970 (for PC). Fig. 3.7.2(2)-5 shows analytical results on lateral sliding displacement and vertical displacement for the primary and the secondary lids. The analytical results were in good agreement with the test results.

Fig. 3.7.2(2)-6 shows time history of the lid opening displacement as a result of the numerical analysis. The opening displacement of the primary lid at the initial stage of the impact showed a displacement as small as about 0.02 mm and gradually the vibration attenuated to the initial value. On the other hand, the opening displacement of the secondary lid showed a displacement as large as about 0.4 mm and still exceeded 0.1 mm even at the end of the analysis of 10 msec. Generally speaking, the spring back displacement of the aluminum gasket is 0.08 mm, which indicates possible leakage due to the lid opening at the impact. On the other hand, the integrity of the secondary lid

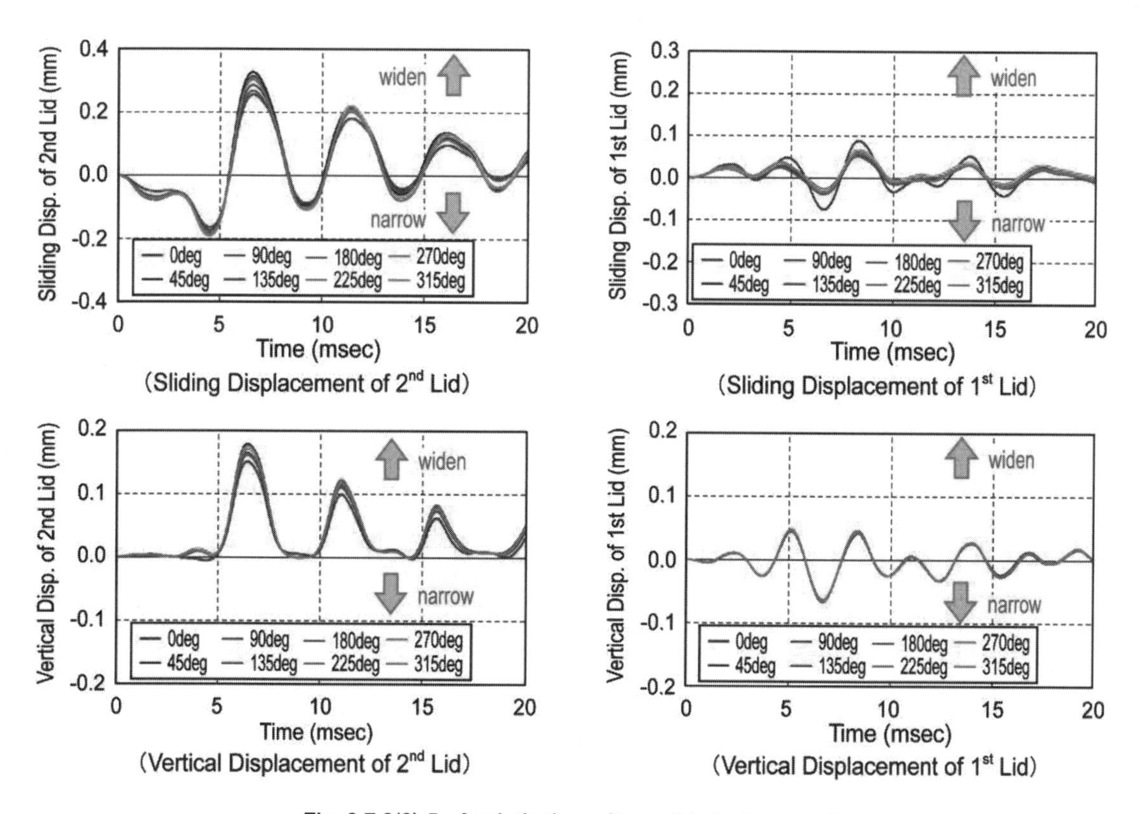

Fig. 3.7.2(2)-5 Analytical results on lid displacement

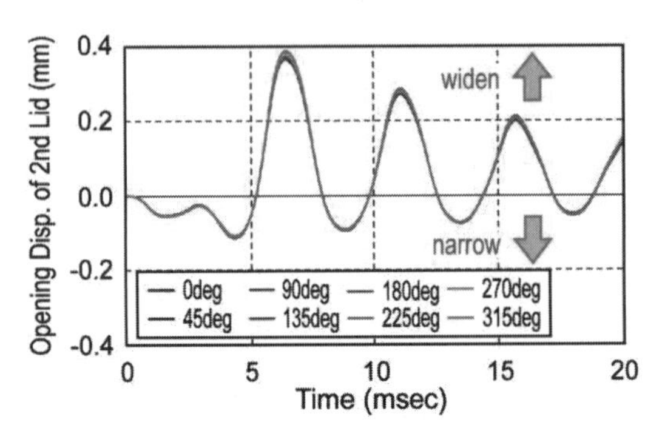

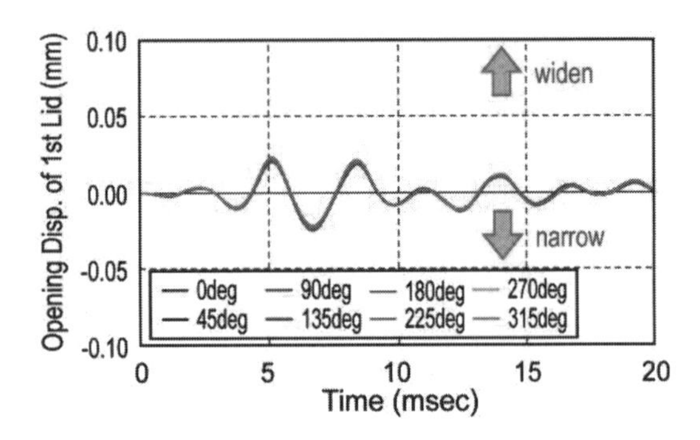

Fig. 3.7.2(2)-6 Analytical results of lid opening at the vertical impact test (LS-DYNA)

bolts had been confirmed and the opening displacement vibration would converge to the initial position. Therefore, we can conclude that the pressure decrease due to the lid opening is small and negligible for the containment performance of the cask. These observations are proof of the fact of the rapid increase of the leak rate right after the impact and the leak rate was below 1×10^{-6} Pa · m^3/sec three hours after the impact test.

4) Evaluation of leak rate

The analytical results showed that the lid opening displacement did not affect the containment performance of the cask and the structural materials of the flange for the secondary and the primary lid were within the elastic region.

The numerical analysis showed the lateral sliding displacement of the primary lid was 0.336 mm at maximum at 0 degree position and that of the secondary lid was 0.958 mm at maximum at 45 degrees position (Fig. 3.7.2(2)-5). While the measurements (Fig. 3.7.2(2)-4) of the sliding displacement of the primary lid was 0.724 mm at maximum at 0 degree position and that of the secondary lid was 1.018 mm at 270 degrees position. The analytical results showed less value for the primary lid at 0 degree position, but the other results were almost in good agreement with the measurements.

Finally, we can estimate the leak rate from the accumulated sliding displacement and the previous section 3.7.2(1)-4. The leak rate was 3.07×10^{-7} Pa · m^3/s for the primary lid and 1.15×10^{-6} Pa · m^3/s for the secondary lid. These values were almost equal to those of the measurements. Thereby, we could confirm the applicability of the evaluation method.

References

1) Central Research Institute of Electric Power Industry: Fiscal year 1991 report on "Verification Test on Spent Fuel Storage Technology" (contract from Ministry of International Trade and Industry), March 1992.
2) B. R. Thomauske: Realization of the German concept for interim storage of spent nuclear fuel -Current situation and prospects-, WM'03 Conference, Tucson, AZ, USA, 2003.
3) K. Namba, K. Shirai, and T. Saegusa: "Evaluation of sealing performance of metal cask subjected to horizontal impact load due to aircraft engine", J. Japan Society of Civil Engineers, Vol.66, No.2, p.177-193 (2010).

4) K. Namba, K. Shirai, and T. Saegusa: "Evaluation of sealing performance of metal cask subjected to vertical impact load due to aircraft engine", J. Atomic Energy Society of Japan, Vol.9, No.2, p.183-198 (2010).
5) P. Barr: Guidelines for the Design and Assessment of Concrete Structures subjected to Impact, Wigshaw Lane, Culcheth Cheshire,WA3 4NE, May 1990.
6) Atomic Energy Society of Japan: "Standard for Safety Design and Inspection of Metal Casks for Spent Fuel Interim Storage Facility" (AESJ-SC-F002: 2010).
7) K. Namba, K. Shirai, T. Saegusa, and K. Shimamura: "Numerical analysis on containment performance of metal cask subjected to impact from air craft", Journal of structural engineering, Vol.55A, 2009.3.

3.8 Interaction between transportation and storage

3.8.1 Influence of mechanical vibration in transport on containment performance of metal gasket in storage

Transport casks of spent nuclear fuel will receive mechanical vibration during transport. It has been known that the containment performance of metal gaskets is influenced by large external load or displacement[1]. Quantitative influence of such vibration during transport on the containment performance of the metal gasket has not been known, but is crucial information particularly if the cask is stored as it is after the transport. The standard for safety design by Atomic Energy Society of Japan[2] stipulates that the dual purpose metal cask shall not lose its containment function for the successive storage by vibration and external force in the normal transport.

The purpose of this section is to find experimentally the influence of mechanical vibration during transport of transport/storage cask with metal gasket on the performance during storage. For the experimental conditions, acceleration values measured in an actual sea transport of transport cask employed. More detailed information is found in the literature[3].

(1) Method

1) Experimental apparatus and specimen

In order to obtain a relationship between the amount of lateral sliding (displacement) of the lid and the leak rate, a 1/10-scale model of a lid structure of metal cask with a metal gasket of double O-ring type was fabricated and assembled as shown in Fig. 3.8.1-1. The gasket had a diameter of 10 mm and was coated with aluminum sheet. The scale model consists of three flanges bolted together and helium gas was installed in a groove of one of the outer flanges.

Eddy current displacement sensors (accuracy of ±0.01 mm) were used to measure displacement of the flanges. Sliding load and relative displacements were applied to the middle flange using loading test equipment. In order to simulate the thermal ageing effect of the metal gasket due to the heat from spent fuel loaded inside the cask, the flanges with the metal gasket were heated for 20 hours at 180 °C inside an oven prior to the tests. The temperature and the time conditions were assumed to simulate the heat history of the gasket after spent fuel loading before transport, with the aid of Larson Miller Parameter equation.

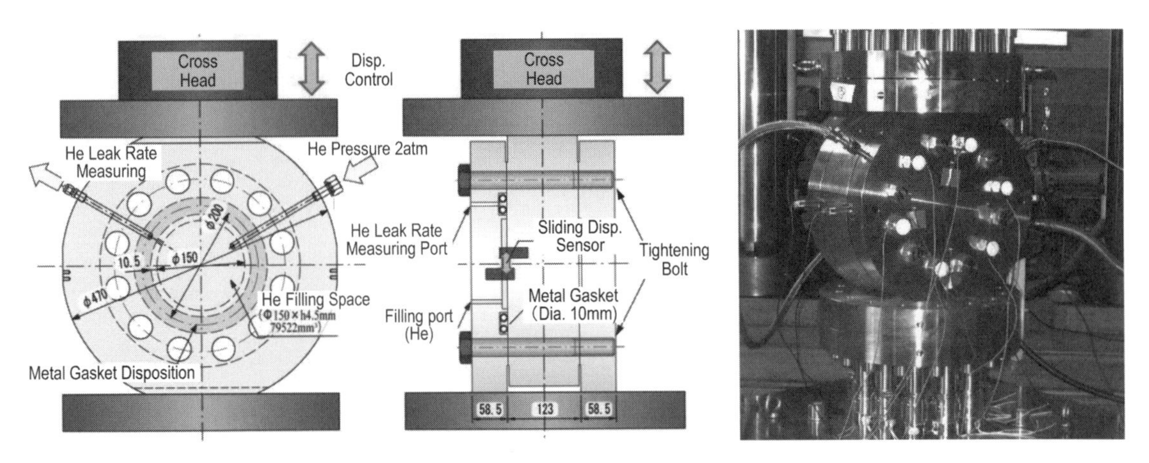

Fig. 3.8.1-1 Scale model (1/10) of a lid structure of metal cask with a metal gasket

2) Experimental conditions

Acceleration wave in transport was assumed by the time history of acceleration measured at a trunnion supports of the spent fuel shipping cask transport frame as shown in Fig. 3.8.1-2[4)].

For analysis, a dynamic analysis code LS-DYNA was employed for the dual purpose cask shown in Fig. 3.8.1-3 (storing 21 PWR spent fuel assemblies) designed by JNES[5)]. Using the time history of the acceleration, we calculated the time history of a lateral sliding of the secondary lid. As the results, the maximum sliding displacement was found in the vicinity of the trunnion, of which the amplitude was approximately 0.014 mm at maximum.

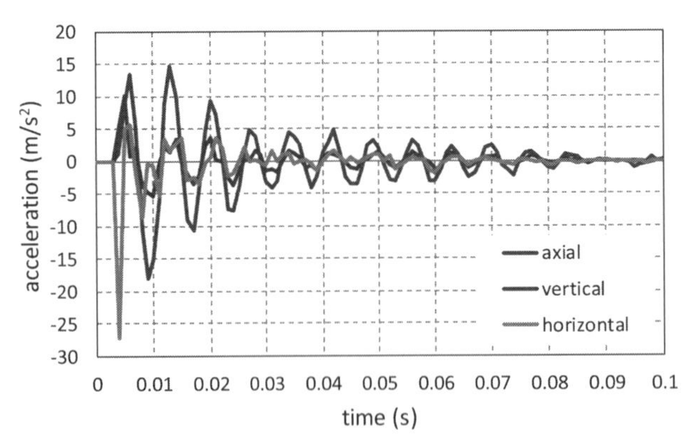

Fig. 3.8.1-2 Time history of acceleration measured at a trunnion supports of the cask transport frame

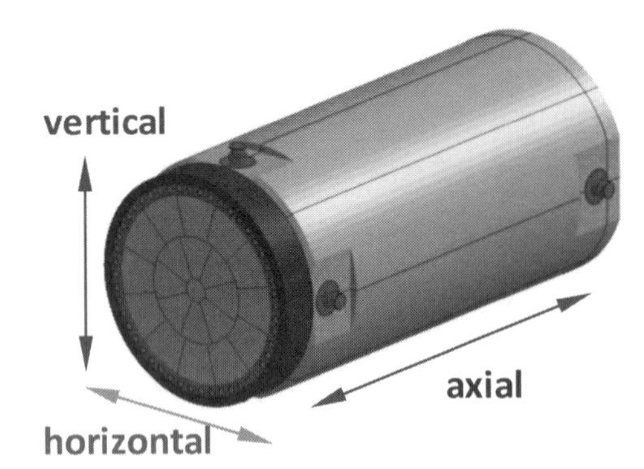

Fig. 3.8.1-3 Cask model for the analysis

(2) Experimental result

The frequency of the cyclic displacement experiment was 0.125 Hz and the nominal displacement was ± 0.02 mm. Fig. 3.8.1-4 shows a representative result showing leak rate and displacement as a function of time. The initial leak rate was 1×10^{-10} Pa · m^3/sec. The leakage started when the radial displacement exceeded ±

0.022 mm and increased as the displacement increased. When the cyclic displacement stopped, the leak rate recovered until the initial leak rate if the cyclic displacement was less than ± 0.025 mm. The leak rate did not recover to the initial leak rate if the cyclic displacement was more than ± 0.035 mm. Nevertheless, the leak rate is still less than 1×10^{-8} Pa · m^3/sec. Corresponding leak rate of a full-scale cask lid model would be less than 1×10^{-7} Pa · m^3/sec taking account of the scale factor[6].

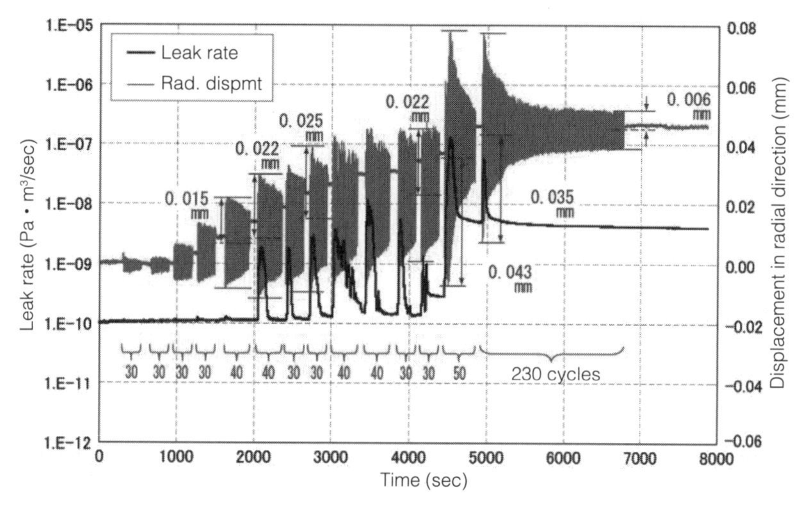

Fig. 3.8.1-4 Measurements of leak rate and radial displacement with elapsed time under cyclic loading

3.8.2 Evaluation of containment performance of metal gasket in transport by ageing of metal gasket under long-term storage

Transport and storage dual purpose metal casks of spent nuclear fuel used in nuclear power plants will be stored in interim storage facilities for 40 to 60 years. The casks usually use metal gaskets for safety and steady containment performance for the long term as shown in Fig. 3.8.2-1. Metal gaskets generally consist of helical metal spring, inner and outer metal jackets. The outer jacket is made of flexible metal material, such as aluminum or silver, and aluminum has been adapted to outer jackets for adhesion between the gasket and lids or cask body surface in Japan. As high temperature due to decay heat of spent nuclear fuel will impose to the seal area of cask lid systems during

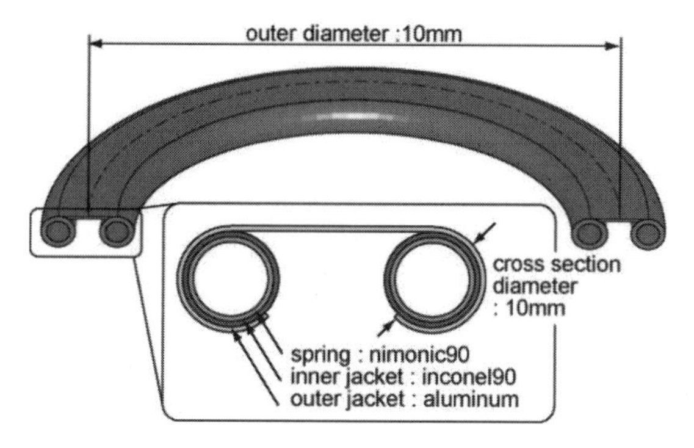

Fig. 3.8.2-1 Overview of a metal gasket

storage, the high temperature can accelerate ageing, such as creep deformation of the outer jacket and corresponding relaxation of the linear load of the gasket complex[7]. Therefore, when the cask and the heated gasket received the vibration force during normal transport operation on land or sea after storage, the containment performance of metal gaskets might be affected.

To evaluate the containment performance of the metal gasket used for long term at high temperature, it is important to comprehend creep characteristics of aluminum at high temperature under compressive loading. In this section, the compressive creep characteristics were obtained by compressive creep tests using aluminum. In order to establish the numerical methodology of the ageing phenomena on containment performance of the metal gasket over the long term, the compressive creep characteristics of the aluminum were introduced in the Finite Element Method code (ABAQUS).More detailed information is found in the literature[8].

(1) Numerical method for gasket containment performance

1) Compressive creep tests of aluminum

To evaluate the relaxation characteristics of the metal gasket complex, a compressive creep tests at high temperature were executed using a 99.5 % pure aluminum with heat treatment at about 250 °C for 4 hours. The test specimen and the test equipment are shown in Fig. 3.8.2-2.

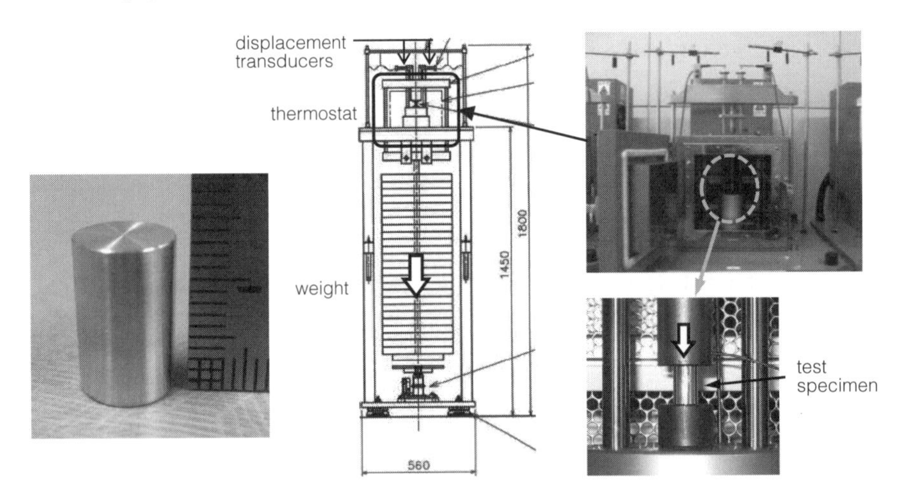

Fig. 3.8.2-2 Test specimen and test equipment of compressive creep test

The creep tests were performed at 150-200 °C under compressive stresses (17-33 MPa). Based on the results of the experimental creep strain curves, strain hardening creep equation were expressed, referring the Sassoulas's creep equation, as follows [9].

$$\dot{\varepsilon}_c = C_1 \cdot e^{C_2 \cdot \sigma} \cdot \varepsilon_c^{C_3} \cdot e^{-C_4/T} \quad , \quad \varepsilon_c = (\alpha t)^{1/(1-C_3)}$$

$$C_1 = 8.974\times10^{-7}, \quad C_2 = 0.657, \quad C_3 = -1.770, \quad C_4 = 10351$$

where, $\dot{\varepsilon}_c$: creep strain rate (/hour), ε_c : creep strain (-)

σ : Mises stress (N/mm^2), t : time (hour), T : Kelvin temperature (K) ,

$\alpha = C_1 \cdot e^{C_2 \cdot \sigma}(1-C_3)/e^{C_4/T}$, e : Napier's constant

2) Method of relaxation analysis

In the relaxation analysis, non-linear 2D axis symmetric model was used as shown in Fig. 3.8.2-3. Table 3.8.2-1 shows the procedure of loading to the metal gasket. The relaxation analysis was executed at time history of temperature as shown in Fig. 3.8.2-4 and holding time was 60 years.

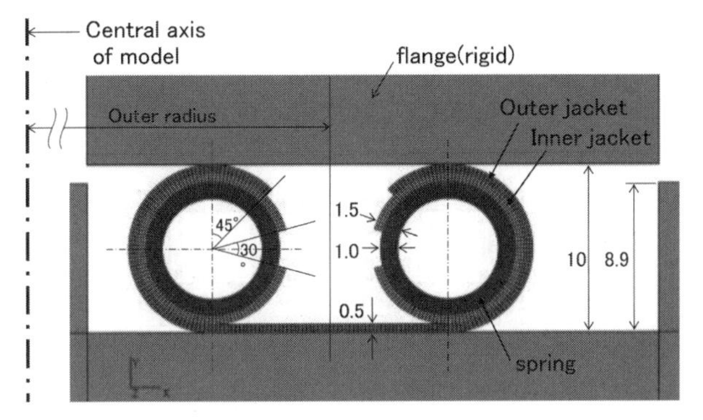

Fig. 3.8.2-3 Non-linear 2D axis symmetric model used in the relaxation analysis

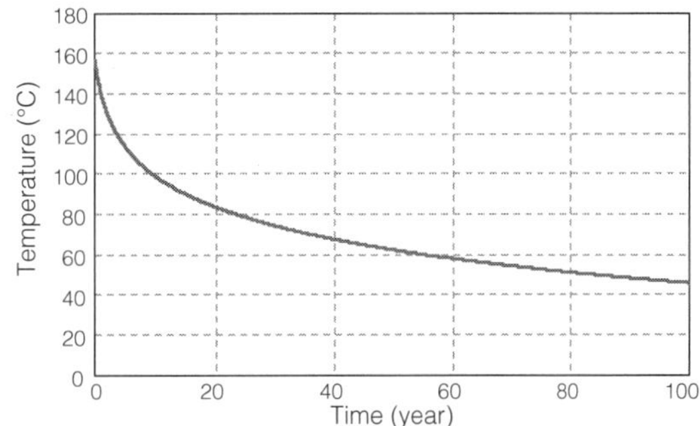

Fig. 3.8.2-4 Time history of temperature used in the analysis

Table 3.8.2-1 Procedure of loading to metal gasket

Step	Temperature	Process	Creep
Step1	Constant (22.6 °C)	Compressive loading	Non
Step2	Heating to test temperature	Considering thermal deformation and material temperature dependency	Considered
Step3	Case1: 160 constant Case2: time history of temperature (Fig. 3.8.2-4)	Heat holding period	Considered
Step4	Temperature on the last time of Step3 (Fig. 3.8.2-4)	Unloading	Non

The CRIEPI carried out relaxation tests using the metal gasket with aluminum outer jacket in a 1/10-scale flange model. As a result, the residual linear load decreased to 212 N/mm from 350 N/mm due to the relaxation of heated gasket complex. Measurements of the residual linear load corresponding to leak rate over 10^{-8} Pa · m^3/s (Y_1) and the effective spring back distance (r_u) were 12 N/mm and 0.12 mm, respectively. Thus, the threshold value of residual linear load Y_1 to maintain containment performance was determined to 12 N/mm. For this experiment, the analytical method showed a good agreement. Using this analytical method we calculated the residual linear load and effective spring back distance of a metal gasket used for storage of 60 years. Fig. 3.8.2-5 shows a result of the analytical result. The external load would not affect the containment performance if vertical displacement of gasket due to an external load is smaller than the spring back distance, r_u=0.09 mm.

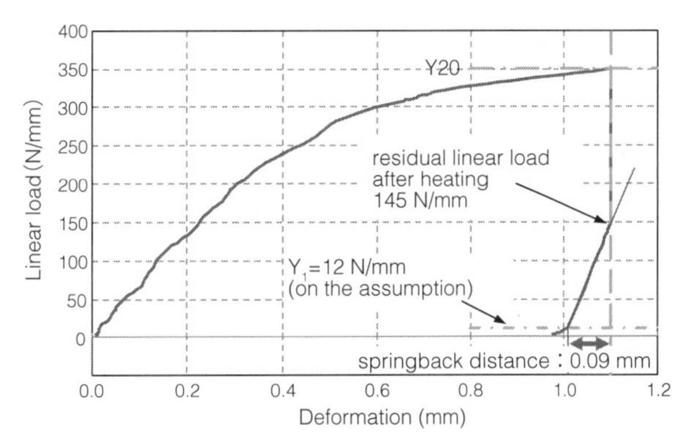

Fig. 3.8.2-5 Relationship between deformation and linear load (analytical result)

(2) Effect of vibration in sea transport on the aged metal gasket

It has been known that the containment performance of a metal gasket is influenced by external load or displacement. Especially, the aged gasket with reduced residual linear load and spring back distance subjected to vibration force during transportation is concerned.

In this section, in order to evaluate the influence of vibration during normal transportation on containment performance of the gasket, the opening displacement perpendicular to the flange surface of the gasket during normal transportation was compared with the spring back distance r_u of the gasket used for 60 years. The displacement of the gasket due to vibration force was calculated by dynamic analysis using Finite Element Method code LS-DYNA.

The cask used in dynamic analysis is a full scale transport and storage dual purpose metal cask as shown in Fig. 3.8.1-3. Time history of acceleration employed in the analytical model was the measurement by gauges set on support frame of cask during actual marine transportation. The measured accelerations as shown in Fig. 3.8.1-2 were of three directions including axial, horizontal and vertical directions. The duration of accelerations used in the dynamic analysis was 10ms comprising maximum accelerations of about 3G in the horizontal direction.

The input data was the measurement without wave filter processing. In the dynamic analysis, the accelerations were input at the trunnion of the cask model and the gravity acted on over-all analytical model.

Fig. 3.8.2-6 shows time history of opening displacement at primary and secondary lid. In primary lid, the peak displacement was very small, less than 0.001 mm. In secondary lid, the peak displacement was about 0.003 mm and the amplitude of the displacement gradually decreased. The results revealed the displacements were much smaller than the spring back distance. Therefore, the containment performance will not be lost by lid opening during the sea transportation within the acceleration measured.

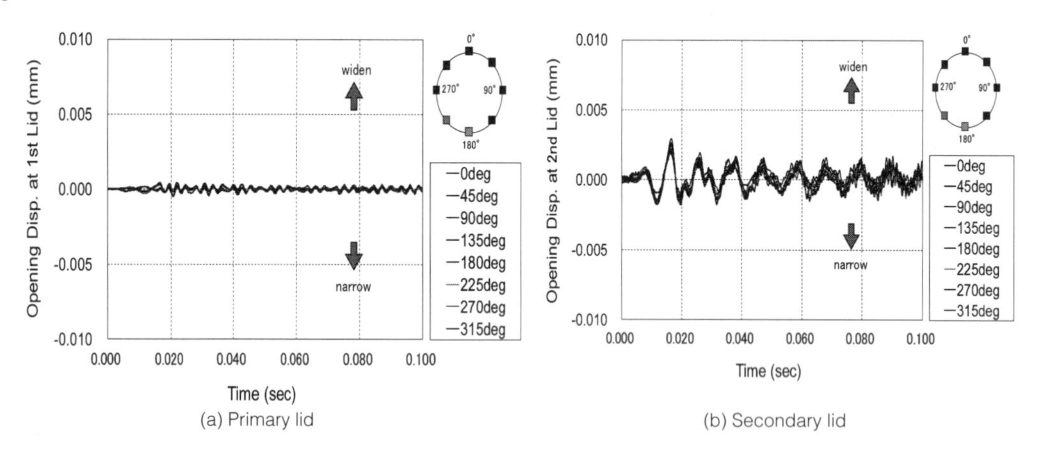

Fig. 3.8.2-6 Dynamic analysis result on lid opening displacement

References

1) H. Yamamoto, et al: "Gasket Performance Test", Kyoto University Research Reactor Institute Report TURRI-TR-293, 80, 1987 (in Japanese).

2) Atomic Energy Society of japan: "Standard for Safety Design and Inspection of Metal Casks for Spent Fuel Interim Storage Facility", AESJ-SCF002: 2010 (in Japanese).

3) T. Saegusa, K. Shirai, H. Takeda, M. Wataru, K. Namba, "Influence of Mechanical Vibration in Transport on Leak-Tightness of Metal Gasket in Transport/Storage Cask for Spent Nuclear Fuel", Proc. PATRAM2010, Oct.4-8, 2010, London.
4) K. Shirai, N. Kageyama, K. Kuriyama: "Estimation of Lid Behavior of Transport/Storage Cask During Transport", 2006 Autumn Meeting at Hokkaido University, Atomic Energy Society of Japan (in Japanese).
5) JNES: Fiscal year 2002 report on "Investigation of spent fuel storage technology (Verification tests of metal cask storage technology)", March 2003 (in Japanese).
6) O. Katoh and C. Ito: "Proposal of an estimation method for leakage from contact surfaces in cask containment system", CRIEPI Report U01006 (2001).
7) M. Wataru, et al.: "Long-term Containment Performance Test of Metal Cask for Spent Nuclear Fuel Storage", 11th International Probabilistic Safety Assessment and Management Conference, PSAM11, Helsinki, Finland, (2012).
8) K. Namba, K. Shirai, M. Wataru, T. Saegusa: "Evaluation of Sealing Performance of Metal Gaskets Used in Dual Purpose Metal Cask under Normal Transport Condition Considering Ageing of Metal Gaskets under Long-Term Storage", Proc. PATRAM2013, San Francisco, August 18-23, 2013.
9) H. Sassoulas, et al.: "Ageing of Metallic Gaskets for Spent Fuel Casks: Century-long Life Forecast from 25,000-h-long experiments", Nuclear Engineering and Design, Vol. 236, p.2411-2417 (2006).

3.8.3 Holistic approach for safety evaluation of post-storage transportation

The metal casks are to be shipped to the reprocessing facilities, etc. after storage in the interim storage facilities.

Spent fuels contained in the shipped metal cask are those whose integrity as spent fuel assemblies at the time of being placed into the metallic casks was confirmed, which is properly contained in the metal casks and properly transported from the power plants to the interim storage facilities. In addition, in the interim storage facilities, the fuel is stored in the environment which is designed in such a manner that safety for long-term storage is sufficiently considered, based on the Safety Review Guidelines, and it is confirmed that there is no abnormality in the pressure between the lids, surface temperature, and external force.

Therefore, shipment from the interim storage facilities to the reprocessing facilities should be studied as follows on the prerequisite of these conditions of the contents[1)-4)].

(1) Items to be confirmed before the start of interim storage

When shipping out the casks after storage, safety as a package (metal cask and its contents) must be assured from the point of transportation regulations, which means that it is necessary to confirm that the metallic casks containing spent fuel must satisfy necessary conditions in terms of securing public safety.

Therefore, the administrative regulatory body need to review in the original approval of design and containers, based on the transportation regulations for the metal casks (Article 59 of the Reactor Regulation Law and transportation rules outside the plants), so that required safety is secured, considering not only transportation from the power plants to the interim storage facilities, but also transportation from the interim storage facilities to the reprocessing facilities. (Specifically, to confirm that the storage casks are designed to allow the necessary items of inspection before shipment can be surely performed (including that the casks are designed in such a manner the leak rate against the sealing boundary of the package can be measured, and if such leak rate fails to satisfy the criteria, the seal part can be repaired/replaced), by appropriately assuming the natures of spent fuel as the contents at the end of the storage.)

Also in the safety review of the storage activities, it is necessary to confirm that the metal casks can be safely shipped out of the facility after the end of storage. Thus, also in the safety reviews related to storage operation, it is appropriate for the administrative regulatory body to confirm that the metal casks should be equipped with functions to the extent necessary for transportation according to the stages of regulation, and that responsibility for shipment of the reactor licensees' (who are the owners of the stored spent fuel) should be clarified. (It is stipulated in the chapter 1. Subject of application of the former safety review guidelines "Spent fuel assemblies are to be placed and delivered in the dry metallic casks, which act as transportation containers for shipping out spent fuel assemblies outside the plant site, then to be stored without being removed to other containers, and after storage, they are to be shipped out. Thus these casks are to be used in the interim storage facilities for spent fuel." Consequently, it is a prerequisite that the casks in concern can be transportable. The purpose of this sentence is to further clarify this point.)

(2) Items to be confirmed during interim storage period

The metal casks, which are confirmed to have functions required for transportation prior to the start of the interim storage operation, need to maintain such functions until they are shipped out after the interim storage actually ends.

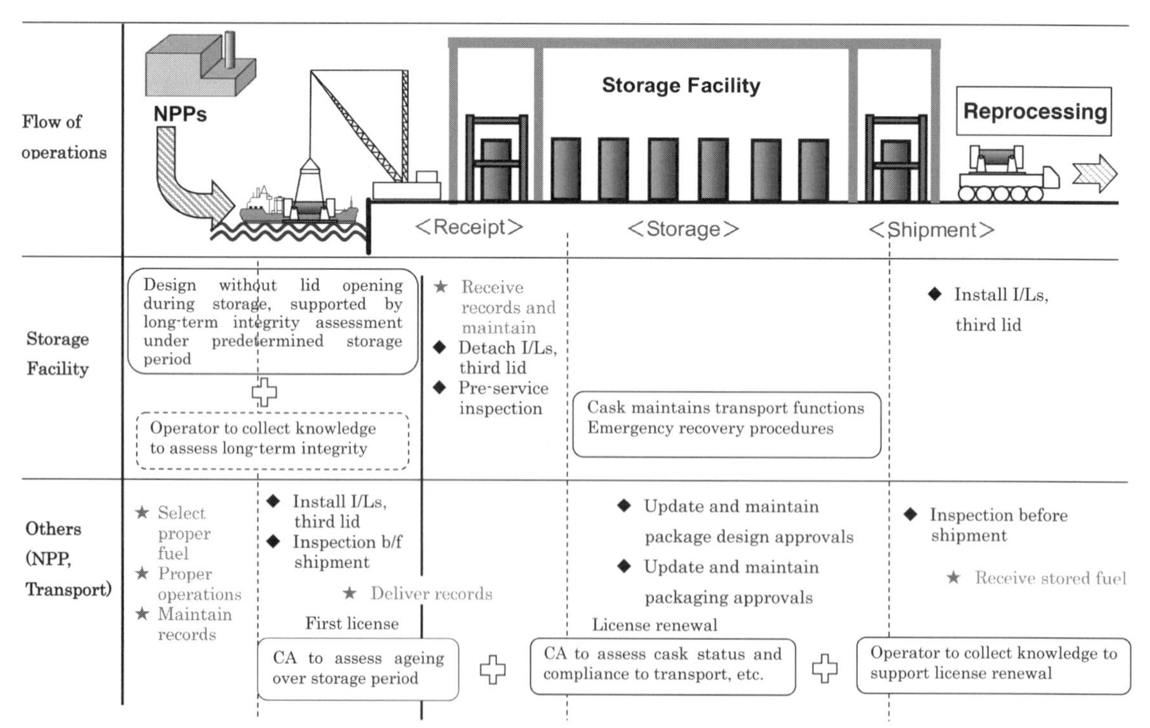

Fig. 3.8.3-1 Example of management of licensing and inspections in spent fuel transport and storage[5]
(CA: Competent Authority)

According to the practice of current transportation regulations, approval for design and containers must be reviewed for renewal generally not later than 5 years, in order to confirm that the packages satisfy the latest regulations and that they still maintain their required performance. Therefore, approval of the metal casks should also be renewed during storage.

Although this approval should be conducted based on the transportation regulations, it is appropriate for the administrative regulatory body to confirm that the renewal of approval is surely conducted by the means such as the licensees' security rule. At the renewal, it is confirmed by the records kept by the licensees that the metal casks, which also act as packages (for transportation), maintain the required performance. The administrative regulatory body should clarify the related procedures so that these records are smoothly and surely delivered and received between the storage licensees and the entities which carry out procedures for renewal (Fig. 3.8.3-1[5)]).

(3) Securing safety during transportation after end of storage (pre-shipment inspection)

1) Items to be confirmed before transportation of spent fuel from power plants

The stored spent fuel is to be shipped out to reprocessing facilities after the end of the storage.

At present, when spent fuel is transported from power plants, generally the following 10 items are inspected (Atomic Energy Society of Japan, "Standard for Safety Design and Inspection of Metal Casks for Spent Fuel Interim Storage Facility", AESJ-SC-F002:2008). (Specific items and methods of inspection shall be stated in the application for design approval of nuclear fuel transportation packages as the items related to handling.)

a. Exterior inspection
b. Airtight leakage inspection
c. Pressure measurement inspection
d. Inspection of dose equivalent
e. Subcriticality inspection
f. Temperature measurement
g. Lifting inspection
h. Weight inspection
i. Contents inspection
j. Inspection of surface contamination density

2) Inspection items which cannot be conducted visual checks etc.

In the above listed inspection items (which are generally carried out), inspections which need visual check of inside the metal casks are included. Therefore, in interim storage facilities which are not equipped with fuel refilling system, it includes the inspection items which cannot be conducted exactly the same as the above, for transportation after the end of storage.

Among those general inspection items for shipment from power plants, the following three inspection items can be specified that cannot be conducted visual checks the same manner for shipment from the interim storage facilities that have no the fuel refilling system, like these inspection items include visual check of the contents and measurement of the interior environment of the containers.

a. Subcriticality inspection

Inspection to confirm that there is no deformation or breakage of baskets, which may affect the function to prevent criticality.

In case shipping out from power plants, the inspection is carried out as visual check of the basket structure appearance.

b. Contents inspection

Inspection to confirm that spent fuel assemblies satisfy the requirements specified at the time of design (specification, quantity and layout when contained in the containers). In case shipping out from power plants, the inspection is carried out confirmation of the specification and quantity by the check of operation records, and also visual inspection of the layout and appearance of the contents.

c. Pressure measurement inspection

Inspection to confirm that atmosphere inside the metal casks is maintained within design requirements.

In case shipping out from power plants, inspection are carried out in such a way that remaining moisture is checked by the degree of vacuum after vacuum drying or by humidity after refilling gas into the casks while constituents and refilling quantity of gas and pressure are checked by the record of the metal cask preparation work.

3) Concept of alternative inspections

In order to confirm safety of transportation at the end of storage, integrity of the metal casks and contents should be checked, for example, by visual check of the condition of the contents and baskets by opening the lid of the metal casks, or by inspecting atmosphere inside the metal casks before shipment in the interim storage facilities. On the other hand, following facts should be considered; (i) interim storage facilities are very stable and static (For example, Paragraph 5.14 of the IAEA Safety Series DS371 (Draft) states, "Under normal operation, spent fuel storage facilities have no sources for a fast increase of nuclear reactivity and as such there are relatively few credible mechanisms for a sudden excursion followed by release of radioactive material."), and (ii) radioactivity of spent fuel contained in the metal casks gradually decays by releasing heat, and (iii) just for visual check. it is required to break the containment boundary of the metal casks. This action is undesirable not only from the viewpoint of reduction of workers' exposure or leak prevention of radioactive materials, but also that it may cause another accident, and would rather increase risks. (At the end of storage, if it is decided to actually measure the pressure in the atmosphere inside the metallic cask, a vent valve installed in the primary lid must be opened after opening the secondary lid. This action is regarded as an operation to break the containment boundary of the cask, just like opening of the primary lid.) Consequently, when the same level of safety can be assured as visual check, it is more desirable to perform inspections based on alternative approaches.

The following are the alternative inspection methods related to the above mentioned inspection items.

a. Subcriticality test

Baskets are exposed to decay heat and radiation of the contained spent fuel, and also to external forces caused by vibration due to earthquake and transfer, etc.

Therefore, the Safety Review Guidelines demand that baskets must be designed (especially selection of materials) and manufactured so as to achieve long-term integrity during storage. The environment which are presumed to be achieved at the time of sealing should be retained by the following way: when metal casks are prepared in the power plants, moisture of the space (baskets with spent fuel are installed) must be completely removed, and then the space are filled with inert gas, and finally sealed by multiple lid structure.

The baskets, to which are treated and managed as above by design and manufacture, are exposed to the same environment as spent fuel. Therefore, it can be judged there is no deformation and breakage if the following points are confirmed:

(a) Baskets are manufactured following the design in the factory.

(b) Moisture is removed and inert gas is filled according to the design requirements during preparation of the metal casks in the power plant.

(c) Casks passed subcriticality inspection for transportation from the power plant to the interim storage facility, and there is no abnormal external force during transportation.

(d) The inert gas environment inside the cask is maintained during the storage period.

As a consequence, when the metallic casks are shipped out from the interim storage facilities that have no fuel reloading equipment, subcriticality inspection during the pre-shipment inspection can be substituted by the documents that proves above listed items from (a) to (d) .

b. Contents inspection

The specification, quantity and layout in the cask for the spent fuel (the contents of the cask) are already confirmed by operational records, etc., when the casks are originally shipped out from the power plant to the interim storage facility. As there exists no factor that will change these conditions, the above mentioned items can be reconfirmed by using the same records for the shipping of the casks from the interim storage facility. Concerning the appearance, although the contents cannot be visually checked, as they are contained in the metal casks which have necessary heat removal function during storage, and sealed with inert gas, it are considered that integrity of spent fuel is not impaired due to chemical, thermal and radiological degradation.

Specifically, if the following points are confirmed, it can be judged that no abnormal change occurs on the appearance of spent fuel:

(a) Moisture is removed and inert gas is filled in the way that satisfies the design condition during preparation of the metal casks in the power plant.

(b) Casks pass the inspection of contents for transportation from the power plant to the interim storage facility, and there is no abnormal external force added during transportation.

(c) There have been no incidents that may damage integrity of contained spent fuel during storage.

As a consequence, when the metal casks are shipped out from the interim storage facilities which have no fuel reloading equipment, the inspection of the contents during the pre-shipment inspection can be substituted by the documents that prove above listed items from (a) to (c) .

c. Pressure measurement inspection

It should be confirmed by pressure measurements inspection that atmosphere inside the metal casks is within the range of design requirements. However, as the inside space of the metal cask is completely dried, filled with inert gas and sealed by the multiple lid structure, it can be considered that the conditions originally assumed are maintained, if the containment function is confirmed to be maintained throughout the storage period.

Specifically, if the following points are confirmed, it can be judged that atmosphere inside the metal casks is within the range assumed at the design stage:

(a) Moisture is removed and inert gas is filled in the way that satisfies the design requirements during preparation of the metal casks in the power plant.

(b) Casks pass the pressure measurement inspection test for transportation from the power plant to the interim storage facility and there is no abnormal external force acted during transportation.

(c) Containment function of the metal casks was confirmed by the acceptance test in the interim storage facility.

(d) Containment function of the metal casks has not been lost during storage.

As a consequence, when the metal casks are shipped out from the interim storage facilities that have no fuel reloading equipment, pressure measurement inspection during the pre-shipment inspection can be substituted by the documents that prove listed items from (a) to (d).

Fig. 3.8.3-2 shows an example of the above-mentioned inspections[4).

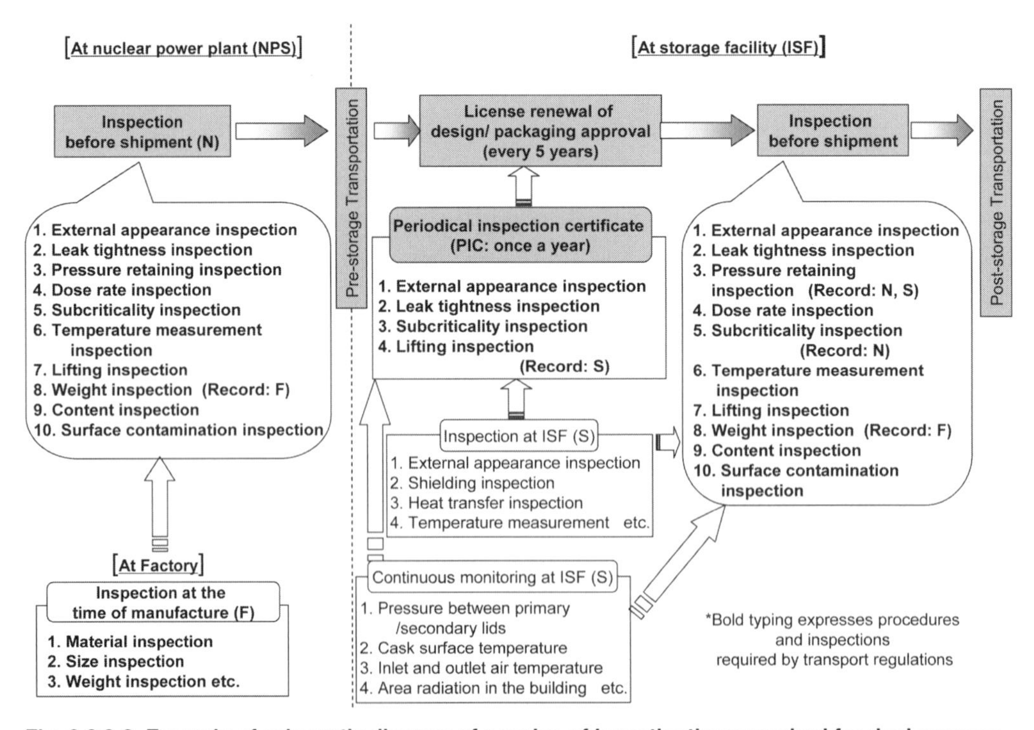

Fig. 3.8.3-2 Example of schematic diagram of a series of investigations required for dual purpose casks (from the viewpoint of transportation) (Provided by Tokyo Electric Power Co, Inc.)

References

1) Nuclear and Industrial Safety Subcommittee of the Advisory Committee for Natural Resources and Energy, Nuclear Fuel Cycle Safety Subcommittee, Interim Storage Working Group and Transport Working Group: "Long-term Integrity of the Dry Metallic Casks and their Contents in the Spent Fuel Interim Storage Facilities", (2009.6.25) (http://www.meti.go.jp/committee/materials2/downloadfiles/g90723b02j.pdf)

2) I. Hanaki : "Interface Issues Arising between Storage and Transport for Storage Facilities Using Storage/Transport Dual Purpose Dry Metal Casks", International Conference Management of Spent Fuel from Nuclear Power Reactors, IAEA, Vienna, (2010). (http://www-ns.iaea.org/meetings/rw-summaries/vienna-2010-mngement-spent-fuel.asp)

3) S. Kojima: "The New Approach to Regulating Long-Term Storage of Spent Fuel", US NRC RIC.2011,(2011). (http://www.nrc.gov/public-involve/conference-symposia/ric/past/2011/docs/abstracts/kojimas-h.pdf)

4) T. Takahashi, M. Matsumoto, and T. Fujimoto: "Confirmation of Maintenance of Function for Transport after Long-term Storage Using Dry Metal Dual Purpose Casks", Proc. PATRAM 2010, London (2010).

5) Nuclear and Industrial Safety Subcommittee of the Advisory Committee for Natural Resources and Energy, Nuclear Fuel Cycle Safety Subcommittee, Interim Storage Working Group (29th) and Transport Working Group (23rd): "Status of Regulatory Framework Development on Interim Storage - Actions for Reactor Operator and Storage Facility Operator -", (2010.7.15)

CHAPTER 4 # CONCRETE CASK STORAGE

4.1 Outline of concrete cask storage

4.1.1 Background and needs

In Japan, metal cask storage is now preceding for commercial use. In the near future, however, as the needs for spent fuel storage expand, the supply of metal casks will not be enough. As will be described in the section 4.1.3, concrete cask storage has merits of economy, contribution to local industry, etc. In addition, prevalence of concrete cask storage over metal cask storage in USA (see Fig. 4.1-1) is suggesting us to make it commercially available as one of multiple storage options other than metal cask storage technology for Japan.

Fig. 4.1-1 Examples of concrete casks in USA (Left: Copyright NAC, Right: Copyright Holtec)

4.1.2 Design concept

Spent fuel is encapsulated in a canister, a cylinder made of a thin stainless steel, and stored vertically in a concrete container. Fig. 4.1-2 shows an example of design concept of concrete cask. The canister is 1.7 m in diameter, 4.6 m in height, and concrete container is 4 m in diameter and 6 m in height, approximately. The total weight is approximately 180 t including spent fuel, canister and concrete container. Containment is secured by welding lid on the canister after spent fuel is installed at a reactor pool. Helium gas is filled in the canister as for the metal cask. The canister is transferred to a concrete container for storage via transport cask to a storage facility. The concrete cask for storage has air inlet and outlet so that air flows naturally along the canister surface from bottom to top of

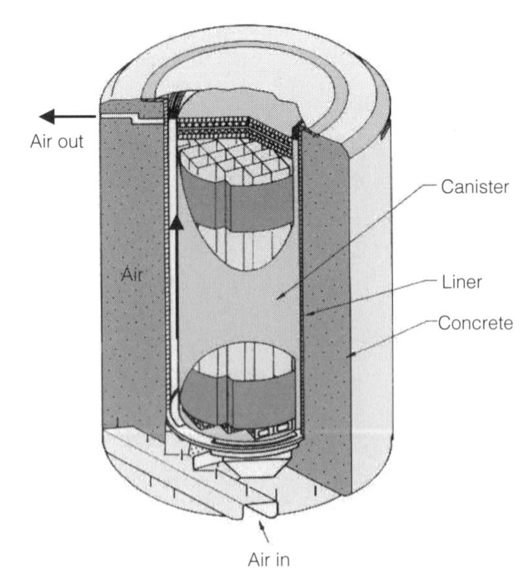

Fig. 4.1-2 Example of concrete cask

the cask to remove the heat form the spent fuel. Radiation shielding is secured by canister and concrete (including inner liner of the concrete cask). Concrete cask storage is considered more economical than metal cask storage[1)] and prevailing method in USA. The concrete cask storage is not yet commercially available in Japan, because of issue of stress corrosion cracking (SCC) of the welded canister made of stainless steel. Research and development is being carried out to overcome the issue. The SCC issue is also recognized in USA and the research is going on.

4.1.3 Comparison with metal cask storage

Table 4.1-1 shows comparison between metal cask storage and concrete cask storage to be assumed for commercial use in Japan.

Table 4.1-1 Comparison between metal cask storage and concrete cask storage

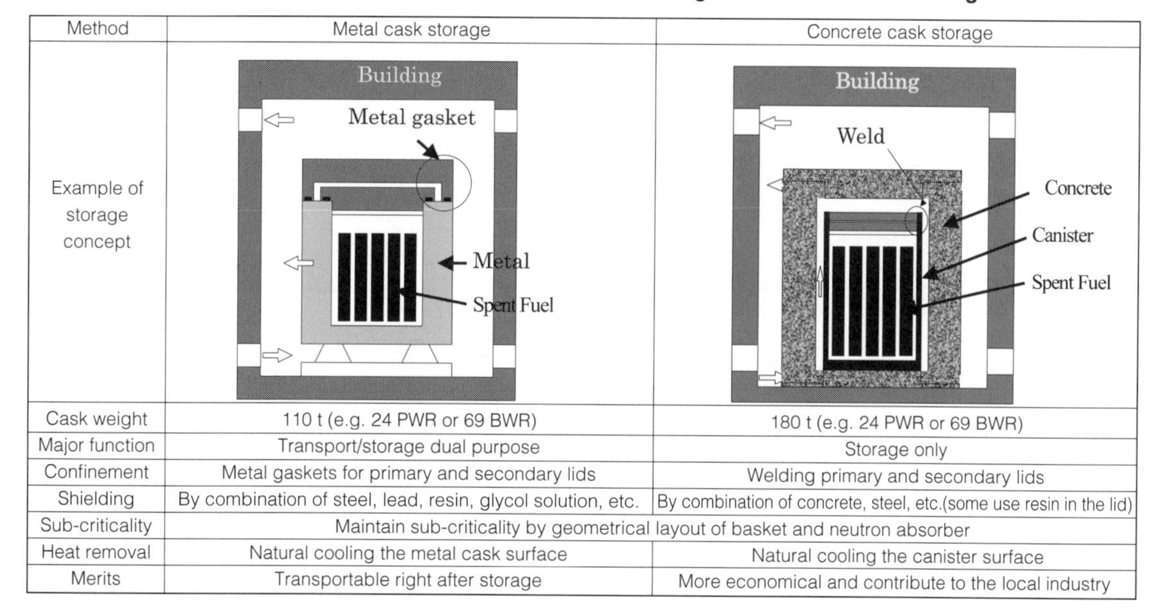

Method	Metal cask storage	Concrete cask storage
Example of storage concept		
Cask weight	110 t (e.g. 24 PWR or 69 BWR)	180 t (e.g. 24 PWR or 69 BWR)
Major function	Transport/storage dual purpose	Storage only
Confinement	Metal gaskets for primary and secondary lids	Welding primary and secondary lids
Shielding	By combination of steel, lead, resin, glycol solution, etc.	By combination of concrete, steel, etc.(some use resin in the lid)
Sub-criticality	Maintain sub-criticality by geometrical layout of basket and neutron absorber	
Heat removal	Natural cooling the metal cask surface	Natural cooling the canister surface
Merits	Transportable right after storage	More economical and contribute to the local industry

Reference

1) R.W. Lambert, D.K. Zabransky, and J.V. Massey, : "Evaluation of Comparative System Economics and Operability of Concrete Casks for Fuel Storage", Proc. Waste Management, 1992

4.2 Design and production of concrete cask

For obtaining validity evaluation data and methods necessary for license of the concrete cask storage method, we made examination and evaluation with validation tests using full-scale or reduced-scale models.

Concrete cask storage facilities consist of casks and a building. The cask structure and the materials shall have strength necessary for loads expected in the designed storage period and safety functions shown in Fig. 4.2-1, under use conditions specific to each country (e.g. site area, seashore location, and quake protection).

The four safety functions are ensured by the following.

(1) Confinement function shall be ensured by the canister in the concrete storage container.

(2) Shielding function shall usually be ensured by both the concrete storage container and the building of the site area is small. Although it depends on the burnup and cooling years of spent fuels, it is usually necessary to make the walls and ceiling of the building thicker as the site area is smaller. There are no buildings built in many cases in the US.

(3) Criticality prevention function shall be ensured by the basket in the canister. For high-burnup fuels, materials containing boron that absorbs neutrons are used.

(4) Heat removal function shall be ensured by both the concrete storage container and the building. Usually, natural air cooling is used for heat removal from dry storage. The building has an exhaust stack for a chimney effect if necessary.

In this section we describe the specifications require for concrete casks and show an overview of two kinds of full-scale casks that we built based on two kinds of structures we chose and basic designs we made.

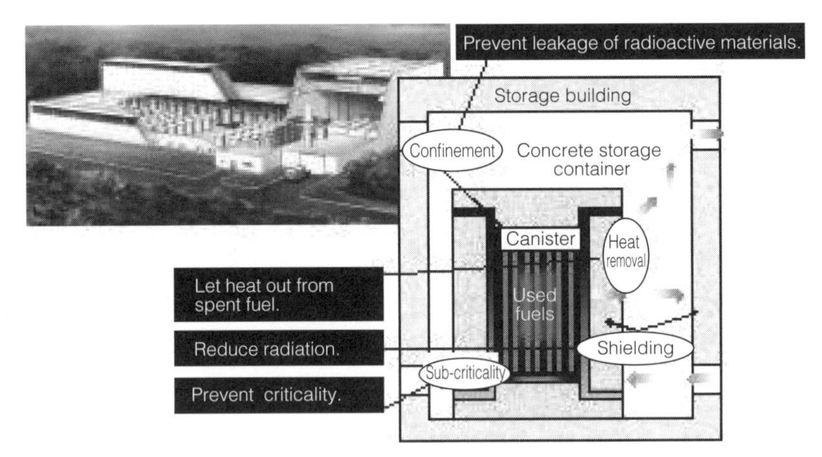

Fig. 4.2-1 Overview of concrete cask storage facilities and safety functions

4.2.1 Basic design requirement

Table 4.2-1 shows basic design items and fuel specifications.

We chose two major structures, reinforced concrete (RC) and concrete filled steel (CFS), shown in Fig. 4.2-2. The RC casks support its load by reinforced concrete, support parts, and support legs and use steel liners as shielding materials. On the other hand, the CFS casks support its load by

Table 4.2-1 Basic design items and fuel specifications

Function	Operation status	Evaluation item
Heat removal performance	Normal condition Abnormal condition	Heat value, natural cooling air volume, temperature of various parts, and clad tube performance
Confinement performance	Normal condition	Quality assurance of welding structure (Fracture mechanical evaluation)
Shielding performance	Normal condition	Exposed dose Radioactivation of cooling air
Sub-criticality performance	Normal condition	When the canister is filled with water. In dry condition.
Structural strength	Normal condition	Durability, temperature stress, pressure capacity, and quake resistance
	Abnormal condition	When the canister falls. When the cask drops. When the air supply inlet is closed.
Fuel specifications	Target fuel	17×17 high burnup PWR
	Initial condensation	4.9 % or lower
	Maximum burnup	55 MWd/kgHM
	Cooling years	10 years

steel parts and use the concrete as shielding materials. Both types are supposed to be used in storage buildings.

A basic design is made for these two kinds of casks. Fig. 4.2-3 shows an evaluation example of heat removal performance of concrete cask and ageing deterioration during the designed period (neutralization, salt damage, etc.). We evaluated the heat removal, confinement, shielding, sub-criticality, and structural strength properties in both ordinary and abnormal conditions and found a positive prospect to practical realization.

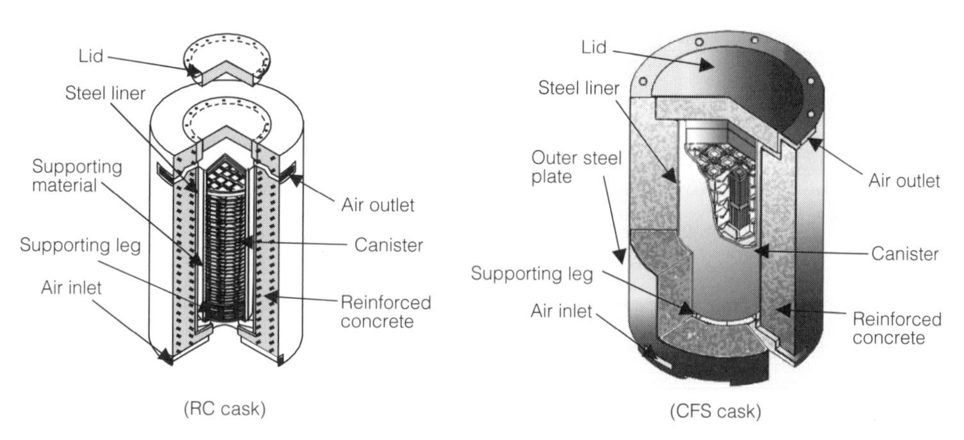

Fig. 4.2-2 Basic structure of concrete cask

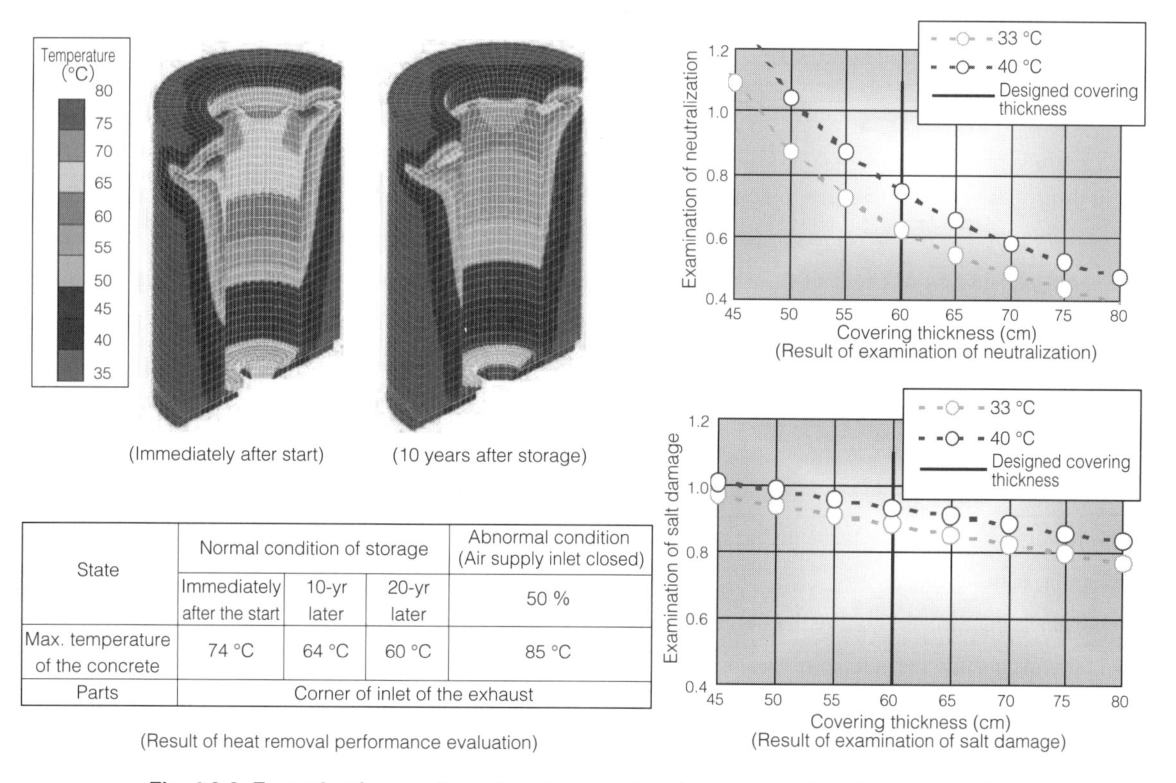

State	Normal condition of storage			Abnormal condition (Air supply inlet closed)
	Immediately after the start	10-yr later	20-yr later	50 %
Max. temperature of the concrete	74 °C	64 °C	60 °C	85 °C
Parts	Corner of inlet of the exhaust			

(Result of heat removal performance evaluation)

Fig. 4.2-3 Example of evaluation of heat removal performance and ageing degradation

4.2.2 Producing test models

(1) Storage container made of concrete

Table 4.2-2 shows the standard and criteria to which we referred to when making test models. In the validation test, we produced two kinds of concrete storage containers shown in Fig. 4.2-2. The construction and management of the concrete parts were based on "Construction work standard specifications and description for reinforced concrete construction of JASS 5N nuclear power plant (2001)" (Architectural Institute of Japan, hereafter referred to as JASS 5N) and various relevant standards of Architectural Institute of Japan and Japan Society of Civil Engineers.

The temperature restriction of the concrete of the RC storage container was designed to allow local temperature up to 90 °C assuming that the concrete would receive only a temperature load in the long term and an inertia force from its weight in earthquakes in the short term and that high-quality concrete would be used for the casks. The temperature restriction of the concrete of the CFS storage container was designed to allow local temperature up to 90 °C in order to ensure the concrete's shielding quality.

Table 4.2-2 Load, standards, and criteria referred to in the production

Storage container		RC	CFS
Loads to be considered	Long-term	Weight and temperature load	
	Short-term	Quake load B (S_2)	
Concrete part	Material	JASS 5N*	Part of JIS (A 5308,1998) JASS 5N*
	Design	Notice 452 RC standard*	(Temperature restriction Note)
	Manufacturing	JASS 5N equivalent*	
Steel part	Material	Design and construction standards**	
	Design	Class 1 support structure	
	Manufacturing	JASS 6*	
*: Architectural Institute of Japan, **: Japan Society of Mechanical Engineers			

In order to suppress cracks due to temperature loads from stored spent fuels, we employed an allowable value of crack width due to corrosion of steel materials designated in Japan Society of Civil Engineers' Standard Design Specifications of Concrete (crack width 0.21 mm with covering depth of 60 mm in a particularly severe corrosion environment). As a result we set the diameter and interval of longitudinal steels to be 38 mm and 125 mm respectively and those of transverse steels to be 38 mm and 150 mm respectively.

Table 4.2-3 and Fig. 4.2-4 show major specifications and production process of RC and CFS storage container models.

Table 4.2-4 shows the concrete components. For each cask, we used high-quality concrete (W/C 50 %) where high-performance AE water reducing agent was used. We also used moderate-heat concrete for the suppression of cracks due to heat generation.

When placing the concrete, we did not make construction joints in the height direction, keeping high workability and finishing quality, and completed the construction as required in the basic design.

(2) Canister

Two kinds of canisters of different materials and structures were made. Twenty-one (21) PWR fuel assemblies

Table 4.2-3 Major specifications of concrete storage container model

Item		Specifications	
Outer diameter		ϕ 3940 mm	ϕ 3890 mm
Inner diameter		ϕ 1850 mm	ϕ 1832 mm
Total height		5787 mm	6030 mm
Lid	Metal thickness	50 mm	30 mm
	Concrete thickness	525 mm	520 mm
Body	Steel liner thickness	40 mm	25 mm
	Concrete thickness	980 mm	980 mm
	Outer steel plate thickness		12 mm
Bottom concrete thickness		460 mm	644 mm
Major material		Reinforced concrete	Concrete filled steel
Weight	Empty weight	About 146 t	About 157 t
	Weight with canister	About 181 t	About 187 t

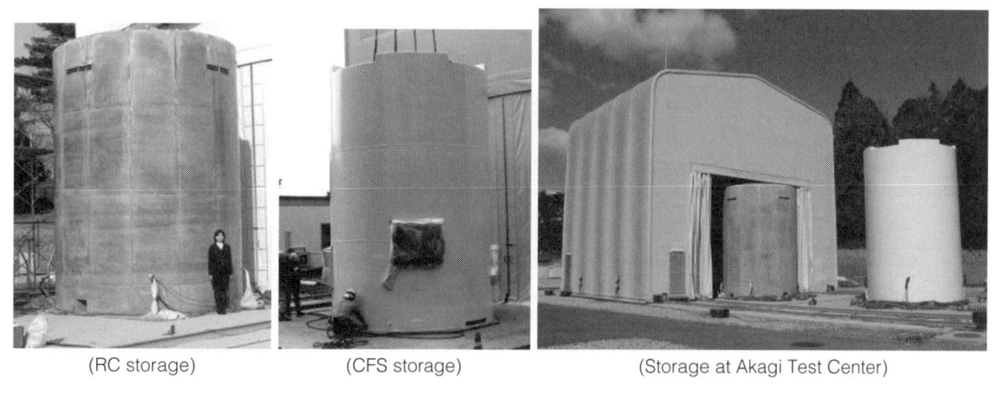

(RC storage) (CFS storage) (Storage at Akagi Test Center)

Fig. 4.2-4 Full-scale model of concrete storage container

Table 4.2-4 Concrete component of concrete storage containers

Container type	Design standard strength (N/mm²)	Water-cement ratio (%)	Fine aggregate ratio (%)	Cement (kg/m³)	Water (kg/m³)	Fine aggregate (kg/m³)	Coarse aggregate (kg/m³)	AE agent (g)	High-performance AE water reducing agent (g)	Slump (cm)	Air volume (%)
RC	30	48.0	49.3	355	170	870	992	13.2	3.20	21±1.5	4.5±1.5
CFS	24	49.5	51.2	344		909	960	1.38	3.10		

were stored in each canister, which had a safety function of criticality prevention or sealing during transport or storage period. The type I canister was made of stainless steel and a square tube called a guide tube was inserted to a space plate to fix the canister. An aluminum heat transfer plate was attached to the space plate to improve the heat removal performance. The canister body was made of highly corrosion-resistant super stainless steel (YUS270). On the other hand, the type II canister's basket was made of aluminum alloy with a lattice structure of hollow aluminum plates. The canister body was made of highly corrosion-resistant 2-phase stainless steel (SUS329J4L).

Fig. 4.2-5 shows basic structure of canisters of types I and II. Table 4.2-5 and Fig. 4.2-6 show major specifications and production process of the canisters.

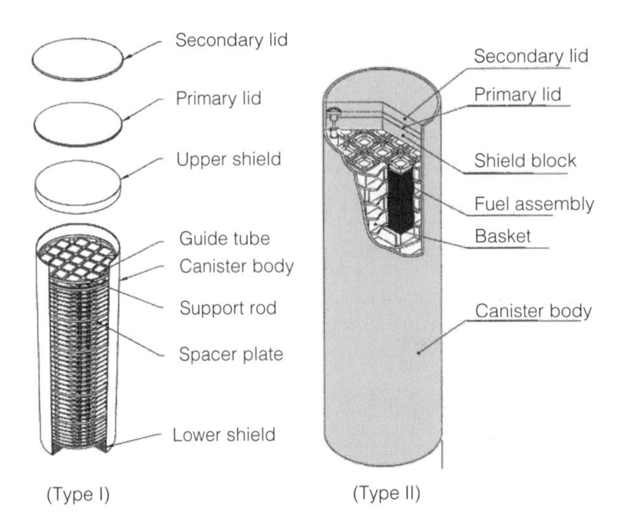

Fig. 4.2-5 Basic structure of canister

(Canister body of type II)

(Basket of type I)

(Basket of type II)

Fig. 4.2-6 Production process of canister

Table 4.2-5 Major specifications of canister

Canister	Type I	Type II
Total length	4630 mm	4470 mm
Outer diameter	1676 mm	1640 mm
Body plate thickness	16 mm	19 mm
Empty weight	19 ton	16 ton
Max weight	35 ton	30 ton
Material	Super stainless steel	2-phase stainless steel
Basket	Stainless steel + carbon steel	Aluminum

4.3 Heat removal performance

4.3.1 Fundamental test for natural convection

A concrete cask is considered to be very reliable and safe because it is cooled by an independent natural convection. This study is provided to establish a refined design method enabling the concrete cask to safely store high-power spent fuel from a nuclear power plant. Fig. 4.3.1-1 illustrates the concrete cask. This cask draws in ambient air from the inlet, and the air is used for cooling the canister which stores the spent nuclear fuel. Then, the heated air is ejected from the outlet. An annular cooling path is thus created between the outer side of the canister and the inner side of the concrete vessel. A shield plate to prevent radiation heat transfer from the canister to the concrete vessel is installed in the middle of the annular cooling path. It divides the cooling path into two paths,

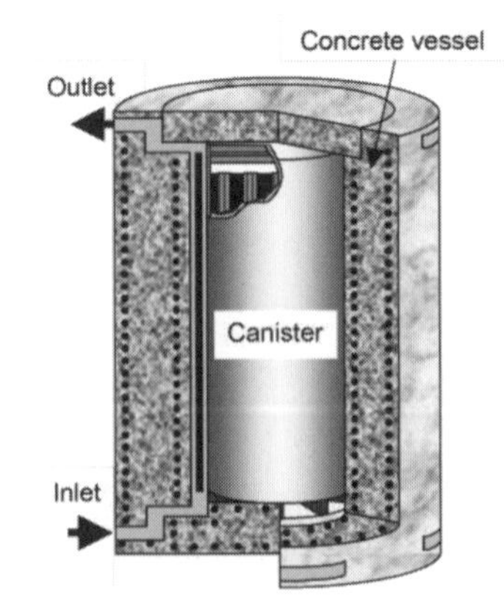

Fig. 4.3.1-1 Cutaway view of concrete cask

the thermal shield plate side and the liner side. Heat from the canister warms up inhaled air and gives buoyancy force to it. The air begins to flow independently from obtaining this buoyancy force. Flow rate of the air is determined by a balance between pressure loss in the concrete cask and buoyancy force given by effluent heat from the canister. The concrete cask is cooled by this kind of self-sustaining flow. Geometry of the cooling path is very simple. At first, estimation of its heat removal seems very easy. In reality, the exact evaluation of the flow is not so easy because of the characteristic features accompanying the buoyancy rising flow.

(1) Experimental apparatus

Fig. 4.3.1-2 is a schematic view of the experimental apparatus composed of a test section, an inlet flow regulator, forced circulation devices, instrument and a traverse mechanism of the test probe, etc. A thirty- degree sector of the annular flow path was chosen as the test section and simulated by a simple flow path with a rectangle shape. Diameter of the canister is about 1.7 m and its curvature is considered to be large enough for the thirty-degree sector model so as to be simulated by a rectangular shape. The test section thus forms a partial model of the concrete cask and its dimension is in full scale except for the circumferential direction.

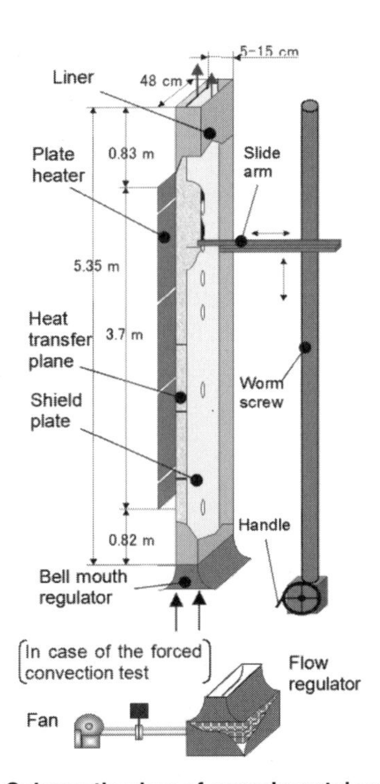

Fig. 4.3.1-2 Schematic view of experimental apparatus with changeable inlet and outlet structures

(2) Test results

1) Kind of flow

We examined what kind of flow occurred in the annular cooling path. Here we employed Reynolds number (Re) and Grashof number ($Gr_{\Delta T}$) as representative parameters on the similarity rule for the natural convection. By our experiments, those numbers could be estimated respectively as 4.9×10^3 and 1.1×10^7 for the design condition. Tanaka has proposed the following flow discrimination criteria to the buoyancy force for $Gr_{\Delta T}>3\times10^5$ (Tanaka et al., 1985)[1].

Natural convection regime:

$$\mathrm{Re} < 16.5 \cdot Gr_{\Delta T}^{8/21} \quad ..... (1)$$

Combined convection regime:

$$16.5 \cdot Gr_{\Delta T}^{8/21} \leq \mathrm{Re} \leq 50 \cdot Gr_{\Delta T}^{8/21} \quad ..... (2)$$

Forced convection regime:

$$\mathrm{Re} > 50 \cdot Gr_{\Delta T}^{8/21} \quad ..... (3)$$

The foregoing values of the Re and $Gr_{\Delta T}$ in our experiment satisfy Eq. (1). Therefore the flow in the concrete cask can be classified as a natural convection.

2) Heat transfer coefficient

Maximum temperature rises from the inlet air temperature at the heat transfer plane surface are plotted on Fig. 4.3.1-3. The highest temperatures were found at almost the same vertical point of the heat transfer plane. The flow rate of the forced convection experiment depicted here is in the range of 1.1 ~ 1.5 times the flow rate of the natural convection experiment. Most of the data obtained in experiments for different gap widths and two flow types are shown on one line in Fig. 4.3.1-3. Temperature rise of the heat transfer plane surface is directly concerned with the heat transfer coefficient. Hence it suggests that the heat transfer coefficient on the heat transfer plane surface might be influenced by the heat flux and not by the gap width, the flow rate nor the flow type. The flow could be clearly categorized as a free convection because of its recognized trend. A unique natural convection flow dependent on the heat flux might occur in the concrete cask.

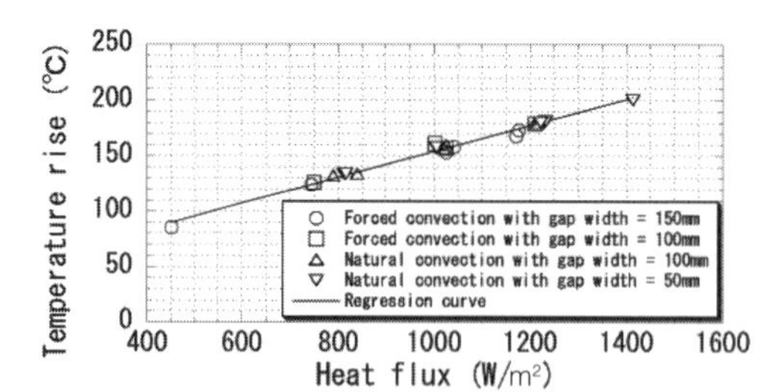

Fig. 4.3.1-3 Maximum canister surface temperature rise at the heat transfer plane surface vs. heat flux

By using surface temperature of the heating plane (T_w), inlet temperature (T_{in}) and heat flux (q), local heat transfer coefficient (h) was estimated by Eq. (4):

$$h = \frac{q}{T_w - T_{in}} \quad ..... (4)$$

Heat transfer coefficients for three different heat fluxes obtained in the study are shown in Fig. 4.3.1-4 together with predictions by Miyamoto's formula (Miyamoto et al., 1982)[2)]. Heat transfer coefficients are nearly constant except in the inlet region.

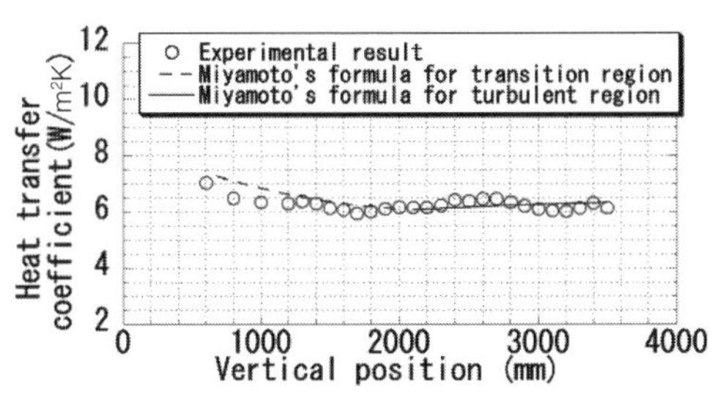

Fig. 4.3.1-4 Heat transfer coefficient vs. vertical position

Miyamoto's formulas are written as follows:

In transition turbulent region ($4\times10^{12}<Gr_z \cdot \mathrm{Pr}<1.5\times10^{13}$):

$$Nu_z = 0.724\cdot(Gr_z \cdot \mathrm{Pr})^{0.208} \quad \ldots\ldots (5)$$

In turbulent region ($1.5\times10^{13}<Gr_z \cdot \mathrm{Pr}<2\times10^{14}$):

$$Nu_z = 0.104\cdot(Gr_z \cdot \mathrm{Pr})^{0.272} \quad \ldots\ldots (6)$$

Our results well agree with Miyamoto's formulas that were obtained for a free convection along a vertical plane. It turned out that the heat transfer coefficient of the natural convection in the concrete cask could be estimated by a formula for free convection.

The flow in an annular flow path of the concrete cask seems to be categorized as a confined natural convection. But it was found that the flow could be treated as a free convection. However, the existing formula applied to a confined natural convection did not agree with the experimental result. The natural convection flow in the concrete cask is one-side heated and its gap width is large. Those features might make the flow similar to a free convection.

3) Friction loss coefficient in the heating zone

We considered the following force balance:

$$\Delta H = \Delta P \quad \ldots\ldots (7)$$

where ΔH is total buoyancy force and ΔP is total pressure loss. Buoyancy force is obtained by integrating air density along one natural convection circuit. But pressure losses, except friction loss in the heating zone, can be calculated by using a measured flow rate and hydraulic resistances given in the handbook. Here, we refer to the handbook issued by Idelchick (Idelchick, 1966)[3)].

Pressure losses are calculated at pressure loss parts drawn on Fig. 4.3.1-5. A single-channel path was used to simplify our estimation. Flow rate was estimated by traversing the hot wire anemometer at the inlet. Velocity at each pressure loss element could be predicted by the flow rate.

By applying the handbook to obtain pressure loss coefficients and using each flow velocity, pressure losses other than friction loss in the heating zone were calculated. Here, we compared it with the pressure loss coefficient of isothermal flow. Friction factor ratio given by the following equation is plotted in Fig. 4.3.1-6 and Fig. 4.3.1-7.

$$f = \xi_{non-isothermal} / \xi_{isothermal} \quad \ldots\ldots (8)$$

Two cases of gap width, 50 mm and 90 mm, were examined, where uniform heat fluxes were given on the heating surface. In their estimations, friction loss in the heating zone accounted for 45 ~ 50 % of the total pressure drop. The horizontal axis shows Gr/Re, a unique parameter applied to the natural convection.

As shown in Fig. 4.3.1-6 and Fig. 4.3.1-7, the friction loss coefficient of the heating surface in the annular cooling path could be expected to be 1.4 ~ 2.0 times the value of an isothermal flow for $2.8\times10^4 < Gr/Re < 3.8\times10^4$ and 2.5 ~ 1.5 times the value of an isothermal flow for $1.6\times10^5 < Gr/Re < 2.4\times10^5$.

The figures also compare experimental results with the value estimated by Polyakov's formula (Polyakov, 1979)[4)]. Polyakov proposed a simple relational expression, Eq. (9), to predict the friction loss coefficient for a vertical channel by theoretical study.

$$\xi = 120 \cdot \sqrt{Gr/(\Pr \cdot \mathrm{Re}^4)} \quad \ldots\ldots (9)$$

The experimental result well agrees with estimated value by Polyakov's formula. An original experiment on the friction loss coefficient of the buoyant force was carried out in the present study. By this study, previous theoretical study could also be validated.

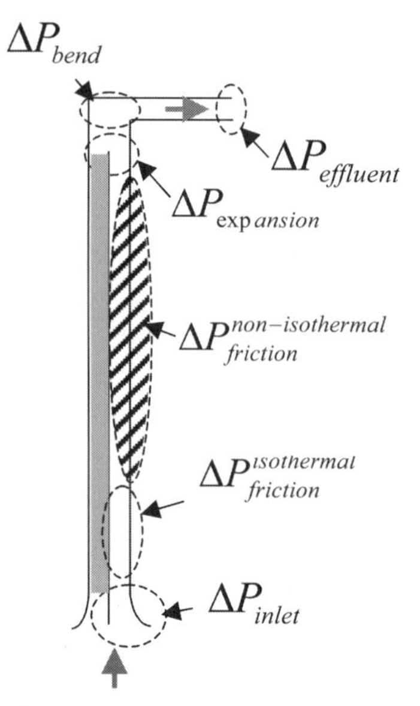

Fig. 4.3.1-5 Estimated pressure losses in the flow path

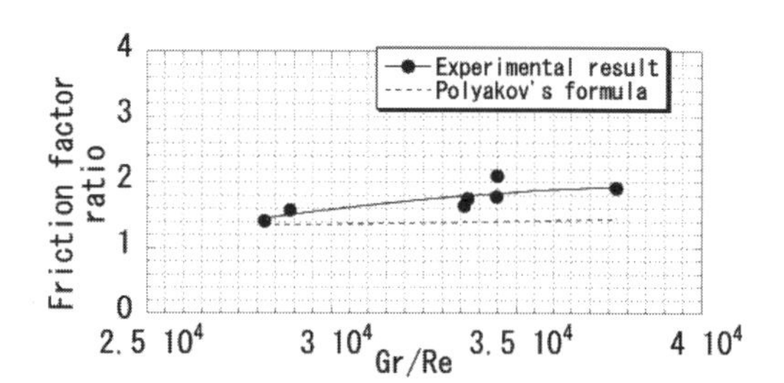

Fig. 4.3.1-6 Friction factor ratio ($\xi_{non-isothermal} / \xi_{isothermal}$) vs. Gr/Re in case of 50 mm gap

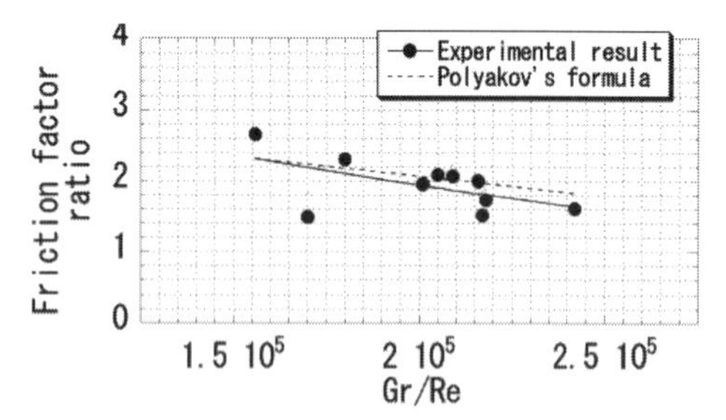

Fig. 4.3.1-7 Friction factor ratio ($\xi_{non-isothermal} / \xi_{isothermal}$) vs. Gr/Re in case of 100 mm gap

(3) Summary

Experiments of heat removal with a natural convection in an annular flow path of the concrete cask have been carried out to clarify its characteristic features. The following results were obtained by the study:

A natural convection other than a combined convection was found to occur in the annular cooling path of the concrete cask. The flow could be estimated as a type of free convection. The natural convection in the concrete cask is one-side heated and its gap width is large. Those features might make the flow similar to a free convection.

The heat transfer coefficient in the heating zone obtained in the experiment well agreed with the estimation by a formula proposed for a free convection. We recommend applying Miyamoto's formula to estimate the heat transfer coefficient in the annular cooling flow of the concrete cask.

Heat from the canister warms the air near its surface. This accelerates the rising speed of air in the vicinity of the wall. As a result, a skewed distribution of the rising speed and a large shear stress occur on the heating surface. This produces a larger friction loss than in an isothermal flow. The friction loss coefficient of the heating surface in the annular cooling path could be predicted to have 2 ~ 2.5 times the value of an isothermal flow for the rated condition.

Nomenclature

$Nu_z = qd/(\lambda \Delta T)$	Nusselt number (—)
Pr	Prandtl number (—)
$Gr_{\Delta T} = g\beta d^3 \Delta T/\nu^2$	Grashof number (—)
$Gr_z = g\beta q z^4/(\lambda \nu^2)$	Grashof number (—)
$Gr = g\beta d^4 q/(\lambda \nu^2)$	Grashof number (—)
$Re = vd/\nu$	Reynolds number (—)
$\Delta T = T_{air} - T_{in}$	Temperature rise of air from the inlet (K)
ΔH	Buoyancy force (Pa=kg m^{-1} s^{-2})
ΔP	Pressure loss (Pa=kg m^{-1} s^{-2})
L	Height of the outlet (m)
d	Equivalent length (m)
f	Friction factor ratio (—)
g	Gravity force (m s^{-1})
h	Heat transfer coefficient (W m^{-2} K^{-1})
q	Heat flux (W m^{-2})
z	Vertical position (m)
β	Thermal expansion coefficient (K^{-1})
λ	Thermal conductivity (W m^{-1} K^{-1})
ν	Kinematic viscosity (m^2 s^{-1})
ξ	Friction loss coefficient (—)

4.3.2 Heat removal verification tests using concrete casks at normal condition

Heat removal verification tests taking account of cask storage periods were performed using a full-scale RC cask and CFS cask, and steady state data were obtained. In the tests, decay heat was simulated by electric heaters. The decay heat at the initial stage of storage (0 year), middle stage (20 years) and final stage (40 years) was 22.6 kW, 16 kW, and 10 kW, respectively[5)].

(1) Test method

Fig. 4.3.2-1 shows a house where heat removal tests were performed. The house is movable and double-wall structure is used. Furthermore, the inner wall is made of an insulator. On the ceiling of the house, a ventilator that controls a flow rate of air in the house is installed. The heat removal tests of the casks were performed inside the house.

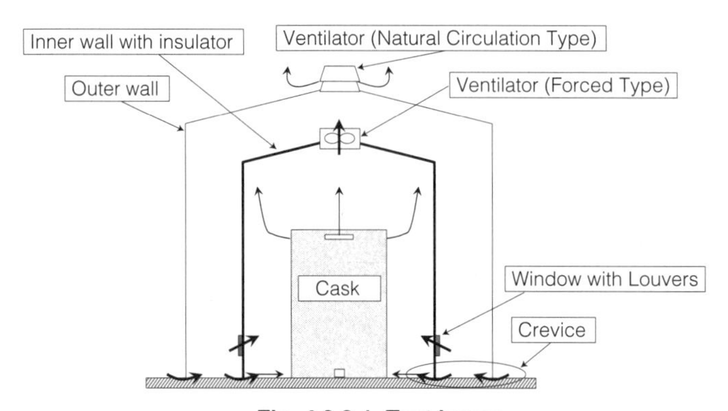

Fig. 4.3.2-1 Test house

Temperature measurement for the steady state data was carried out at midnight because the change of the outside temperature is small.

During the heat removal test, a ventilation flow rate was controlled so that a thermal stratification boundary in the house was kept above outlets of the cask.

It took about two weeks to reach a steady state for each test case.

(2) Test results of RC cask

1) Air flow rate

Four air inlets are provided at the lower part of a concrete container. For measurement of air velocity, a hot wire anemometer was used. Fig. 4.3.2-2 shows instruments of velocity measurement. A duct with a bell-mouth shape

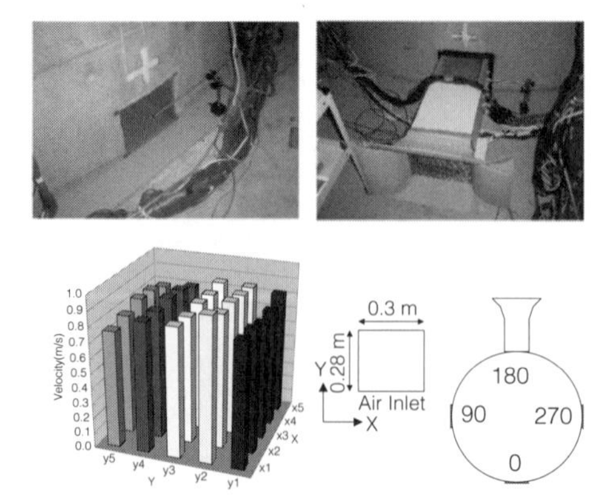

Fig. 4.3.2-2 Flow rate measurement at air inlet

was installed on the inlet in order to regulate three-dimensional flow into one-dimensional flow, and the velocity distribution in a cross section of the inlet could be measured.

The velocity distribution in the case of 22.6 kW is shown in this figure. There was not large velocity difference in the cross section. The mean velocity in the cross section was 0.837 m/s and the total flow rate was 0.281 m^3/s.

2) Velocity and temperature distributions at outlet

Fig. 4.3.2-3 shows the velocity at the outlet. In an outlet duct, steel plates for radiation shielding are placed and divide each outlet into four areas.

In order to obtain the heat rate conveyed by cooling air accurately, enthalpy at the inlet and outlet should be obtained. Then, the velocity and temperature distributions at the outlet were measured. Especially the velocity measurement for the outlet is difficult because air temperature at the outlet is high and fluctuating. A propeller meter was used for the velocity measurement at the outlet. The velocity was measured at the center of each of the four areas. The velocity is faster at the upper areas than at the lower areas, because air at outlet flows under the influence of buoyancy force.

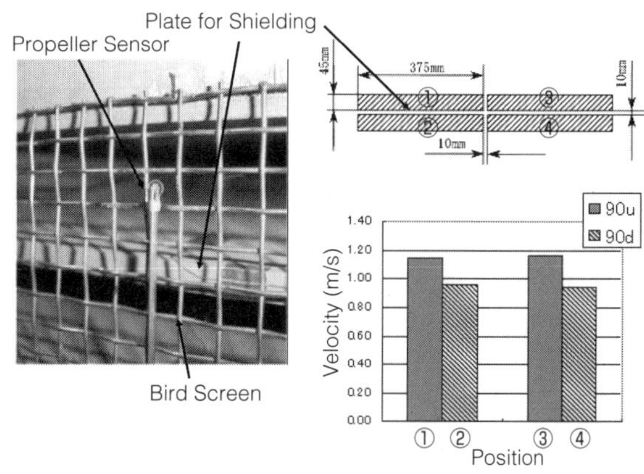

Fig. 4.3.2-3 Velocity measurement at air outlet

Fig. 4.3.2-4 shows the air temperature at the outlet. At the inside of the outlet, About 60 °C temperature difference was observed between the upper side and the under side.

The air temperature of the upper side depends on the hot air rising along the canister surface. The air temperature of the under side depends on cold air rising up through a flow space between the thermal shielding plate and the inside of the concrete cask. At the outside of the outlet, air temperature difference between the upper side and the under side was reduced to 5 °C, because hot air and cold air were mixed in the outlet duct.

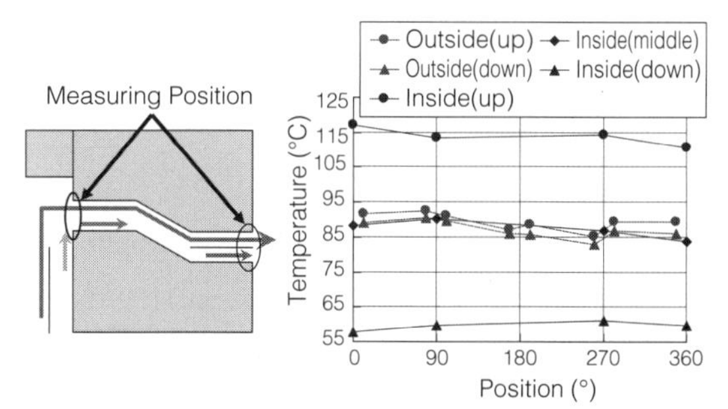

Fig. 4.3.2-4 Outlet air temperature

3) Heat balance

Heat discharged from the cask to the environment is distributed into heat convection of cooling air (Q1), heat transfer from the side of the concrete container (Q2), heat transfer from the top of the concrete container (Q3), and heat transfer from the bottom of the concrete container to the floor surface (Q4).

The heat removed ratio of the cask was calculated using test data. The amount of the heat convection by the cooling air was calculated using the enthalpy at the air inlet and outlet. The amount of the heat removed from the surface of the cask was calculated using temperature gradient in the concrete container. Fig. 4.3.2-5 shows the ratio of the heat transfer. It was found that 80 % of the total heat is removed by the cooling air.

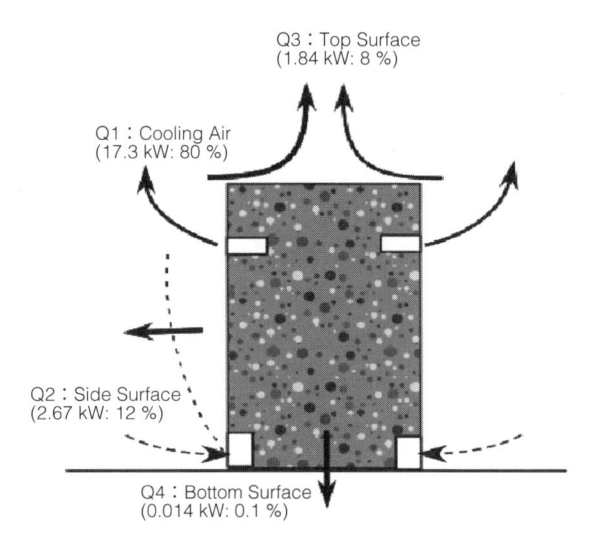

Fig. 4.3.2-5 Heat balance

4) Concrete temperature

Fig. 4.3.2-6 shows temperature distributions inside the concrete container under the condition of an inlet air temperature of 33 °C.

In the case of 22.6 kW, concrete temperature around the outlet was very high. The maximum concrete temperature was 91 °C. This value exceeds specified temperature of concrete (90 °C). Thus, it is necessary to modify cask design.

On the other hand, the maximum concrete temperature in the case of 10 kW is below 65 °C.

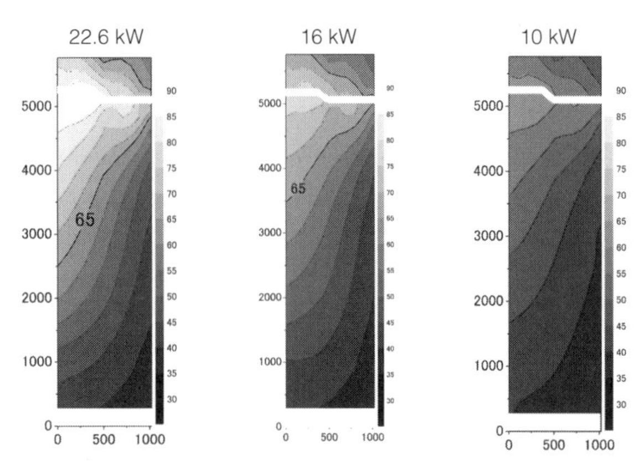

Fig. 4.3.2-6 Concrete temperature (RC cask)

5) Canister temperature

Fig. 4.3.2-7 shows surface temperature distributions in an axial direction of the canister. These depended on an angle in a circumferential direction of the canister. The reason of the temperature difference was the gap between the

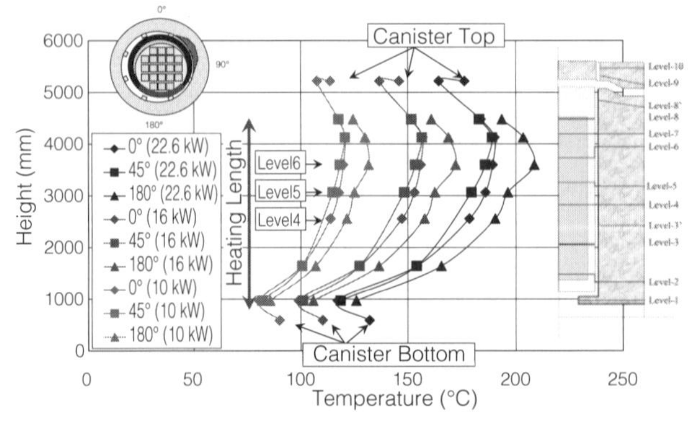

Fig. 4.3.2-7 Canister surface temperature (RC cask)

canister and a guide rail on a steel liner of the cask. At a 45 degree direction, the canister was contact with the guide rail and the heat of the canister was conducted from the canister surface to the guide rail. Then, the temperature at this direction decreased.

Furthermore, it was considered that a basket contacted with the canister. At the 180 degree direction, the basket contacted with the canister and the heat of the basket was conducted to the canister. It is necessary to take account of such contact conditions in the design and evaluation of the cask. In the final period of storage, when the heat rates is 10 kW, the area where the temperature of the canister surface does not reach 100 °C will appear. Thus, the point of view of stress corrosion cracking (SCC), it is important to evaluate the cold part of the canister surface.

Fig. 4.3.2-8 shows such a temperature distribution in a radial direction of the canister. Shape of the lines are almost symmetry. In the canister, the temperature distribution is not affected by the contact condition mentioned above.

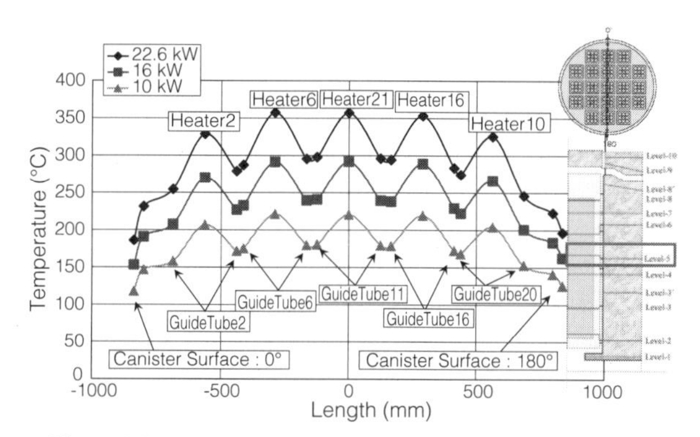

Fig. 4.3.2-8 Internal temperature of canister (RC cask)

6) Evaluation

Fig. 4.3.2-9 shows temperature distribution of the cask components in the cases of 22.6 kW and 10 kW.

Although temperatures of the heater and the canister surface differed largely, the difference in concrete surface temperature was few.

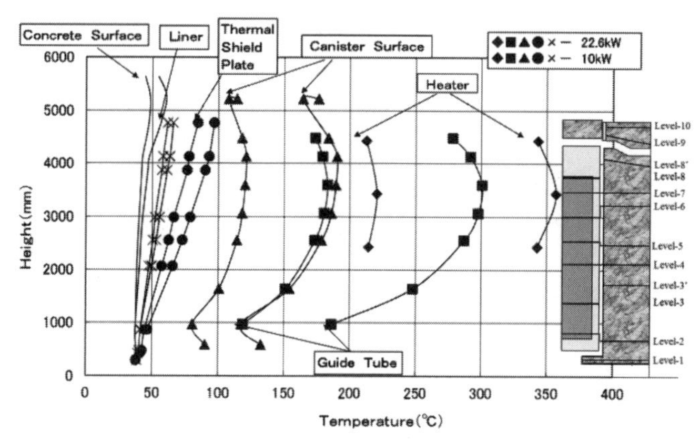

Fig. 4.3.2-9 Temperature distribution of each components of cask (RC cask)

Moreover, Table 4.3.2-1 shows the temperature and flow rates from the test results in various cases of decay heat.

Design modification is necessary because the area exceeding the specified temperature of concrete (90 °C) was

Table 4.3.2-1 Temperatures and flow rates (RC cask)

	22.6kW	16kW	10kW
Temp. Inlet Air(°C)	33	33	33
Temp. Concrete Container (°C)	Max. 91	Max. 78	Max. 65
Temp. Canister Surface (°C)	Max. 209	Max. 171	Max. 132
	Min. 89	Min. 77	Min. 66
Temp. Guide Tube (°C)	Max. 301	Max. 243	Max. 183
Temp. Increase of Coolind Air (°C)	65	51	36
Flow Rate (kg/s)	0.335	0.363	0.271

observed inside the outlet in the case of 22.6 kW.

Under the condition of final period of the storage (10 kW), the area on the canister surface, which did not reach 100 °C, appeared. From the point of view of SCC, it is important to evaluate such a cold area of the canister surface.

(3) Test results of CFS cask

1) Cask temperature

Fig. 4.3.2-10 shows temperature distributions inside the concrete container under the condition of an inlet of 33 °C. In the case of 22.6 kW, the maximum concrete temperature was under the specified temperature of concrete. The maximum temperature in the case of 10 kW was below 65 °C.

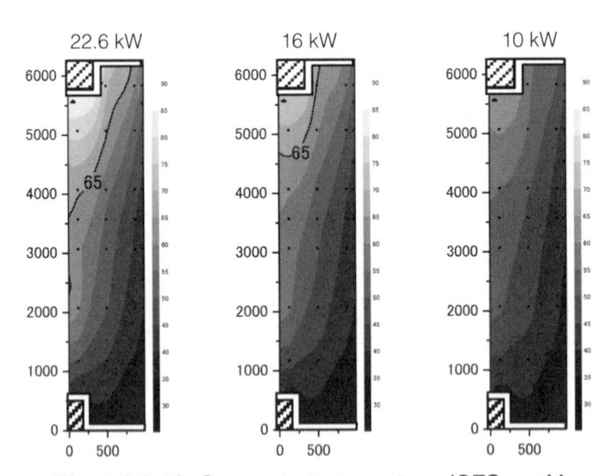

Fig. 4.3.2-10 Concrete temperature (CFS cask)

2) Canister temperature

Fig. 4.3.2-11 shows the vertical temperature distribution in the axial direction of the canister surface.

The temperature of the canister bottom was high. It was caused by a flow channel shape which has a tendency to stagnate air flow at the canister bottom. In the case of 10 kW, the area where the temperature was below 100 °C was observed although it was small. Fig. 4.3.2-12 shows temperature distribution in the radial direction of the canister.

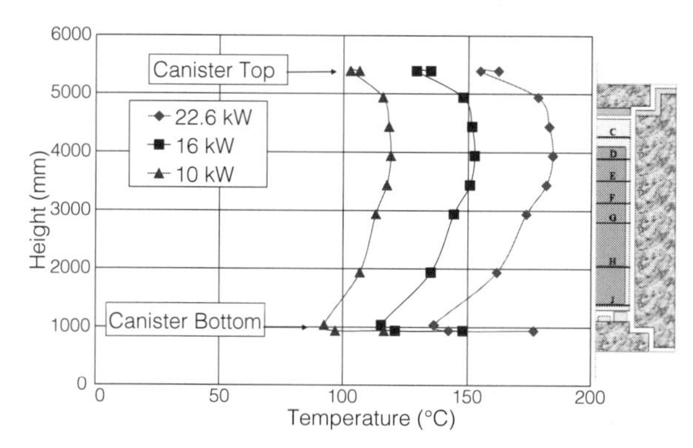

Fig. 4.3.2-11 Canister surface temperature (CFS cask)

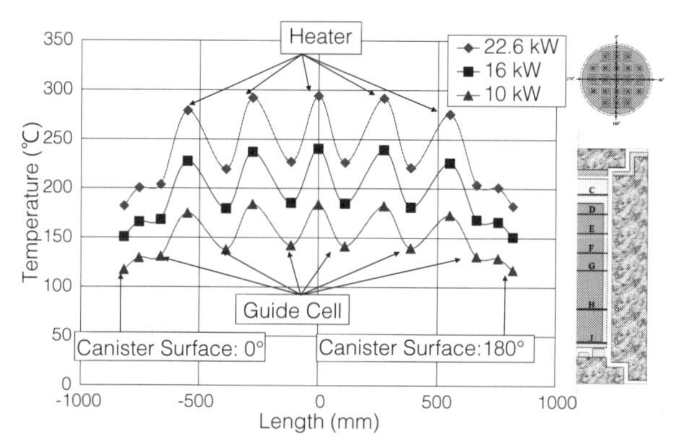

Fig. 4.3.2-12 Internal temperature of canister (CFS cask)

3) Evaluation

Table 4.3.2-2 shows the temperature and flow rates obtained from the test results in various cases of decay heat. In the case of 22.6 kW, the maximum concrete temperature was under the specified temperature of concrete (90 °C), and the guide tube temperature was lower than that of small RC cask.

Table 4.3.2-2 Temperatures and flow rates (CFS cask)

	22.6kW	16kW	10kW
Temp. Inlet Air(°C)	33	33	33
Temp. Concrete Container (°C)	Max. 83	Max. 74	Max. 63
Temp. Canister Surface (°C)	Max. 192	Max. 158	Max. 123
	Min. 123	Min. 106	Min. 85
Temp. Guide Tube (°C)	Max. 228	Max. 186	Max. 143
Temp. Increase of Coolind Air (°C)	52	42	30
Flow Rate (kg/s)	0.363	0.385	0.344

(4) Comparison between RC cask and CFS cask

Fig. 4.3.2-13 shows axial direction temperature distribution of the canister surface, and Fig. 4.3.2-14 shows axial direction temperature distribution of the guide tube.

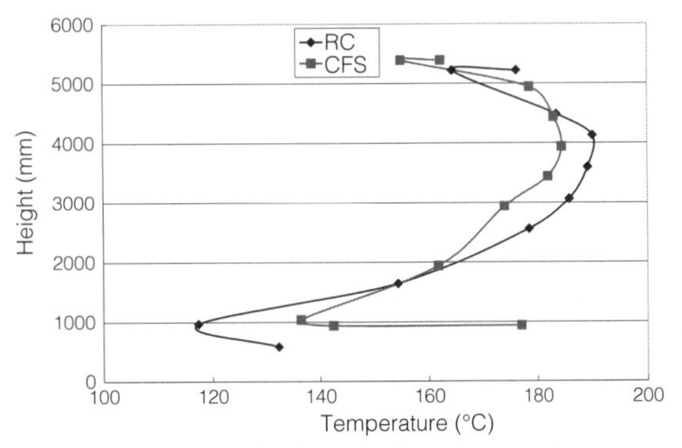

Fig. 4.3.2-13 Canister surface temperature (RC cask & CFS cask)

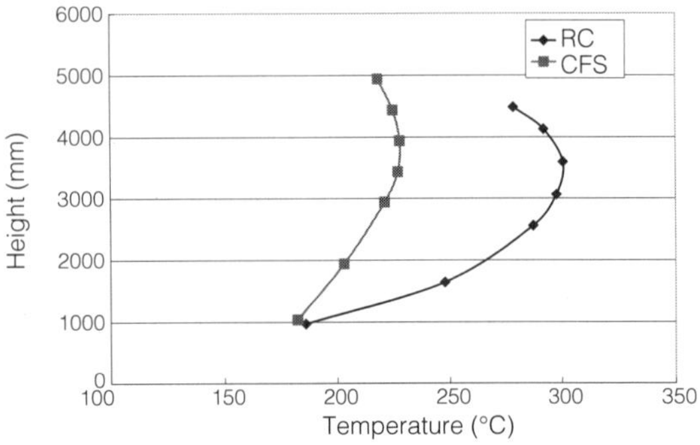

Fig. 4.3.2-14 Guide tube temperature (RC cask & CFS cask)

Especially, the temperature of the canister bottom of the RC cask was lower than that of the CFS cask. The reason is the difference in flow channel shape. In the RC cask, the air coming through the inlet collides with the bottom of the canister directly, and the bottom is cooled. The guide tube temperature of the CFS cask (Type II canister) was 73 °C lower than that of the RC cask (Type I canister), because an aluminum alloy is used for the basket in the Type II canister.

Fig. 4.3.2-15 shows axial direction temperature distributions of the steel liner installed in the concrete container.

In the CFS cask, the thermal shielding plate is installed at only the upper part of an annulus gap, so that the lower liner is heated by radiation. On the other hand, the upper liner is protected from radiation.

Fig. 4.3.2-16 shows axial direction temperature distributions of the concrete surface.

In the RC cask, the outlet is located at a part lower than the top of the cask, so that the temperature around the outlet is higher than the other parts.

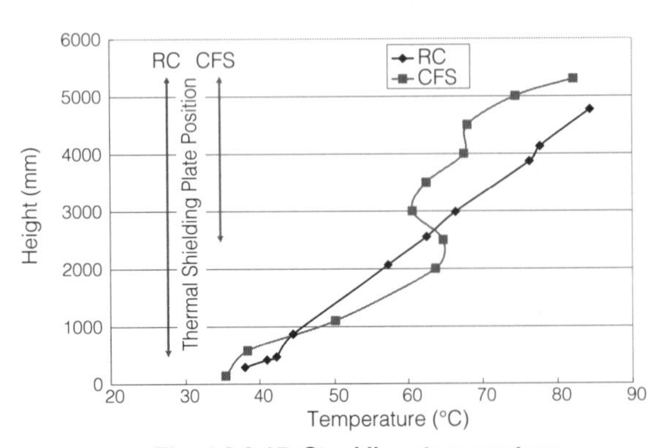

Fig. 4.3.2-15 Steel liner temperature (RC cask & CFS cask)

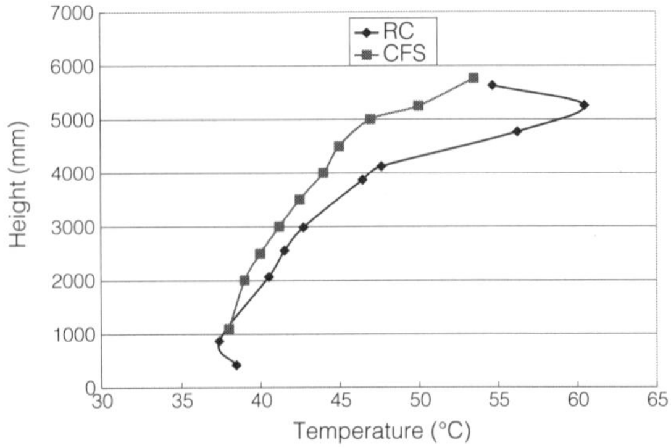

Fig. 4.3.2-16 Concrete surface temperature (RC cask & CFS cask)

(5) Summary

The heat removal verification tests were performed using the two types of full-scale concrete casks. In the tests, the heat rate was selected as a test parameter in order to simulate the condition of the concrete casks during the various storage periods.

The quantitative data were obtained for safety evaluation. Furthermore, the data of heat discharged from the cask to the environment were obtained.

The design modification is necessary for the RC cask, because the area exceeding the specified temperature of concrete (90 °C) was observed inside the outlet under the condition of the initial storage period.

The concrete cask has such a characteristic cooling system that the air goes through the annulus gap between the concrete container liner and the canister surface. From the point of view of the heat removal, the design of the flow channel is important for the concrete cask.

Finally, it was found that the aluminum alloy is thermally effective for the basket of the canister.

4.3.3 Heat removal analysis at normal condition

For interim storage of spent fuel, metal casks have been used mainly and number of examples of heat removal analyses including experimental data have been published [6)-11)]. On the other hand, concrete casks have been used mainly in USA, and their experimental data and verification of their analytical methods might not be enough[12)-14)].

For analytical evaluation by numerical calculation, there is an evaluation method by thermal hydraulic analysis and the other method by coupling a thermal hydraulic analysis for the cooling air and a heat conduction analysis for the solid parts (inside the canister and cask body). For the evaluation by the thermohydraulic analysis, the amount of the cooling air is obtained by the balance of the buoyancy force and the resistance of flow path, and a heat transfer boundary will be assumed at a location of heat transfer with air. The thermal hydraulic evaluation of natural convection needs verification by experimental data, but the flow rate in the concrete cask is relative slow and hard to coincide with the experimental data. Therefore, it is necessary to obtain experimental data.

We carried out heat removal demonstration tests using full-scale concrete casks, and established and verified the analytical method based on the experimental data[15),16)]. More detail information is found in the literature[16)].

(1) Analytical method

1) Object of analysis

The objects of the analysis are the two kinds of concrete casks used for the demonstration tests as described in the previous section. Fig. 4.3.3-1 shows the analysis area. Because geometry is symmetrical and it is necessary to evaluate air inlet blockage event in which two inlets among four are blocked, analysis area was 1/4 sector of the cask.

2) Analytical model and method

The analysis model is composed of two parts. One is inside the canister and another is outside the canister. The canister has a cylindrical shape with a basket of lattice structure. Because its shape is important for the heat removal evaluation, we employed x-y-z coordinate.

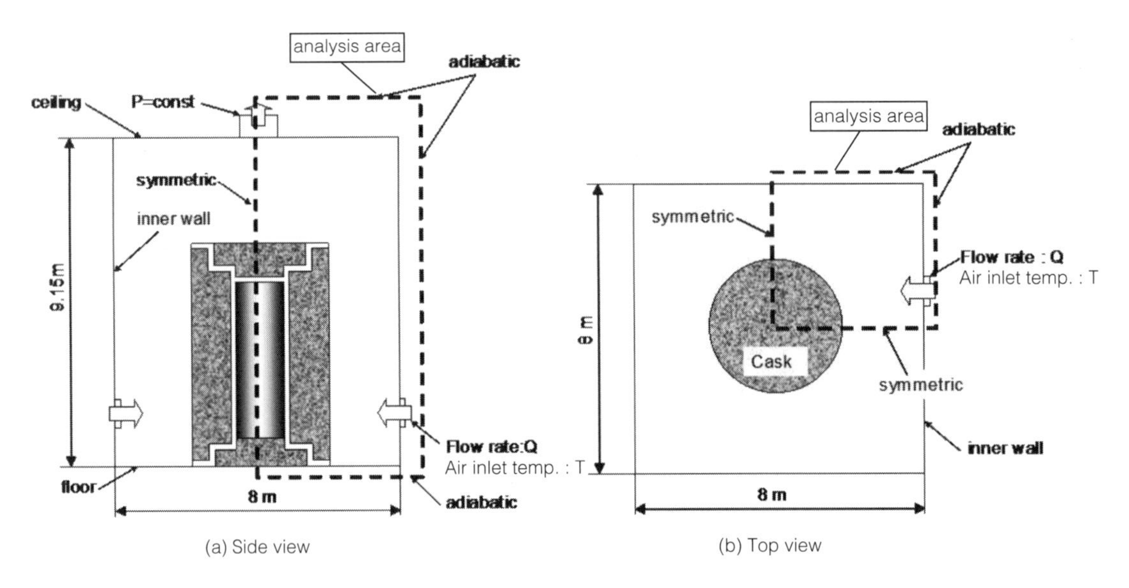

(a) Side view

(b) Top view

Fig. 4.3.3-1 Thermal analysis area

On the other hand, for the outside the canister the natural convection of the cooling air is important and the flow path surrounded by the canister and the liner is represented by mesh model of r-θ-z coordinate. However, the air inlets composed of rectangular ducts, which cannot represented by the r-θ-z coordinate and a boundary-fitted coordinate (BFC) was employed. Analysis method is the same for both types of casks. In this section, analysis method and results for CFS cask are described .

a. Inside the canister

In the canister, helium gas is filled. Helium is inert gas that has good heat conductivity. Supposing that the influence of natural convection on temperature distribution is small, heat conduction analysis is adopted for this area.

Taking account of radiation effect, equivalent value of heat conductivity is used for helium. The heat radiation was replaced by the equivalent heat conductivity rate in the calculation of this section. The heat transfer rate of helium gas was represented by the heat conductivity with the equivalent heat conductivity for radiation[17),18)].

Analytical code for the inside the canister was FIT-3D. The FIT-3D code is thermal hydraulic analysis code but, in this section, function of only heat conduction and radiation were used.

b. Outside the canister

Outside the canister, natural convection of the air transfers the heat between the cooling air path and the outer surface of the canister. Inside the concrete cask body, heat conduction delivers the heat from inside to outside. These heat transfer by the air flow and the heat conduction in the cask were analyzed by one model. The analytical code was PHOENICS, a general thermal hydraulic analysis code. Time differential was based on semi-implicit method (SIMPLEST).

3) Conditions for analysis

The analysis was carried out for the normal condition of the CFS cask. Table 4.3.3-1 shows the conditions for analysis. The analysis was steady state analysis.

Table 4.3.3-1 Conditions for analysis

Concrete cask for analysis	CFS cask
Heater power (kW)	22.6
Temperature of inlet air (°C)	27.5
Flow rate at the inlet of the tent (kg/s)	1.086

(2) Result of analysis

The overall analysis with the models for inside and outside the canister reproduced the image of flow rate vector in a storage building showing a high temperature air climbing from the air outlet by its buoyancy force.

Temperature in the cooling air path inside the cask was relatively higher on the canister surface and the flow rate due to natural convection was high. Although most of the air flows near the canister surface, a weak air flows outside the thermal shield plate in the cooling air path. There was no circular flow inside the cooling air path.

In the following, comparison between analytical results and experimental results. Fig. 4.3.3-2 shows axial temperature distribution outside the canister.

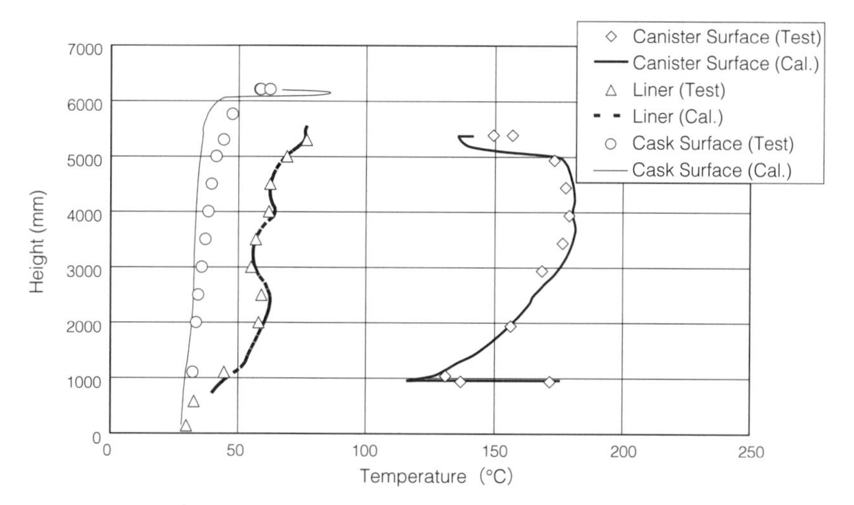

Fig. 4.3.3-2 Axial temperature distribution outside canister

Temperature distributions in the axial direction are shown in Fig. 4.3.3-2. Canister surface temperature is in good agreement with test results except for the top area. In the top area, the calculation results are lower than the test results. In the calculation, convection effect in the cavity between the canister top and cask lid is overestimated. There is an inflection point (at 2.5 m from the bottom) in the liner temperature. It is caused by the thermal shielding plate. The plate exists from the middle (at 2.5 m from the bottom) to the top and temperature decreases at the start point of the plate. The plate cuts off the direct heat of radiation. The model simulates this effect.

Concerning the cask surface temperature at the top, the calculation results are higher than the test results. The calculated cooling air flow in the cask was 0.326 kg/s, which was lower than the test result 0.363 kg/s. Then, the calculated outlet air temperature becomes higher. In fact, the temperature difference between air inlet and outlet was 52 °C in the calculation and 56 °C in the analysis.

Fig. 4.3.3-3 shows temperature of the basket. The calculation results were lower than the test results. The tendency of temperature distribution in the axial direction was similar. The reason why the calculation results were lower would be that the canister temperature was estimated lower in the calculation than the test results, which affected the calculated temperature of the basket.

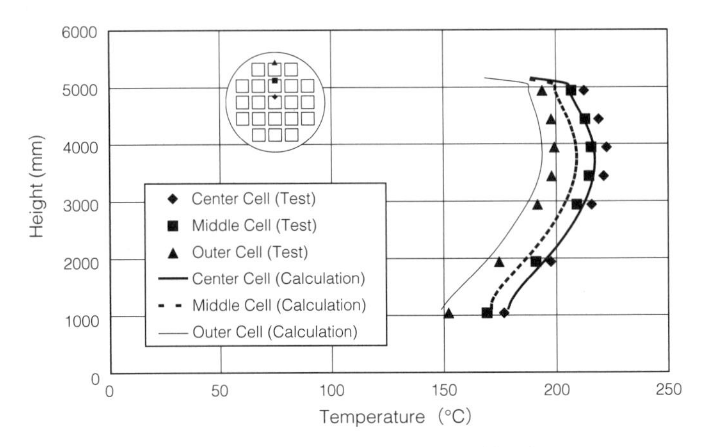

Fig. 4.3.3-3 Temperature of basket

Fig. 4.3.3-4 shows calculated heat removal allocation at surfaces of canister and cask. For CFS cask, about 80 % of the heat inside the canister was removed by the cooling air to the outside.

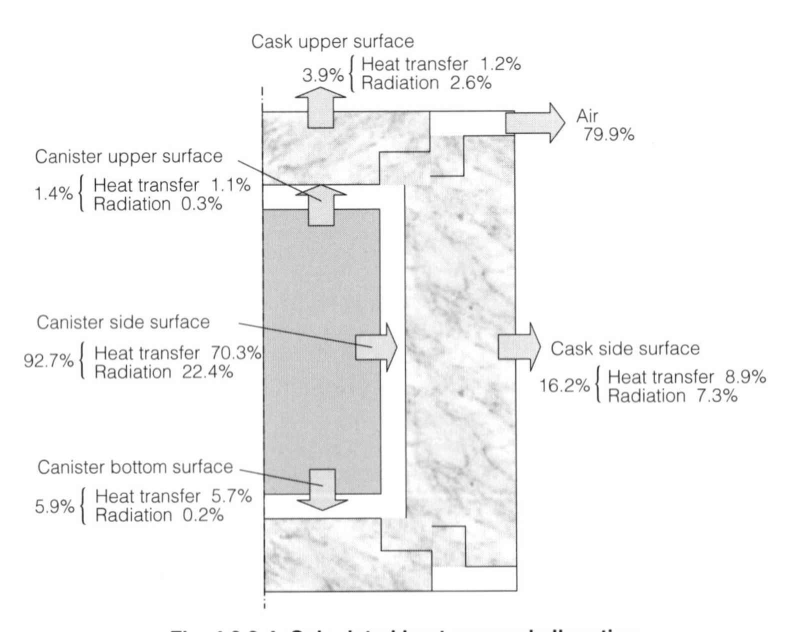

Fig. 4.3.3-4 Calculated heat removal allocation

(3) Summary

1) The analysis with the independent canister model showed a good agreement with the test results. However, conservative evaluation is necessary based on conservative assumption or experimental data for the gaps and contacts between the canister inner surface and basket, etc.

2) An overall analysis linking models inside and outside the canister was carried out and the results were compared with the test results. The differences were within about 10 % for the amount of air flow and about 7 % for the maximum temperature inside the canister. From these results, the adequacy of the analytical method was verified. It should be noted, however, the analysis did not always give conservative result and safety margin should be carefully considered.

4.3.4 Heat removal tests under abnormal conditions

Following the heat removal test at normal condition, we carried out heat removal tests under abnormal conditions in order to find heat removal characteristics under the abnormal conditions.

Conventionally, heat removal characteristics have been evaluated mainly by analysis and little test data have been available[19),20)]. Concrete cask storage is a method to remove heat by natural convection and verification of the analysis required by test data. Particularly, events such as blockage of air inlet are hardly evaluated only by analysis.

We conducted tests using full-scale concrete casks (RC type cask and CFS type cask) for the following abnormal conditions[21)].

a. Air inlets are blocked by 50 % (two out of four inlets are blocked)

b. Air inlets are blocked by 100 % (all of four inlets are blocked)

(1) Test method

The casks and equipment for the tests are as shown in the section 4.2.2. Heat power inside the cask was 22.6 kW as designed. The blockage events will be detected by daily patrol, thereby 24 hours will be the maximum period of the event. In fact, temperature limit of the concrete cask for short term is assumed 24 hours in the code by JSME[22)]. The current tests were continued for 48 hours to obtain longer period of time[21)].

The heat removal tests were performed changing parameters. The test parameters are closure rate of air inlet. For 50 % closure test of CFS cask, there are two patterns of inlet closure. In one case, inlets at angle positions (see Fig.

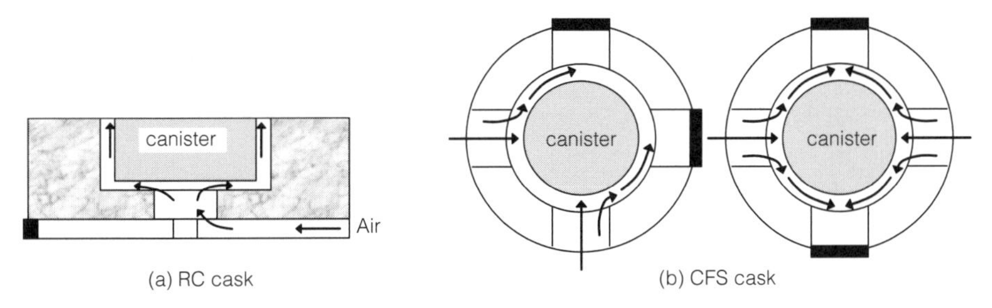

(a) RC cask (b) CFS cask

Fig. 4.3.4-1 Air flow patterns of different air inlet shapes under 50 % blockage

4.3.4-1(b)) of 0° and 90° are closed. In another case, those at 0° and 180° are closed. For RC cask, it is prospected that influence of closure pattern on temperature distribution is negligible. This is because, in case of RC cask, inlet air goes into cavity below the canister and is distributed uniformly to the side area of the canister. Table 4.3.4-1 shows the test cases.

Table 4.3.4-1 Test cases

No.	Cask Type	Cask Orientation	Cavity Gas	Total Heat Power	Closure rate of the air inlet	Situation
1	RC	Vertical	He	22.6 kW	50 %*1	Steady state
2	RC				100%	Transient
3	CFS				50 %*1	Steady state
4	CFS				50 %*2	Steady state
5	CFS				100 %	Transient

*1: Inlet of 0° and 90° were closed. *2: Inlet of 0° and 180° were closed.

(2) Test results

1) Blockage of 50 % condition

a. Change of temperature and air flow rate

Table 4.3.4-2 shows all the results of the tests. Maximum temperature increase is about 5 °C for RC cask and 14 °C for CFS cask due to the 50 % blockage. When two air inlets among four are closed, decreased rates in centigrade temperature of air flow are about 5 % for RC cask and 22 % for CFS cask. This difference means that increase of flow resistance in case of 50 % blockage for CFS cask is bigger than that for RC cask. It is concluded that the influence of the 50 % blockage on the temperature is small and the safety under 50 % blockage condition is verified.

Table 4.3.4-2 Temperature and flow rate at 50 % blockage condition

Cask type	RC	CFS	
Closure angle (°)	0,90	0,90	0,180
Concrete body (°C)	96 (+5)	93 (+10)	94 (+11)
Canister surface (°C)	214 (+5)	200 (+8)	199 (+7)
Center cell of basket (°C)	306 (+5)	235 (+7)	235 (+7)
Temperature difference of air between inlet and outlet (°C)	70 (+5)	66 (+14)	66 (+14)
Flow rate (kg/s)	0.321 (-4.2%)	0.280 (-22.9%)	0.290 (-20.1%)

Note( ): Difference between 50% blockage and normal condition

b. Temperature of concrete body

Figs. 4.3.4-2 and 4.3.4-3 show isothermal image of the concrete bodies. It can be seen that temperature increased at around air outlet by the 50 % blockage of the air inlet as compared with the temperature profile at the normal condition.

2) Blockage of 100 % condition

a. Result for RC type cask

For RC cask, it is observed that cooling air does not go out from the outlet. Fig. 4.3.4-4 shows the flow pattern. Fig. 4.3.4-5 and 4.3.4-6 show the temperature distribution. Temperature of the canister surface increases

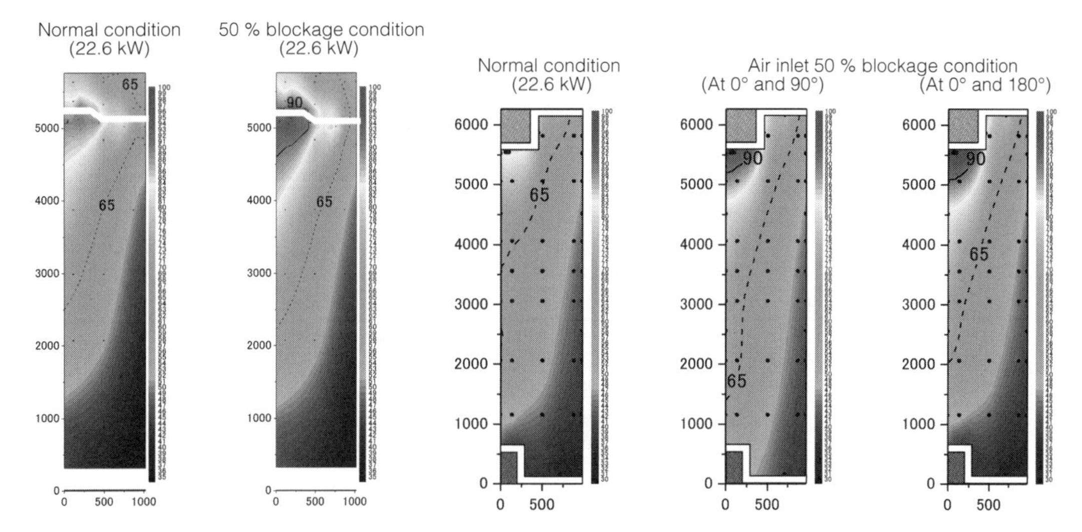

Fig. 4.3.4-2 Temperature distribution of concrete body (Cross section of RC cask at 90 °)

Fig. 4.3.4-3 Temperature distribution of concrete body (Cross section of CFS cask at 90 °)

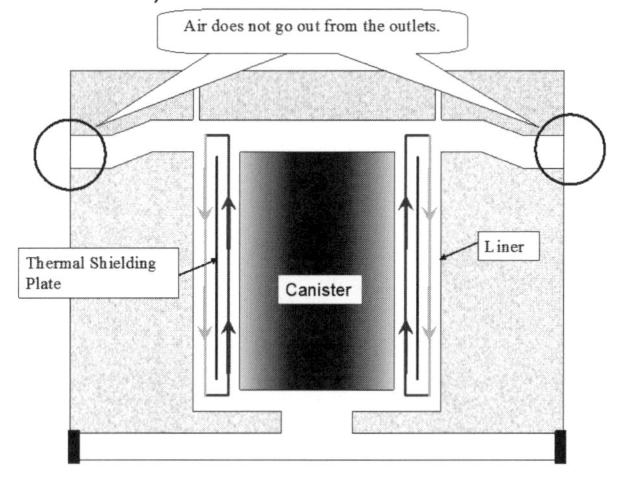

Fig. 4.3.4-4 Air flow at 100 % blockage condition (RC cask)

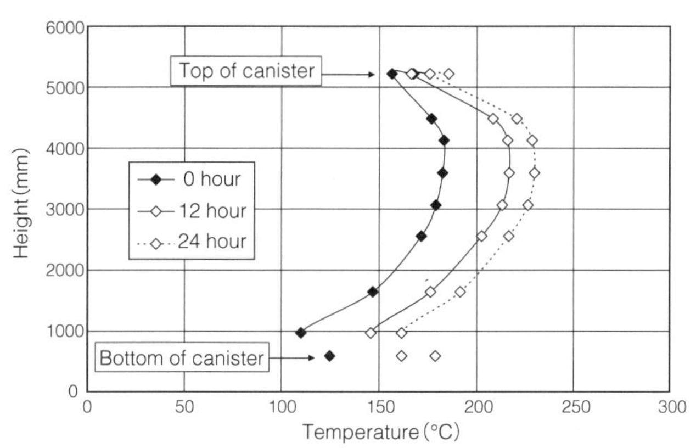

Fig. 4.3.4-5 Temperature distribution in axial direction (Canister surface, RC cask, 100 % blockage)

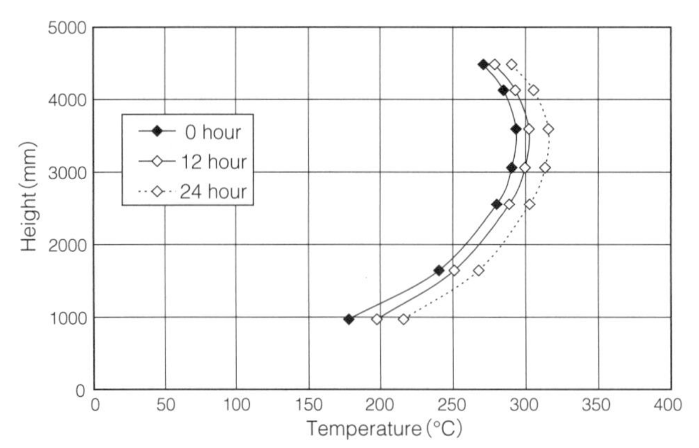

Fig. 4.3.4-6 Temperature distribution in axial direction (Center cell of basket, RC cask, 100 % blockage)

uniformly at all angle positions. This result shows that flow pattern near the canister surface does not change after inlets are closed. Judging from temperature distribution, cooling air circulates in the area between canister surface and liner (see Fig. 4.3.4-4). The reason why air does not go out is the shape of outlet duct. Its shape is down step. Hot air is not likely to go downwards because of buoyancy force.

Fig. 4.3.4-7 shows changes of the isothermal images of the concrete body with time, which show expansion of the high temperature region.

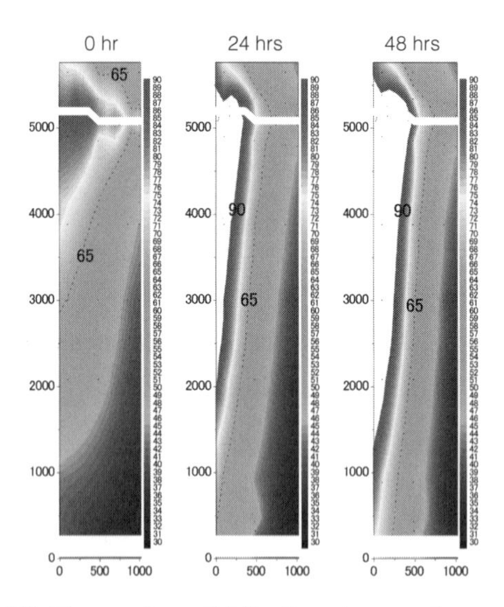

Fig. 4.3.4-7 Temperature distribution change of concrete body (Cross section of RC cask at 90 °)

b. Result for CFS type cask

For CFS cask, outflow and inflow are observed at the outlet. Fig. 4.3.4-8 shows the flow pattern where cooling air goes out from the outlet at angle positions of 90° and 180° while it goes in at the angle position of

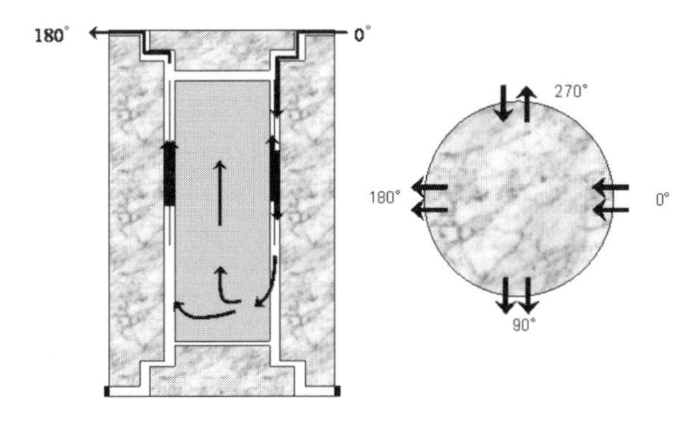

Fig. 4.3.4-8 Air flow with 100 % blockage at air inlet (CFS cask)

0°. At an angle position of 270°, in the half part of outlet, it goes out and in another half part, it goes in. The outlet duct of CFS cask is divided into right and left part by a steel plate. The shape of outlet is up-step for CFS cask. It is easily to go out for hot air. During the test period, this flow pattern did not change. Figs. 4.3.4-9 and 4.3.4-10 show the temperature distribution. It is observed that temperature of canister surface at the angle position of 0° is affected by cool air which comes in through the outlet. From these results, it is concluded that the shape of air outlet affects the flow pattern during the 100 % inlet blockage accident.

Fig. 4.3.4-11 shows the isothermal image of the concrete body in the 90°direction. Although the test continued for 48 hours, some data were missing. Fig. 4.3.4-11 show the image after 42 hours, which showed expansion of the high temperature region.

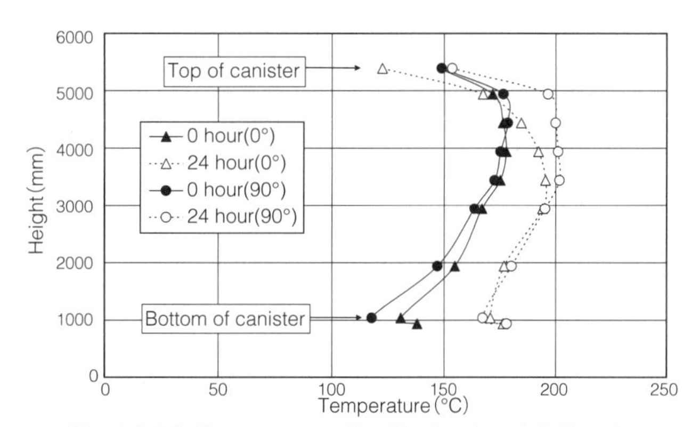

Fig. 4.3.4-9 Temperature distribution in axial direction (Canister surface, CFS cask, 100 % blockage)

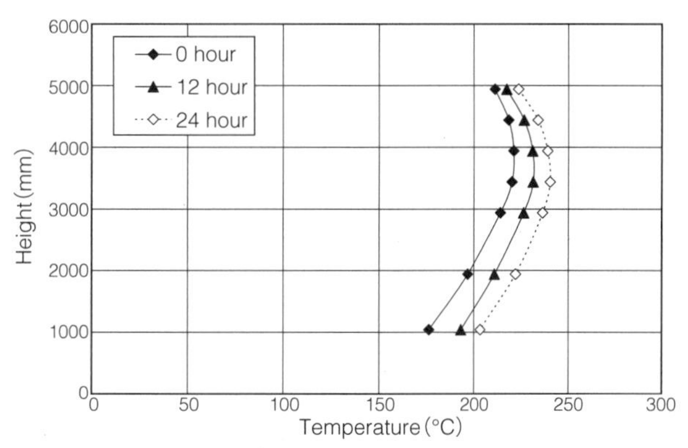

Fig. 4.3.4-10 Temperature distribution of axial direction (Center cell of basket, CFS cask, 100 % blockage)

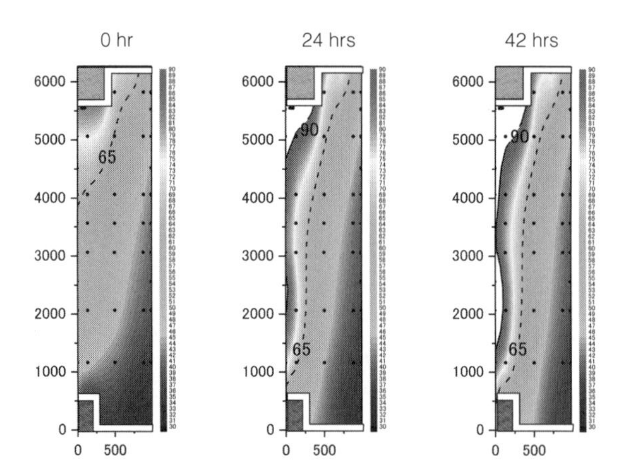

Fig. 4.3.4-11 Temperature distribution change of concrete body (Cross section of CFS cask at 90 °)

c. Results of temperature changes for RC and CFS casks

Table 4.3.4-3 shows maximum temperature of the casks after 24 hours from the beginning of the 100 % blockage test. Temperature increase of RC cask is bigger than that of CFS cask. For RC cask, heat energy accumulates inside and is transferred slowly by heat conduction. On the other hand, for CFS cask, cooling air coming in through the outlets removes heat energy and goes out from the other outlets.

Table 4.3.4-3 Maximum Temperature at 100 % blockage of air inlet after 24 hours

	RC	CFS
Concrete body (°C)	113 (+25)	101 (+19)
Canister surface (°C)	250 (+44)	209 (+18)
Center cell of basket (°C)	321 (+23)	247 (+19)

Note ( ): Difference between 100 % blockage and normal condition

(3) Summary

For 50 % blockage of air inlet, temperature increase in the concrete cask is small. When two air inlets among four are closed, decreased rates of air flow are about 5 % for RC cask and 22 % for CFS cask. The safety under 50 % blockage condition is verified.

For 100 % blockage of air inlet, different types of flow pattern were observed according to the shape of outlet. In case of outlet with down step shape, air does not go out from the outlet. On the other hand, in case of outlet with up-step shape, air goes out and in from different outlets. In the former case, temperature increase is bigger than that in the latter case. If air does not go out from the outlet, heat energy accumulates inside and temperature continues to increase.

These data are useful for thermal evaluation under accident conditions.

References

1) H. Tanaka, et al.: "Study on heat transfer of forced vertical pipe flow and combined natural convection", Proc. 22nd National Heat Transfer Symposium, pp.422-424 (1985).

2) M. Miyamoto, et al.: "Development of turbulence characteristics in a vertical free convection boundary layer", Proc. 7th Int. Heat Trans. Conf. 2, 323-328 (1982).

3) Idelchick, I.E., 1996. Handbook of hydraulic resistance, Begell House, New York.

4) A. F. Polyakov: "Turbulent force flow and heat exchange in vertical channels in conditions of free convection", J. Eng. Phys. 3.2.5, 801-811 (1979).

5) H. Takeda, et al.: "Heat removal verification Tests using concrete casks under normal condition", Nucl. Eng. Design, Vol. 238, pp. 1196-1205 (2008).

6) H. Yamakawa et al.: " Demonstration test for a shipping cask transporting high burn-up spent fuels - Thermal test and analyses -", Proceedings of PATRAM'98, Vol.2, pp. 659-666, 1998.

7) M. Greiner et al.: " Response of a spent fuel nuclear fuel transportation package to regulatory format thermal events", Proceedings of PATRAM'95, Vol.2, pp.664-671, 1985.

8) D. J. Burt et al.: " Ullage temperatures in 'wet' spent fuel transport flasks", Int. J. of Radioactive Materials Transport, Vol.13, No.3.3.1, pp.263-268, 2002.

9) H. Yamakawa, Y. Gomi, S. Ozaki, and A. Kosaki: "Establishment of Cask Storage Method for Storing Spent Fuel - Evaluation on Heat Transmission Characteristic of Storage Cask -", CRIEPI Report U92038 (1993).

10) J. M. Creer et al.: "The TN-24P PWR spent-fuel dry storage cask: Testing and analyses", PNL-6054, 1987.

11) T. E. Michener et al.: "Thermal-hydraulic analysis of the TN-24P cask loaded with consolidated and unconsolidated spent nuclear fuel", PATRAM'89 Proceedings, pp.299-307, 1989.

12) M. Sakai, et al.: "Concrete storage cask for interim storage of spent nuclear fuel - evaluation of thermal performance -", IHI technical report, Vol. 42, No.1, pp.47-55 (2001).

13) H. Tsuji, et al.: "Thermal evaluation method for concrete cask is established by full-scale mock-up tests", MES technical report, No.180, pp.37-42 (2003).

14) J. C. Lee et al.: "Thermal-fluid flow analysis and demonstration test of a spent fuel storage system", Nucl. Eng. Design, Vol. 239, p. 551 (2009).

15) M. Wataru et al.: "Thermal hydraulic analysis compared with tests of full-scale concrete casks", Nucl. Eng. Design, Vol. 238, p. 1213 (2008).

16) T. Saegusa, K. Shirai, H. Takeda, M. Wataru, et al.: "Development of Concrete Cask Storage Technology for Spent Nuclear Fuel", CRIEPI Report N09 (2012).

17) R. D. Manteufel and N. E. Todreas:"Analytic Formulae for the effective conductivity of a square or hexagonal array of parallel tubes", HTD-Vol.207, Fundamental Problems in Conduction Heat Transfer, ASME, pp.43-54 (1992).

18) M. Wataru, H. Yamakawa, K. Nakajima, and T. Nagashima: "Development of Heat Transfer and Thermal-Analysis Code for Radioactive-Material Packages (CRISCAT)", Proc. PATRAM '92, pp.1347-1354 (1992).

19) Topical safety analysis report for the ventilated storage cask system (Rev.1), Pacific-Sierra nuclear associate,

1990
20) NISA, Sub-committee of safety of nuclear fuel cycle, "Technical report on spent fuel storage facility using concrete casks", 2004.
21) M. Wataru et al.: "Heat removal verification tests of full-scale concretes casks under accident conditions", Nucl. Eng. Design, Vol. 238, p. 1206 (2008).
22) JSME: "Codes for Construction of Spent Nuclear Fuel Storage Facilities - Rules on Concrete Casks, Canister Transfer Machines and Canister Transport casks for Spent Nuclear Fuel-" JSME S FB1-2003.

4.4 Shielding performance

4.4.1 Streaming from air inlet/outlet

For the evaluation of shielding performance of concrete casks, it is necessary to take account of long-term durability in comparison with metal casks and shielding loss due to cracks in the concrete. Also, since a cooling air path needs to be secured, it is necessary to take account of streaming and other design requirements specific to the concrete casks. However, safety design and evaluation methods of concrete casks with these being taken account of are not well established.

Radiation dose evaluation using streaming from inlet/outlet of concrete casks is mostly conducted with two-dimensional code (e.g. DOT). In the US, evaluation with three-dimensional code (e.g. MCNP) that utilized a Monte Carlo method is often conducted.

In order to evaluate the shielding performance of the concrete casks, we made a streaming test and test analysis by using a model that had an inlet and outlet and using a radiation source. We then examined the analysis code and made a streaming evaluation of actual casks. For details, see Reference[1].

(1) Streaming test

1) Test system

Fig. 4.4.1 shows an overview of the test model. The dimensions of the test model are width 900 × length 1700 × height 1600 mm (approximate dimensions of whole body) and the model consists of a concrete cask model with an inlet/outlet duct and radiation source shielding equipment. For the test, a radiation source was placed near one of the two openings of the duct and a detector was placed near the other. We made four types of ducts for the test: Straight type, twice-curved type, thrice-curved type, and pipe assembly type. The radiation sources were 252Cf for neutron rays and 137Cs and 60Co for gamma rays and the detectors were 3He proportional counter for neutrons and NaI scintillation detector and gamma ray survey meter for gamma rays.

The position of the radiation source and the detector was changed in the test as a parameter by placing them at or far from the duct openings. The obtained measurement values were corrected to eliminate background and converted to dose equivalent rate by using conversion coefficients of ICRP-51.

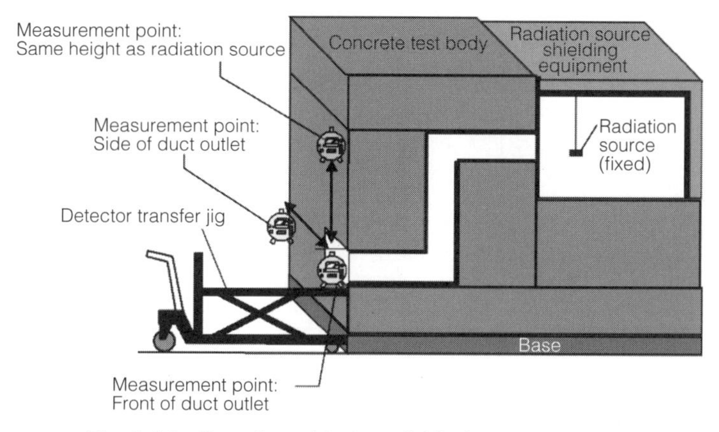

Fig. 4.4.1 Overview of test model (twice-curved duct)

(2) Test analysis

1) Analysis method

For the analysis, a Monte Carlo method based three-dimensional code MCNP-4B was used. The analysis model simulates details of the model shape, radiation source, and measurement points. Also, componential analysis was made for a concrete sample taken when the concrete part was manufactured and the result was reflected on the analysis model.

2) Analysis results

For the evaluation of the validity of the analysis results, the ratio C/E of the value obtained in the analysis to the value obtained in the test was evaluated. Fig. 4.4.2 compares the test results and analysis results of the neutron case by showing C/E and Fig. 4.4.3 compares those of the gamma ray. Fig. 4.4.4 compares the test and analysis using gamma ray spectrum. The largest difference between the test and the analysis was about 60 % for the neutron and about 40 % for the gamma ray.

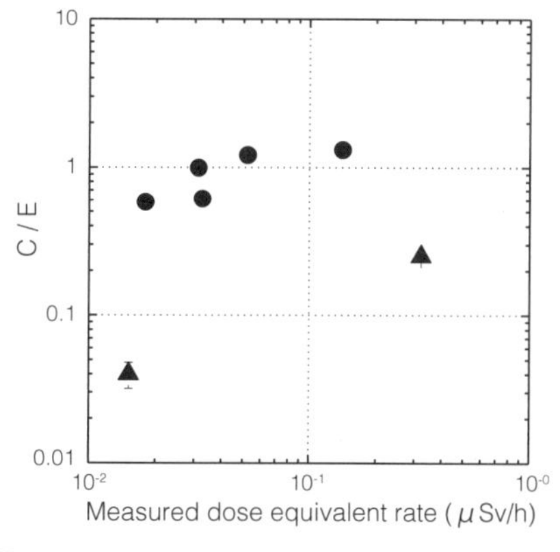

Fig. 4.4.2 Comparison between test and analysis using neutrons

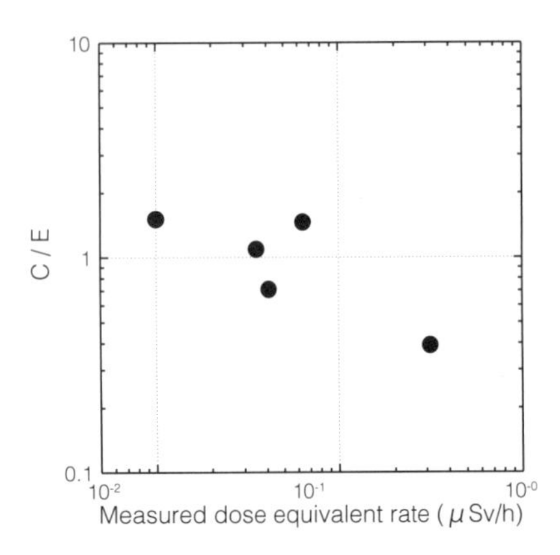

Fig. 4.4.3 Comparison between test and analysis using gamma ray

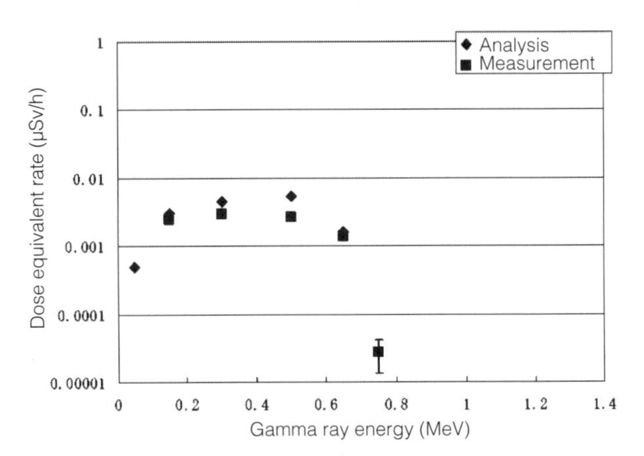

Fig. 4.4.4 Comparison between test and analysis using gamma ray spectrum (Case G-1)

(3) Analysis with actual casks

Based on the analysis method used for the streaming test, MCNP-4B was used to analyze an actual concrete cask.

1) Evaluation criteria

The target evaluation criteria were set as in the following.

(i) Cask surface: 2000 μSv/h

(ii) 1 m from cask surface: 25 μSv/h

Here, the evaluation criteria for the position 1 m from the cask surface was set to 25 μSv/h for the following reasons.

* It is necessary to reduce radiation exposure as much as possible since workers may access the air inlet at the lower body of the concrete cask. Contribution of the surrounding four casks to the dose equivalent rate in the storage area should be taken account of.
* Dose control section of the storage area should be a basis for the facility shielding design and the exposure dose of workers. The dose equivalent rate of workers designated by laws[2)] is 50 mSv/h. The evaluation standard shall be about half of this restriction value.
* For the determination of the criteria of the dose equivalent rate in the storage area, one year is considered to be 50 weeks and the access frequency to the storage area to be 2 hours a week.
* Safety margin is set to 2.5, according to the shielding test and analysis results (analysis value and measurement value) obtained last year or before.

2) Analysis conditions and analysis method

Table 4.4.1 shows specifications of fuels stored in the cask. The fuel assembly of average burnup fuels and maximum burnup fuels stored in the cask were treated as a single radiation source in the canister. Since the fuel effective part, upper nozzle, and upper plenum may contribute to the dose in different ways, they were treated not as a single source but as separate radiation source areas. Since the canister lid was one of the shielding materials, it was not included in the source. Gamma rays and neutrons were evaluated separately. For the gamma rays, fuel effective part's rays and activated rays were analyzed separately. Therefore, we assumed three radiation sources, activated gamma rays, fuel effective part gamma rays, and fuel effective part neutrons, and analyzed and evaluated them separately. In the dose equivalent rate evaluation, separately-calculated dose equivalent rates were summed up. We used a cutoff energy and weight window for the analysis and set the default cutoff energy of MCNP to the following values.

* Neutron source: 0.01 eV

* Gamma ray: 0.05 MeV

In the MCNP, calculation was made until FCD output became 10 % or lower so that the statistical error of the calculation results was 10 % or lower.

Table 4.4.1 Specifications of fuels stored in cask

Item		Average burnup	Max burnup
Concentration	wt%	4.7	4.7
Burnup	MWD/t	50000	55000
Specific power	MW/t	38.4	
Cooling period (year)		10	
Burnup distribution		Upper: 1/12	PF=1.0
		Central: 10/12	PF=1.15
		Lower: 1/12	PF=1.0
Number of fuels stored		21 assemblies	
Position in cask		12 assemblies outside	9 assemblies inside

3) Analysis system and evaluation position

The whole system was divided into 8 parts by dividing it into two in the axis direction and into 4 in the circumferential direction in order to take account of the axial and circumferential symmetry of the inlet/outlet shapes of the concrete cask (include canister).

The analysis was made with three-dimensional MCNP-4B for both neutrons and gamma rays.

The evaluation points, shown in Fig. 4.4.5, are set on the side of the cask, surface of the top, and 1 m away from the surface.

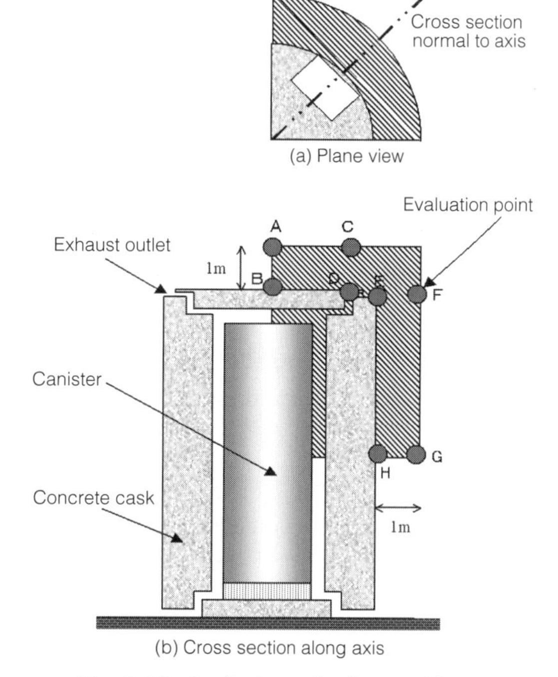

Fig. 4.4.5 Analysis evaluation position

4) Analysis results

Table 4.4.2 shows the analysis result at each evaluation point.

Among the evaluation points on the surface, point B in the lid center gave the highest dose, about 77 μSv/h, still lower than the target criteria value. Among the points 1 m away from the surface, point A above the lid center and G at the fuel center level gave higher doses than the target standard value but almost equivalent to the value. The dose rate distribution above the cask gave a higher dose distribution than the target standard value but it is almost equivalent to the value. The distribution on the side of the cask showed a similar behavior.

Table 4.4.2 Analysis result (dose equivalent rate)

(Unit:μSv/h)

Position	Surface				1 m from surface			
Evaluation point	B	D	E	H	A	C	F	G
Result	77	64	4.5	56	27	19	4.6	29
Criteria	2000				25			

(4) Summary

1) As a result of comparison between the test result and the analysis result, we can conclude that the MCNP code can be used for appropriate evaluation of the contribution of the streaming from the air outlet to the dose equivalent rate.

2) In the actual cask, the dose equivalent rate near the outlet met the target criteria value.

References

1) T. Saegusa, K. Shirai, et al.: “Development and evaluation of technology of storing used fuels in concrete,” Central Research Institute of Electric Power Industry, General Report NO 9, May, 2010 (in Japanese).

2) Prime Minister’s Office: “Laws and regulations on prevention of radiation damages due to radioactive elements,” ministerial ordinance No. 8, March 31, 1998 (in Japanese).

4.5 Structural integrity

4.5.1 Fracture toughness of welded part of canister

Canister stores spent fuels inside and maintains its containment function. The canister has a cylindrical shape by welding, of which size is about 1.7 m in diameter, 15 to 20 meters in height, and 20 mm in thickness. In this section testing and evaluation methods of fracture toughness applicable to the stainless steels, and the actual fracture toughness data are described, in order to be possible to evaluate the integrity of welded structural canister against impact at a drop accident, etc.

(The unit system)

At the following section (1), the metric system is used partially. Comparison with mutual original values in the plural figures and tables may be easier by the other units. In case of conversion to SI unit, the following formul a can be used.

Load : 1 (kgf)= 9.8 (N), Stress : 1 (kgf/mm^2) = 9.8 (N/mm^2)= 9.8 (MPa)

Stress intensity factor K and Linear elastic fracture toughness K_{IC}: 1 ($kgf/mm^{3/2}$)= 0.3101 ($MPa \cdot m^{1/2}$)

J integral and non linear (elastic plastic) fracture toughness J_{IC} : (N/m) or (k J/m^2)

where, 1×10^3(N/m) = 1 (k J/m^2)

(1) Fracture toughness of welded part of conventional stainless steels

Candidate materials for canister in Japan at present include types 304, 304L, 316L, and 316LN for BWR (conventional stainless steels that have been used in the nuclear power plants). In addition, types 329J4L, YUS270, etc., are also candidates that have not been used in service, but high corrosion resistant.

In this section, the testing and the evaluation methods of fracture toughness for welded part of the conventional stainless steels types 304, 304L, and 316L, are described.

Generally, austenite stainless steels have high ductility and are dealt as materials that need not fracture toughness evaluation in MITI Code 501[1)] that provides codes for material, design and inspection of the nuclear power plants in Japan.

However, the canister of concrete cask dose not undergo solid solution treatment from the point of view preventing deformation, because it is made of welding structure with thin shell thickness of about 15 to 20 mm in thickness. Then, it would be possible that the welded part gets sensitized by heat and loses the natural properties of stainless steel, or it is used with existence of residual stress. In addition, the canister is exposed to neutron irradiation during the storage periods from spent fuel inside.

With the above consideration, the testing and the evaluating method of fracture toughness applicable to the stainless steels, and the actual fracture toughness data will be necessary, in order to evaluate the integrity of welded structural canister against impact at a drop accident, etc.

1) J_{IC} (JQ) of conventional stainless steels

Static fracture toughness tests by the single specimen method, namely the unloading compliance method, were conducted for base metal, deposited metal, and heat affected zone (HAZ) of type 304, 304L, 316LN, in accordance with ASTM Code[2)], using 0.5 inch thick compact tension (0.5T-CT) test specimens with appropriate side grooves.

The single specimen method uses only one specimen with repeated frequency of load and unload conditions. The J-integral value is calculated using its stress-strain curve. The J_{IC} value can be determined using the J-integral value and extrapolated crack extension length. Fig. 4.5.1-1 shows the results of type 304 and 304L. The obtained JQ values increased gradually with rising temperature, and reached peak values within the temperature range from about 20 °C to 50 °C, and then decreased. This decreasing trend after the peaks is similar to that of carbon steels for the pressure vessel of nuclear power plant. All the fracture surface of the test specimens showed dimples.

The J values in Fig. 4.5.1-1 obtained by the unloading compliance method was compared with the other data in the literature. The J values in this study showed the same level or remarkably higher values in comparison with J_{IC} values of austenite stainless steels as shown in Table 4.5.1-1 in the JSME Code S NA1-2000[3)].

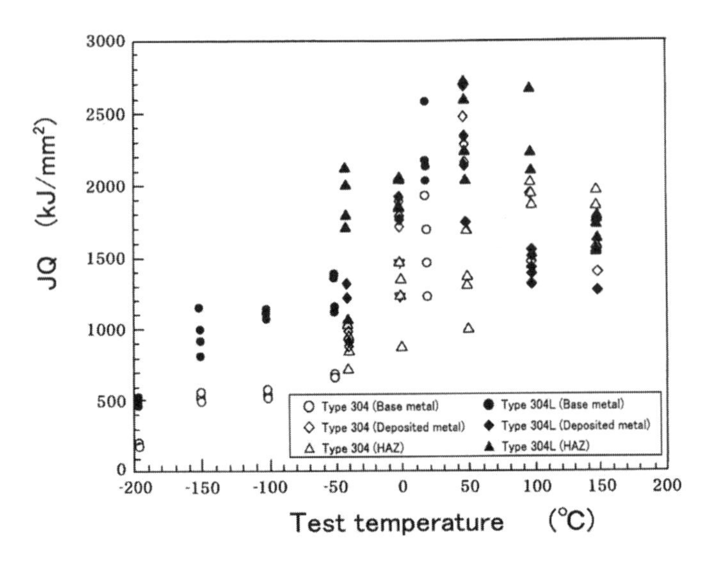

Fig. 4.5.1-1 Temperature dependence of fracture toughness of conventional stainless steels

Table 4.5.1-1 Fracture toughness of austenite stainless steels [3)]

Welding process etc.	Test temperature (°C)	Domestic data		Foreign data	
		J_{IC} (kJ/m^2)	Number of data	J_{IC} (kJ/m^2)	Number of data
SAW, SMAW	0°C～R.T.	49～218	4 [2), 3)]	81～259	4 [4)]
	288°C	191	1 [3)]	47～168	4 [4)]
TIG	R.T.	–	–	195	1 [4)]
	288°C	–	–	558	1 [4)]
HAZ (Base metals are type 304 and 316)	R.T.	–	–	263～788	6 [4)]
	288°C	–	–	613～753	2 [4)]
Base Metal (Type 304, 316)	0°C～R.T.	500～686	4 [2), 3)]	795～1148	2 [4)]
	288°C	400	1 [3)]	473～832	3 [4)]

Note) R.T.: Room Temperature.

Because it is very important to confirm if the obtained J can be accepted as J (Ji) of crack initiation correctly or not, crack opening displacement (COD) tests were conducted as shown in the next section to confirm J values of crack initiation.

2) Ji value obtained from COD test

Crack opening displacement (COD) tests by the multiple specimens method were conducted for base metal, deposited metal, and heat affected zone (HAZ) of type 304, 304L, and 316LN, in accordance with British Standard BS5762-1979[4)], etc., using 0.5T-CT test specimens with appropriate side grooves. We obtained the fracture mechanics parameter δi, that is a crack tip opening displacement and gives the stable crack initiation. Also, correlation between δi and J was investigated and J (Ji) of crack initiation was obtained. The multiple specimens method is a method to obtain the initiation point of stable crack using multiple specimens and method of extrapolation.

Measurement procedure of δi by multiple specimens method was based on BS5762-1979 mainly, but, the calculate formula of δ by compact tension (CT) specimen was based on WES 1108-1995[5], etc.

Using the relation of $J = m \cdot \sigma_y \cdot \delta$ [6], J of ductile crack initiation, Ji, can be obtained from δ.

Further we assumed non-dimensional Φ normalized from J as the parameter of fracture mechanics to be used finally as an evaluation method. This parameter Φ enables to evaluate both J-integral value and COD values.

Namely,

$\Phi = J/(2\pi E \cdot \varepsilon_y^2 \cdot a) = \delta/(2\pi a \cdot \varepsilon_y)$ (an example case of m = 1) ..... (1)

here, E : Young's modulus, ε_y : yield strain, δ : crack tip opening displacement (COD),

a : crack length

Using a testing apparatus (electro-hydraulic servo fatigue testing machine), load- displacement-curves were obtained, which composed of load and clip gage displacement. During the test, the specimen was unloaded at several loads as shown in Fig. 4.5.1-2. After unloading, fracture surface was colored by the heat tincture method and the fracture surface was revealed by force, and then the length of pre-crack by fatigue a_0 and the propagation length of stable crack Δa were measured.

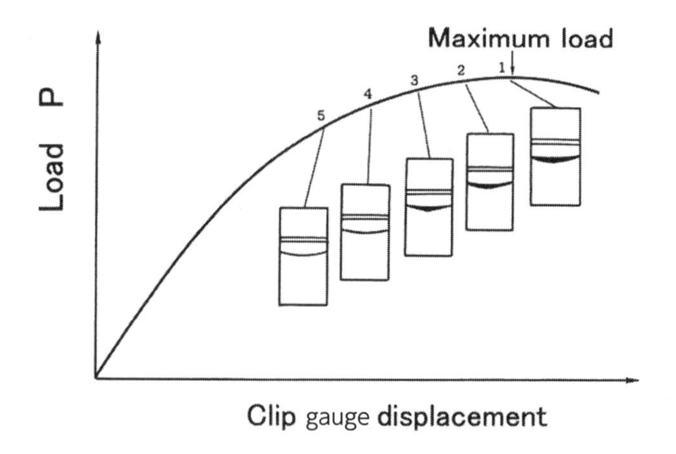

Fig. 4.5.1-2 Load and clip-gauge-displacement curves [4]

Plotting the measured each propagation length of stable crack Δa and COD (δ) as shown in Fig. 4.5.1-3, COD value was obtained by extrapolation up to the point of Δa = 0 that is defined as COD value (δi) at the point of stable crack initiation.

Applying the experimental data (J integral obtained from load-displacement-curve and δ measured by experiment) to the general formula of $J = m \cdot \sigma_y \cdot \delta$, the coefficient m was determined. Namely, we plotted the relation between J and $(\sigma_y \cdot \delta)$ and approximated these plotted data using straight line passing (0, 0) point of the co-ordinates, then coefficient m was determined as inclination of this approximated straight line.

Here, σ_y is effective yield strength and is the mean value between 0.2 % proof strength and tensile strength at each

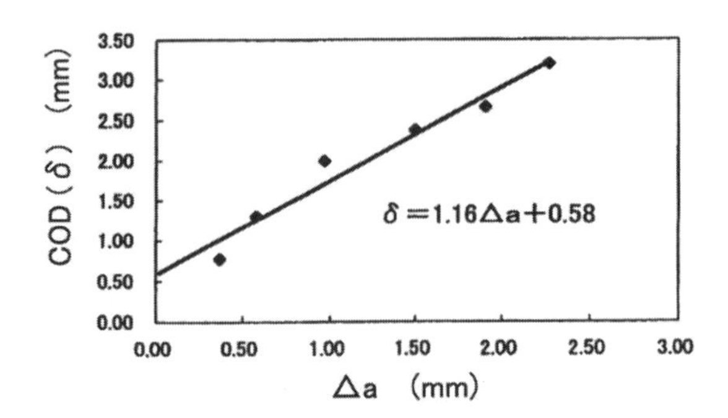

Fig. 4.5.1-3 An example of COD (δ) -△a curve (Type 304L base metal, at 150 °C)

temperature.

The value m is non-dimensional coefficient and known as value from 1 to 3 [6] at plane-strain condition. Within the temperature range from -40 °C to 150 °C, Ji values of heat affected zone (HAZ) were lower than those of base metal and deposited metal in case of type 316LN. It indicated that fracture toughness of HAZ shows the lowest value.

Namely,

Ji (HAZ) < Ji (deposited metal) < Ji (base metal) ..... (2)

The Ji value of the base metal of type 304 and 304L showed much smaller values than JQ values obtained in the past within the temperature range from -40 °C to 150 °C. Similarly, Ji values of the deposited metal of type 304 and 304L showed smaller values than Ji values of the base metal within the same temperature range from -40 °C to 150 °C and showed the same relative tendency as type 316LN.

Fig. 4.5.1-4 to Fig. 4.5.1-6 show Ji values obtained by COD tests and arranged for each material type. The Ji value of base metal of type 304 and 304L are smaller than JQ values of the same material type obtained by the single specimen method, and smaller than Ji values of type 316LN base metal.

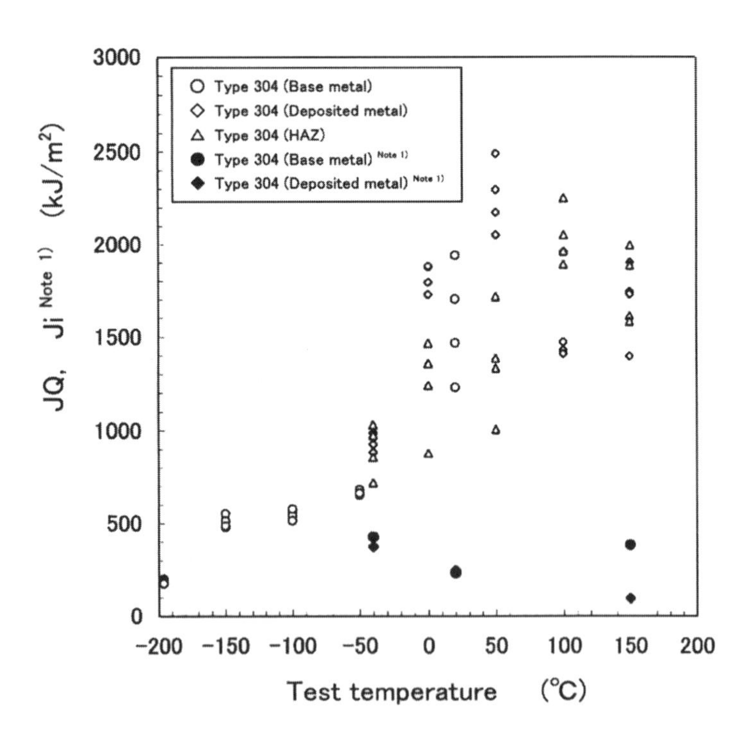

Fig. 4.5.1-4 Temperature dependence of fracture toughness of type 304 by 0.5 T-CT specimens

(Plotted data include both JQ values by the compliance method (the single specimen method) and Ji values by COD tests (the multiple specimens method).)

Note 1) Symbols of which inside are painted black in this figure are Ji values, which show J integral of stable crack initiation.

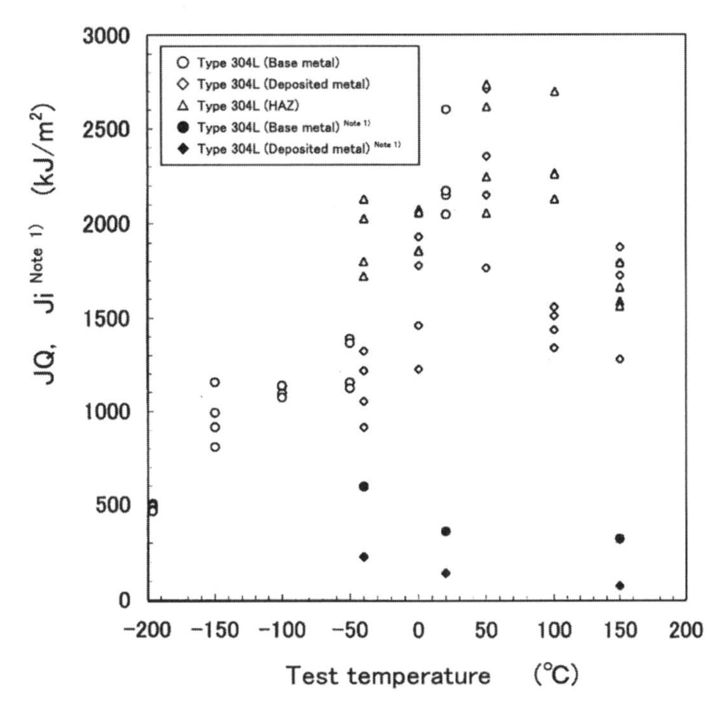

Fig. 4.5.1-5 Temperature dependence of fracture toughness of type 304L by 0.5 T-CT specimens

(Plotted data include both JQ values by the compliance method (the single specimen method) and Ji values by COD tests (the multiple specimens method).)
Note 1): Symbols of which inside are painted black in this figure are Ji values, which show J integral of stable crack initiation.

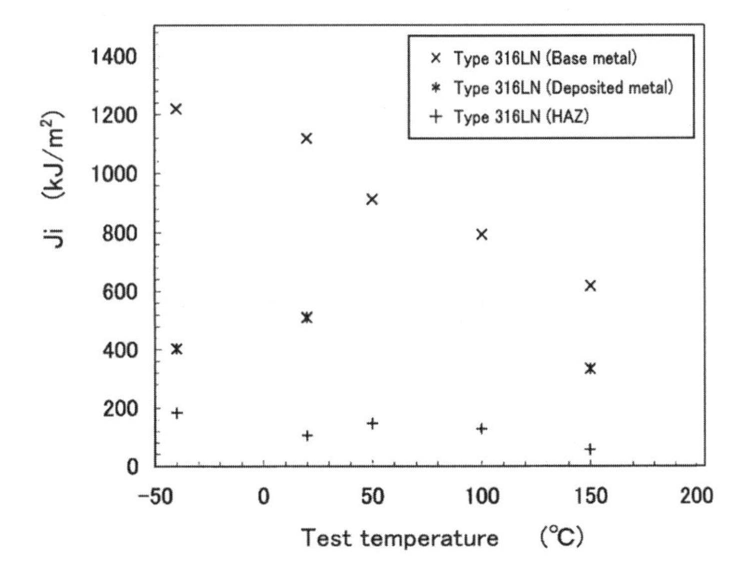

Fig. 4.5.1-6 Temperature dependence of fracture toughness of type 316LN by 0.5 T-CT specimens

(All of plotted data are Ji values by COD tests (the multiple specimens method).)
Note 1) All of plotted data in this figure are Ji values, which show J integral of stable crack initiation.

3) Fracture toughness for evaluating crack initiation (JQ and Ji)

In order to evaluate the crack initiation of canister made of stainless steel, fracture toughness data of materials are necessary. The JQ value can be obtained by the single specimen method (the unloading compliance method) specified in American Society for Testing and Materials (ASTM) Code, etc. On the other hand, Ji can be obtained by the crack opening displacement (COD) tests by the multiple specimens method in accordance with British Standard BS 5762-1979. In this case, use of the Ji value as fracture toughness value of materials will likely indicate the crack initiation more correctly and will give more conservative evaluation as compared with that using the above JQ values.

4) Fracture toughness of material $J_{0.2}$ that permits microscopic crack initiation

Stainless steels for canister have high ductility and microscopic crack initiation at crack tip may be permitted as long as it is allowed, the fracture toughness of material at crack initiation point. Referring to the ASTM Code that defines industrial crack initiation point J_{IC} by the cross point with 0.2 mm offset line, the fracture toughness of material $J_{0.2}$ for evaluation of crack initiation can be obtained by assumption of crack propagation length $\Delta a = 0.2$ mm. The $J_{0.2}$ value instead of Ji value may be used for stainless steels for canister with high ductility.

5) Noteworthy point in case of using JQ

The JQ values obtained by using the crack opening displacement (COD) tests by the multiple specimens method may include large values of crack propagation length that may include crack propagation up to the length of full thickness. Accordingly, it is not appropriate to apply such JQ values. In case of using JQ value, it should be confirmed that the crack initiation length is not so large with problem.

Comparison of several J integral as the fracture toughness of material is shown in Fig. 4.5.1-7, schematically.

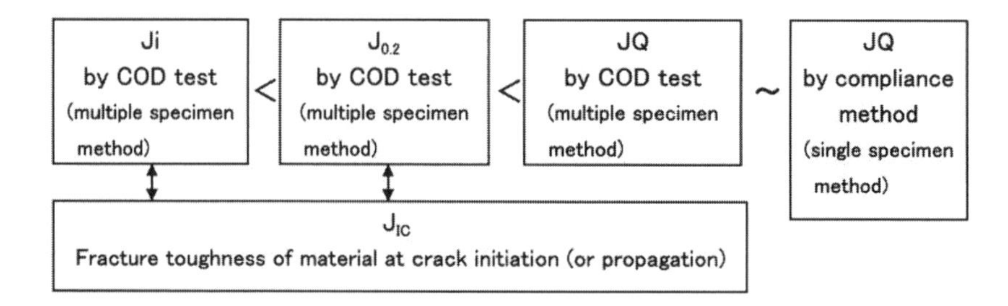

Fig. 4.5.1-7 Schematic figure of several J integral as the fracture toughness of material

6) Assessment of ductile crack propagation

In case of evaluating integrity of the welded part of canister when external force is loaded, J integral-evaluate-curve (Φ or J integral), as an index of external force can be obtained by analysis in consideration of the welding residual stress and the restrictive force. The Φ or J integral value as external force is compared with fracture toughness of the material Φ (J_{IC}, Ji, or $J_{0.2}$ of material). Ductile crack initiation, or propagation, can be assessed by judgment of which value is bigger or smaller.

Namely, ductile crack will not initiate or propagate under the following condition.

Φ (or J_{IC}, or Ji, or $J_{0.2}$) as fracture toughness of material > Φ as external force (or J integral) ..... (3)

where,

Φ: non-dimensional J integral (see to formula (1). Obtained by J integral analysis, etc. [7),8),9)])

References

1) Ministry of International Trade and Industry (MITI), present METI, MITI Code 501, (1980).

2) ASTM Standards, ASTM E813-89 "Standard Test Method for J_{IC}, A Measure of Fracture Toughness", (1989).

3) JSME Codes for Nuclear Power Generation Facilities - Rules on Fitness-for-Service for Nuclear Power Plants -, JSME S NA1-2000, (2000).

4) British Standard Institution, BS5762-1979 "British Standard Methods for Crack opening displacement (COD) testing", (1979).

5) WES 1108-1995 "Standard test method for Crack-Tip Opening Displacement (CTOD) fracture toughness measurement", The Japan Welding Engineering Society, (1995).

6) S. Machida, "Ensei Hakai Rikigaku (Ductile Fracture Mechanics)", Nikkan Kogyo Shimbun Ltd., (1984).

7) A. Kosaki, N. Urabe, "Ductile Fracture Evaluation of Ductile Cast Iron and Forged Steel by Nonliner-fracture-mechanics -Proposal of J-integral Design Curves-", The 7th National Symposium on Power and Energy Systems (SPES 2000), JSME[No.00-11], B201, pp.321-326, (2000).

8) A. Kosaki, "Brittle Fracture Tests of Spent Fuel Storage Cask Materials", 1998 Fall Meeting of the Atomic Energy Society of Japan, AESJ, E67, p.327, (1998).

9) A. Kosaki, N. Urabe, K. Shirai, "Ductile Fracture Evaluation of Stainless Steel by Nonliner-fracture-mechanics", The 8th National Symposium on Power and Energy Systems (SPES 2002), JSME [No.02-7], P23-02, pp.439-444, (2002).

(2) Fracture toughness of highly corrosion resistant stainless steel

In Japan, storage facilities may be placed near the sea and chloride-induced external stress-corrosion cracking (ESCC) due to sea salt contained in cooling air could occur. Therefore, not only conventional stainless steel of SUS304 or SUS316 but also highly corrosion resistant stainless steel (SUS329J4L, YUS270) can be used[1)-3)]. In order to use these types of highly corrosion resistant materials, it is necessary to show that the canister would not be damaged even in a falling accident or others when an object is handled in the operation. Since the canister lid is welded to the body after spent fuels are set in the canister, the welding would be partial penetration single-side welding and one cannot check the penetration from the other side of the lid. So it is necessary to assume that there could be a defect in the primary layer of the welding, make fracture-mechanical evaluation of the welded spot of the lid to confirm the structural integrity. Fracture toughness property of the canister materials is one of major parameters necessary for fracture-mechanical evaluation. Since the fracture toughness could depend on production method, heat treatment condition, and plate thickness of the materials even if they were made under the same standards, it is important to evaluate the materials which were made under the same manufacturing conditions of actual canisters. In particular for the welded parts whose fracture toughness property largely depends on the welding method and condition, it is crucial to make evaluation with a reproduction of actual canister's welded parts. However the fracture toughness property of the welded parts of canisters made of these highly corrosion resistant materials has not been clarified. Therefore in order to clarify the fracture toughness property of the welded parts of two kinds of highly corrosion resistant materials, we performed elastic-plastic fracture toughness tests with a test temperature parameter for base metal, welding heat affected zone, and weld metal of weld joints which simulated the canister lid's welded parts[4].

1) Experimental method

(i) Material under test and test body

We made weld joints that simulated secondary lid welded parts and used them for the test.

* SUS329J4L weld joint

Base metal: SUS329J4L (JIS G4304-1999[5]), weld metal: SUS329J4L equivalent material

* YUS270 weld joint

Base metal: YUS270 (ASME SA240 S31254[6] equivalent), weld metal: Alloy 625 equivalent material (JIS Z3334 YNiCrMo-3[7])

The chemical composition of the materials under test is shown in Table 4.5.1(2)-1 and Table 4.5.1(2)-2.

Table 4.5.1(2)-1 Chemical composition of SUS329J4L joint material (wt%)

		C	Si	Mn	P	S	Cu	Ni	Cr	Mo	W	N
Base metal	Spec.	≦0.030	≦1.00	≦1.50	≦0.040	≦0.030	-	5.5 ～ 7.5	24 ～ 26	2.5 ～ 3.5	-	0.08 ～ 0.3
	Sample	0.01	0.41	0.45	0.024	0.001	0.49	6.88	25.67	3.33	0.4	0.23
Weld metal	Sample	0.016	0.3	0.5	0.008	0.002	0.49	9.06	25.28	3.06	1.96	0.22

Table 4.5.1(2)-2 Chemical composition of YUS270 joint material (wt%)

		C	Si	Mn	P	S	Cu	Ni	Cr	Mo	Fe	N	Nb+Ta	Al	Ti
Base metal	Spec.	≦0.020	≦0.80	≦1.00	≦0.030	≦0.015	0.5 ～ 1	17 ～ 19.5	19 ～ 21	5.5 ～ 6.5	-	0.16 ～ 0.24	-	-	-
	Sample	0.013	0.51	0.55	0.023	0.001	0.62	17.84	19.84	6.92	-	0.19	-	-	-
Weld metal	Sample	0.02	0.13	0.11	0.009	0.001	0.14	55.88	21.59	8.34	0.19	-	2.88	0.3	0.31

The same TIG welding used for actual canisters was used in the test and the welding conditions of the test were also the same as those for actual canisters. The root shape was almost the same as that of actual canister's welded part. However, in order to be collected from a heat affected zone (HAZ), the root shape on one side was perpendicularly set up. The welding conditions and root shapes of each weld joint are shown in Table 4.5.1(2)-3 and Fig. 4.5.1(2)-1.

Table 4.5.1(2)-3 Welding condition

Welding method	TIG	TIG
Welding wire	DP3WT	YNiCrMo-3
Welding current (A)	250 ～ 300	150 ～ 250
Welding voltage (V)	11.5	9
Number of pass	35	20

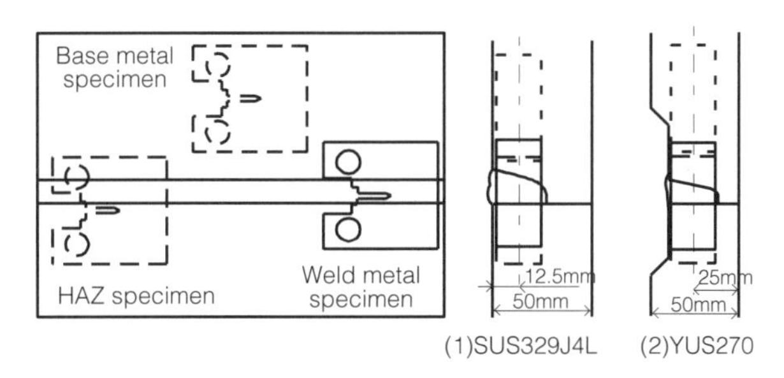

Fig. 4.5.1(2)-1 Method of sampling specimen from weld joint

A 1-inch thick compact tension type (1TCT) specimen with side grooves was used for the test. Specimens of each material were sampled from the base metal, weld metal, and HAZ with the method shown in Fig. 4.5.1(2)-1. The specimen sampling location in the thickness direction was the weakest area of the base metal identified in a Charpy impact test conducted in advance.

(ii) Test method

For the fracture toughness test, we employed the unloading compliance method based on ASTM E1820-99[8]. In this method, the unloading process is repeated in the middle of the loading process to obtain the relation between the energy J given to the specimen and the crack extension Δa from the relation of the load P and the load line displacement V_{LLD}. The J-Δa relation obtained from the P-V_{LLD} relation measured in the test is approximated to an exponential function by using the least-square method and a crack extension resistance (J-R) curve was obtained. A blunting line was obtained with the least-square method from the linear part of the J-Δa relation and shifted by a 0.2 mm offset for V_{LLD}. J value at the crossing point of the obtained line and the J-R curve was the fracture toughness J_Q. When J_Q met all the requirements about the specimen dimensions, uniform crack propagation, and data reliability, the value of J_Q was regarded as effective fracture toughness J_{IC}.

(iii) Test condition

Since the canister temperature depends on stored used fuels, cooling air conditions, and storage period, it is important to know the fracture toughness property in the expected temperature range. Therefore, the following temperature conditions were chosen.

Test temperature: 233 K, 298 K, 373 K, 423 K, 473 K.

The tests were repeated twice or more for each specimen collected from each position.

2) Test results

The propagation of ductile cracks was observed in all the specimens of any materials, sampled from any positions, at any test temperatures. Since no unstable failure was observed, the high toughness level of the materials were indicated. Fig. 4.5.1(2)-2 shows an example of the relation J-Δa (circles in the figure) and J-R curve. Although disturbance in the J-Δa relation in a small Δa region was observed in some specimens, stable J-Δa relations could be observed in most cases as shown here. The relation between J_Q and the temperature of SUS329J4L and YUS270 is shown in Fig. 4.5.1(2)-3 and Fig. 4.5.1(2)-4. J_Q of the base metal of SUS329J4L weld joint reached the minimum,

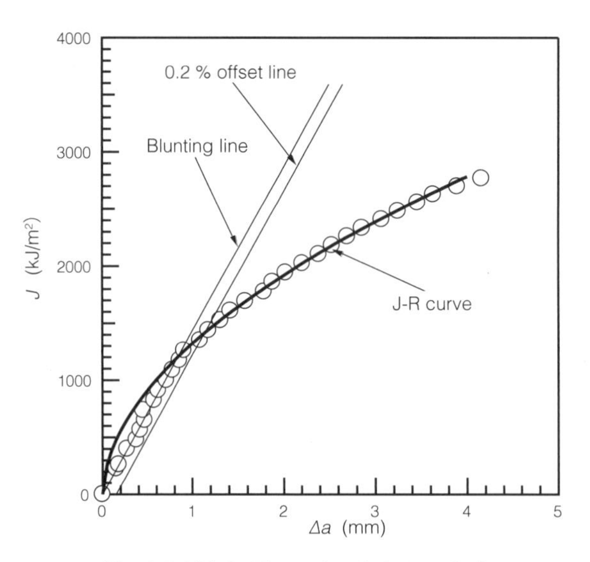

Fig. 4.5.1(2)-2 Example of J-Δa relation

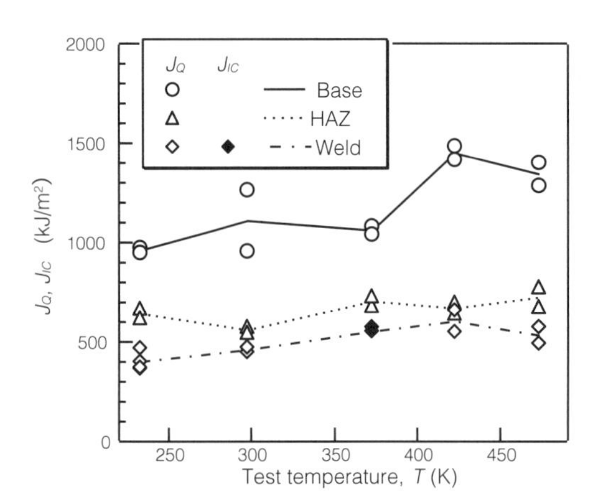

Fig. 4.5.1(2)-3 Relation between fracture toughness and temperature of SUS329J4L joint material

Fig. 4.5.1(2)-4 Relation between fracture toughness and temperature of YUS270 joint material

i.e. about 600 kJ/m^2, at 233 K and the maximum, i.e. about 1800 kJ/m^2, at 298 K. It slightly decreased with the increase of the temperature and reached about 1500 kJ/m^2 at 473 K. The temperature dependence of J_Q of HAZ showed the same tendency as that of the base metal and the J_Q value was almost the same as, or slightly lower than, that of the base metal. J_Q of the weld metal was almost the same as that of the base metal at 233 K, increased with the temperature, and reached the maximum value (about 1000 kJ/m^2) at 423 K.

Fig. 4.5.1(2)-4 shows the result of YUS270 weld joint, where the average of J_Q for the base metal and weld metal is about 900 kJ/m^2 and 420 kJ/m^2 at 233 K respectively and monotonically increases with the temperature, reaching about 1400 kJ/m^2 and about 550 kJ/m^2 at 473 K respectively. J_Q of HAZ was slightly larger than that of the weld metal at any temperature. It was the weld metal that gave the smallest J_Q at any test temperature.

J_{IC} was obtained only in the weld metal of SUS329J4L weld joint at 233 K and 298 K and in the weld metal of YUS270 weld joint at 373 K. Under the other conditions, the specimen dimension requirement for the validity of ASTM E1820 could not be satisfied and J_{IC} could not be obtained. Larger specimen dimensions are necessary to obtain J_{IC}, but it is difficult to collect a specimen of the dimensions that meet this requirement because of the plate thickness of actual canister's welded part. It was reported[9)-12)] that the crack extension-resistance *J-R* and fracture toughness J_{IC} measured for a small CT specimen of 304 stainless steel and low-alloy steel were the same as or smaller than those for larger CT specimens. Also, since the obtained J_{IC} values were all within the variation range of J_Q, the J_Q values obtained in the present test should represent J_{IC} equally or conservatively. The fracture toughness value of 304 stainless steel that has been used for canisters in the US has not been obtained. Also, that of the welded parts of light-water reactor pipes has not been obtained. However, it was reported[13)-16)] that J_{IC} of the base metal s of 304 and 316 stainless steels was about 500-1150 kJ/m^2 and that of the TIG welded parts was 200-600 kJ/m^2 at 273 K to room temperatures. Since J_Q of the present specimens at room temperatures was equal to or higher than these values, the present specimens should have the same or higher toughness than 304 or 316 stainless steel.

3) Summary

(i) For SUS329J4L base metal and HAZ, the fracture toughness J_Q reached the minimum at 233 K and maximum at room temperature and then slightly decreased as the temperature increased. For weld metal, J_Q increased with the test temperature. For YUS270, J_Q increased with the test temperature irrespective of where the specimen was collected.

(ii) Fracture toughness value of the welded parts of the both specimens was equal to or smaller than that of the base metal or HAZ. It was therefore found that the fracture toughness value of the weld metal should be used for the evaluation of the welded parts of the canister.

The canister design standard[17)] of the Japan Society of Mechanical Engineers used the above results as reference data for selecting an evaluation method.

References

1) T. Saegusa, M. Mayuzumi: "Material and environment of interim storage vessel of recycled atomic-energy fuels," Airyo-to-Kankyo, 53(2004), 246 (in Japanese).
2) G. Abe, H. Ito, and H. Kajimura: Proceedings of Autumn Meeting 2002, Atomic Energy Society of Japan, Tokyo, (2002), 570.
3) H. Nakayama, T. Hirano, S. Kobayashi, and T. Sakaya: 48th Material and Environment Symposium, Japan Society of Corrosion Engineering, Tokyo, (2001), 143.
4) T. Arai, M. Mayuzumi, L. Niu, K. Takaku: Iron and Steel, Vol.91, No.5 (2005).
5) Japan Industrial Standards JIS G4304-1999, Japanese Standards Association, 1999.
6) ASME Boiler and Pressure Vessel code Sec. II, ASME, 1999.
7) Japan Industrial Standards JIS Z333.4.1.1999, Japanese Standards Association, 1999.
8) ASTM standard, designation E1820-99, ASTM, 1999.

9) E.M. Hackett and J.A. Joyce: Nuclear Engineering and Design, 134(1992).
10) J.D Landes and D.E. McCabe: EPRI NP-4768, EPRI, 1986.
11) V. Papaspyropoulos: NUREG/CR-4575 BMI-2137, US NRC, 1986.
12) R.A Hays: NUREG/CR-4538, Vol.1, US URC, 1986.
13) M.F. Kanninen: EPRI NP-2347, EPRI, April, 1982.
14) P.C. Paris: NUREG-0311, US NRC, August, 1977.
15) P.C. Paris and R.E. Johnson: ASTM STP803, 11, ASTM, 1983.
16) H. Itoh, T. Shige, K. Matsunaga, K. Murakami, K. Ohnishi, H. Okunishi: ISSF Seminar 2003, Tokyo, 2003.
17) Standards of the Japan Society of Mechanical Engineers, JSME S FB1-2003, The Japan Society of Mechanical Engineers (2003).

4.5.2 Ultrasonic test to detect flaw in welded parts of canister

As stated in Sec. 2.2.1, "Technological requirements for used fuel storage facilities (interim storage facilities) with concrete casks[1]" (hereafter referred to as technological requirements) by the Nuclear and Industrial Safety Agency, Ministry of Economy, Trade and Industry requires non-destructive tests of the canister lid welded parts by using a multi-layered PT (multi-layer penetrant test) and UT (ultrasonic test). On the other hand, the current concrete cask structure standard designated by the Japan Society of Mechanical Engineers (JSME)[2] in Sec. 2.2.3 employed only multi-layered PT (and UT as alternative) as volume inspection, following the double-lid welded structure inspection method based on US ASME Code Case N-595[3].

For early establishment of the concrete cask technology in Japan, it is necessary to make a safe and practically-rational standard by taking account of the standardization in the US which is a pioneer of the canister storage, current reliability of the UT technology for stainless steel, and practical restriction of the canisters. As shown in Fig. 4.5.2-1, the canister lid welding was single-side welding from the outer side and therefore partial penetration welding. UT inspection methods of the lid welded parts of this structure were designated in a welding standard for disposal vessels made of carbon steel[4], but only a few for similar lid structures made of austenite stainless steel.

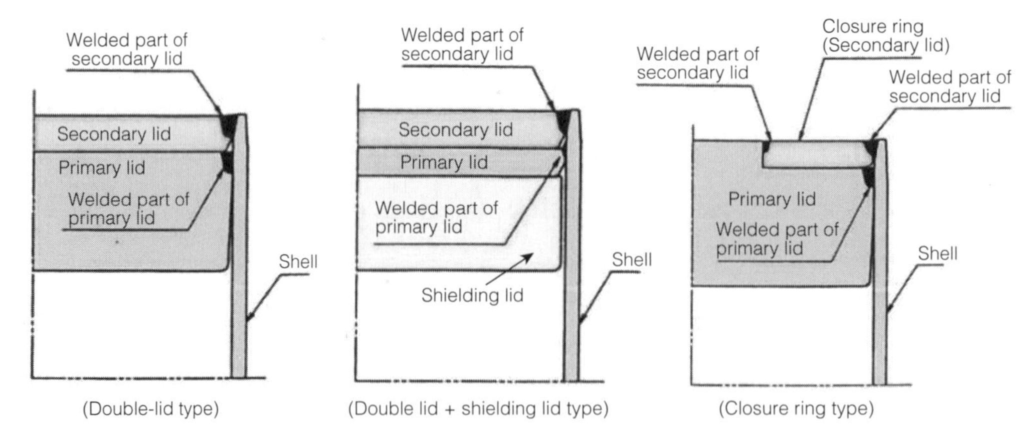

Fig. 4.5.2-1 Lid types of canister (double closure weld detail)

The UT for the lid welded parts could be used for searching for defects from the wall side and from the lid side as in Fig. 4.5.2-2. In this section, for the evaluation of the validity of the UT for the lid welded part structure of the canister, we made a full-size lid model test body made of austenite stainless steel Type 304L (UNS S30403) shown in Fig. 4.5.2-3 and estimated minimum defect size and defect-search area (to find the UT test applicability) by using the latest UT technology. Below are the evaluation results. For details, see Reference[5].

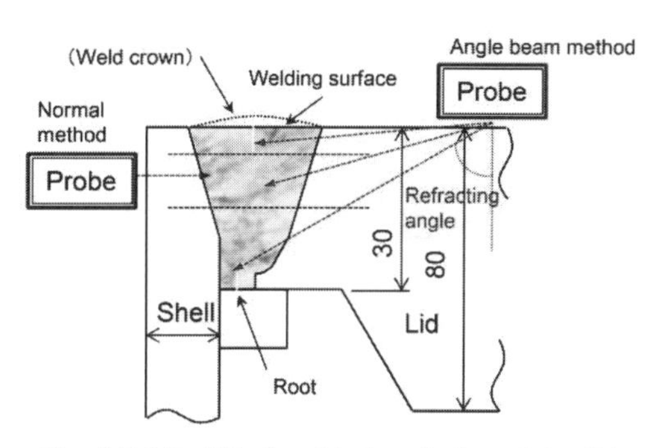

Fig. 4.5.2-2 UT of welded part of canister lid

Fig. 4.5.2-3 Full-scale canister lid model

(1) Manufacturing lid model

As shown in Fig. 4.5.2-4, the canister lid model that we manufactured consisted of a shell and a lid. The lid was placed on the spacer ring (backing strip and supporting lug) inside the shell and welded. Then the shell was cut to manufacture two canister lid model test bodies (models A and B).

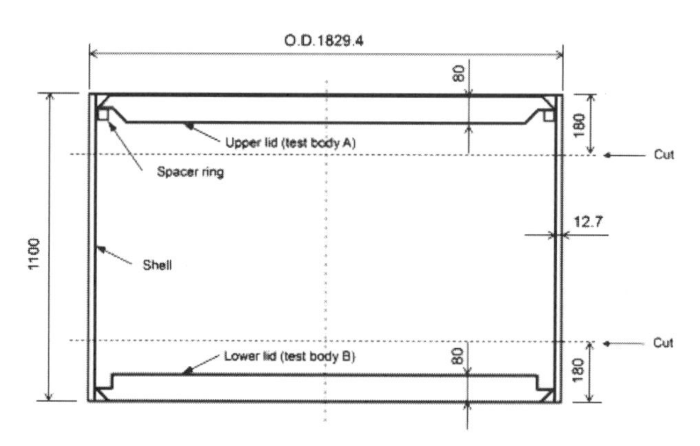

Fig. 4.5.2-4 Shape and size of lid model test body before cutting

The weld crown was removed with a grinder. The influence of excess metal on the defect detection area was taken account of in the analysis of UT defect detection data.

An electro-discharge machined defect (hereafter referred to as EDM defect) was made on model A and artificial welding defects were made on model B. Fig. 4.5.2-5 shows the shape and dimensions of the EDM defect made on model A. Since the detection target was a crack on the welding root, a defect that could work as a reflection source similar to this natural crack was created by EDM. The depth target was set to 1/8 t_w (t_w: welding depth) by taking account of the initial defect's depth 1/4 t_w assumed in JSME Structure Standards and of safety ratio 2. Since the test body's welding depth was 30 mm, the size of the target defect was 3.7 mm and the standard defect depth was 4 mm. To check the sensitivity of the defect size detection, defects of various shapes were made at depths of 2, 4, and 6 mm. On the other hand, four artificial welding defects, including fusion failure and blow hole, were created on the welding root of model test body B.

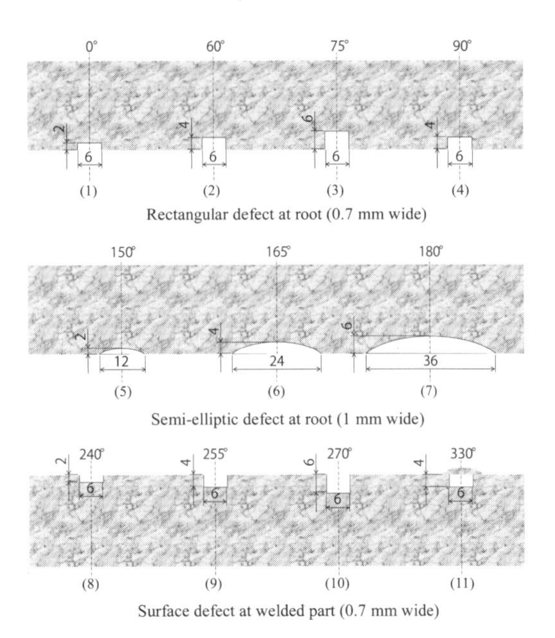

Fig. 4.5.2-5 Shape and dimensions of EDM defect

(2) Minimum detection size with UT

In the test, defect detection equipment with an automatic scanner accessible from top of the lid was used. The equipment is shown in Fig. 4.5.2-6. To search for defects, we employed three methods: Phased Array method (PA method) and conventional methods such as the fixed angle beam method and normal beam method. "Conventional methods" are defined as the ones that combine an ultrasonic defect detector that displays defect search waveforms (A scope) and a probe that has one or two vibrators.

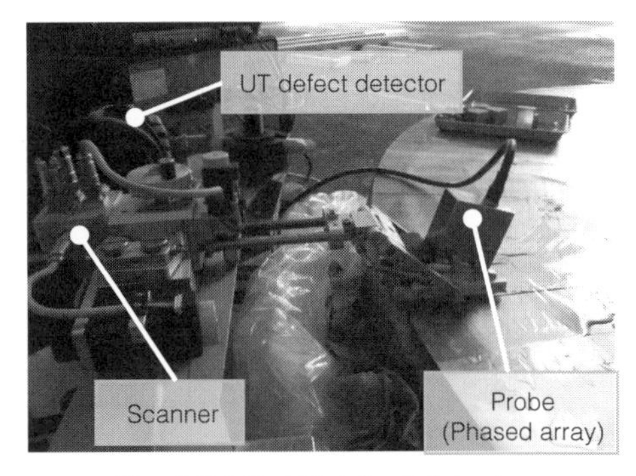

Fig. 4.5.2-6 UT defect detector

1) EDM defect detection result

Table 4.5.2-1 shows the detection results of the artificial defects on the welding root of test body A. In both the PA method and conventional methods using transverse waves, the detection performance was better at the frequency 2 MHz than at 5 MHz and at the reflecting angle 45° than at 60°. In particular, in the defect detection from above the lid at the frequency 2 MHz and reflecting angle 45°, all the defects could be detected at DAC100 % or higher. Therefore this condition is the best for the detection. In the PA method with longitudinal waves at 2 MHz or 5 MHz, the image analysis of defects viewed from above the lid was almost always difficult due to pseudo-echoes from the surroundings. It was also difficult from the side of the shell due to the pseudo-echoes. These test results indicate that UT under the best conditions could detect a defect at a depth 2 mm from the root. This 2 mm depth defect is 1/15 of the welding depth 30 mm of the test body, smaller than 1/4 of the welding depth assumed in the JSME Structure

Table 4.5.2-1 Detection result of EDM artificial detect on welding root (model A)

UT technique				Target defect (number, shape, depth / length)							
Method	Mode	Frequency	Defect face	Reflecting angle	(i) Rectangular 2 mm / 6 mm	(ii)Rectangular 4 mm / 6 mm	(iii) Rectangular 6 mm / 6 mm	(iv) Rectangular 4 mm / 6 mm	(v) Semi-elliptic 2 mm / 12 mm	(vi) Semi-elliptic 4 mm / 24 mm	(vii) Semi-elliptic 6 mm / 36 mm
PA	Transverse	2 MHz	Lid top face	45°	◎	◎	◎	◎(▲)[2]	◎	◎	◎
				60°	◎	○	○	○(▲)[2]	○	◎	◎
				Arbitrary[1]	◎	◎	◎	○(▲)[2]	◎	◎	◎
		5 MHz		45°	○	○	○	○(▲)[2]	◎	◎	◎
				60°	○	○	○	○(▲)[2]	○	○	◎
				Arbitrary[1]	○	○	○	○(▲)[2]	◎	◎	◎
	Longitudinal	2 MHz		45°	▲	▲	○	×	▲	▲	○
				60°	▲	▲	○	×	▲	▲	○
				Arbitrary[1]	▲	▲	○	▲	▲	▲	○
		5 MHz		45°	▲	▲	▲	▲	▲	▲	◎
				60°	▲	▲	▲	▲	▲	▲	○
				Arbitrary[1]	▲	▲	▲	▲	▲	▲	◎
			Side of body	–	×	△	△	×	×	×	△
Conventional method	Transverse	2 MHz	Lid top face	45°	◎	◎	◎	◎(▲)[2]	◎	◎	◎
				60°	▲	○	◎	○(▲)[2]	○	◎	◎
		5 MHz		45°	○	▲	○	○(▲)[2]	○	○	◎
				60°	▲	▲	▲	○(▲)[2]	▲	△	○

1) The largest echo height (%CRT) at an arbitrary reflecting angle
2) Distinguishable but many similar indications in the surroundings

Standards. Therefore the UT method was found to be sufficiently valid for the detection of defects in the welded parts of the lid and should be able to be applied to actual facilities.

Fig. 4.5.2-7 shows an example of detecting the defect (i) (transverse wave, 2 MHz, refracting angle 45° or around). The defect could be clearly recognized by the visualization.

Since not only the given defect but also many pseudo-echoes were detected around the defect (iv), judgment criterion (was put down in the list.

On the other hand, in actual canister lids, there is always weld crown existing in welded parts and the defect scanning study from above the lid cannot be made over the weld crown. As mentioned in (1) above, weld crown was removed from the test body in this report. Fig. 4.5.2-8 shows an extended defect scanning area and indicates the area where defects cannot be detected due to the weld crown in the UT defect detection data analysis image. The figure shows the influence of the weld crown on the welding root and surface in the test with the PA and conventional methods.

Defects in the root can be detected with the conventional methods which use smaller probes than the PA method, but those on the surface cannot be detected even with the conventional methods. Therefore, the conventional methods are effective for the detection of defects in the root in the presence of excess metal. For the detection of defects on the surface with the PA and conventional methods, another measure needs to be taken against the excess metal.

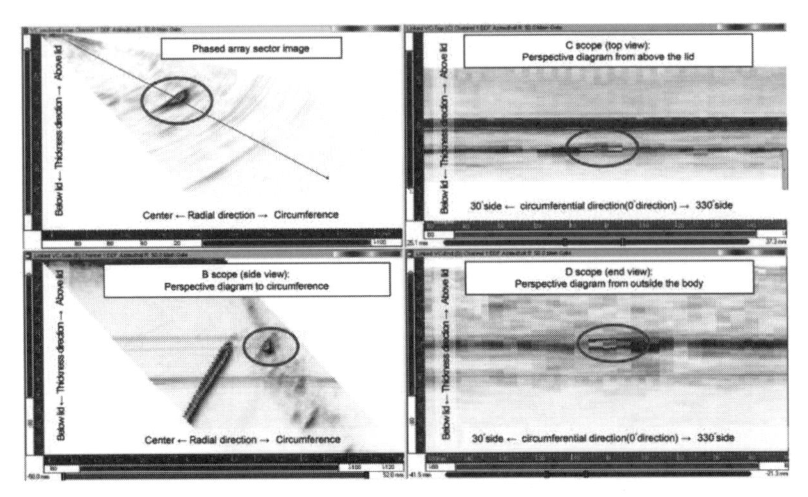

Fig. 4.5.2-7 Example of detection of defect (i) on the welding root (transverse wave, 2 MHz, reflecting angle of around 45°)

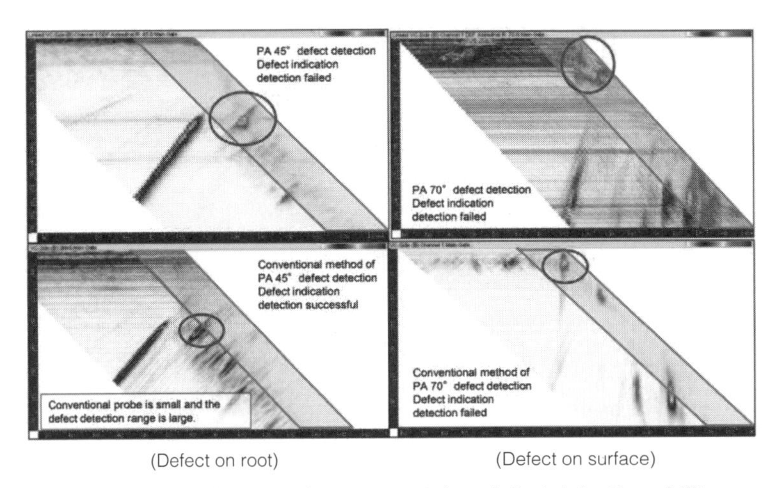

(Defect on root) (Defect on surface)

Fig. 4.5.2-8 Influence of excess metal on defect detection ability

2) UT detection ability of artificial welding defects

Four kinds of artificial welding defects, which could actually occur, were created on the welding root on test body B as shown in Table 4.5.2-2. Macroscopic image of the cross section with the defects (at 45°, 135°, 225°, 315°) on test body B were taken and the UT and RT were conducted at the defect positions. The fusion failure at 45°, blow hole at 225°, and crack at 315° observed on the macroscopic cross section had the shape that we expected. The high-temperature cracking at 135° should be fusion failure on the observation cross section.

Fig. 4.5.2-9 shows macroscopic image of the cross section where fusion failure was created and the UT detection images. The fusion failure defects in the root could be distinguished with both transverse and longitudinal waves. On the other hand, small spherical defects like blow balls could be distinguished if the size was larger than 2 mm. Some of the spherical defects of the size smaller than 2 mm could not be identified.

Table 4.5.2-2 Creation method of artificial welding failure defects

Creation point	Defect type	Attachment gap	Defect creation method	Circumferential direction position
Welding root	Lack of fusion	0 mm	Only welding rod is melted by a low electric current and fuse failure is caused on the base material by melting only the welding rod with a low electric current.	45°
	Wrong welding material (High temperature cracking)	3 mm	Inconel material is used for welding and a high temperature crack is created on the second layer due to the wrong component.	135°
	Porosity (Blow hole)	6 mm	Oxidation of the back bead is promoted by reducing the gas amount.	225°
	Crack	3 mm	The bead is cracked by applying an impact load on the groove.	315°

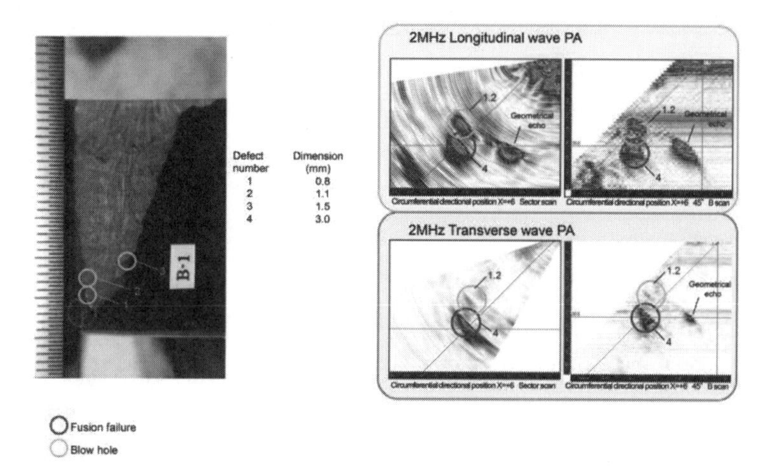

Fig. 4.5.2-9 Macroscopic picture of cross section of fusion failure area and results of UT

(3) Minimum detection limit evaluation of defect dimensions in UT

The study on test bodies A and B showed that crack type defects of the size 2 mm or larger could be detected. The size 2 mm was 1/15 of the welding depth 30 mm of the test bodies and smaller than 1/4 of the welding depth assumed in the JSME Structure Standards. Therefore, the UT method was shown to have high enough ability of detecting defects on the welding parts of the lid.

References

1) Nuclear and Industrial Safety Agency: Technical requirement for used fuel storage facilities (interim storage facilities) with concrete casks, April 2006.

2) The Japan Society of Mechanical Engineers: Used fuel storage facility standards, Concrete cask, Canister refilling equipment and canister transport cask, Structure Standard, JSME S FA1-2003, 2003.12.

3) ASME : Case of ASME Boiler And Pressure Vessel Code, Case N-595-4, Requirements for Spent Fuel Storage Canisters Section III, Division 1, 2004.5.

4) The Japan Welding Engineering Society: Marginal depth disposal vessel welding standard, WES 7901, 2011.2.

5) M. Goto, H. Shoji, and K. Shirai, "Study on Spent Fuel Storage by Concrete Cask for Practical Use -Applicability

Evaluation of Ultrasonic Test with Imaging Analysis for Type304L Stainless Steel Canister Lid weldment-,"
Central Research Institute of Electric Power Industry, Research Report N11057 (2012).

4.5.3 Canister drop test and analysis

Canister is transferred from transfer cask to concrete cask at a storage facility vertically or horizontally. In USA, NRC published safety evaluation of canister assuming that impact load works on the canister by hypothetical drop or tip over event from view point of defense in depth. On the other hand experimental case study is rarely reported in domestic and overseas countries. Accumulation of basic data on deformation and damage of full-scale canister is necessary.

We performed drop tests based on non-mechanical events using the two full-scale canisters used for heat removal tests with welded double lid structure in order to evaluate the canister's integrity (namely, there is no significant deformation or damage for maintaining containment performance). The drop tests were performed at the drop test yard in Akagi test yard of CRIEPI as shown in Fig. 4.5.3-1[1].

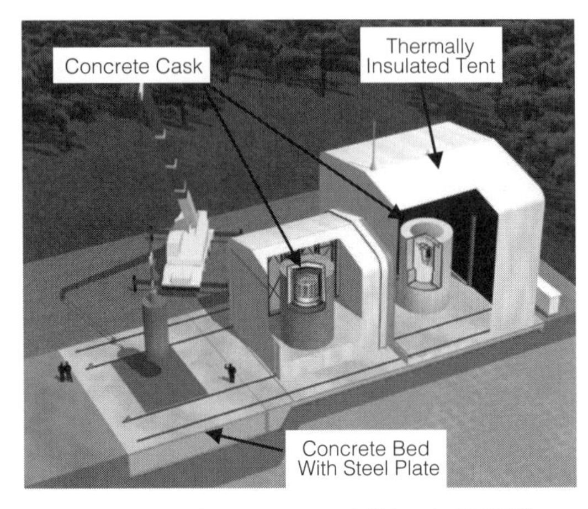

Fig. 4.5.3-1 Drop test yard (Akagi, CRIEPI)

(1) Drop test specimens

Two types of canisters were used in the drop tests as shown in Fig. 4.2-5 and Table 4.2-5. Simulated weight for spent fuel was installed in the canisters for the drop tests.

Type I canister consists of upper shield plate, a pile of primary and secondary lids. Both lids are partially welded to the canister body. The basket consists of spacer tube, guide tube, and support rods. Type II canister consists of a thick primary lid and secondary lid. Both lids are partially welded to the canister body. The basket is assembled of aluminum alloy plates to form lattice shape.

(2) Drop tests

Table 4.5.3-1 shows the drop test conditions. Two drop tests in horizontal and vertical orientations were conducted considering non-mechanistic drop or impact events during handling, and the drop heights were 1 m and 6 m, respectively. The drop height of 1 m was determined as a reference and unit height. The drop height of 6 m was determined as a lifting height of the canister when the canister is transferred to concrete cask.

As for the object target, the hard target, namely, the 5 cm thickness steel plate attached to the concrete block (widh 13 m, thickness 2 m, length 10 m, total weight about 550 ton) was applied. In order to monitor the impact response of the target block during the tests, the accelerometers were set inside the concrete block. Containment tests of the canisters were prformed before and after the drop tests.

Table 4.5.3-1 Canister drop test conditions

Canister	Type I	Type II
Non-mechanical events during handling	Tip over	Drop
Orientation	Horizontal	Vertical
Height	1 m*	6 m**

* Equivalent drop height for rotational velocity caused by tipping-over from height of GC
** Drop height from cask height

1) Horizontal drop test

Fig. 4.5.3-2 shows photographs of the test canister before and after the drop test. The acceleration waves were treated with a 1 kHz low-pass filter. The test canister was slightly deformed near the impacted area. It also shows time histories of acceleration at various points in the test canister, measured in the drop test. The average deceleration value was about 436 G at the top of the lids.

Measured leakage rates shows the integrity of sealability at lids and canister shell, as all values are under 1.0×10^{-9} $Pa \cdot m^3/s$ (Design criteria : 1.0×10^{-6} $Pa \cdot m^3/s$).

In Fig. 4.5.3-3, photographs of the cut section of the directly impacted welded part during horizontal drop test are shown through microscope with magnified by 5.7 times. Crack initiation could be found in this figure due to the impulsive moment around the top corner of the test canister. However, the initiated crack was arrested in the first welded layer.

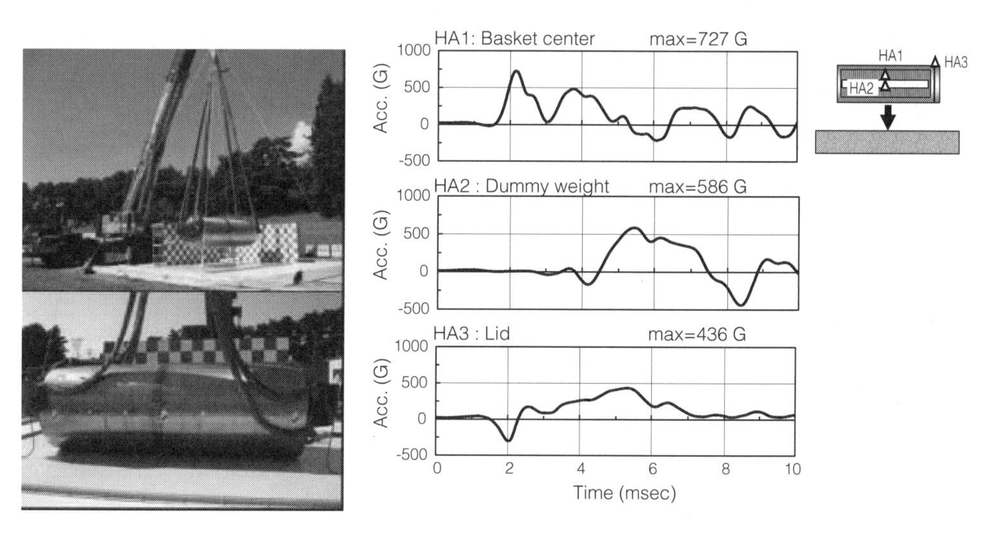

Fig. 4.5.3-2 Example of measured time histories of acceleration during horizontal drop test

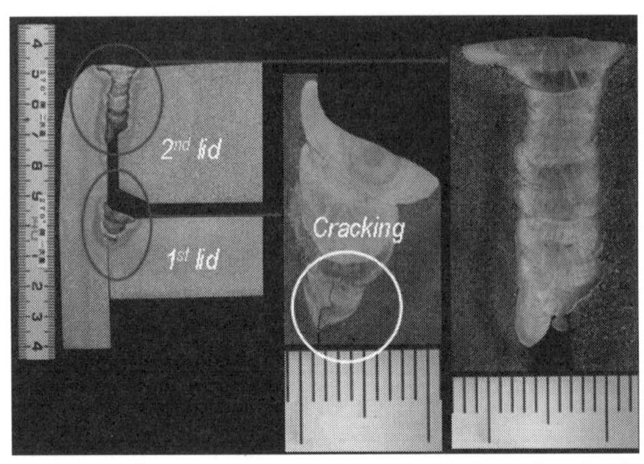

Fig. 4.5.3-3 Magnified view of cut section of welded part after horizontal drop test

2) Vertical drop test

Fig. 4.5.3-4 shows photographs of the test canister before and after the drop test. The bottom plate of the test canister was deformed by the force of inertia of the contents. However, the basket was slightly deformed near the impacted area. It also shows time histories of accelerometers and strain gauges at various points in the test canister, measured in the drop test. The average deceleration value was about 1153 G at the center of the shell.

He-leak tests were performed before and after the drop tests to confirm the integrity of leak-tightness of the test canister (especially welded lids) against impact loads. Measured leakage rates shows the integrity of sealability at lids and canister shell, as all values are under 1.0×10^{-9} Pa*m^3/s.

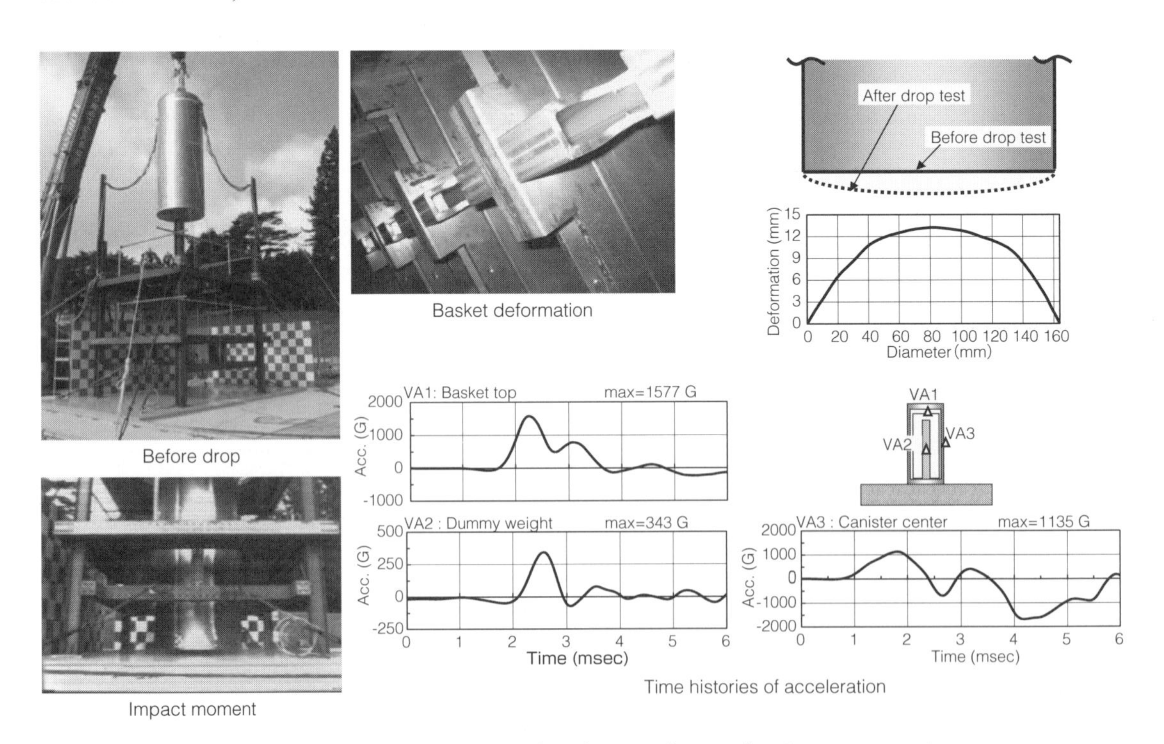

Fig. 4.5.3-4 Vertical drop test of canister and example of measurements

(3) Drop test analysis

To investigate the impact loads and strain distributions yielded in the canister's bodies and inner structures during drop tests, impact analyses were executed using LS-DYNA code.

Table 4.5.3-2 shows the material properties of the canister used in the analyses. These properties were obtained from static tensile tests of specimen machined from the base material. The stress-strain relation of these materials were expressed by bilinear approximation and based on the isotropic hardening rule. For failure criterion, the Von-Mises yield criterion was applied, but the strain rate effect was not considered by the first order approximation. In Fig. 4.5.3-5, the test results and the analytical results were compared as to the acceleration time history for each drop test. In the analysis, the maximum acceleration is in good agreement with the test results in each drop orientation.

Table 4.5.3-3 shows the summary of the drop analysis. An important plastic strain yielded in the vicinity of the welded joints, however the corresponding strains (8.4 % at maximum) were considerably less than the ultimate strain (more than 20 %) of the material in each drop orientation.

Table 4.5.3-2 Material properties of canister

Material	Super stainless steel	Austenitic-ferritic stainless steel
Density	8.0 g/cm^3	8.0 g/cm^3
Elastic modulus	192 GPa	213 GPa
Hardening modulus	1012 MPa	786 MPa
Yield stress	407 MPa	664 MPa
Poisson ratio	0.31	0.27

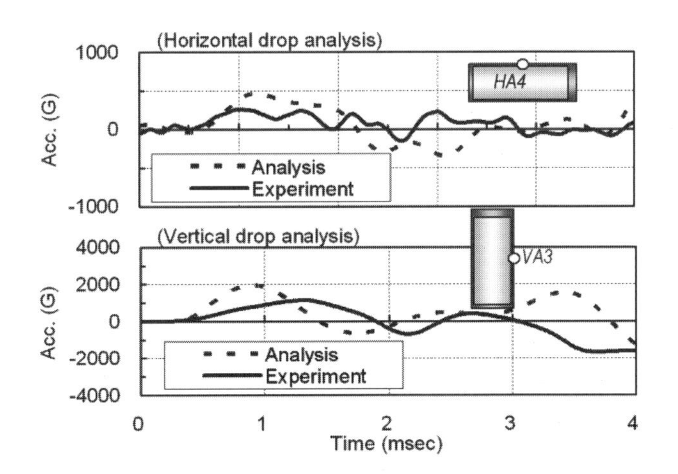

Fig. 4.5.3-5 Comparison between analysis and measurement of drop test

Table 4.5.3-3 Summary of analysis for drop tests

Part	Plastic strain %	
	Horizontal drop	Vertical drop
Shell body	8.4	5.0
Primary lid	5.3	2.6
Secondary lid	5.4	7.0

(4) Summary

Two drop tests in horizontal and vertical orientations with the full-scale canisters were conducted onto the hard target, and each drop heights were 1 m and 6 m, respectively. In the drop tests, containment performance of the canisters were maintained and the yielded strain at the local part of the canisters were less than 10 %, which is much less than the ultimate strain (more than 20 %) for fracture. In order to investigate the impact loads and strain distributions occurred in the canister bodies and inner structures during drop tests, impact analyses were executed using LS-DYNA3D code. According to these investigations, the structural and containment integrities of the canisters were verified, even if subjected to extreme loads related to the non-mechanical drop or impact events.

Reference

1) K. Shirai and T. Saegusa: "Demonstrative drop tests of transport and storage full-scale canisters wit high corrosion-resistant material", Nucl. Eng. Design, Vol.238, pp.1241-1249 (2008).

4.5.4 Temperature-caused crack test of concrete container

A crack could occur on the periphery of a concrete storage container due to thermal expansion or temperature stress since the container is exposed to a high temperature by the heat generation of used fuels. As will be mentioned below, the local concrete temperature limit of the reinforced concrete storage containers was set to 90 °C in the design, by assuming that there will be only a temperature load in the long term and an inertia force of the weight during an earthquake in the short term and that high-quality concrete is used for the casks. Therefore we could expect a large local temperature increase near the exhaust outlet of the concrete cask (Fig. 4.5.4-1).

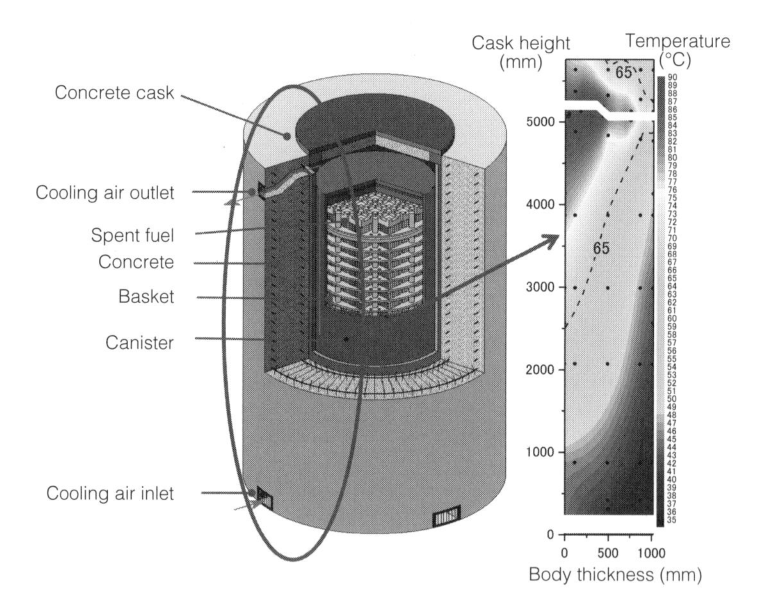

Fig. 4.5.4-1 Example of temperature distribution of concrete cask

In this section, a method of analytical evaluation of crack width is explained for the evaluation of crack generation of concrete casks due to temperature loads. A high temperature fracture toughness test was conducted with a non-reinforced concrete joist test body having a cutout to study the temperature dependence of the fracture mechanics parameters. We also performed a heat transfer test for an RC cylindrical structure object and made numerical analysis to quantitatively investigate crack width. For details see Reference[1].

(1) High temperature fracture toughness test

1) Temperature dependence of toughness value and destruction energy

Fig. 4.5.4-2 shows the temperature dependence of fracture toughness value K_{IC} and destruction energy G_F. The K_{IC} remains almost the same at temperatures up to 65 °C but decreases at 90 °C and then increases with the temperature. On the other hand, G_F increases with the temperature up to 65 °C but slightly decreases at 90 °C and then increases with the temperature. It was pointed out[2] that the mineral composition of concrete materials changes due to thermal influences at around 65 to 90 °C and the fracture mechanical parameters largely change with the change of the mineral composition.

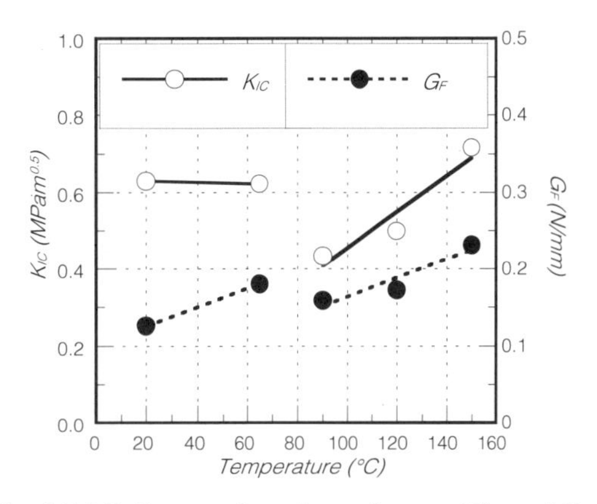

Fig. 4.5.4-2 Temperature dependence of K_{IC} and G_F

2) Temperature dependence of tensile softening characteristic

From a load-CMOD curve obtained in the fracture toughness test, we estimated the tensile softening characteristic by using the multi-line approximation method[3] proposed by the Japan Concrete Institute. Fig. 4.5.4-3 shows the temperature dependence of the tensile softening characteristic. The tensile strength gradually decreases with the increase of the temperature up to 90 °C and the limit virtual crack width increases. On the other hand, at around 120-150 °C, the tensile strength is recovered a little and the limit virtual crack width is larger than at room temperature.

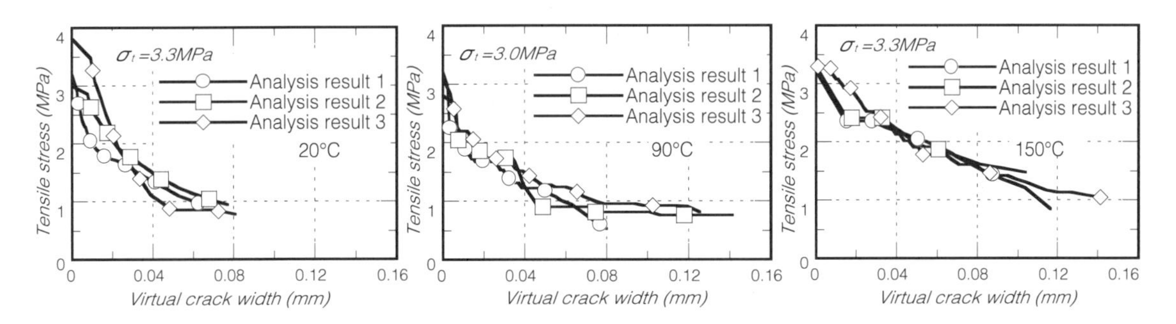

Fig. 4.5.4-3 Temperature dependence of tensile softening characteristic

(2) Heat transfer test of RC cylindrical structure

1) Test body and test method

We manufactured a cylindrical test body shown in Fig. 4.5.4-4 and performed a heat transfer test. The test body dimensions were outer diameter 1200 × inner diameter 590 × height 1000 mm. Inside the cylindrical test body, a 9.5 mm-thick steel liner plate was placed. To prevent interference with the concrete, two Teflon sheets with about 0.25 mm thick grease in between were inserted between the concrete and the steel liner plate. The concrete composition was the same as that used in the fracture toughness test. The D16 reinforcing bars were used for hoops and D13 for vertical reinforcement. The reinforcement ratio was 1.07 % in the circumferential direction and 0.48 % in the axial direction. Wedge-shape cutouts of width 10 mm and height 20 mm were introduced to the concrete. A clip gauge was mounted on each cutout to measure the change of the size during the test. A strain gauge and thermocouple were attached to the surface of the test body and a thermocouple to the inside of the test body to measure stress and temperature distribution.

The test body was placed in a thermostat bath and the atmospheric temperature of the test body and the thermostat bath was increased to and kept at 38 °C. Then a heater was used to increase the temperature of the inner surface of the test body to 90 °C. The temperature increase speed was about 2 °C/h. Then, the temperature was kept for a certain period of time and the test body was naturally cooled. The cutout size change, strain and temperature were measured during this process and the cracks of the test body were observed after the test.

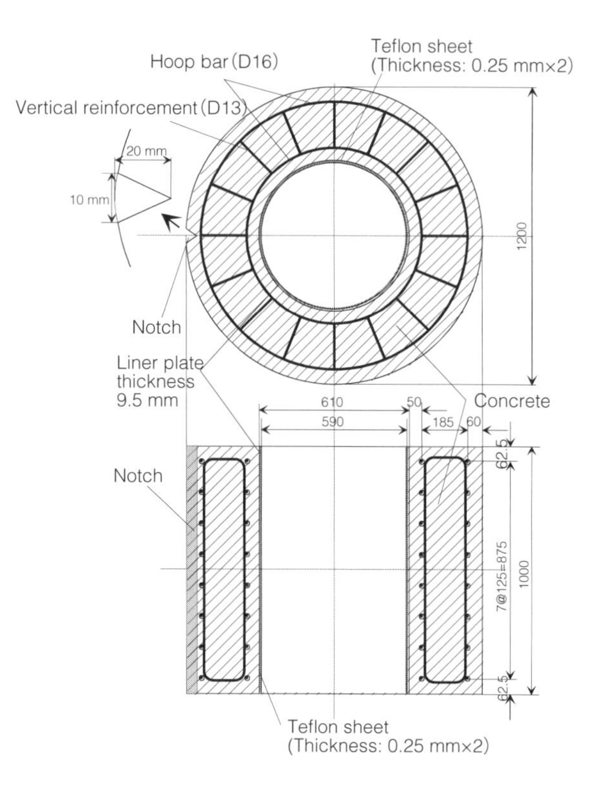

Fig. 4.5.4-4 Shape and dimensions of the test body

Fig. 4.5.4-5 shows crack conditions of the test body after the test. The cracks occurred not only near the cutouts but also in other areas. On the top face, a crack penetrated the concrete to the liner plate. Most of these cracks were found near the vertical reinforcement.

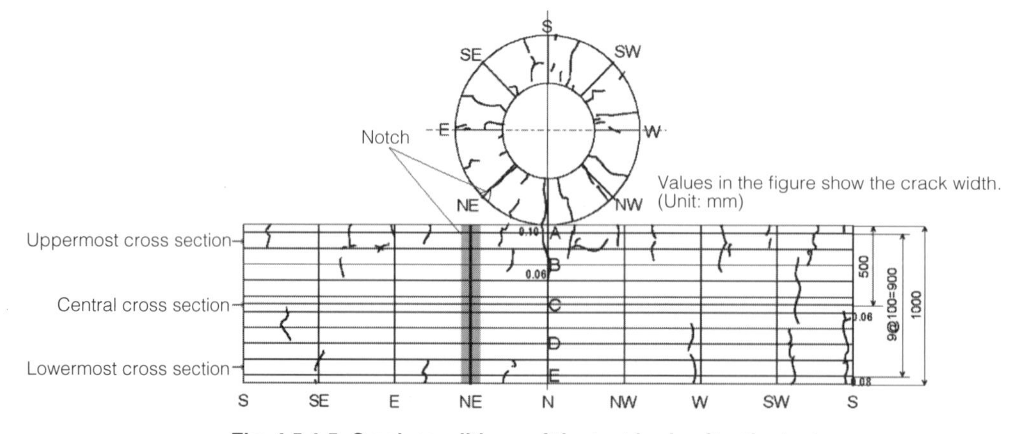

Fig. 4.5.4-5 Crack conditions of the test body after the test

2) Test result

The first crack occurred at the cutouts after 190 hours and the temperature difference was about 7 °C at the initiation of the crack. The largest crack width at the end of the notch in the test was 0.2 mm.

(3) Crack analysis

1) Method

For the analysis, the two-dimension finite element method program CRANCYL was used. CRANCYL was an improved code based on CRAN[4)], an analysis program of cracks of gravity dams. For the linear elastic body only, a discrete crack model based on the linear fracture mechanics can be used to evaluate crack conditions and crack extension.

A plane strain condition was used for the analysis. After the initiation of the cracks, a contact spring with equivalent stiffness of the reinforcement was loaded on the surface of the cracks to model the reinforcing bar pull-out. In the crack model, actual measurement data were used for the tensile strength, fracture toughness, tensile softening curve, and other properties of the concrete and the temperature at the boundaries. Also, the same linear expansion coefficient as the concrete was used for the liner plate in order to introduce the effect of the Teflon sheets that relaxed the interference.

The elastic modulus, tensile strength, and fracture toughness were assumed to form normal distributions (with variation coefficient of 0.1) by taking account of the variation.

2) Analysis result

Fig. 4.5.4-6 shows crack initiation conditions. The result shows penetrating cracks and small cracks, and is in relatively good agreement with the test result.

Fig. 4.5.4-7 shows crack width at the end of the notch in comparison with the measured data. The crack occurred at the end of the notch 199 hours after (temperature difference: 11 °C) and penetrated the concrete 215 hours after. The crack initiation time obtained in the analysis was in good agreement with the measured time. The largest crack width was 0.2 mm in the measurement and 0.32 mm in the analysis. Namely, the analysis gave 1.5 times as large width, which was accurate enough in practice.

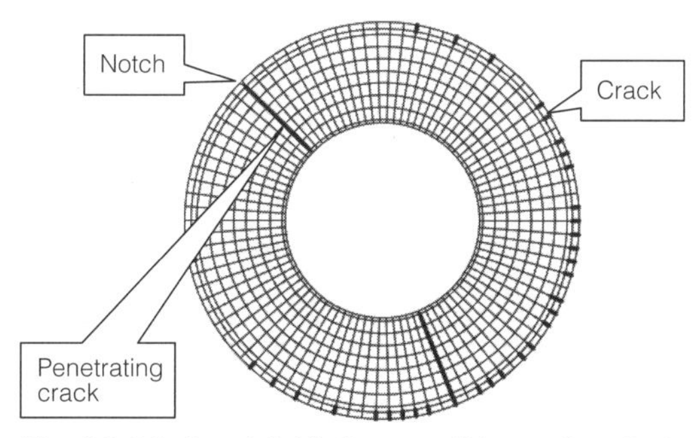

Fig. 4.5.4-6 Crack initiation conditions of analysis

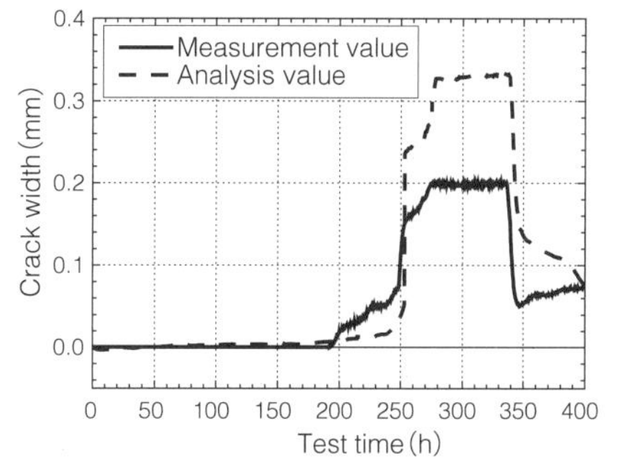

Fig. 4.5.4-7 Comparison between test and analysis of crack width

(4) Summary

For the evaluation of the temperature crack behavior of the concrete casks with temperature loads, a cylindrical reinforced concrete test body with cutouts was manufactured and the heat transfer test was conducted to compare the test result with the analysis result of the crack propagation. From the result we found that both results were in good agreement with respect to the crack initiation and propagation condition.

References

1) T. Saegusa, K. Shirai, et al.: "Development and evaluation of concrete storage technology of used fuels," Central Research Institute of Electric Power Industry, General Report No.9, May 2010.

2) A. Atkinson and J.A. Hearne, "The Hydrothermal Chemistry of Portland Cement and its Relevance to Radioactive Waste Disposal", UK Nirex Ltd. Report, NSS/R187, 1989.

3) Japan Concrete Institute: Survey Research Committee Report on Concrete Fracture Characteristic Test Method, May 2001.

4) M. Irobe and S.Y. Peng, Proceedings FRAMCOS-3, pp.1605-1614, 1997.

4.6 Earthquake resistance

4.6.1 Seismic tipping and sliding test using scale model

A concrete cask storing spent fuel is required to be designed according to safety importance so that the surrounding general public is not exposed to radiation at the time of an earthquake. Also, a measure that the concrete cask is vertically placed with being unfixed can be assumed in terms of rationality. In the case of the unfixed placement, it is required to verify that the concrete casks do not tip over or interfere with each other and the spent fuel is not affected by setting proper design seismic force acting on the concrete cask. Furthermore, because the concrete cask has such a structure that a concrete storage container (Seismic Class B, maintaining an Ss function) contains a canister (Seismic Class S) storing the spent fuel, the canister is required to maintain criticality prevention and containment functions for the spent fuel at the occurrence of a scenario earthquake.

Fig. 4.6.1 shows a flow of seismic capacity evaluation of the concrete cask that is vertically placed with being unfixed. In this evaluation flow, first, the evaluation is performed on the occurrence of tipping and sliding of the concrete cask in reaction to assumed seismic loads, and seismic tipping stability is evaluated based on comparison with allowable values. A heat removal function using natural circulation between the canister and an inner surface of the concrete cask, which is a characteristic of the concrete cask, is especially expected. In this structure, such a space as a flow channel can be maintained even during an earthquake. Thus, on the ground that dynamic interaction occurs to the unfixed canister and concrete storage container when dynamic force due to an earthquake is applied to the concrete cask, it is required to evaluate acceleration generated on the concrete cask and the initiation stress of the canister.

In this chapter, real shapes and interaction behaviors of the concrete storage container and the canister during

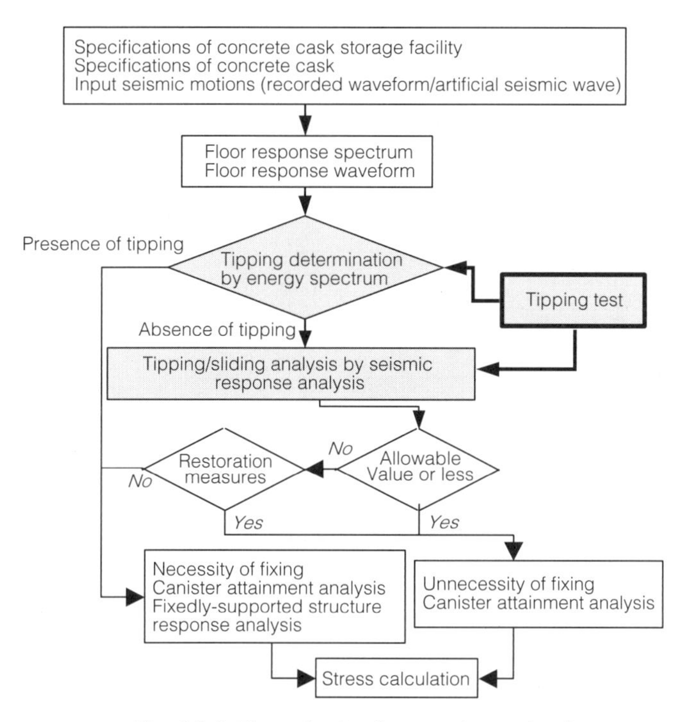

Fig. 4.6.1 Flow of seismic capacity evaluation

an earthquake are treated as dynamic vibration problems between cylindrical rigid bodies, and a tipping stability evaluation method using an energy spectrum is described while a tipping test on a scale similar model of the concrete cask is conducted by using a large-sized oscillation stand. (Refer to the reference[1] for details)

(1) Test body for tipping test

The concrete cask as an object was a reinforced-concrete (RC) cask. The main specifications are shown in Sec. 4.2. Fig. 4.6.2 and Table 4.6.1 show a shape, dimensions, and the main specifications of the similar model of the concrete cask.

Its representative fraction was 1/3 in consideration of similarity with an actual cask and a performance limit of the 100 t-class large-sized oscillation stand.

Table 4.6.1 Main specifications of similar model of concrete cask

Part	Dimension		Material
Storage container	Overall height	1901 mm	Shell part : carbon steel Leg part : reinforced concrete
	Outer diameter of upper structure	1230 mm	
	Inner diameter of upper structure	952 mm	
	Mass	8.17 ton	
Canister	Overall height	1543 mm	Carbon steel
	Maximum outer diameter	559 mm	
	Outer diameter of center part	420 mm	
	Mass	1.95 ton	

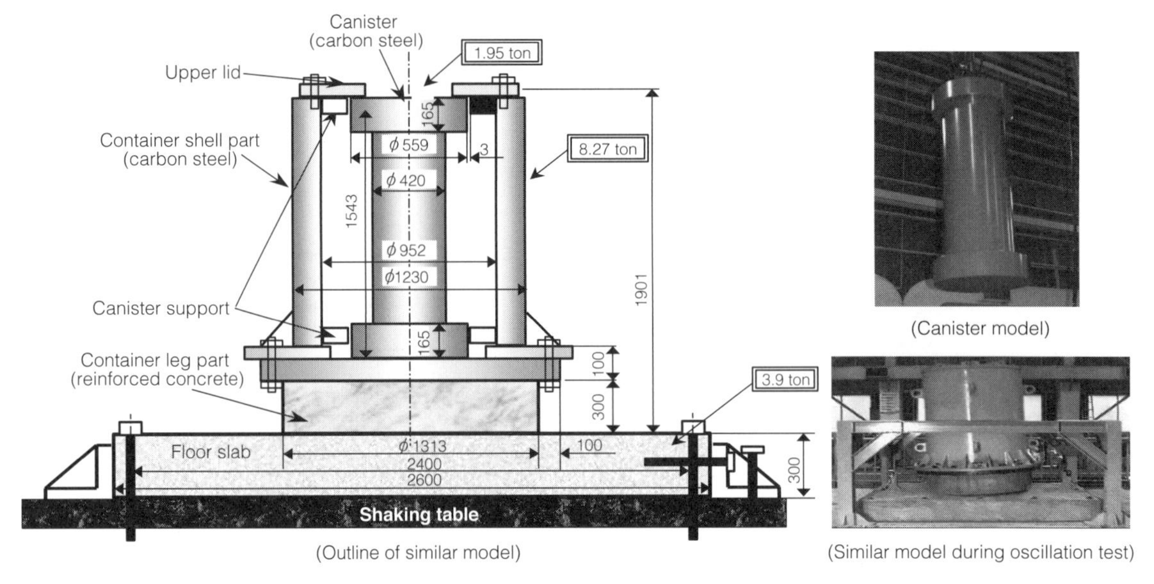

(Outline of similar model) (Similar model during oscillation test)

Fig. 4.6.2 Shape and dimensions of similar model

(2) Tipping test

1) Test conditions

Table 4.6.2 shows seismic input waveforms used in a vibration test. Two kinds of recorded waves (El Centro, and JMA Kobe) and two kinds of artificial seismic waves (low seismic area Ss, and high seismic area Ss) were used in the vibration test.

Table 4.6.2 Seismic input waveforms used in vibration test

Input waveform		Wave information	Maximum acceleration value
Recoded wave	El Centro	Imperial Valley Earthquake in 1940	NS 342 gal, UD 206 gal
	JMA Kobe	Southern Hyogo Prefecture Earthquake in 1995 : Kobe Local Mythological Office	NS 821 gal, UD 333 gal
Artificial seismic wave	Seismic committee No.1 : low seismic area Ss	Magnitude: 6.5, Epicentral distance: 7.2 km	H 259 gal, UD 168 gal
	Seismic committee No. 2 : high seismic area Ss	Magnitude: 8.5, Epicentral distance: 68 km	H 204 gal, UD 124 gal

2) Test result

When the waveform of JMA Kobe of the Southern Hyogo Prefecture Earthquake in 1995 was input horizontally and vertically at the same time, the vibration behavior of the scale model was rocking vibration associated with three-dimensional spinning vibration, which was generated after uplift. The maximum response angle was 0.417 rad, the maximum uplift displacement at the gravity center of the scale model was 26.5 mm, but residual sliding displacement was about 5 mm.

Fig. 4.6.3 shows an example of the relation between the input magnification of the maximum input acceleration and the maximum response angle. The result showed that in the artificial seismic waves for design, the rocking vibration and sliding were not generated at one-fold of an input level. Also, when the input levels of the artificial seismic waves were amplified three-fold or more, the large rocking vibration was rapidly generated on the scale

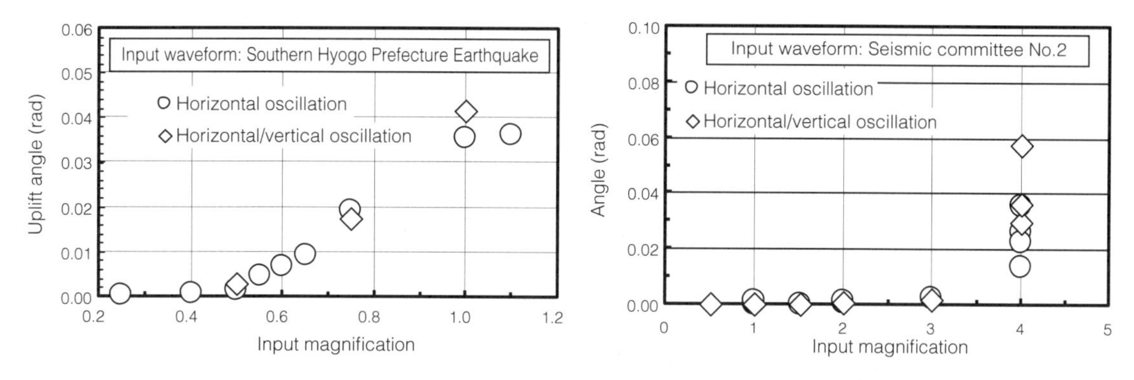

Fig. 4.6.3 Relation between maximum input acceleration and maximum response angle

model test body of the cask, but the test body did not tip over. In the vibration behavior of the test body, the rocking behavior was dominant, and the sliding was not generated alone.

A variation in response of the rocking vibration due to repetitive oscillation became marked as the rocking vibration of the test body increased in size, and it almost doubled at a maximum. Also, in the influence of a vertical motion, the variation increased by about 20 % at a maximum compared with the response only to the horizontal input.

Fig. 4.6.4 shows the relation between the maximum input acceleration (artificial seismic wave: Type 2) and the maximum response angle in the case where a gap of 3 mm equivalent to one of the actual cask was provided between the storage container model and the canister model. The result showed that the rocking vibration can be suppressed if the gap is provided between the cask and the canister.

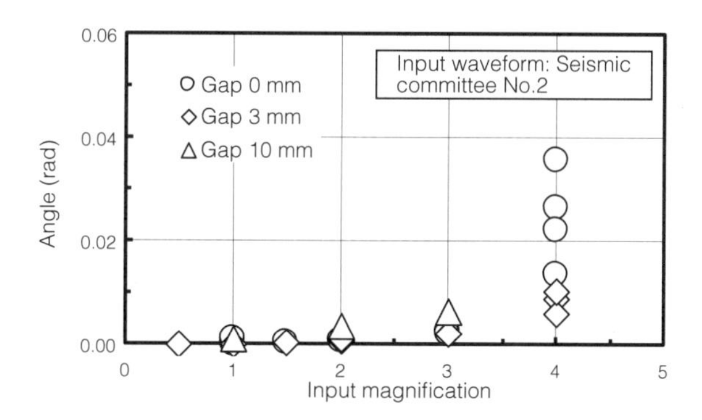

Fig. 4.6.4 Influence of gap amount on maximum response angle

(3) Tipping determination method by energy spectrum

In this section, a tipping evaluation method proposed as a method for determining the presence of the tipping by Akiyama et al.[2)] is verified based on the test result shown in the previous chapter, while its outline is described.

Akiyama et al. proposed the tipping evaluation method based on the energy spectrum with a two-dimensional model of a homogeneous cuboid placed on a rigid floor as a target. Its tipping evaluation formula is expressed as follows.

$$ {}_{ou}V_E(a) < V_{Ereq} $$

${}_{ou}V_E(a)$ expresses a tipping energy spectrum, and V_{Ereq} expresses a value corresponding to tipping limit energy velocity required for the tipping. Fig. 4.6.5 shows the energy spectrum of the waveform of JMA Kobe.

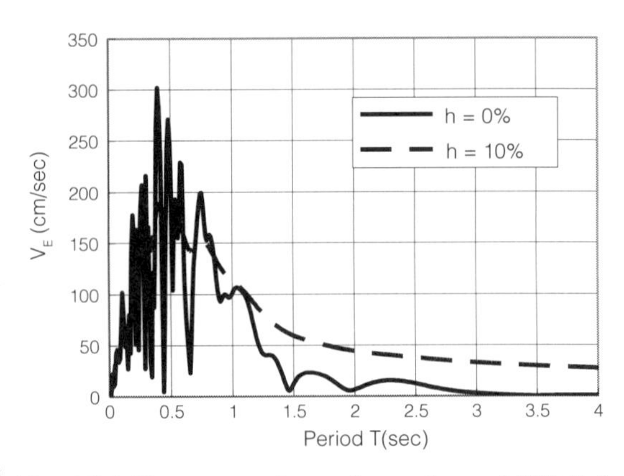

Fig. 4.6.5 Energy spectrum of waveform of JMA Kobe

Fig. 4.6.6 shows the comparison of a prediction result (input energy) obtained by the tipping energy spectrum with a test result (response energy) in the result of the vibration test using the wave of JMA Kobe.

The tipping energy spectrum $V_{EH,EV}$ in the case of the horizontal and vertical input was calculated based on the following formula using the input energy E_H, E_V in the horizontal and vertical directions.

$$ V_{EH,EV} = \sqrt{2(E_H + E_V)/M} $$

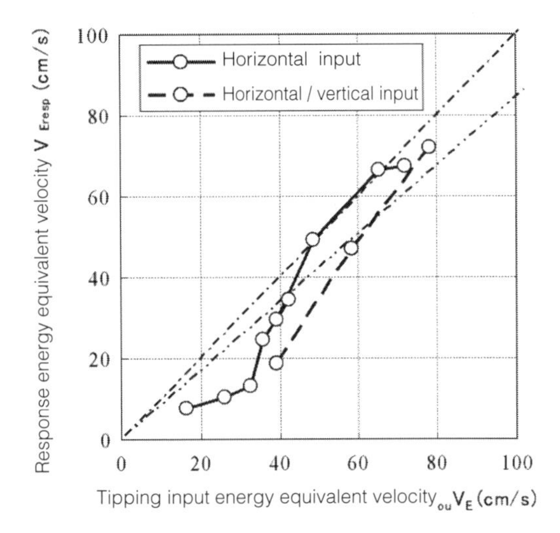

Fig. 4.6.6 Input and response energy (JMA Kobe)

The wave of JMA Kobe has a characteristic waveform which has large amplitude in the early period of the seismic motion. The amplitude has sufficient acceleration for the start of the rocking. In this case, there is a good correlation between the input energy obtained by the tipping energy spectrum and the response energy calculated based on the amount of the uplift.

Thus, the occurrence of tipping of the concrete cask during an earthquake can be determined with a sufficient margin of safety by setting a proper safety factor in the comparison of the energy input during an earthquake with the energy require for the uplift.

(4) Seismic response analysis

When the occurrence of the tipping is not predicted, the generation amount of the uplift and sliding is required to be properly estimated in order to determine the necessity of fixing of the concrete cask.

Seismic response analysis considering the tipping, sliding, and rotation was performed based on two and three-dimensional models whose gravity center, mass, and rotatory inertia were equivalent to the similar model by using two general-purpose analysis codes: TDAP, a two-dimensional finite element method code; and ABAQUS, a three-dimensional finite element method code (Explicit version).

Fig. 4.6.7 shows the comparison of analysis result values with experimental values in the case of inputting the wave of JMA Kobe. Fig. 4.6.8 shows examples of the three-dimensional analysis model and the analysis result. The

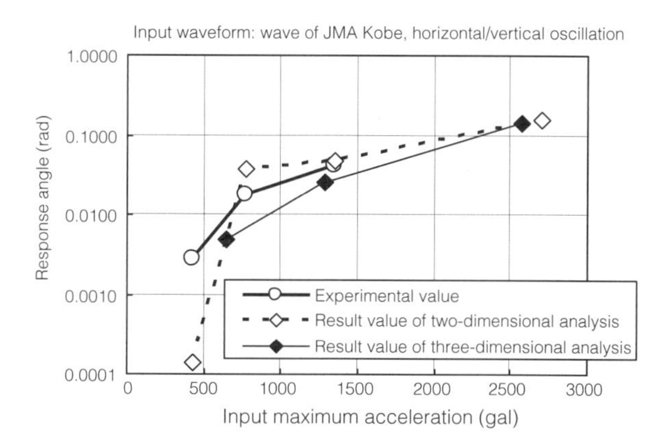

Fig. 4.6.7 Comparison of analysis result values and experimental values (wave of JMA Kobe)

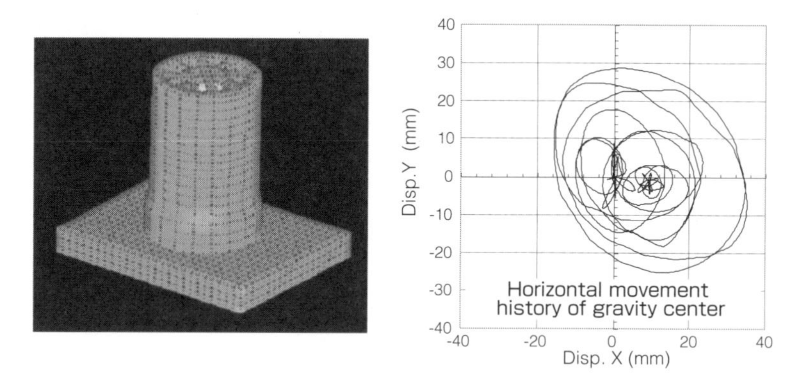

Fig. 4.6.8 Seismic response analysis considering rocking, sliding, and rotation (analysis code: ABAQUS)

three-dimensional analysis reproduced a state where the rocking vibration and sliding were generated in association with rotational vibration. Furthermore, the result of either two- or three-dimensional analysis showed that the maximum response angle obtained by the test was generally evaluated on the conservative side, so that tracking with practically-sufficient accuracy was possible. In the future, stress and strain generated at each part of the actual concrete cask can be directly evaluated by performing the detailed three-dimensional analysis.

4.6.2 Seismic test of full-scale concrete cask

(1) Outline of the test

Vibration tests[3)] of a full-scale concrete cask model (180 ton) were conducted using a large shaking table as follows.

* A full-scale concrete cask model, dummy spent fuel assemblies, a floor of storage facility, etc. were designed and produced.
* The 3D shaking table of E-defense of National Research Institute for Earth Science and Disaster Prevention

was used for frequency characteristic tests and vibration tests with three kinds of earthquake waves (El Centro, JAMA Kobe, artificial earthquake waves) (see Figs. 4.6.9 and 4.6.10)

* We obtained three dimensional seismic response of the concrete cask accompanied with rocking, sliding, and spinning. We also found that the concrete cask will not tip over even if we input waves with increased amplitudes exceeding the critical acceleration obtained from the formula of critical tip over evaluation.

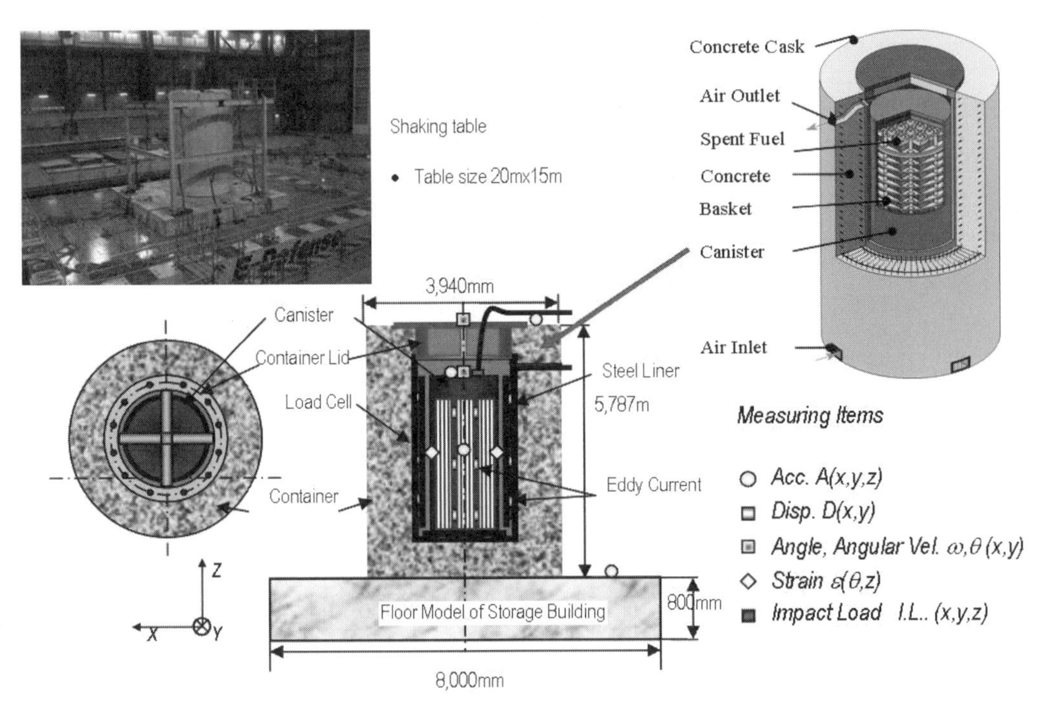

Fig. 4.6.9 Test of full-scale concrete cask on concrete floor using 3D shaking table

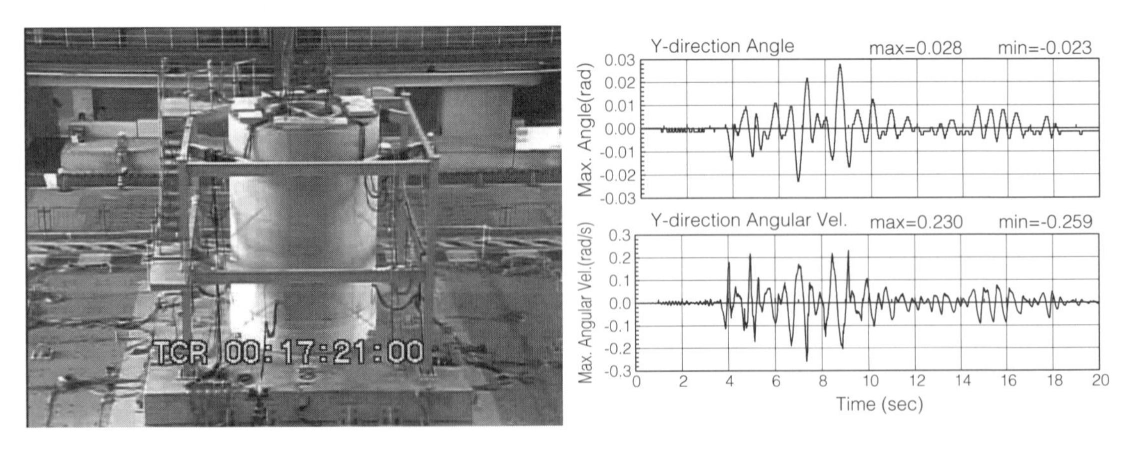

Fig. 4.6.10 Example of seismic response of cask rocking (JMA Kobe wave)

* Because of the gaps between the cask/canister and the canister/dummy spent fuel, the strain yielded at the dummy spent fuel did not increase proportionally. The response of the fuel was within the elastic region for the input earthquake waves.

From these results, we confirmed that the unfixed concrete cask maintains its stableness for tip over and the spent fuel maintains the integrity if it were subjected to a large earthquake (the earthquake wave observed at the 1995 South Hyogo earthquake with 1285 gal at maximum).

(2) Proposal of evaluation method of stability against earthquake

We proposed window energy that contributes cask rising from the ground focusing a relationship between input energy effective for cask rising[2] and instantaneous input energy within a limited time span (window width: T_{window}) [4]. If a value of $T_1 = 0.3\sqrt{a}$ is used for the evaluated time span T_{window}, the maximum value of the window energy will give approximately double the value enveloping the experimental value (shown by the result of vibration of the 1/3 scale model in Fig. 4.6.11). The symbols in the figure mean as follows. T0: Minimum response time, T1: Maximum response time, a: Spin radius, H: Height of rigid body, B: Width of rigid body, θ: Rising angle.

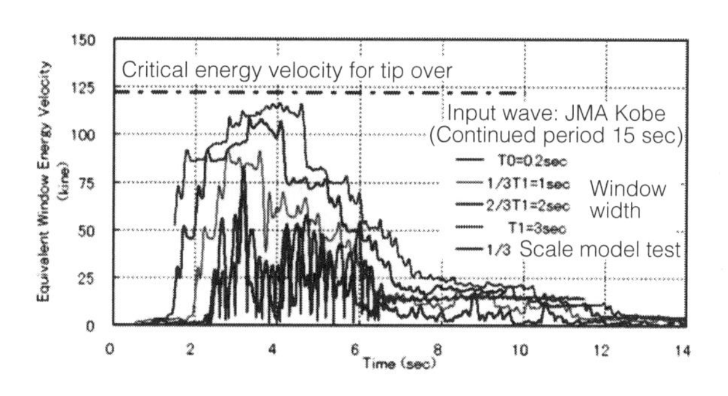

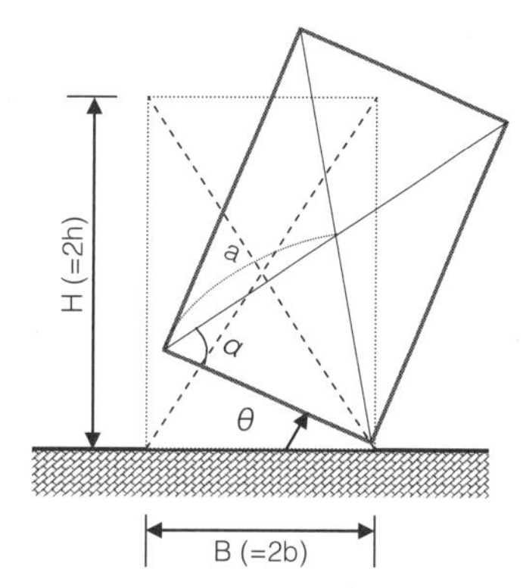

Fig. 4.6.11 Applicability of window energy

References

1) T. Saegusa, K. Shirai, et al.: "Development of concrete cask storage technology for spent nuclear fuel", CRIEPI General Report N09, May 2010.

2) H. Akiyama, et al.: "Tip over prediction of rigid body using energy spectrum", Collection of Architectural Institute of Japan article reports, No.488, 1996

3) K. Shirai, et al.: "Experimental Studies of Free-Standing Spent Fuel Storage Cask subjected to Strong Earthquake", PATRAM 2007, Miami, Florida, USA (2007).

4) K. Shirai, et al.: "Tip over of spent fuel storage cask evaluated by energy spectrum", L53.5., Spring meeting of Atomic Energy Society of Japan, 2008.

4.7.1 Stress corrosion cracking of canister

(1) Long term reliability against SCC

In the dry storage of spent nuclear fuels using concrete casks, stainless-steel canisters act as an important barrier for encapsulating spent fuels and radioactive materials. According to the spent fuel storage concept, the decay heat of spent nuclear fuels dissipates through the canister wall by air cooling. Hence, the canister wall is in direct contact with air containing sea salt particles and is possibly contaminated by chlorides, because the interim storage facilities for spent nuclear fuels will be built in coastal regions in Japan. Austenitic stainless steels are susceptible to stress corrosion cracking (SCC) in certain environments under tensile stress. SCC induced by sea salt particles and chlorides, for example, has been observed on various structures of chemical plants built in coastal regions.

SCC occurs when three conditions, stress, environment and material, are satisfied. As for the canister, residual stress by weld, deposition of sea salt particles on the surface and using of austenitic stainless steel are causes of SCC. If one of these conditions is removed, SCC will not occur. Removing more than two of them is desirable measure for safety operation.

Process of the SCC, expected on the canister, is schematically drawn like Fig. 4.7.1(1)-1. At first sea salt particles attach on the canister surface. Then sea salt deliquesces with humid in the air. Localized corrosion such as pitting or crevice corrosion occurs by salt water. Finally cracking occurs if tensile stress exists around the corroded pit. Localized corrosion yields rust spot, so cracking do not occur at the portion without rust spot.

On following parts, we will explain on each SCC factors.

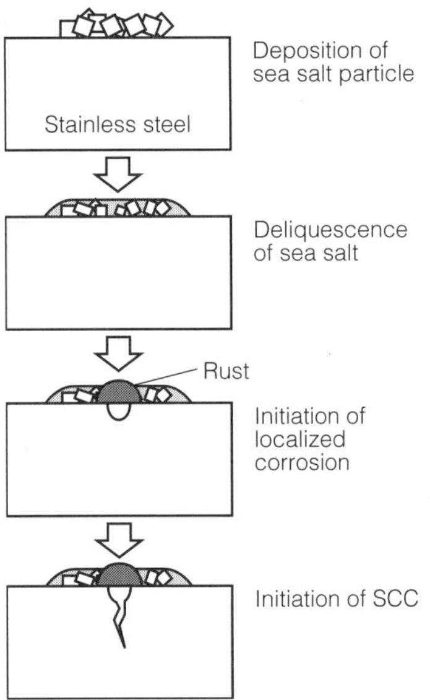

Fig. 4.7.1(1)-1 Schematic drawing of the process of SCC initiation

1) Stress

Most of the metallic materials swell by heating. Small unconstrained plate deforms due to the swelling and shrinking at weld line. As for the constraint structures like the canister, residual stress emerges instead of deformation. Simplified model for residual stress is shown on Fig. 4.7.1(1)-2. Plastic deformation occurs around weld line with heat and then shrinks with cooling, finally tensile stress remains. Residual stress by weld is possible to exceed yield stress.

Surface finishing with hand grinder may cause tensile stress[1]. The grinder makes two axes stress with working direction of the grinding tool and rotational direction of the grinder head. Surface finishing with grinder also increase hardness. It is reported that stress corrosion crack depth increases with hardness on surface of material at operation temperature of light water reactor[2].

Compressed stress does not be a cause of SCC. SCC will be mitigated when the tensile stress lowered to less than zero. Several stress relaxation methods, which are already applied on light water reactors, are applicable on the canister. For example, shot peening, laser peening and water-jet peening are easily apply compression stress. Annealing is also one of the stress relaxation methods. Although the canister, which wall thickness is thin, may deforms by annealing. In order to avoid deformation, retainer is required.

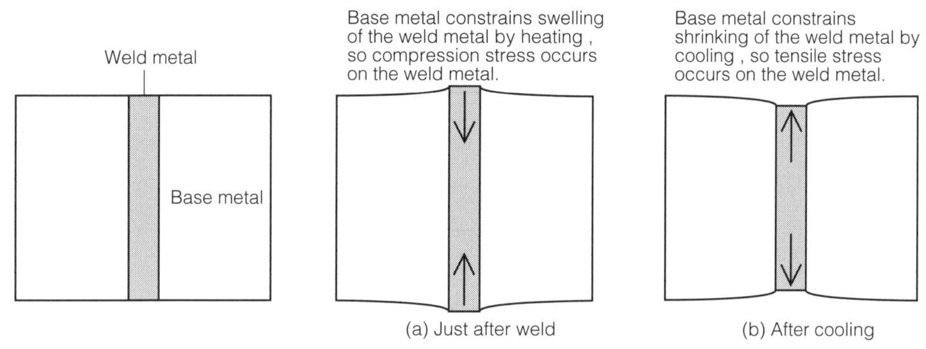

Fig. 4.7.1(1)-2 Mechanism of generating residual stress by weld

2) Environment

Sea salt, transported by cooling air, is major environmental factor on SCC. Sea salt contains 60 % of sodium chloride and 25 % of magnesium chloride by weight. Sodium chloride deliquesces at 75 % relative humidity. Magnesium chloride deliquesces at 30 to 35 % RH. When sea salt is placed in the air with 35 % RH, highly concentrated chloride solution yields between the salt and metal surface. It is reported that SCC sensitivity is highest at 35 % RH[3]. Sea salt contains small amount of salt such as calcium chloride that deliquesces at lower RH so the SCC occurs at lower than 30 % RH. Threshold RH for SCC of type304 stainless steel at 80 °C is 15 %[4].

Relative humidity is derived from temperature and absolute humidity. If absolute humidity is constant, relative humidity decreases with increase of temperature. Maximum absolute humidity in Japan is 23 to 28 g/m^3. Fig. 4.7.1(1)-3 is temperature dependency of relative humidity at 30 g/m^3 of absolute humidity. If surface temperature on the canister is higher than 70 °C, relative humidity is less than 15 % so it is not possible to make the SCC.

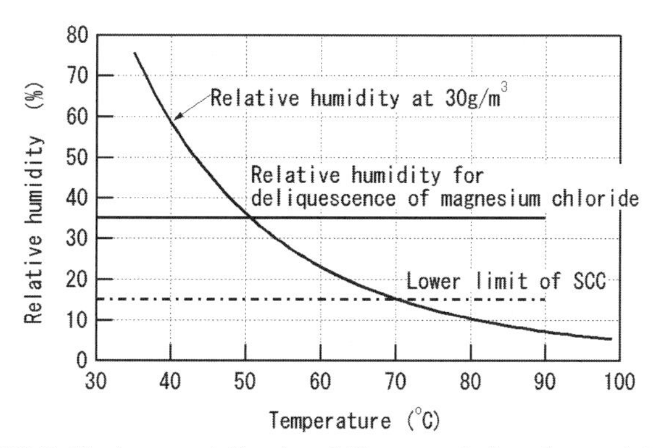

Fig. 4.7.1(1)-3 Maximum relative humidity expected on the canister surface

Sea salt is carried as small particle, so the sea salt deposits on the canister surface as an island. In order to initiate the localized corrosion, certain area of continuous salt is necessary, so it is anticipated that initiation and propagation of the SCC depends on the density of deposited sea salt. Such the threshold sea salt density will be utilized on design of storage building.

3) Material factor

Austenitic stainless steel such as type304 and duplex stainless steel are vulnerable to the chloride induced SCC in the air. Procedure of the SCC shown on Fig. 4.7.1(1)-1 indicates that the pre-process such as pitting or crevice corrosion is necessary to initiate the SCC. So the initiation conditions of the SCC correspond to the initiation conditions of the localized corrosion. Pitting resistant equivalent (PRE) is an index of the pitting resistivity of stainless steel and Ni based alloy. PRE is calculated from the following equation which is consisted with concentration of Cr, Mo and N in alloy. Chemical composition of major alloying elements and PRE values for materials listed on the standard[5] on the structure of the canister are shown on Table 4.7.1(1)-1.

PRE=[%Cr]+3.3[%Mo]+16[%N]

It is assumed that high PRE material has high SCC resistivity. Constant load test for these materials at 80 °C, 35 % RH with enough sea salt on the specimen surface was conducted. As shown on Fig. 4.7.1(1)-4 type304 and type316 failed at around several hundred hours while S31254 and S31260 did not fail for more than 60,000 h[6),7)].

SCC measures are applicable from every aspect of stress, environment and material. Best method should be applied in point view of environment of the facility and management policy of organizers.

Table 4.7.1 (1)-1 Chemical composition for major element of canister materials

	Cr	Ni	Mo	N	PRE
S30403	18.00-20.00	8.00-12.00	-	<0.10	18
S31603	16.00-18.00	10.00-14.00	2.00-3.00	<0.10	23
S31260	24.00-26.00	5.50-7.50	2.50-3.50	0.10-0.30	34
S31254	19.50-20.50	17.50-18.50	6.00-6.50	0.18-0.22	42

(wt%)

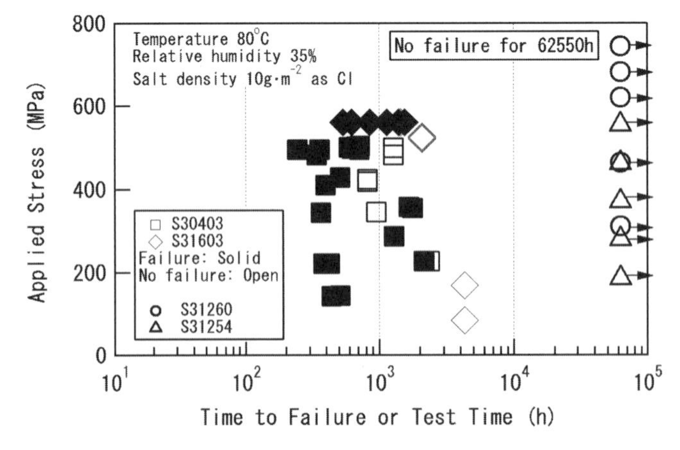

Fig. 4.7.1(1)-4 SCC resistivity of canister materials evaluated with constant load test

References

1) S. Suzuki, et. al., Aturyoku-gijutu, 42, 188 (2004). [Japanese article]

2) M. Tsubota, Y. Kanazawa and H. Inoue, Proceeding of the 7th International Symposium on Environmental Degradation of Materials in Nuclear Power Systems - Water Reactor, p.519 (1995).

3) S. Shoji, N. Ohnaka, Y. Furuya and T. Saito, Boshoku-gijutu, 35, 559, (1986). [Japanese article]

4) M. Mayuzumi et al., Nuclear Engineering and Design 238, 1227-1232 (2008).

5) JSME Codes for Nuclear Power Generation Facilities, Rules on Concrete Casks, Canister Transfer Machines and Canister Transport Casks for Spent Nuclear Fuel 2003 edition, JSME SFB1-2003.

6) J. Tani, M. Mayuzumi and N. Hara, Corrosion, 65, 187 (2009).

7) J. Tani and M. Mayuzumi, CRIEPI report Q08007 (2009).

(2) Criteria to prevent SCC initiation for normal austenite stainless steel

For interim storage of used fuels, there is not only the metal cask storage method, which is a major storage method in Japan, but also the concrete cask method or silo method. In these storage methods, used fuels are confined in a canister and the canister is stored in a concrete module for storage. In particular, the concrete cask method has an advantage in terms of economic efficiency, local employment, and reduction of waste and is therefore a major interim storage method used in the US.

In Japan, technologies, guidelines, and private company standards for concrete cask storage facilities with canisters that can economically store a large amount of used fuels from nuclear power plants for a long period of time have been developed in order to ensure the flexibility of choosing interim storage methods[1)-3)].

On the other hand, naturally-ventilated concrete canister storage facilities have a problem that chloride induced stress corrosion cracking (SCC) on the surface of the canisters could damage the containment function, and designs of preventing SCC have been studied. It is necessary to reflect the study results to the private company standards and the regulation criteria.

In the technology requirements in Sec. 2.2.1, various basic safety functions are required for the concrete casks. For example, a heat removal function and shielding functions are required for concrete module, a containment function for the canisters, and a criticality prevention function for baskets contained in the canisters (Fig. 4.2-1). As shown in Fig. 4.7.1(2)-1, the heat removal function takes in the cooling air by using natural convection of the cooling air from the opening on the lower side of the storage vessel so that the cooling air directly contacts and cools the canister surface. Therefore if the facilities are located near the sea, sea salt particles contained in the air could attach to the canister surface. As given in three factors for SCC in Fig. 4.7.1(2)-2, SCC could occur on the canisters since they are made of austenite stainless steel, since sea salt particles can attach to the canisters, and since tensile residual stress exists at and around the welded parts. Therefore the technology requirements require appropriate countermeasures in the design against SCC such as use of highly corrosion-resistant materials, reduction of the residual stress at the welded parts, and improvement of the salt deposit environment in order to keep the containment function.

In the US, normal austenite stainless steels of Type304, 304L, and 316L (hereafter referred to as normal materials) are commonly used for canisters. However these materials are more SCC sensitive than highly corrosion-resistant

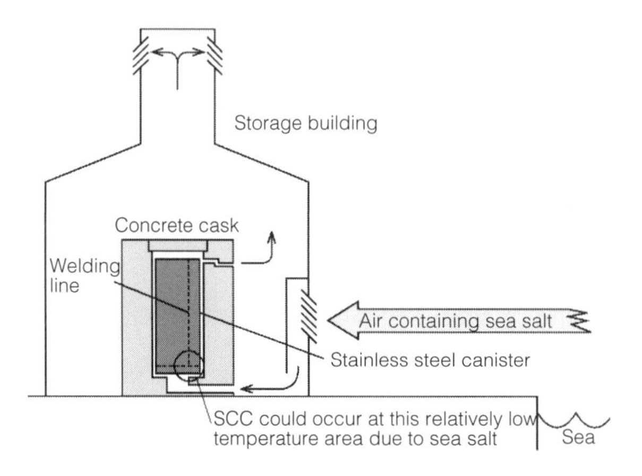

Fig. 4.7.1(2)-1 Concrete cask storage facility structure and SCC risk

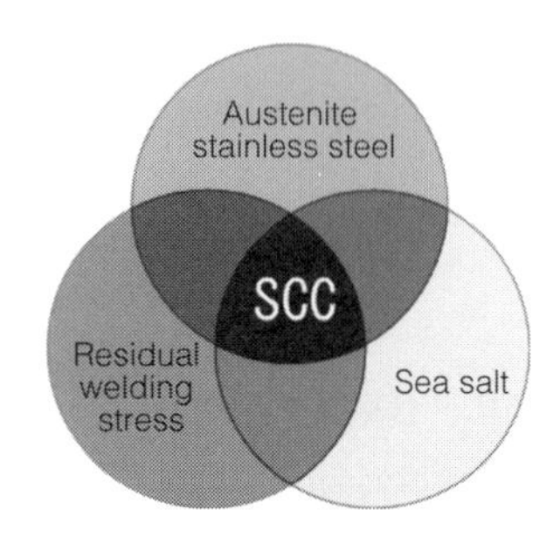

Fig. 4.7.1(2)-2 Three factors for SCC

materials. Therefore SCC could occur if the canisters made of these materials are stored in natural-ventilation-based concrete cask storage facilities near the sea[4),5)]. In the current concrete cask structure standards by Japan Society of Mechanical Engineers, not only these normal materials but also highly corrosion-resistant materials such as super austenite stainless steel materials SA-240 S31254 and SA-182 F44 and duplex stainless steel material GSUS329J4L can be used.

1) Scenario of countermeasure against SCC

With this background, countermeasures against SCC need to be considered in order to have more choices of interim storage methods and aim at practical application of the concrete cask method. Therefore the development of a scenario of not only countermeasures using highly corrosion-resistant materials[6)] but also countermeasures against SCC is important for canisters made of 304L (SUS304L hereafter) or 316L (SUS316L hereafter).

According to the standards of nuclear power facilities for electricity generation, design and construction standards (NC-CC-002) by Japan Society of Mechanical Engineers[7)], which gives a guideline of SCC suppression of light-water reactors, it is preferable to take countermeasures if possible against multiple causes among the three, i.e. material, stress, and environment related causes. In the safety design of countermeasures against SCC, a combination of technological countermeasures against two SCC causes among the three in Fig. 4.7.1(2)-2 (Environment + Material, Material + Stress, Stress + Environment) should be used. Table 4.7.1(2)-1 compares options of method selection with focus on the three causes of SCC[8)].

In Fig. 4.7.1(2)-3 we propose a scenario of countermeasures against SCC on the basis of the evaluation case 2 in Table 4.7.1(2)-1. The scenario consists of two scenarios. In Scenario 1, an attached salt amount measurement is made. If it is found that the salt concentration limit is exceeded, Scenario 2 of crack propagation control starts. With the scenarios, the salt concentration on the canister surface would not exceed the SCC occurrence criterion concentration during the storage period and, even if SCC occurs, the containment function could be maintained by controlling the crack propagation.

In this section, we describe the results of the SCC test for the integrity evaluation of the SCC prevention standards for canister made of normal materials. This test was conducted to evaluate how the SCC occurrence salt concentration limit of SUS304L and 316L, which was obtained in a constant-load SCC evaluation test, and the residual stress improvement processing affect the prevention of SCC.

Table 4.7.1(2)-1 Options of method selection with focus on countermeasures against SCC

<table>
<tr><th>Method</th><th>Material</th><th>Stress</th><th>Environment</th></tr>
<tr><td>Reference case
(no SCC countermeasure)
Employed in the US</td><td>Normal materials</td><td>No countermeasure</td><td>Topographical environmental cause reduction
(Inland)</td></tr>
<tr><td>Evaluation case 1
(no SCC countermeasure)
Following the reference case</td><td>Normal materials</td><td>No countermeasure</td><td>Topographical environmental cause reduction[Note 1)]
(Seashore: Salt monitoring)</td></tr>
<tr><td>Evaluation case 2
Countermeasure
(Stress + Environment)</td><td>Normal materials</td><td rowspan="2">Residual stress improvement measure
. Peening
. Burnishing[Note 1)]
. Laser welding[Note 2)]</td><td>Surface salt concentration control</td></tr>
<tr><td>Evaluation case 3
Countermeasure
(Material + Stress)</td><td>Highly corrosion-resistant materials</td><td>No countermeasure</td></tr>
<tr><td>Evaluation case 4
Countermeasure
(Environment + Material)</td><td>Highly corrosion-resistant materials</td><td>(No countermeasure)</td><td>Surface salt concentration control</td></tr>
</table>

Note 1: A residual stress countermeasure technology for Yucca Mountain Project canisters, called Low Plasticity Burnishing (LPB), was developed at US INL (Idaho National Laboratory). Its applicability in Japan is evaluated.
Note 2: In comparison to SAW or TIG welding, which was commercialized in the US and examined in Japan to be used for canisters, the laser welding can improve the residual stress and shorten the manufacturing period. Therefore the validity and cost of the manufacturing method is to be evaluated.

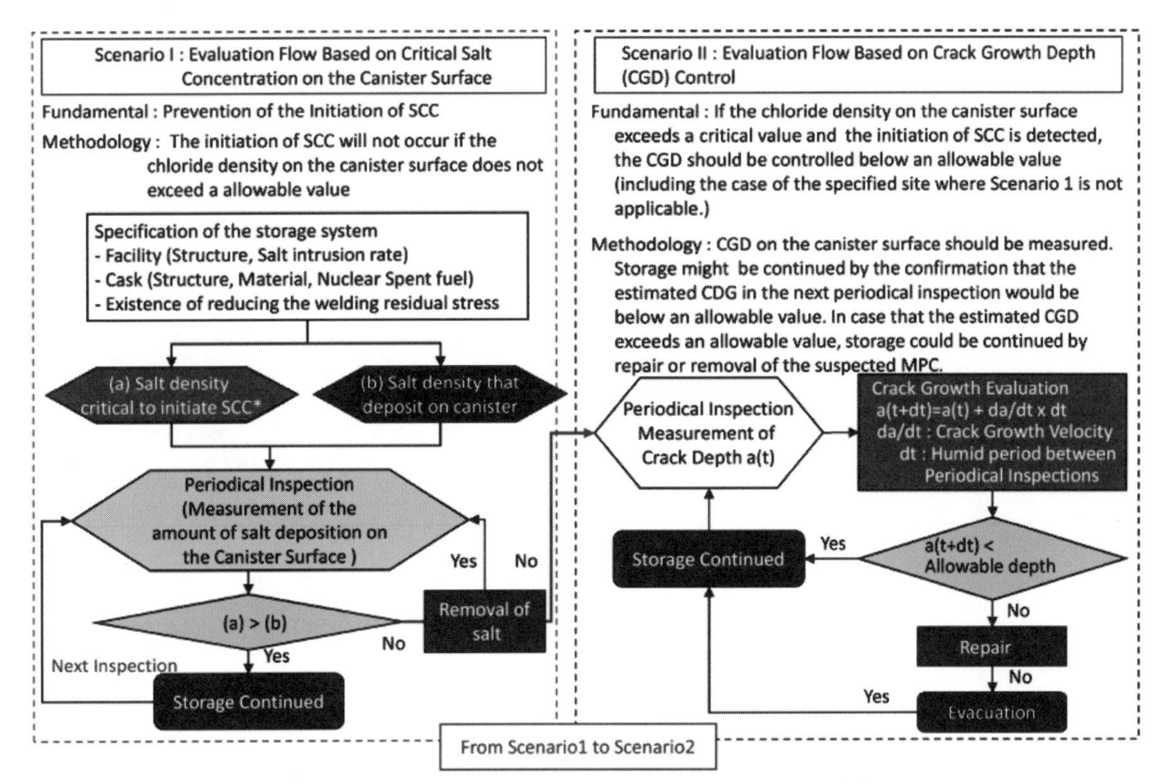

Fig. 4.7.1(2)-3 Canister SCC countermeasure scenario[8)]

2) Constant-load SCC test of SUS304L and 316L materials

(i) Specimen

We used SUS304L and SUS316L materials of 2 mm thickness as test specimens. Fig. 4.7.1(2)-4 shows the test specimens and dimensions used in the constant-load SCC test. To remove influential surface factors other than salt and stress, we wet-polished the surface of the specimens with #600 papers and ultra-sonic washed them with acetone.

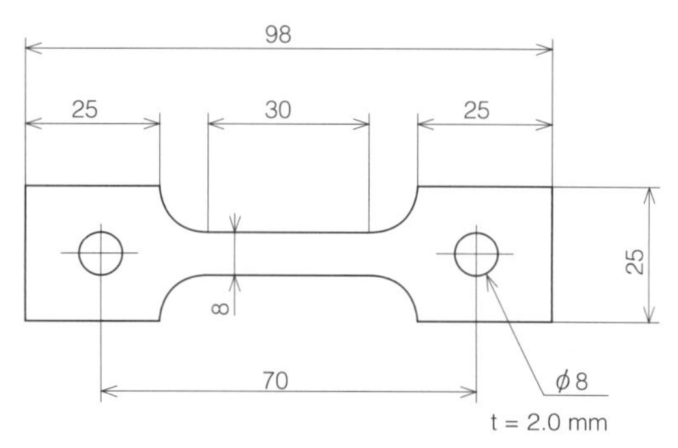

Fig. 4.7.1(2)-4 Shape and dimensions of constant-load SCC test specimen

(ii) Evaluation of SCC initiation criterion

SCC has extendibility since non-extendable small cracks are produced in the initial stage and the produced cracks grow close to each other and combine with each other. Some evaluation studies reported[9),10)] about the SCC occurrence in high temperature water that cracks of the depth less than 50 μm did not progress. Since these cracks had a semi-ellipse shape of the aspect ratio a/c = 1, the presence of surface cracks of the size 100 μm or larger on the test specimen was used as criterion of judging the SCC occurrence, as shown in Fig. 4.7.1(2)-5.

For the observation of surface cracks, we used SEM (Scanning Electron Microscope).

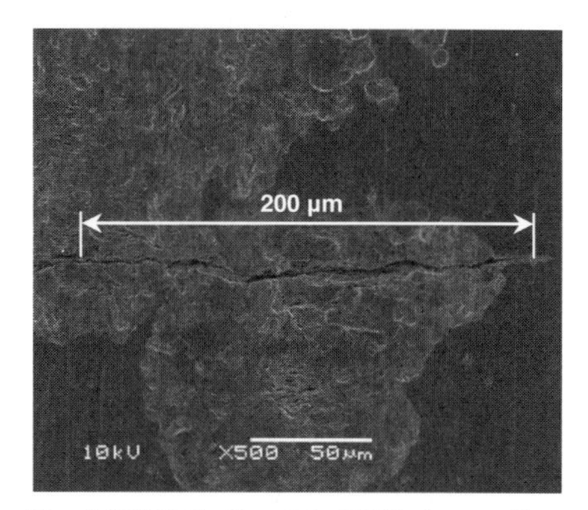

Fig. 4.7.1(2)-5 Crack in SEM observation of rust part (where SCC occurred)

(iii) Constant-load SCC test method

In the SCC test, we employed a constant-load SCC test method using a spring. Fig. 4.7.1(2)-6 shows constant-load SCC test jigs. A test specimen was attached to the jig and the spring was pressed with a compression tester until the specified load was obtained. Then it was fixed with a nut to apply a tension load. Salt was supplied intermittently by using a nozzle which could spray fine particles. Salt water was sprayed so that the salt water droplets on the specimen surface would not combine with each other.

Fig. 4.7.1(2)-6 Appearance of test jigs

The test temperature and relative humidity were determined by taking account of the actual environment based on the domestic climate observation data. Fig. 4.7.1(2)-7 shows the relation between the critical temperature and the critical humidity. The critical humidity of SCC is relative humidity at which sea salt deliquesces. The domestic maximum absolute humidity is about 30 g/m^3 and the relative humidity and temperature at which sea salt causes SCC in an atmospheric environment SCC are 35 % and 50 °C respectively. We therefore set the temperature to 50 °C and the relative humidity to 35 % in the test.

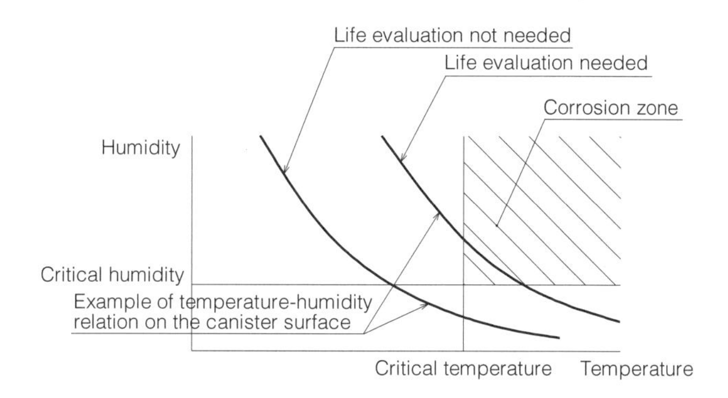

Fig. 4.7.1(2)-7 Critical temperature and humidity, and life evaluation

Fig. 4.7.1(2)-8 shows the relation among the SCC test results, stress and deposited salt concentration for each of SUS304L and 316L materials. For the SUS304L material, no cracks or significant load stress dependency was observed if the salt concentration was lower than 0.8 g/m^2 as Cl. So, we set the salt concentration limit of SUS304L to 0.8 g/m^2 as Cl. Similarly, the salt concentration limit of SUS316L was set to 4 g/m^2 as Cl, which is about 5 times larger than the limit of SUS304L. Namely, SUS316L has better performance in terms of SCC than SUS304L.

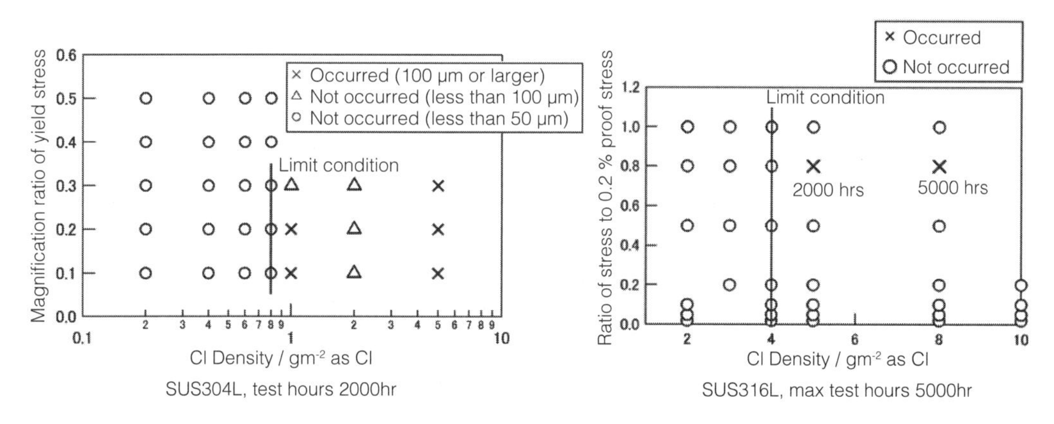

Fig. 4.7.1(2)-8 Relation among crack initiation, stress, and deposited salt concentration

If we take account of the grinder work on the welded part of the actual canister, the salt concentration limit could be lowered to half of the limit for the materials whose surface was wet-ground. Namely, the hardened surface layer could lower the limit.

3) SCC evaluation test of full-scale SUS304L canister

A full-scale SUS304L canister was used to make an SCC examination test and check the effect of the residual stress relaxation processing. Salt was attached on the full-scale canister model test body (with diameter of ϕ1836

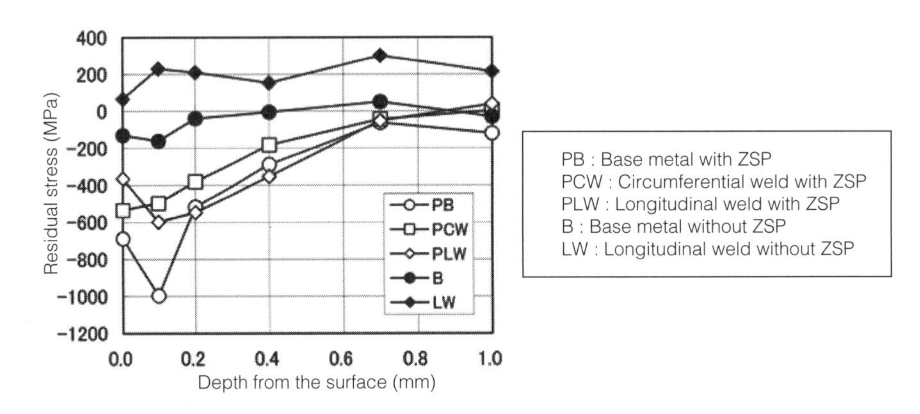

Fig. 4.7.1(2)-9 Residual stress distribution in depth direction

mm, height of 1100 mm, shell thickness of 12.7 mm, and weight of about 2.1 tons; manufactured by laser welding) where residual stress relaxation processing was applied a part of the welded area and the test body was placed in a constant-temperature bath at a temperature of 80 °C and with humidity of 35 %RH to study the influence of the residual stress on the SCC occurrence[11].

(i) Residual stress processing

To reduce the residual stress of the test specimen, the residual stress mitigation processing (ZSP: zirconia shot peening) was conducted for the area ±100 mm from the center of the canister welding line after the manufacturing[12),13)].

Fig. 4.7.1(2)-9 shows the residual stress distribution in the depth direction near the welded part. In the welded parts with no ZSP processing, almost uniform tensile stress (of about 200 MPa) occurred down to the 1 mm depth. However with the ZSP processing, the stress became compressional stress from the surface to a depth of 0.7 mm.

In the base materials where no ZSP processing was applied, compressive stress that could be caused by the rolling process exists down to a depth of 0.2 mm. In order to clearly understand the influence of the stress of SCC, a part (0.37 mm deep) of the surface of the welded parts where no ZSP was applied was ground by electrolytic grinding to identify the position of the welding tensile residual stress.

(ii) SCC evaluation test

The concentration of the salt water sprayed to the welded part on the circumferential direction and longitudinal direction of the canister was set to be about 5 times larger than the criterion value of SUS304L, namely to 4 g/m^2 as Cl. The salt water (artificial sea water: Yashima Chemicals; Aquamarine) was sprayed with a nozzle to the top of the canister test body which was placed sideways on a rotary table. Before the salt water spraying, masking tape was applied to the surrounding area around the area to which the salt water is sprayed, and the inside was warmed with a stove to prevent the sprayed artificial sea water from dripping and to make the salt concentration uniform.

After the salt was sprayed, the canister test body was placed in a constant-humidity, constant-temperature bath. After the temperature and humidity reached 80° and 35 %RH, the canister test body was kept for 2000 hours in that environment and then removed from the bath.

(iii) Surface and cross section observation of specimen

On the surface of the welded parts of the test body taken out of the constant-temperature bath after heating and curing, there was rust occurring over the area where salt water was sprayed. In order to find SCC, we used a rust remover to remove rust and performed a penetrant test (PT). For the test, a fluorescent penetrant test (fluorescent PT) method which could detect small cracks was employed. The surface was studied in the fluorescent PT according to the current criterion of the PT (JIS Z 2343-1)[14] and no penetrant flaw indication of defects was found. However it was difficult to judge visually or by PT whether a SCC crack (of the size bout 100 μm) occurred since grinder scratches remained on the surface or since the rust reacted with a color former and produced false indication. As a result of microscope observation of the test body surface, a lot of SCCs of the length 100 μm or more were found near the welding line where no ZSP processing was applied. The cross section of the As Weld part where SCC was found was observed.

Fig. 4.7.1(2)-10 shows observation results of a cross section of the As Weld part. The crack in the Fig. was located near (within 5 mm) the welding line of the welded parts in the circumferential direction. Its maximum depth was about 4 mm. Many cracks occurred on the surface of the welded parts where no ZSP processing was applied but no cracks were found in the welded parts where ZSP processing was applied. Therefore we confirmed the validity of the welding residual stress mitigation processing.

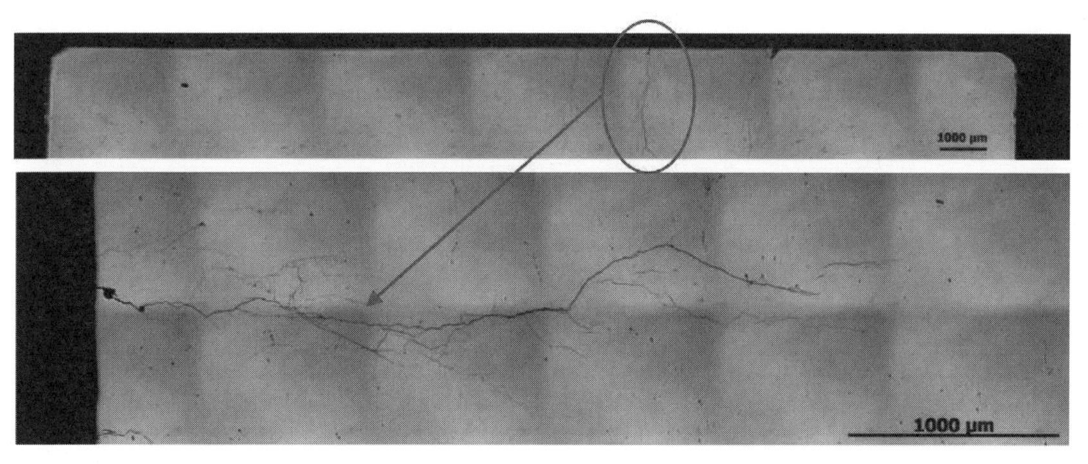

Fig. 4.7.1(2)-10 Maximum crack dimensions (without residual stress mitigation processing)

References

1) Atomic Energy Society of Japan: Atomic Energy Society of Japan Standards, safety design and inspection standards of concrete cask and canister refilling equipment for used fuel interim storage facilities, AESJ-SC-F009, June 2007.

2) Nuclear and Industrial Safety Agency: Technical requirements for used fuel storage facilities (interim storage facilities) with concrete casks, April 2006

3) The Japan Society of Mechanical Engineers: Used fuel storage facility standards, Concrete cask, Canister refilling equipment and canister transport cask, Structure Standard JSME S FB1-2003, 2003.12.

4) K. Sorenson, et al., "Long Term Storage of Used Nuclear Fuel in the U.S.", Proc. PATRAM 2010, London, 2010.
5) L. Caseres and T.S. Mintz, "Atmospheric Stress Corrosion Cracking Susceptibility of Welded and Unwelded 304, 304L and 316L Austenitic Stainless Steels Commonly Used for Dry Storage Containers Exposed to Marine Environments", NUREG/CR-7030, Oct.2010.
6) K. Shirai: "Future recycled fuel storage technology development", Central Research Institute of Electric Power Industry Forum 2010, October 2010.
7) The Japan Society of Mechanical Engineers: Standards of nuclear power facilities for electricity generation, design and construction standards, case example standards, "consideration of suppression of stress corrosion cracking" of nuclear power facilities for electricity generation, JSME S NC-CC-002, 2005.
8) K. Shirai et al.: "Study on Interim Storage of Spent Nuclear Fuel by Concrete Cask for Practical Use -Feasibility Study on prevention of Chloride Induced Stress Corrosion Cracking for Type 304L stainless steel Canister-," Central Research Institute of Electric Power Industry, Research Report, N10035, 2011.5.
9) M. Akashi: "CBB Test Method for Assessing the Stress Corrosion Cracking Susceptibility of Stainless Steels in High-Temperature, High-Purity Water Environments," in "Localized Corrosion -Current Japanese Materials Research," Vol.4, F. Hine, K. Komai, K. Yamakawa, Eds., Soc. Mat. Sci. Jap., Elsevier Applied Science, London and New York, pp. 175-196, 1988.
10) Corrosion and anti-corrosion handbook CD-ROM, 2nd ed. Japan Society of Corrosion Engineering, Maruzen, 2000.
11) M. Goto, J. Tani, and K. Shirai: "Study on practical application of used fuel storage with concrete cask -Evaluation of chloride stress corrosion cracking sensitivity of 304L, 316L stainless steel-", Central Research Institute of Electric Power Industry, Research Report, N12023, 2013.5
12) Japan Nuclear Technology Institute: Preventive preservation construction method guideline [Peening method], JANTI-VIP-03-2nd ed., January 2008.
13) M. Takemoto: Stress corrosion cracking (SCC) for on-site engineers, 2011 Revised edition.
14) Japanese Society for Non-Destructive Inspection: Nondestructive inspection -Penetrant testing- Part 1: General rules: Penetrant testing method and penetrant indication classification, JISZ2343-1, 2001.4.

(3) Evaluation of salt concentration in air and deposit on surface for SCC countermeasures

There are three conditions for SCC countermeasures of canister, i.e. environment, material, stress. For the environmental condition, it is important to confirm salt deposit on surface does not exceed the threshold value for SCC initiation by periodical inspection during storage period. In addition, even if the salt amount exceeded the threshold value, it is necessary to control the crack so that the containment shall be maintained. Therefore, it is important to measure the salt deposit on the canister surface and evaluate, realistically. Fig. 4.7.1(3)-1 shows a concept of salt particles driven by wind.

We carried out tests to measure salt deposit on a metal surface placed in the air containing salt particles in order to find relationship between salt particle in the air and salt deposit. In the beginning, we have conducted laboratory tests that found influence of specimen temperature and flow rate on salt deposit by spraying salt water in a wind tunnel with a large salt concentration in the air[1]. In order to obtain relationship between salt deposit data of laboratory test and those of field tests and accumulate salt deposit data under many weather conditions as much as possible, we investigated change in salt deposit during a relative short period (1 to 25 days).

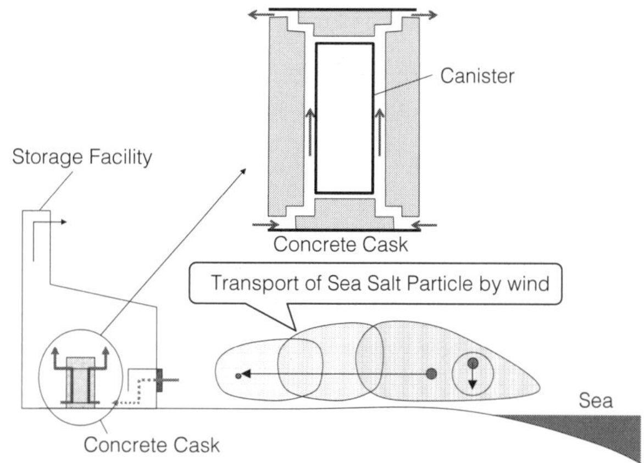

Fig. 4.7.1(3)-1 Concept of sea salt particle transport to cask storage facility

1) Existing studies on salt damage

Salt deposit rate on bridges in the existing studies[2] decreases exponentially with distance from the coastal line. The maximum rate at 200 to 500 m from the coastal line is about 0.5 g/m^2/h. While, a measured result at domestic exposure test center by dry gauze method was about 0.09 g/m^2/day at maximum. Using the latter velocity, the time to reach to a rusting condition 0.1 g/m^2 that were obtained by Tsujikawa[3] will be about 1.1 days.

The salt deposit on the canister surface in the concrete cask is different from the existing studies will be different in the following points.

a. The concrete casks are placed inside a building.

b. The air containing salt particle will flow up vertically along the canister surface.

c. The canister surface temperature is high (about 200 °C).

Considering the above points we obtained test data in order to evaluate salt deposit on the canister surface. We also developed a device to measure salt concentration in the air and obtained measured data. More detail is found in the literature[4].

2) Development of device to measure salt concentration in the air

Conventionally ejector-type sampler[5] has been used to measure salt concentration in the air. This method has high efficiency to collect sea salt particle and the device structure is simple, durable and adequate for field use. On the other hand, it needs pure water, which may evaporates and scatters, and should be refilled for long term measurement. We developed a devise for long term measurement overcoming the weakness.

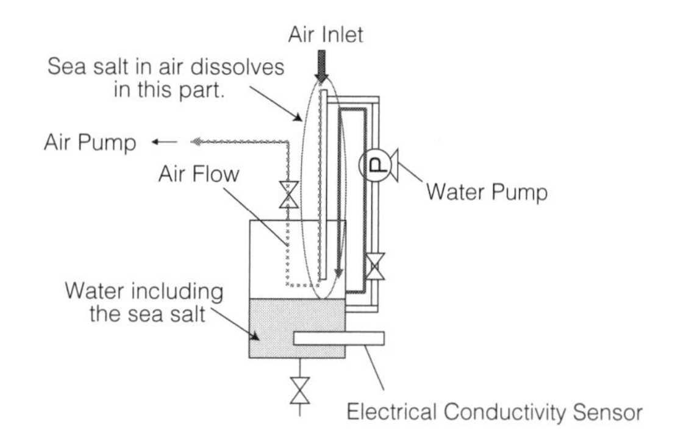

Fig. 4.7.1(3)-2 System concept of device to measure salt concentration in the air

Fig. 4.7.1(3)-2 shows the system concept, which consist of suction pipe and pump, sampling container, flowmeter, electric conductivity meter, level sensor, supplement container for pure water.

This device enable to measure continuously for 60 days without operator by setting 10 days per cycle that starts measurement for a certain period until collection of water in a sampling container. In the meantime, it requires refilling pure water to avoid shortage of pure water once during the period.

3) Simple analytical method for salt concentration in the air

Table 4.7.1(3)-1 shows measurement results of salt concentration in the air using the ejector-type salt sampler in the air at the Choshi exposure center of the Japan weathering test center (about 4 km from the coastal line).

Table 4.7.1(3)-1 Measurements of salt concentration in the air by ejector-type sampler

Measurement Period	Measurement Result (a) (Indoor)	Measurement Result (b) (Outdoor)	(b) ／ (a)
2009.9.1~10.1	2.4	3.9	1.6
2009.10.1~10.31	1.7	10.0	5.9
2009.10.31~11.2	—	6.4	—
2009.12.31~1.30	0.8	3.4	4.3
2010.1.30~2.28	1.6	9.7	6.1
2010.7.3~7.31	—	3.9	—
2010.7.31~9.4	—	2.6	—
2010.9.4~10.3	—	6.5	—
2010.10.3~10.30	—	5.0	—
2010.10.30~11.27	—	4.5	—
2010.11.27~12.31	—	2.3	—
2011.11.21~12.4	1.7	4.2	2.5

(unit :μg/m^3 as Cl)

We calculated the salt concentration in the air using a simple model for salt scattering prediction that was developed by CRIEPI and compared with the measurements. The calculation method is based on the Guassian plume equation, complete deposit model in consideration of gravity sedimentation. More detail information is available in the literature by Katoh, et al[6].

Fig. 4.7.1(3)-3 shows the calculated values on the horizontal axis and measurements by the ejector-type sampler for the salts in the air on the vertical axis. The figure demonstrates that the measurements are within about twice the calculated values. We propose to use a factor of two with the measurements for conservative results.

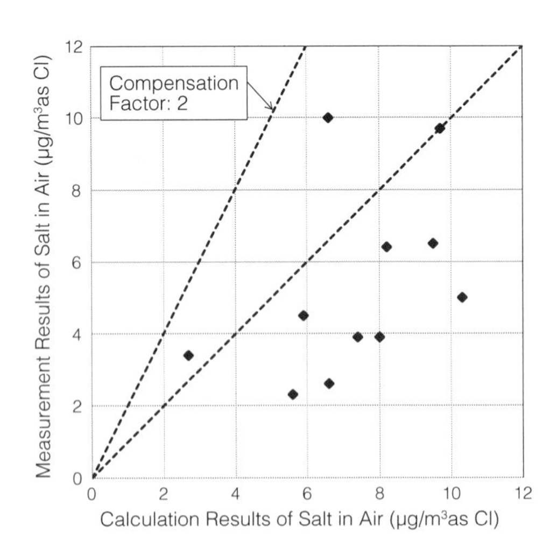

Fig. 4.7.1(3)-3 Salt concentration in the air (measurements and analysis)

4) Salt deposit tests in a real field

For salt deposit tests in a real field, we placed stainless steel specimens (SUS304) after cleaning in five wind tunnels with certain operating conditions The test period was 1 to 25 days since a relatively long period measurements have been conducted in the past. The test parameters were flow rate in the tunnels and specimen temperature. Representative flow rate was assumed considering the flow rate in the cooling path in the concrete cask[7)].

The specimen temperatures were 30 °C as near the room temperature and 100 °C as high temperature[8)].

Fig. 4.7.1(3)-4 shows results of salt deposit on the specimens in the salt deposit test device in the Choshi exposure test center together with the measurements at laboratory obtained in the past. The results show the measurements for less than one month are similar to those for several months. Namely, there was not time and temperature dependence on the salt deposits, which was about 3 mg/m^2 (as Cl) in the initial period and stayed afterwards. This value is significantly small as compared with the threshold value (800 mg/m^2)[9)] for the SCC initiation of SUS304L that is a candidate material for the canister. This means the air flows in parallel with the canister surface and the salt deposits less. In the future, measurement on the salt deposit should be conducted at a real site for evaluation.

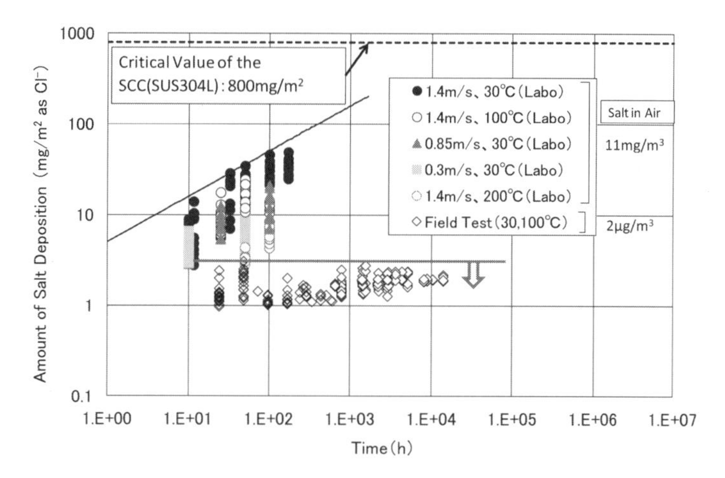

Fig. 4.7.1(3)-4 Measured salt deposit on specimen surface (Accelerated test in laboratory and field test)

Fig. 4.7.1(3)-5 shows a summary of measured results of salt concentration in the air and salt deposit at Choshi Test Center. The real concrete cask storing spent fuel may be radioactive on the canister surface and hard to access for measurement of the salt deposit. In that case we may be able to estimate the salt deposit on the canister surface from the other data near the concrete cask. Relationship between the data on the real canister and the other should be established in advance. For instance, we can estimate the salt deposit from measurements of a specimen for inspection put near the concrete cask. The location and the posture of the specimens are very important. We should properly estimate the salt deposit change with time by a relationship between the data of specimen and that on the canister surface. It should be noted that the deposit depends on the angle of the specimen to the air flow. The deposits are different for the horizontal and vertical surfaces.

Fig. 4.7.1(3)-5 Measurements of salt concentration in the air and salt deposit at Choshi Test Center

References

1) M. Wataru, et al: “Evaluation of the salt deposition on the canister surface of concrete cask - Measurement test of the salt deposition in the laboratory and the field -”, CRIEPI Report N09023, 2010.

2) K. Kishitani, et al.: “Durable concrete structure series - Salt damage -”, Gihodo Shuppan, P.12, 1986.

3) S. Tsujikawa: "Rusting of stainless steel", Corrosion-center news, No.009, P.1-9, 1995.

4) M. Wataru: “Evaluation of the salt deposition on the canister surface of concrete cask (Part 2) - Measurement test of the salt concentration in air and salt deposition in the field -”, CRIEPI Report N11028, 2012.

5) H. Kato, et al.: “Improvement of the prediction method for wind-driven sea salt: (1) Verification of the performance of an ejector-type sampler and improvement of a simple prediction model for wind-driven sea salt”, CRIEPI Report T03019, 2004.

6) H. Kato, et al.: “A Simple Model for Estimating the Amount of Wind-driven Sea Salt over Coastal Area”, Journal of Agricultural Meteorology, Vol. 57, P.79-92, 2001.

7) M. Wataru et al., Thermal Hydraulic analysis compared with tests of full-scale concrete casks, Nuclear Engineering and Design, Vol.238, Issue5, P. 1213-1219, 2008

8) H. Takeda et al., Heat removal verification tests using concrete casks under normal conditions, Nuclear Engineering and Design, Vol.238, Issue 5, P.1196-1205, 2008

9) K. Shirai, et al.: “Study on Interim Storage of Spent Nuclear Fuel by Concrete Cask for Practical Use -Feasibility Study on Prevention of Chloride Induced Stress Corrosion Cracking for Type304L Stainless Steel Canister-”, CRIEPI Report, N10035, 2011.

(4) Application of salt particle collection device for preventing SCC on Canister

Spent nuclear fuel generated at nuclear power plants is safely managed and stored in a spent nuclear fuel storage facility (hereafter called the storage facility) after being cooled in pools for a specified period. Now, in Japan, while a metal cask is used for storing, a practical use of concrete casks which is a storage method using a canister is under review because of its cost effectiveness and procurement easiness. In the review of the practical use, stress corrosion cracking (SCC) of a canister container in the concrete cask becomes an issue which is needed to be resolved soon. A natural ventilation system is generally adopted for the storage facility. Especially in Japan, the storage facility will be built near coasts, so that cooling air includes sea salt particles. Thus, the occurrence of SCC is concerned when the sea salt particles adhere to welded parts of the canister.

It has been reported through laboratory experiments that the threshold of chloride density for rust of SUS304L is 0.2 g/ m^2 as Cl and the threshold of chloride density for the occurrence of SCC is 0.8 g/m^2 as Cl under the condition of 50 °C and 35 % of relative humidity[1)]. These threshold values were not obtained through discussions of electrochemical mechanism as generally acknowledged by Macdnald[2)]. In this study, the chloride density on the canister surface is expressed as the average amount of chloride deposited per canister surface. However, our microscopic observation revealed that the deposited salt particles distribute like small islands on the canister surface. When the salt particles deliquesce with moisture in the air, a thick chloride solution is produced at interfaces between the individual salt particle and metal parts of the canister. Thus, local corrosion proceeds at those interfaces. Further electrochemical investigation has not been performed yet on this kind of atmospheric corrosion.

Saegusa, et al[3)]. estimated sea salt concentration in the air and the length of time when occur rust and SCC on each of stainless steels (SUS304, LSUS329J4L, S31254) as a canister material. In the case of SUS304L, the length of time which occur rust is about 20 years and the length of time when occur SCC is about 120 years at an airborne salt concentration of 20 μg/m^3. On the other hand, the length of time when occur rust and SCC is doubled when the airborne salt concentration is reduced from 20 μg/m^3 to10 μg/m^3. Consequently, it will be effective SCC prevention measures to reduce the amount of salt flowing into the storage facility and the concrete cask. We proposed a salt particle collection device (hereafter called the device) with low pressure loss which does not influence heat removal performance[4)]. We have been investigated about the pressure loss and a salt particle collection rate of the device based on the test results, and evaluated the applicability of the device to the concrete cask[5),6),7)]. Particularly, the effect of electric field on the particle collection rate was evaluated in this paper.

1) Test apparatus

The device is constructed with multiple stainless plates. Fig. 4.7.1(4)-1 shows a sample of the device composed of ten plates which have 0.5 mm thickness, 300 mm length and are arranged parallel at even intervals. H in this figure is height between plates. Particle collection tests were been performed by the apparatus in which the device was installed. The scale of acryl duct which is arranged the device are a width of 100 mm (B) and a hight of 100 mm (H_0). A filter (Nitta Co. Filter SE-98-X) which can collect particles over 98 % [8)] was arranged at downstream of the device. The tests with dry plates and the tests with wet plates were performed. In the tests using the wet plates, pure water in the tank was circulated using a bellows pump so as to keep the plates wet as shown in Fig. 4.7.1(4)-2. The tests using dry plates were performed in three forms : Ten normal dry stainless plates, ten electret filter (ERIA100EF10 by Breast Co.) plates in place of the stainless plates as shown in Fig. 4.7.1(4)-3, and ten dry stainless plates with a 1kV electric field as shown in Fig. 4.7.1(4)-4. The electret filter is made of permanent charged fiber.

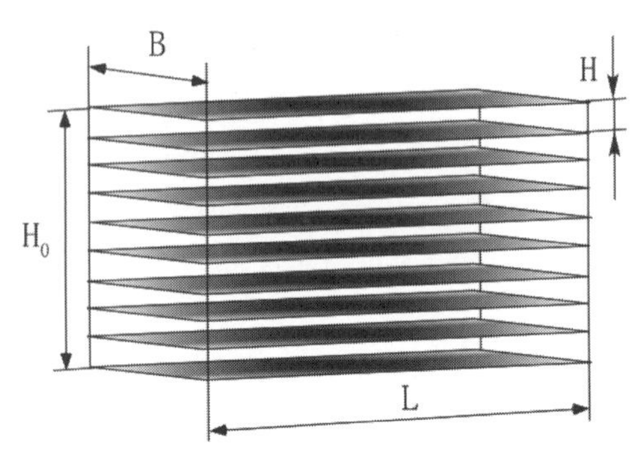

Fig. 4.7.1(4)-1 Salt particle collection device

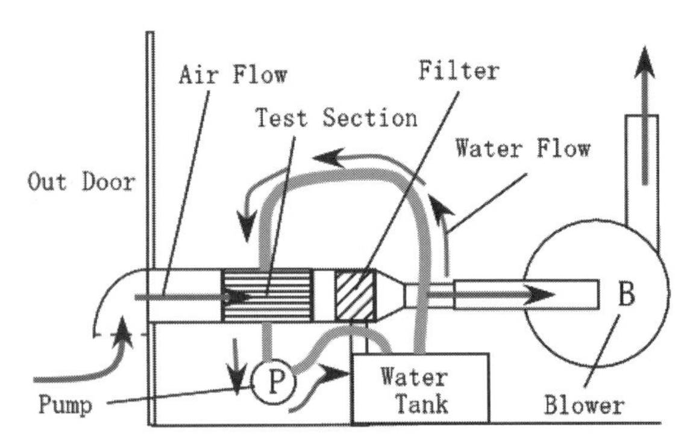

Fig. 4.7.1(4)-2 Apparatus for salt particle collection

Fig. 4.7.1(4)-3 Electret filter plates

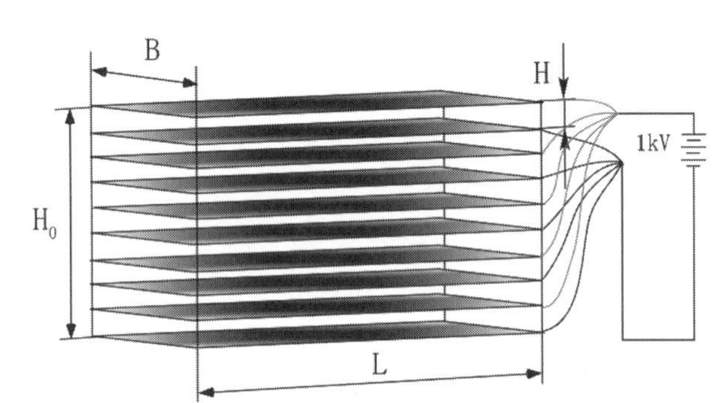

Fig. 4.7.1(4)-4 Electric field (1kV)

2) Pressure loss test results and discussion

Pressure loss tests were performed on the device with the each plate length (L) of 300 mm, and with the number of the plates of ten. In this test, the device was installed in the above-mentioned acryl duct, and the flow rate measurement and the differential pressure measurement between the upward and downward part of the device were performed while the flow rate was changed by the blower. For comparing the differential pressure with a commercial filter with low pressure loss, the differential pressure of the above-mentioned commercial filter, SE-98-X by Nitta Co [8]., was also measured. In selecting filters, the pressure loss of commercial filters were compared, and the Nitta filter which has the smallest value under the same condition was employed. A pressure loss coefficient ξ is expressed in the formula (1) based on the measured differential pressure and flow rate.

$$\xi = \frac{2\Delta P}{\rho_f U_0^2} \quad \ldots\ldots (1)$$

Here, ΔP is the differential pressure, U_0 is average velocity at the cross section of the duct, and ρ_f is air density (considering temperature dependence)

A Reynolds number Re_0 is expressed in the formula (2).

$$Re_0 = \frac{U_0 L_m}{\nu} \quad \ldots\ldots (2)$$

where ν is a coefficient of kinematic viscosity of air, and L_m is characteristic length (duct height (or width) of 100mm was used).

Fig. 4.7.1(4)-5 shows the relation between the Reynolds number Re_0 and the pressure loss coefficient ξ of the device (L: 300 mm, number of the plates: 10) and the Nitta filter. It was found that the pressure loss coefficient of the device was 1/30 to 1/20 compared with that of the commercial filter (Nitta filter). To quantitatively evaluate the pressure loss coefficient ξ of the device, the pressure loss coefficients at each part of the device were obtained by a quantity survey based on the values written in the handbooks of the pressure loss [9),10)], as expressed in the formula (3). In this formula, ξ_{in} expresses the pressure loss coefficient at an inlet part of the device, ξ_{out} expresses the pressure loss coefficient at an outlet part, λ expresses a friction coefficient of the duct, and u expresses the velocity in the duct with the height H between the plates.

$$\Delta P = \frac{1}{2}\rho_f \xi U_0^{\ 2}$$

$$= \frac{1}{2}\rho_f \left[\xi_{in} + \left(\frac{\lambda L}{De}\right) + \xi_{out} \right] u^2 \quad \cdots\cdots (3)$$

A hydraulic mean diameter; De is represented by $De = \dfrac{2HB}{(H+B)}$

As shown in Fig. 4.7.1(4)-6, the test values completely coincide with the calculated values.

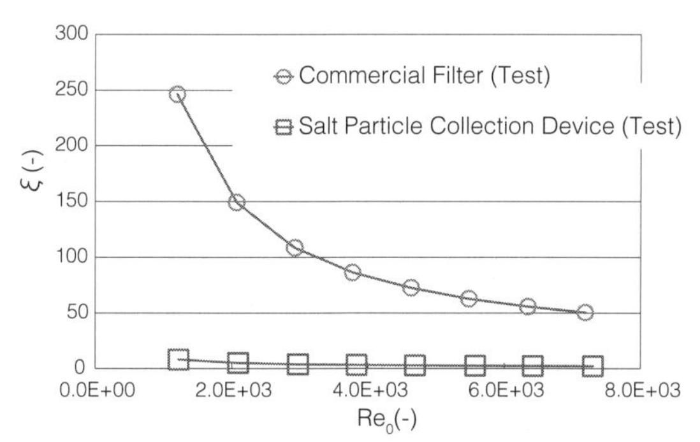

Fig. 4.7.1(4)-5 Pressure loss coefficient with Reynolds number comparing commercial filter

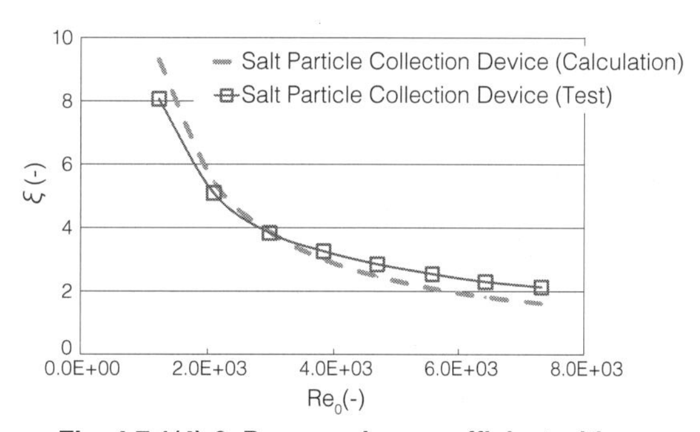

Fig. 4.7.1(4)-6 Pressure loss coefficient with Reynolds number comparing calculation

3) Particle collection field test results and discussion

The test apparatus was arranged about 50 m away from the Pacific coast and at a height of about 10 meters in Japan, and field tests of salt particle collection were conducted. In this test, the air including the sea salt particles was sucked into the device with the blower. Some particles collected by the device, and the other particles went into the filter at the duct arranged downstream of the device. The amount of the both particles was measured about every month, so that the collection rate of the device was calculated. For the calculation of the particle collection rate, the electrical conductivity of the sample solution was used. The components of particles in air were also analyzed by chromatography.

a. Particle collection field test results

The influence of an electric charge on the particle collection rate in the case of using the dry plates is shown in Fig. 4.7.1(4)-7. In this test, the ten dry stainless plates with and without the 1 kV/m electric field and the ten electret filers were used, and the particle collection rates with various velocities were measured. The measurement values has dispersion, however the particle collection rate decreases according to the increase in velocity. The particle collection rate of the ten dry stainless plates with 1 kV/m is almost the same as the rate of the electret filter, and was about 4 % higher than that of the ten dry stainless plates without electrical field.

Fig. 4.7.1(4)-8 shows the comparison of the particle collection rate between the wet plates and dry plates. When the tests using the wet plates were conducted, the water was circulated using the pump in a manner to keep the plates wet as shown in Fig. 4.7.1(4)-2. The particle collection rate with wet plates was 40 % higher than that of the dry plates. The reason could be because the re-entrainment of the particles which had dropped on the device was prevented.

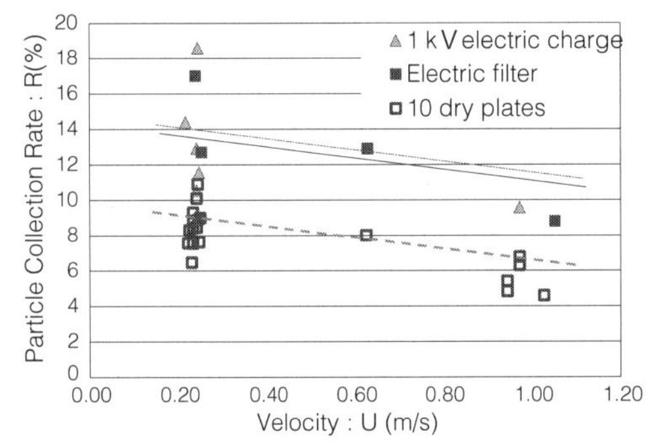

Fig. 4.7.1(4)-7 Particle collection rate with velocity (dry test)

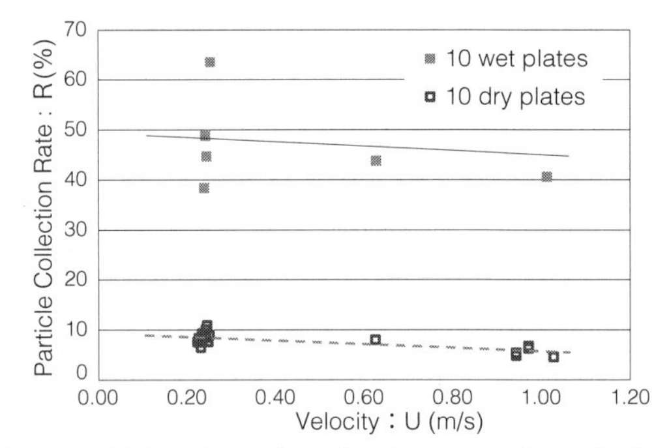

Fig. 4.7.1(4)-8 Particle collection rate with velocity (dry and wet tests)

Fig. 4.7.1(4)-9 shows the particle concentration in the air obtained by the measurement about every month. During the measurement, the particle concentration in the air was 9-24 μg/m^3, the mean value was 16 μg/m^3. Fig. 4.7.1(4)-10 shows ion ratios in the air compared with that of the sea water.

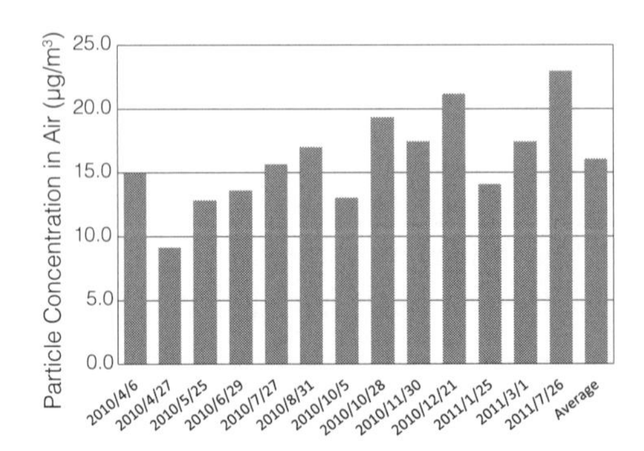

Fig. 4.7.1(4)-9 Particle concentration in the air

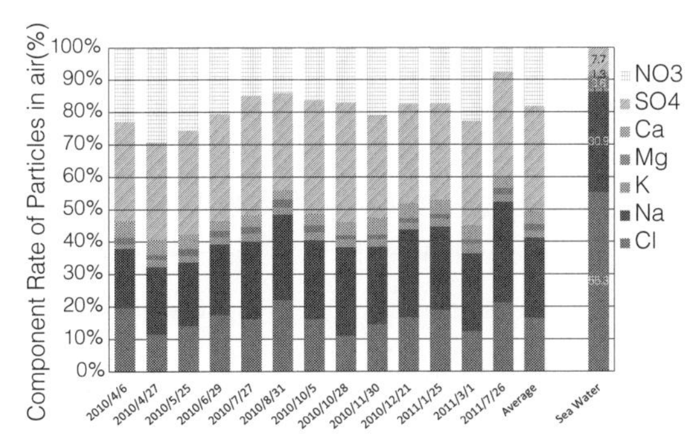

Fig. 4.7.1(4)-10 Ion ratios in the air

Na and Cl account for about half of the ions, and the remaining half ions are NO_3 and SO_4. Generally, NO_3 is almost not included in the sea water. On the other hand, SO_4 accounts for 8 % of the sea water; however the ratio of SO_4 in the air is extremely large. Accordingly, NO_3 and SO_4 could be a part of air pollutants.

The particle distribution for diameter shown in Fig. 4.7.1(4)-11 was obtained by the measurement using a ten-line sampler. The measurement was performed during three days, and it was found that there are two peeks in the particle distribution. One peek of the particle distribution is 2.5 μm or less. The other peek is 10 μm or more.

Such a tendency like this has been reported in other studies [11),12)].

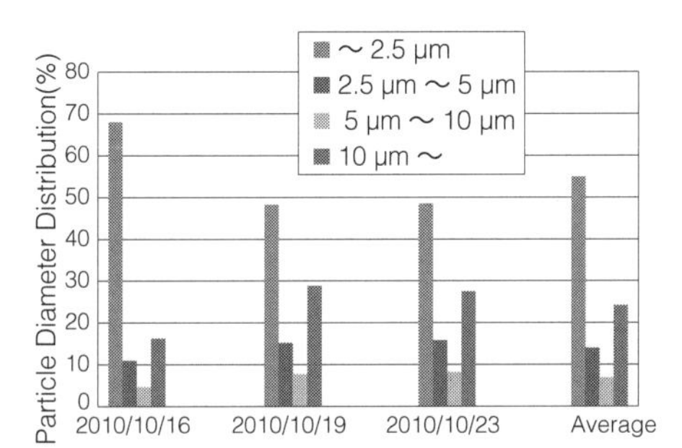

Fig. 4.7.1(4)-11 Particle distribution for diameter

b. The effect on collection ratio by gravity

The particle re-entrainment could not occur by using the wet plates, therefore the particles are collected mainly by gravity deposition. The diameter of the minimum particle which is collected by the device is expressed in the formula (4). In the formula, L is the plate length, H is the height between the plates, u is the velocity of the particle, η is the viscosity coefficient of the air, ρ_p is the density of the particle, and ρ_f is the density of the air.

$$d_p = \sqrt{\frac{18\eta Hu}{Lg(\rho_p - \rho_f)}} \quad \text{..... (4)}$$

It was found that the distribution of particle diameter of the salt near the sea has the two peeks as shown in Sec. 3)-a. Based on the result of experiment using the wet plates, we supposed normal distribution of particle diameter, which has two peaks; One peak has mean diameter of 2.5 μm, standard deviation of 0.8 μm and occupies 56 % of all particles, and the other peak has mean diameter of 20 μm, standard deviation of 5 μm and occupies 44 % of all particles. The normal distribution is expressed in the formula (5). In this formula, N(d) is the number of the particles with a diameter (d), N_0 is all of the number of the particles, d_m is a mean diameter of the particle, and σ is a value of a standard deviation.

$$N(d) = N_0 \times \frac{1}{\sigma\sqrt{2\pi}} \exp\left[-\frac{(d-d_m)^2}{2\sigma^2} \right] \quad \ldots\ldots (5)$$

In the case of $d \geqq d_p$, the all particles with a diameter of d can be collected during passing through the device. However, even in the case of $d<d_p$, if the condition of $h<H$ is satisfied, such particles can be collected. The relation of $h \propto d^2$ is formed between d and h. Thus, in the case of $d<d_p$, the distribution of the number of the particles passed through the device is expressed in the formula (6).

$$N(d)_{after} = N_0 \times \frac{1}{\sigma\sqrt{2\pi}} \exp\left[-\frac{(d-d_m)^2}{2\sigma^2} \right] \times \left[1 - \left(\frac{d}{d_p} \right)^2 \right] \quad \ldots\ldots (6)$$

The particle collection rate(R) is obtained from measured mass of the particles, so that it is expressed in the formula (7).

$$R = \frac{\int_0^{d_p} d^3 N(d) \left(\frac{d}{d_p} \right)^2 \delta d + \int_{d_p}^{\infty} d^3 N(d) \delta d}{\int_0^{\infty} d^3 N(d) \delta d} \quad \ldots\ldots (7)$$

When the particle diameter distribution has two peaks, the particle collection rate of the device is obtained by the formula (8). In this formula, R is a total collection rate, $R_{d=2.5}$ is the collection rate calculated from the normal distribution where the mean diameter is 2.5 μm and the standard deviation is 0.8 μm, and $\Phi_{d=2.5}$ is the particle ratio (56 %) in the all particles. Similarly, $R_{d=20}$ is the collection rate calculated from the normal distribution where the mean diameter is 20 μm and the standard deviation is 5 μm, and $\Phi_{d=20}$ is the particle ratio (44 %) in the all particles. The particle diameter distributions before and after the device at the particle velocity of 1.00 m/s are shown in Fig. 4.7.1(4)-12. It was found that in the device model taking account of only gravity effect, the collection rate of the relatively larger particles is affected by the structure of the device and the particle velocity, however, the relatively smaller particles can not be collected.

$$R = R_{d=2.5}\Phi_{d=2.5} + R_{d=20}\Phi_{d=20} \quad \ldots\ldots (8)$$

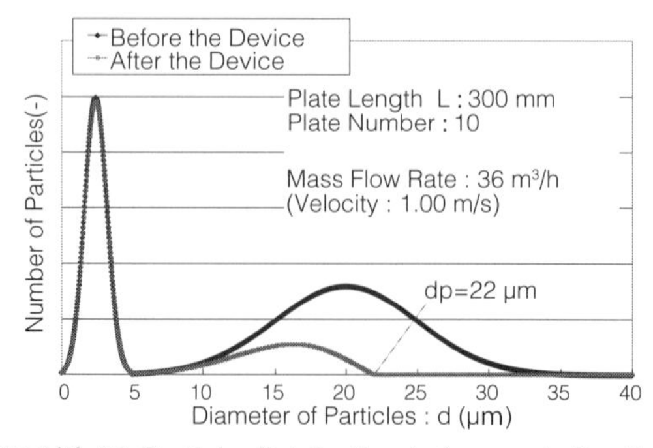

Fig. 4.7.1(4)-12 Particle distribution before and after the device

c. The effect on collection ratio by electric field

The airborne particle concentrations before and after the device at the particle velocity of 1.00 m/s on the condition of using wet plates, electret filter plates and dry normal plates are shown in Fig. 4.7.1(4)-13. The ion ratio of the particles collected by these methods is shown in Fig. 4.7.1(4)-14. As the results, the collection ratio of SO_4 and NO_3 were especially higher in the wet plates method. It is supposed that the wet plates prevented the re-entrainment of the relatively bigger particles such as SO_4 and NO_3. Moreover, it is supposed that in the electret filter plates method, many of the relatively smaller particles including NO_3 were collected.

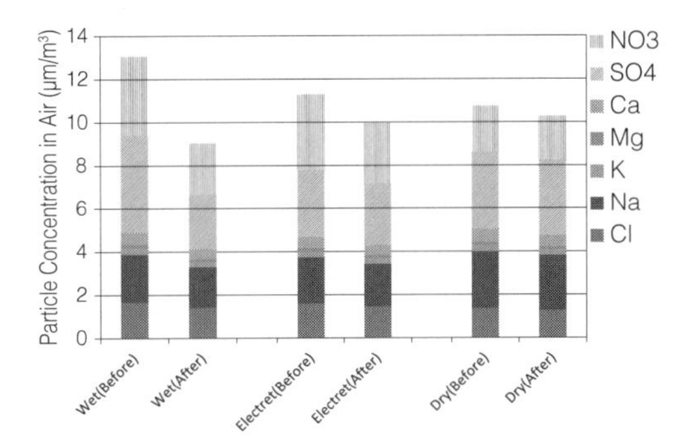

Fig. 4.7.1(4)-13 Airborne particle concentration in air before and after the device

Fig. 4.7.1(4)-14 Ion ratio of collected particles

When an electric field is applied between plats, the vertical velocity of the particle is expressed in the formula (9). In this formula, E (V/m) is an electric filed. According to Blanchard [13)], the charge of the sea salt particles depends on a bubble age (the time until a bubble in the sea bursts on the sea surface) and a radius of the particle. He obtained the formula (10) by experiments.

$$V = V_s(1 - e^{-\beta t}) \quad \ldots\ldots (9)$$

Here,

$$V_s = \frac{d^2(\rho_p - \rho_f)g}{18\eta} + \frac{qE}{3\pi d\eta}, \qquad \beta = \frac{18\eta}{d^2\rho_p} \quad \ldots\ldots (10)$$

$$q = 1.76 \times 10^{-8} t_B^{\,0.41} r^{1.3} \quad (\text{esu})$$

In formula (10), t_B is the bubble age (sec), and r is the radius of the particle (μm).

The time t_f required to pass through the length L (m) at the velocity u (m/s) is expressed in the formula (11). And the condition that the all particles are collected during the time t_f is expressed in the formula (12).

$$t_f = \frac{L}{u} \quad \ldots\ldots (11)$$

$$H < \int_0^{t_f} V dt = V_s\left[t_f + \frac{1}{\beta}e^{-\beta t_f}\right] \approx V_s t_f \quad \ldots\ldots (12)$$

The vertical dropping distance (h) relative to the particle diameter with t_B as a parameter under the condition of 10^2 kV/m between the plates is shown in Fig. 4.7.1(4)-15. It was clarified that the longer t_B becomes and the larger the charge of the particle becomes, the longer vertical dropping distance becomes even the particle having the same diameter.

Next, the vertical dropping distance of the particle relative to the particle diameter with electric field intensity as a parameter was calculated under the condition of t_B of 1sec. The results are shown in Fig. 4.7.1(4)-16. In this test, the length H between the plates of the devise was set to 0.01 m. Consequently, in the case of H<h, the particles can be collected perfectly. In the case of no electric field, the smallest diameter of the particle which can be collected is 22 μm. In the case of using the electric field, the more the electric field intensity increases, the smaller particles the device can collecte. The particle diameter distribution before and after the device under the condition of an electric field intensity of 10^3 kV/m and a duct velocity of 1.00 m/s is shown in Fig. 4.7.1(4)-17. A particle collection rate relative to the particle velocity with the electric filed intensity as a parameter is shown in Fig. 4.7.1(4)-18. In the case of 10^2 kV/m, the particle collection rate is not improved. On the other hand, in the case of the electric field intensity of 10^3 kV/m, it is assumed to be improved.

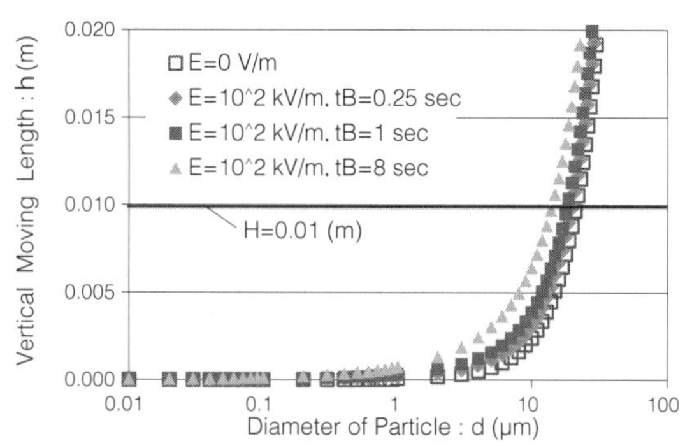

Fig. 4.7.1(4)-15 Vertical dropping distance under E=10^2 kV/m

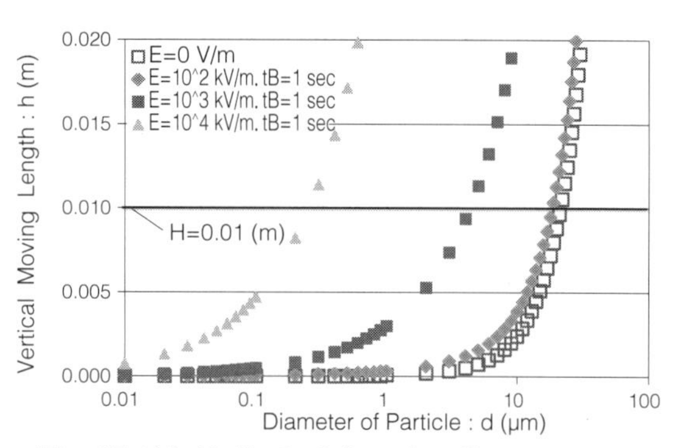

Fig. 4.7.1(4)-16 Vertical dropping distance under each E (V/m)

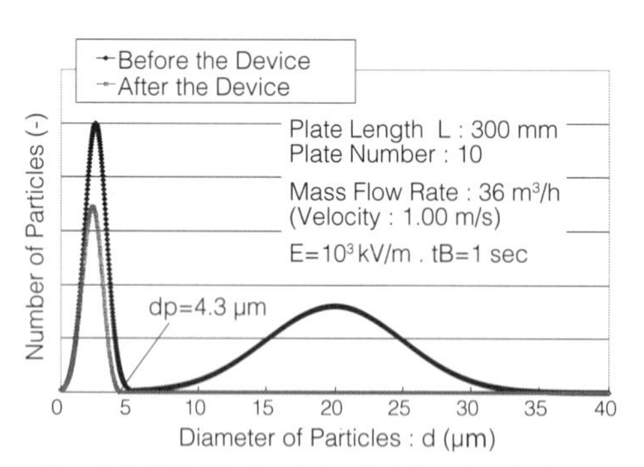

Fig. 4.7.1(4)-17 Particle distribution before and after the device

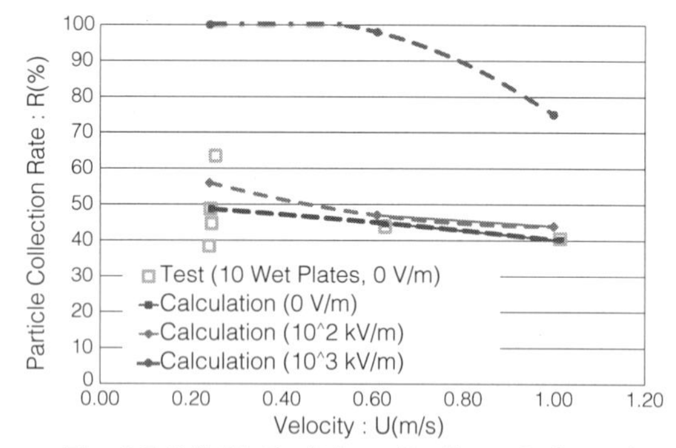

Fig. 4.7.1(4)-18 Particle collection rate in each electric field

4) Applicability of the device for concrete cask

As for the application of the device, there are two ways. One is for an air inlet of a storage facility, and the other is for an air inlet of a concrete cask. Here, we discuss the evaluation of heat removal and the particle collection effect in the case of installation for the air inlet of the concrete cask. The concrete cask which we selected for the evaluation is the full scale cask model which the heat removal test was conducted [14]. The model is shown in Fig. 4.7.1(4)-19. The cask has four air inlets and outlets at each 90-degree angle on its circumferential surface. The specifications of the concrete cask used for the evaluation are shown in Table 4.7.1(4)-1.

Table 4.7.1(4)-1 Specification of concrete cask

Inlet air temperature : T_{in} (°C)	33
Outlet air temperature : T_{out} (°C)	98
Heat rate of Canister : Q (kW)	22.6
Air Velocity at inlet : U_0 (m/s)	0.865
Hight of Inlet : H_0 (m)	0.28
Width of Inlet : B (m)	0.3

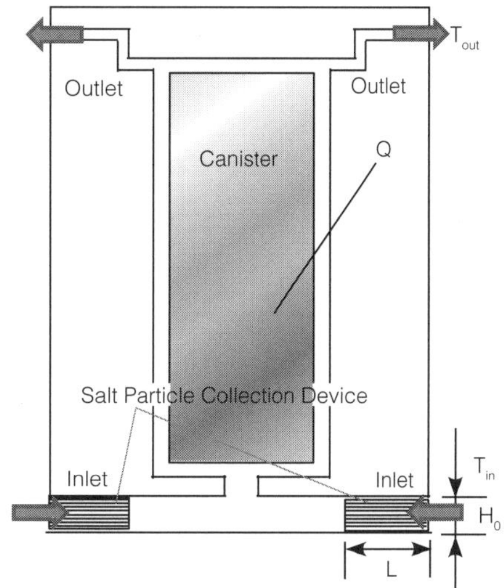

Fig. 4.7.1(4)-19 Concrete cask

It was found that the most of heat of the canister is removed by heat convection on the basis of Table 4.7.1(4)-1, and the formula (13) holds. The balance between the heat and buoyancy is expressed in the formula (14). In this formula, β is the coefficient of thermal expansion, c is the specific heat of the air, and L_D is the characteristic length.

$$Q = \rho_f c \Delta T U_0 H_0 B \times 4 \quad \text{..... (13)}$$

$$\beta g \rho_f \Delta T L_D = \frac{1}{2} \xi \rho_f U_0^2 \quad \text{..... (14)}$$

ξ is calculated by using the formula (13), (14) and the values in Table 4.7.1(4)-1. And the relation between δΔT (the deviation of ΔT) and δξ (the deviation of ξ) is expressed in the formula (15). The effect of the increasing pressure loss by increasing plates on outlet air temperature can be evaluated by using the formula (15).

$$\delta \Delta T = \frac{\Delta T}{3\xi} \delta \xi = 0.649 \times \delta \xi \quad \text{..... (15)}$$

Table 4.7.1(4)-2 shows the relation between an outlet air temperature increase rate and the particle collection rate with the number of the plates in the device as a parameter. The particle collection rate is based on the experimental results under the condition of using the wet plates and no electric charge, as shown in Sec. 3)-b. From Table 4.7.1(4)-2, it was found that the particle collection rate could be expected to be 40 % in the case of using the 15 plates with the length of 0.5 m and the thickness of 0.5 mm and

Table 4.7.1(4)-2 Performance of the device (L=0.5 m)

The number of plates n	Electric Field E (kV/m)	Height of Duct separated by plates H (mm)	Minimum dirmete of particle collected dp (μm)	Pressure loss coefficient ξ	Temperature increse rate δΔT (℃)	Particle collection rate (%)
1	0	279.5	87.7	0.05	0.03	3.3
5	0	55.5	39.2	0.20	0.13	16.6
10	0	27.5	27.7	0.44	0.28	31.7
15	0	18.2	22.6	0.74	0.48	40.0
	10^2		18.3			44.4
	10^3		4.2			76.3
30	0	8.8	16.0	3.13	2.03	45.7
40	0	6.5	13.9	5.86	3.80	46.8

the corresponding outlet air temperature increase rate could be expected to be 0.48 °C which is not disturb the heat removal function. Thus, the salt concentration into the cask could be expected to be half by using the device, and the time to generate SCC on the canister could be prolonged by two times. In the case of providing the electric field with 10^2 kV/m and the re-entrainment prevention treatment on the 15 plates, the minimum diameter of the collected particle could be 18.3 μm, and the particle collection rate could be 44.4 %. Similarly, in the case of the electric field with 10^3 kV/m, the minimum diameter of the collected particle could be 4.2 μm and the particle collection rate could be 76.3 %.

5) Summary

The decrease in concentration of salt in the air is useful as the measure against SCC for the canister in the cask arranged near seaside. Regarding this, we proposed the salt particle collection device which could be installed at air inlet of the concrete cask and quantitatively evaluated the influence of pressure loss and the particle collection rate by using experimental results. It was found that the device could be expected to cut 40 % of particles in the air without disturbing the heat removal function of the cask. On the other hand, it is clarified from the result of researching the ion ratio of collected particles that Cl ion which causes SCC can be removed only 15 % by using the wet plates. It means that the salt particles are small comparatively, so that the salt particles are difficult to collect only by gravity effect. In this test, the wet plates were used for the re-entrainment prevention of the particles and for the evaluation of the particle collection rate. However, adhesive plates would be preferable for the practical usage. And in the case of using the electric field to increase the particle collection rate, the usage of the electret filter would be effective. It is supposed that the electric field of more than 10^3 kV/m or more is required for the specification of the device to get the maximum effect.

References

1) K. Shirai, J. Tani et al., "May, 2011. Study on Interim Storage of Spent Nuclear Fuel by Concrete Cask for Practical Use - Feasibility Study on Prevention of Chloride Induced Stress Corrosion Cracking for Type304L Stainless Steel Canister - ", CRIEPI Report N10035 (in Japanese)
2) D. D. Macdonald, 1999. Passivity- the key to our metals-based civilization, Pure Appl. Chem., Vol.71, No.6, pp.951-978.
3) T. Saegusa, J. Tani et al., Challenge to Overcome the Concern of SCC in Canister During Long Term Storage of Spent Fuel. IAEA International Conference on Spent Fuel Management, May 31-June 4, 2010.
4) J. Tani, H. Takeda et al., June, 2007. Chloride induced stress corrosion cracking of candidate canister materials for concrete cask storage of spent fuel Vol.5 - Evaluation of SCC and development of salt particle collection device - CRIEPI Report Q06014 (in Japanese)
5) H. Takeda, T. Saegusa, April, 2012. Development of Salt Particle Collection Device for Preventing SCC on Canister (Part 2) -Applicability Evaluation for Salt Particle Collection Device - CRIEPI Report N11044 (in Japanese)
6) H. Takeda, T. Saegusa, Evaluation of Salt Particle Collection Device for Preventing SCC on Canister - Effect on Particle Collection Rate by Electric Field - Global 2013 Salt Lake City, Utah, Sept.29-Oct.3, 2013.

7) H. Takeda, T. Saegusa, Salt Particle Collection Device for Preventing SCC on Canister (in Japanese) JSME Conference Sept. 2012.
8) Catalogue of Nitta Corporation, http://www.nitta.co.jp/product/airclean/catalog/fbook.html
9) JSME. December 2004. JSME Data Book Hydraulic Losses in Pipes and Ducts. (10th Edition) ISBN 4-88898-003-9 C 3353
10) I.E.Idelchik, Handbook of Hydraulic Resistance(3rd Edition),JAICO PUBLISHING HOUSE,2008
11) Catalogue of Shinohara Electric Co.,LTD., http://www.shinohara-elec.co.jp/shouhin/filter/air_taien.html
12) Yoshida, Narita et al., 2007. Chemical composition and size distribution of sea fog over the northern North Pacific, Chikyukagaku (Geochemistry) 41, 165-172.
13) D. C. Blanchard, The electrification of the atmosphere by particles from bubbles in the sea, Progress in Oceanography, Vol.1, PP73-112, 1963.
14) H. Takeda, M. Wataru et al., Heat Removal Verification Tests using Concrete Casks under Normal Conditions, Nuclear Engineering and Design, Vol.238,Issue 5, 1196-1205,2008

(5) Measurement technology of chlorine attached on canister surface

Periodic measurement to verify that the concentration of chlorine in salt attached on a canister is below the concentration that induces stress corrosion cracking (SCC) is necessary in the operation of storage to prevent the loss of the confinement function of spent fuels owing to the occurrence of SCC [1),2)]. In this section, we present candidate methods of measuring the concentration of chlorine in salt attached on a canister. We also show the experimental results.

An electrical conductivity (EC) meter or ion chromatography (IC) is commonly used for the measurement of surface salt concentration by wiping the surface of the measurement point with gauze or filter paper. EC reflects the conductivity of water and is related to the concentration of salt dissolved in water. IC can also be used to analyze the elements dissolved in water. The surface salt concentration can be determined directly by EC measurement when the EC measurement device is attached on the surface of a target, and water is filled in the gap between the device and the surface to dissolve surface deposits. The measurement range is generally 0 to 2 g/m^2. EC relates to the sum total of all salts dissolved in water and can be converted to chlorine concentration using the equation implemented in relation to ISO 8502-9. EC is dependent on the water temperature; therefore, it is necessary to compensate the temperature dependence. Some EC measurement devices have a temperature compensation function and can work at temperatures from 0 to 50 °C.

IC has a high sensitivity up to several tens of ppb because it can be used to directly measure each ion concentration. However, it is necessary to wipe away the salts attached on a canister or dissolve them in gauze or water and deliver them to the IC device for measuring the surface salt concentration because IC can be used to analyze only the solution. If the salts attached on the canister are dissolved in water, water vaporization can occur owing to the hot environment of the canister surface. Therefore, the salt sampling procedure is an important issue for adopting IC to the analysis of surface salt.

X-ray fluorescence (XRF) analysis is a method of measuring the concentration of each element in the target by analyzing the X-ray fluorescence, the energy of which depends on the elements, from the target irradiated by the characteristic X-ray produced by a discharge tube. Although this method enables noncontact measurement in principle, the salt sampling procedure from the surface of a canister using, for example, a tube is necessary.

Laser-induced breakdown spectroscopy (LIBS) is a method of measuring the concentration of each element in the target by analyzing the emission spectra of plasma produced by the laser irradiation on the target[3]. A similar method using high-voltage discharge instead of laser irradiation to produce plasma is called spark-induced breakdown spectroscopy (SIBS). LIBS or SIBS can be applied for onsite measurement outdoors because no sample preparation is necessary for the measurement.

A comparison of the methods of surface salt measurement is shown in Table 4.7.1(5)-1. The methods, which can be used for onsite measurement, are shown in the table; an electron probe microanalyzer (EPMA) is not included in the table because of the difficulty of downsizing the device. For all the methods shown in the table, it is necessary to consider their applicability in an actual environment before being used for the quantitative measurement of chlorine in salt attached on a canister. For example, the surface temperature of a canister just after storage of spent fuel was estimated to be up to 200 °C [4]. Therefore, a noncontact method or the use of a heat-resistant probe in a contact method is necessary, and applicability evaluations of the devices in terms of radiation resistance are also necessary because the surface of the canister is in a radioactive environment.

Table 4.7.1(5)-1 Comparison of methods of surface salt measurement

Method	Physical parameters	Identification of elements	Contact to the target
X-ray fluorescence analysis	X-ray fluorescence	Possible	Noncontact
Ion chromatography	Anion, Cation	Possible	Contact
Electric conductivity measurement	Electric conductivity	Impossible	Contact
Laser-induced breakdown spectroscopy	Plasma emission	Possible	Noncontact

Examples of chlorine concentration measurement by LIBS are shown below[5]. Although the applicability evaluation for the measurement under high-temperature and high-radioactivity environments is needed for the measurement of salt attached on the canister that is stored in a concrete body, the sensitivity for chlorine concentration under room temperature was evaluated first because the salt concentration measurement on metal surfaces by LIBS had not been performed. The experimental setup for chlorine concentration measurement by LIBS is shown in Fig. 4.7.1(5)-1. Q-switch Nd:YAG lasers were used for lasers 1 and 2 in Fig. 4.7.1(5)-1, and the second harmonics were irradiated on a stainless-steel plate. The pulses of lasers 1 and 2 were superposed using a polarization beam splitter and by rotating the plane of polarization, and were focused by a plano-convex lens. One of the laser pulses was used for producing the plasma, and the other was used for re-exciting the laser-induced plasma. In comparison with the use of one laser pulse, which is called the single-pulse (SP) configuration, the use of two laser pulses, which is called the double-pulse (DP) configuration, is effective for enhancing the emission intensity of the laser-induced plasma, which contributes to the enlargement of the quantitative measurement range. The emission from the laser-induced plasma was collected on a bundle fiber by a plano-convex lens through a long-wave pass filter, and was detected using a spectrometer and an intensified charge-coupled device (ICCD) camera. A time delay of exposing the ICCD camera

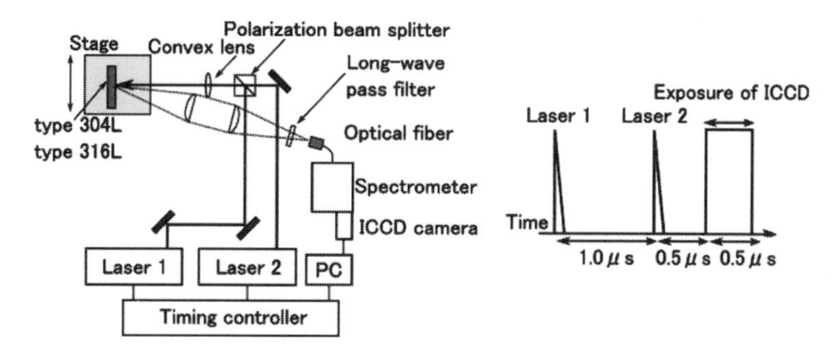

Fig. 4.7.1(5)-1 Experimental setup and time relationship of laser pulse irradiation and detection of emission spectrum

after laser irradiation was set because no emission line was observed just after the laser irradiation owing to the bremsstrahlung of electrons in plasma. The stainless-steel plate was moved perpendicularly to the laser irradiation by 1 mm per shot to irradiate each laser pulse on the fresh surface of the plate because the salt attached on the plate was removed by laser irradiation. Types 304L and 316L stainless-steel plates, which are candidate canister materials, were used for LIBS. Synthetic seawater was sprayed uniformly on the plate, and the chlorine concentration was controlled by adjusting the number of sprayings.

Examples of emission spectra are shown in Fig. 4.7.1(5)-2. Iron, oxygen, calcium and chlorine emission spectrum were observed in the wavelength range from 827 to 852 nm. There was no noticeable change in the spectrum waveform between the chlorine concentrations of 0.1 and 1.0 g/m^2; the spectral intensity, however, decreased at the chlorine concentration of 1.0 g/m^2. Although molybdenum is doped at 2 - 3 % in type 316L, the spectral intensity difference between types 304L and 316L was within the variability of the emission intensity shot by shot, and no noticeable difference in spectral intensity was observed. These results suggest that the presence of molybdenum in stainless steel does not affect the waveform and intensity of emission spectra.

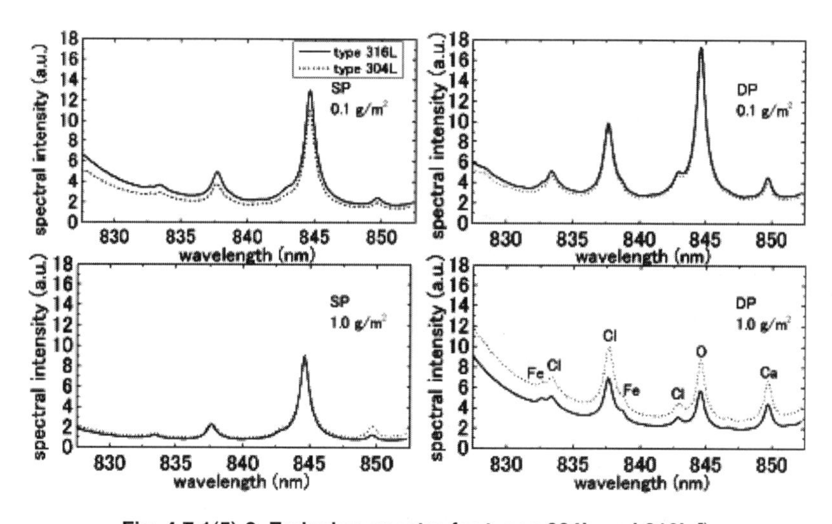

Fig. 4.7.1(5)-2 Emission spectra for types 304L and 316L [5)]

The chlorine concentration dependences on the chlorine emission intensity are shown in Fig. 4.7.1(5)-3. The chlorine emission spectrum at the 837.59 nm wavelength, which has the highest emission intensity in chlorine, was chosen for obtaining the emission intensity. The chlorine emission intensity was 2 - 4 times higher in the DP configuration than in the SP configuration, and it decreased with a chlorine concentration of over 0.4 g/m^2. The results are considered to be caused by insufficient plasma generation in SP and DP configurations, and show that the calibration curve for the chlorine concentration can be obtained using the chlorine emission intensity when the chlorine concentration is from 0.05 to 0.4 g/m^2.

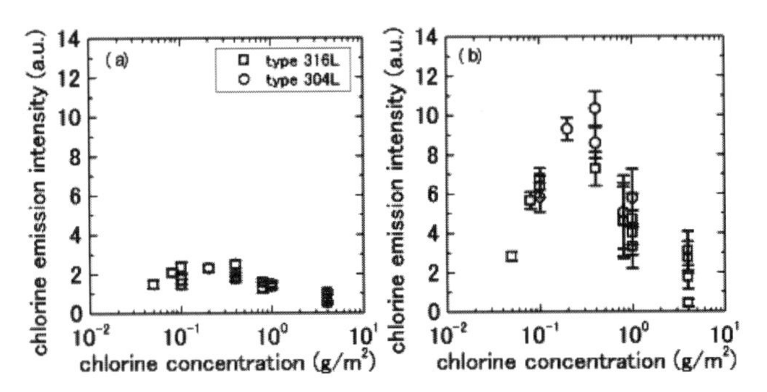

Fig. 4.7.1(5)-3 Chlorine concentration dependence on chlorine emission intensity for (a) SP and (b) DP configurations [6)]

One method of correcting the calibration curve in LIBS is to use the emission intensity ratio of each element. This affects the emission intensity of the elements other than the element concentration; for example, the electron density in plasma can be canceled out using the emission intensity ratio. The emission intensity ratio between oxygen and chlorine was used for the correction in the experiment. The emission intensity ratios between oxygen (844.63 nm) and chlorine are shown in Fig. 4.7.1(5)-4. The chlorine emission intensity was subtracted from the sodium emission intensity, which interferes with the chlorine emission intensity. The emission intensity ratio increased monotonically with the chlorine concentration from 0.05 to 0.4 g/m^2 in DP configration. Although the variability of the emission intensity ratio between oxygen and chlorine was large at the chlorine concentration from 0.4 to 1.0 g/m^2, the chlorine concentration in this range can be estimated from the emission intensity ratio. On the other hand, the

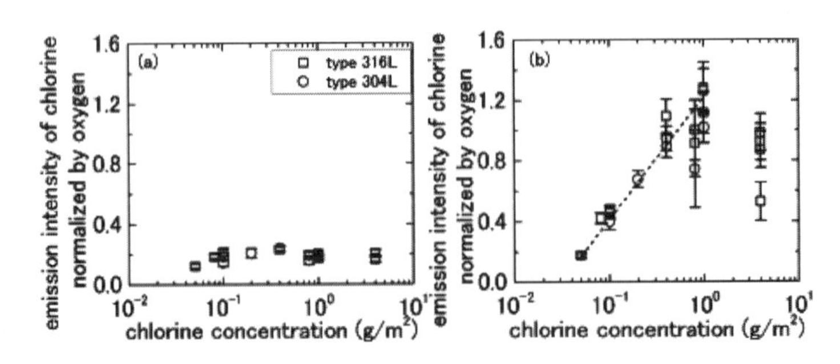

Fig. 4.7.1(5)-4 Chlorine concentration dependence on emission intensity ratio between oxygen and chlorine for(a) SP and (b) DP configurations [5)]

emission intensity ratio between oxygen and chlorine has a weak dependence on the chlorine concentration in the SP configuration. The results suggest that the chlorine in salt was not excited sufficiently by the laser-induced plasma.

References

1) J. Tani, M. Mayuzumi, N. Hara, Initiation and propagation of stress corrosion cracking of stainless-steel canister for concrete cask storage of spent nuclear fuel, Corrosion 65 (2009) 187-194.

2) K. Shirai, J. Tani, M. Gotoh, M. Wataru, SCC evaluation method of multi-purpose canister in long-term storage, Proceedings of PSAM 11 & ESREL, 2012, 16BS-We2-3.

3) W. Miziolek, V. Palleschi, I. Schechter, Laser-induced Breakdown Spectroscopy (LIBS), Cambridge University Press, Cambridge, 2006, pp. 5-17, ISBN 0521852749.

4) H. Takeda, M. Wataru, K. Shirai, T. Saegusa, Study on concrete cask for practical use -Heat removal test under normal condition-, CRIEPI report N04029 (2004) [in Japanese].

5) S. Eto, T. Fujii, K. Shirai, Development of measurement technology of chlorine attached on canister using laser - application of LIBS using collinear geometry-, CRIEPI report H11020 (2012) [in Japanese].

6) S. Eto, T. Fujii, K. Shirai, Measurement of chlorine concentration in salt attached on stainless steel by laser-induced breakdown spectroscopy in collinear geometry, Proceedings of LANE'13, 2013, p. 48.

4.7.2 Visual inspection of canister in service

Idaho National Laboratory (INL), USA has been conducting demonstrative test of VSC-17 concrete cask since 1990. We conducted visual inspection of the canister inside the concrete cask in service[1].

The VSC-17 concrete cask was designed and manufactured by BNFL (British Nuclear Fuel Limited, former Pacific Sierra Nuclear, co.).

The INL installed the spent fuel at the hot cell in the TAN (Test Area North) 607 facility on October 1990, performed heat removal and shielding performance evaluation tests, and started storage on the concrete pad outside the TAN 791 area. The VSC-17 was transferred to the outside of CPP2707 area at INTEC (Idaho Nuclear Technology and Engineering Center) site in the CFA (Central Facilities Area) on October 2004, and being stored at the same area. Fig. 4.7.2-1 shows the VSC-17 concrete cask being transported to the INTEC site from TAN site, and the current status of the cask at INTEC (December 2004).

Fig. 4.7.2-2 shows a basic structure of VSC-17. The spent fuel was originally configured as a 15 × 15-rod array in PWR assemblies. The fuel were repackaged as a part of the Dry Rod Consolidation Technology project into canisters that have the same external dimensions as a PWR assembly (4.3 m in length × 216 mm square), but contain 410 rods from consolidated assemblies with a storage density ratio of 1.8.

The burnup of the spent fuel was 26.8 to 35.4 GWd/MWt and the cooling time 9 to 15 years. The heat generation was 704 to 1050 W (average 875 W) per assembly after rod consolidation, and the total heat generation was 14.9 kW.

The canister was made of carbon steel (currently stainless steel is used for concrete casks in most cases.). The inner and outer surfaces of the canister were coated with a solid lubrication film (coating material Everlune 812). The lid was fixed by bolts with a metal gasket to secure confinement (Currently lids are welded in most cases).

(Transferred from TAN to INTEC site, 2004.10.) (Storage at INTEC site, 2004.12.)

Fig. 4.7.2-1 VSC-17 concrete cask

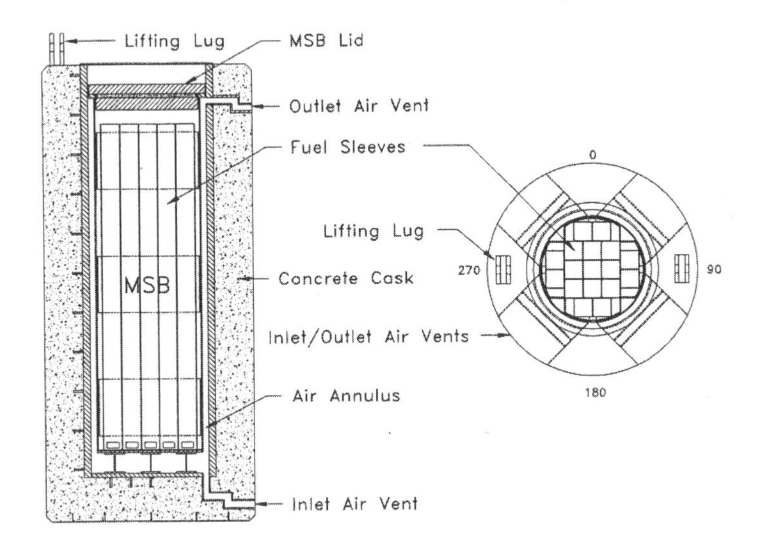

Fig. 4.7.2-2 Basic structure of VSC-17 concrete cask

(Area under weather cover) (Nozzle made by Swagelok for gas sampling)

Fig. 4.7.2-3 Canister surface condition after 15 years of storage

(1) Surface of the canister lid

The cask was modified and installed a more robust upper plate as part of the weather cover in order to secure the strength during handling prior to the transfer from the outside area of TAN 791 to the outside area CPP 2707 on October 2004. Visual inspection was performed when the cover was replaced with respect to corrosion. Fig. 4.7.2-3 shows the canister surface condition and the nozzle for gas sampling after 15 years of storage. The gas sampling has been made periodically even now, which reported nothing abnormal so far. The visual inspection revealed nothing abnormal. The lid surface revealed no significant corrosion maintaining its good condition.

(2) Annulus inspection

We performed video inspection of the conditions of the annulus between the multi-element sealed basket (MSB, canister) and the concrete shield of the VSC-17 cask. Three different borescope cameras were inserted from the upper annulus to downward along the canister surface until the canister support frame of the VSC-17 cask. Fig. 4.7.2-4 shows locations of the inspection.

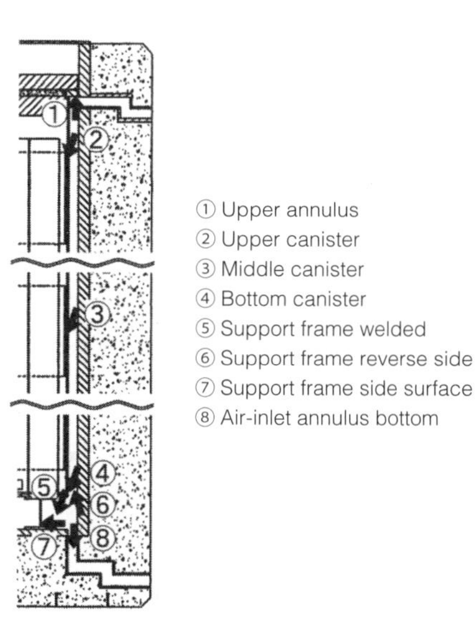

Fig. 4.7.2-4 Locations of video inspection using borescope cameras

Table 4.7.2-1 shows specifications of the borescope cameras used for the visual inspection.

Fig. 4.7.2-5 shows borecope images taken by the two cameras for comparison. The image by Toshiba Model IK-M44H showed condensation stains on the canister surface with relatively better resolution and less noise. The linear stains look like rust stains that appear to be the result of water, either wind-blown or condensed, running down the side of the carbon steel canister.

From these results, we can expect that micro cameras sold in the market may be applicable for the future inspection devices by optimum combination of probe driving function, probe lens diameter, and light source, etc.

Table 4.7.2-1 Specifications of the borescope cameras

Date of inspection	September 2004	March 2007
Borescope camera	Everest/VIT (processor XL Pro/light source used jointly)	Toshiba Model IK-M44H (processor IK-CU44 used jointly)
Probe	With side view optical adapter (PXT850SG：50-degree field of view and 9–160 mm depth of field) 8.4 mm diameter probe	17 mm diameter probe (JK-L75M：7.5-mm lens)
Image sensor	1/6 inch Super HAD™ CCD	1/2 inch IT- CCD
Signal	NTSC/PAL optional	NTSC
Effective pixels	537×505 pixels	768×494 pixels
Camera exterior	Left: Toshiba Model EM-QN42H (IK-M44H) Right: Everest/VIT XL Pro type	

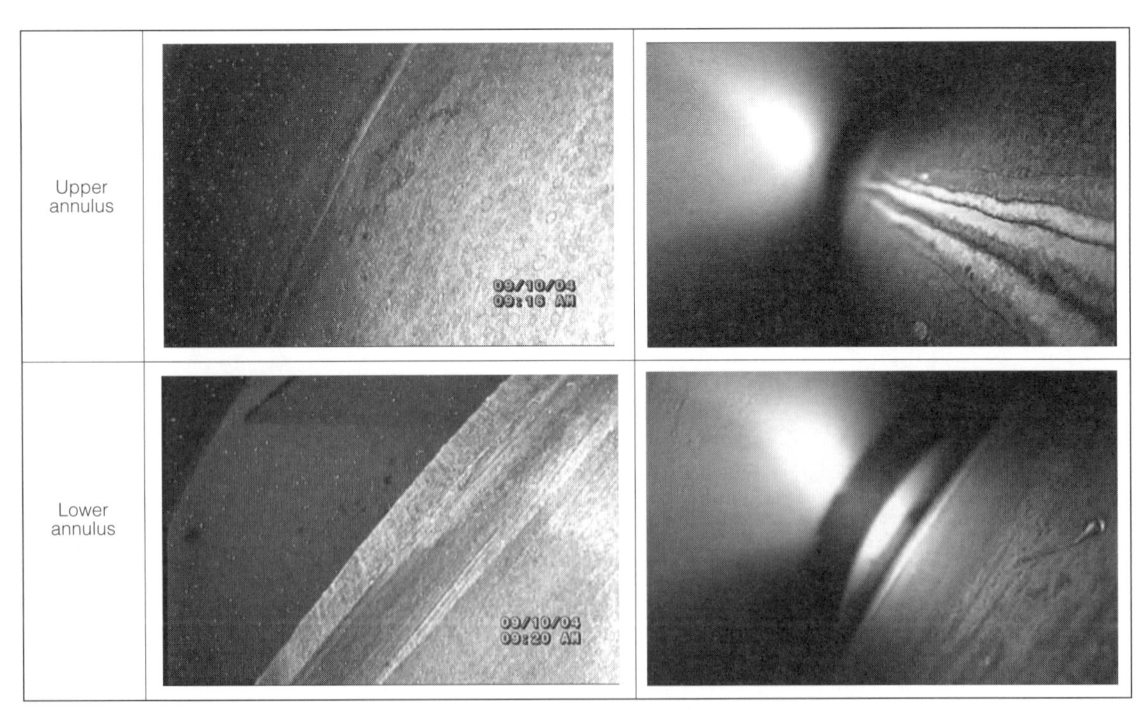

Left: Welch Allyn Everest/VIT borescope images, September 2004
Right: Toshiba Model IK-M44H images, March 2007.

Fig. 4.7.2-5 Comparison of images of annulus

(3) Others

The video of the inspection showed no detectable changes between 2004 and 2007. The only indications of degradation of the external surface of the canister are some rust stains that appear to be the result of water, either wind-blown or condensed, running down the side of the carbon steel canister. On the other hand, the side plate of the canister support frame revealed general corrosion, which indicates the necessity of the similar countermeasures for corrosion for the canister[2].

References

1) K. Shirai, T. Saegusa, A. Sasahara, T. Hattori, T. Matsumura, S. L. Morton and P.L. Winston : "Ageing Characteristics of the VSC-17 Concrete Cask Storing Spent Nuclear Fuel for 15 years", CRIEPI Report N08057, 2008.

2) J. Kessler, "EPRI Initiative in Dry Storage", INMM Spent Fuel Management Seminar XXIX, Jan. 13-15, 2014, Washington DC.

4.7.3 Detection method of helium leak from canister

In a concrete cask, spent fuel is installed in a canister and a lid of the canister is sealed by weld. The canister is filled with helium gas and its containment should be maintained and inspected during storage.

Helium gas enhances the function of heat removal from spent fuel. When the helium gas leaks, the effect of helium convection is weakened in the canister. Then, temperature on a surface of the canister changes. In light of this principle, a detecting method using information on temperature change on a side surface of the canister has been proposed[1].

However, there are the following problems in the proposed method.

1) It is difficult to determine the helium leak by the change of the side surface temperature, because the change is very small and depends on inlet air temperature.
2) Experimental determination is necessary in order to detect the helium leak using transient temperature information on the canister surface.

In this study, we developed an improved helium leak detecting method to detect the helium leak quickly and accurately[2].

(1) Test cases and measurement points

Helium leak tests were performed using two types of concrete casks (see 4.2.2). Table 4.7.3-1 shows the test cases.

Table 4.7.3-1 Test cases of helium leak

CASE No.	Cask Type	Initial Pressure (kPa)	Final Pressure (kPa)	Leak rate (Pam^3/s)
CASE 1	CFS cask	56	5	4.86×10^{-1}
CASE 2	CFS cask	151	1	5.16
CASE 3	RC cask	59	1	3.60×10

In CASE 1 and CASE 3, helium gas of atmosphere pressure (0 kPa) level was filled into the canister before heating. Then, the inner pressure reached 56 kPa in the steady state with the condition of a heating rate of 22.6 kW in a CFS cask. In the same way, it reached 59 kPa in the RC cask. In CASE 2, helium gas of 100 kPa was filled about before heating. Then, the inner pressure reached 151 kPa in the steady state with the condition of a heating rate of 22.6 kW.

In each case, the tests started when a valve was opened to leak the helium gas.

Fig. 4.7.3-1 shows temperature measurement points.

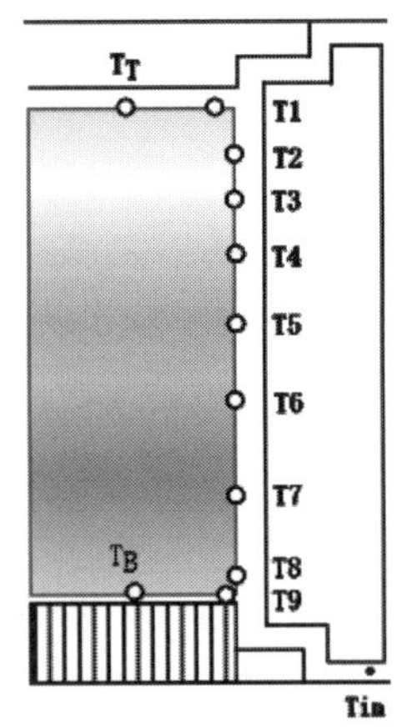

Fig. 4.7.3-1 Temperature measurement points

(2) Test results

1) CASE 1

Fig. 4.7.3-2 shows the relation between the inner pressure of the canister and the canister surface temperature, including inlet air temperature. The initial condition was in such a steady state that the heating rate was 22.6 kW and the inner pressure was 56 kPa.

The valve of the canister was slightly open, so that the helium gas leaked and the inner pressure decreased from 56 kPa to 5 kPa in 4 days. Finally, the valve was opened completely to drop the pressure to 0 kPa.

It was found that the canister surface temperature depends on the inlet air temperature. It is difficult to determine the helium leak by the change of the surface temperature.

Fig. 4.7.3-3 shows canister surface temperature distribution in normal condition, and Fig. 4.7.3-4 shows one in helium leak condition.

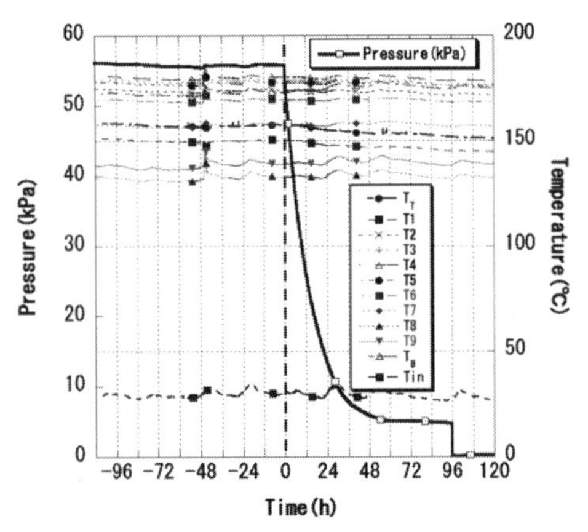

Fig. 4.7.3-2 Canister surface temperature and pressure

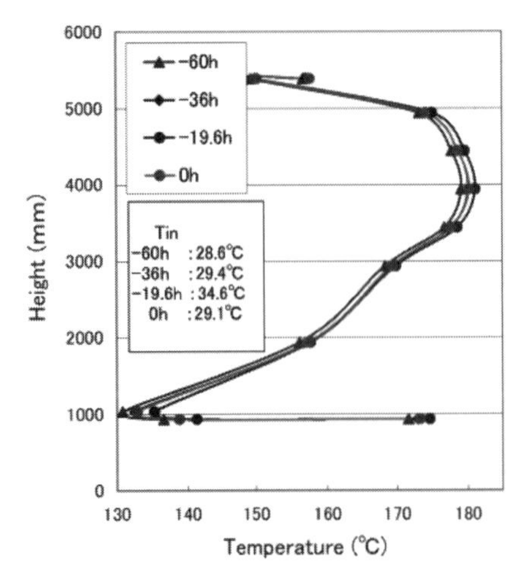

Fig. 4.7.3-3 Canister surface temperature distribution (Normal)

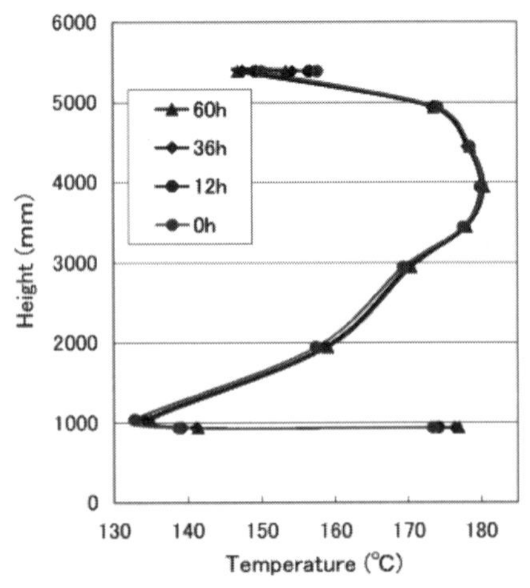

Fig. 4.7.3-4 Canister surface temperature distribution (Helium leak)

In normal condition, the canister surface temperature followed the change of the inlet air temperature.

On the other hand, in helium leak condition, the temperature at the center of the canister bottom increased, but the temperature at the center of the canister top decreased.

Fig. 4.7.3-5 shows the change of the canister surface temperature distribution at each height in normal condition. Fig. 4.7.3-6 shows the change of the canister surface temperature distribution at each height in helium leak condition. It was seen that temperature difference between the center of the canister bottom (T_B) and the center of the canister top (T_T) increased about 8 °C in 60 hours after the test started.

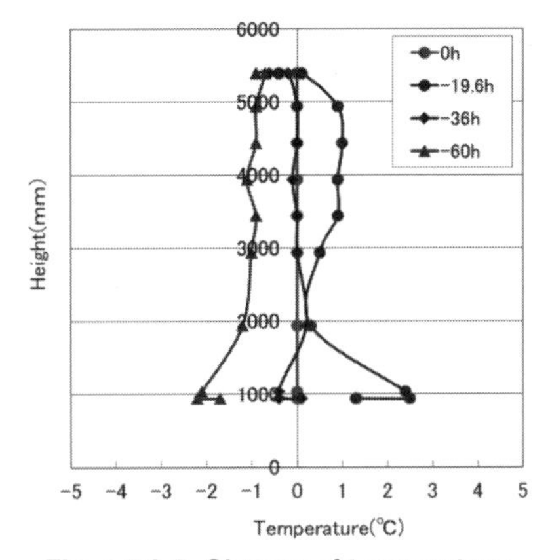

Fig. 4.7.3-5 Change of temperature distribution (Normal)

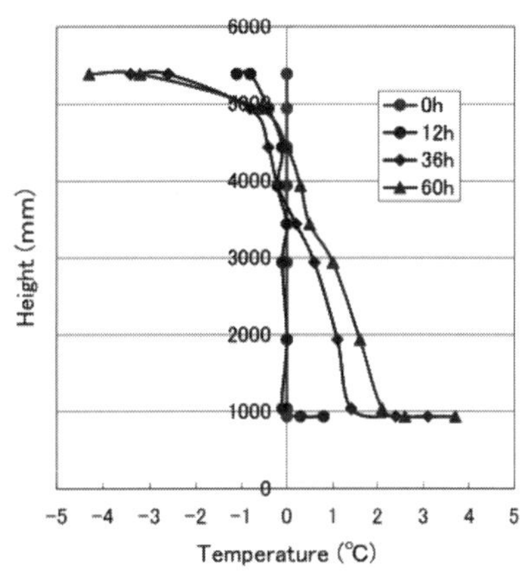

Fig. 4.7.3-6 Change of temperature distribution (Helium leak)

The temperature difference (=T_B - T_T) was defined as ΔT_{BT}, and a detecting method of the helium leak using ΔT_{BT} was examined.

Fig. 4.7.3-7 shows the relationship between the change of ΔT_{BT} and that of the inner pressure of the canister. Fig. 4.7.3-8 shows the relationship between the change of ΔT_{BT} and that of the inlet air temperature.

It was found that ΔT_{BT} increases significantly due to the helium leak.

Fig. 4.7.3-8 shows the relationship between the change of ΔT_{BT} and that of the inlet air temperature. It was found that ΔT_{BT} increases during the helium leak although the inlet air temperature decreases.

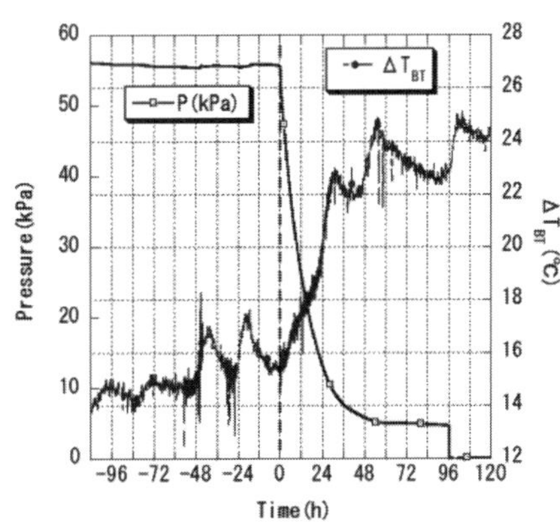

Fig. 4.7.3-7 Change of ΔT_{BT} and pressure (CASE 1)

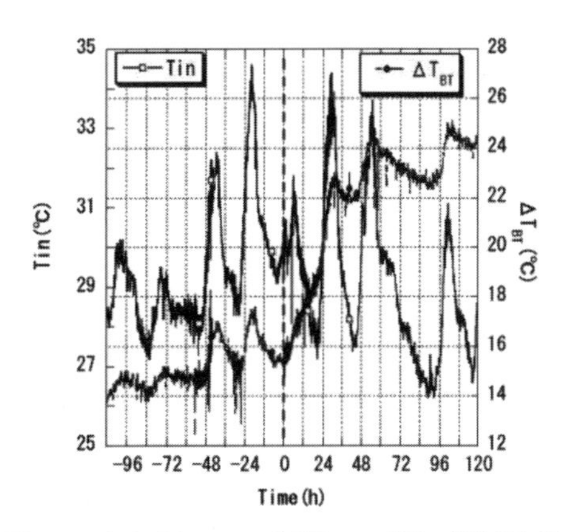

Fig. 4.7.3-8 Change of ΔT_{BT} and Tin (CASE 1)

Fig. 4.7.3-9 shows canister surface temperature distributions at the various heating rates. Values of ΔT_{BT} are 14.7 °C under the condition of 22.6 kW that simulated the initial storage, 13.2 °C under the condition of 16 kW (20-year storage) and 10.2 °C in the condition of 10 kW (40-year storage). The values of ΔT_{BT} decreased with the heating rate decreasing.

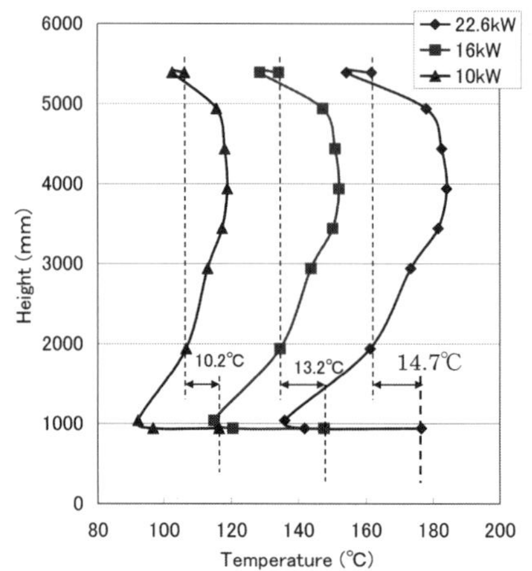

Fig. 4.7.3-9 Surface temperature distribution at various heating rates

On the other hand, when the helium gas leaked, the values of ΔT_{BT} increased. Thus, this detecting method using ΔT_{BT} can accurately detect the helium leak.

2) CASE 2

In this case, the initial inner pressure was 151 kPa. The pressure dropped to 36 kPa in one day. We opened the valve completely, so that the pressure dropped to 1 kPa.

Fig. 4.7.3-10 shows the relationship between the change of ΔT_{BT} and that of the inner pressure. Fig. 4.7.3-11 shows the relationship between the change of ΔT_{BT} and that of the inlet air temperature.

In this case, the helium leak rate was calculated to be 5.16 (Pa · m^3/s).

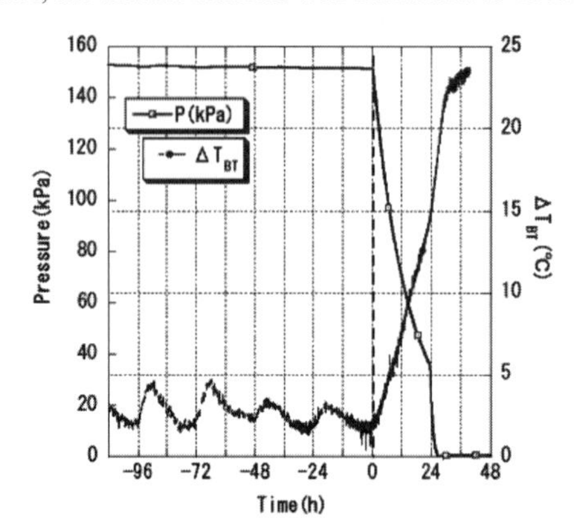

Fig. 4.7.3-10 Change of ΔT_{BT} and pressure (CASE 2)

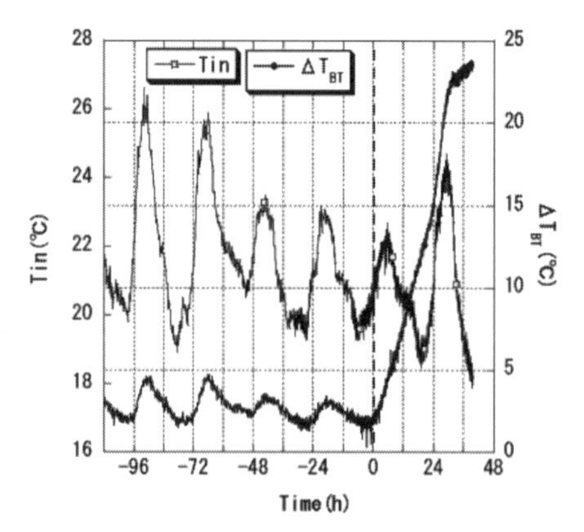

Fig. 4.7.3-11 Change of ΔT_{BT} and Tin (CASE 2)

The values of ΔT_{BT} increased approximately 20 °C due to the helium leak.

Fig. 4.7.3-12 shows the canister temperature distribution at the various inner pressures under the condition that the inlet air temperature is 33 °C. The values of ΔT_{BT} were 1.8 °C at 151 kPa, 17.9 °C at 56 kPa, and 23.9 °C at 0 kPa. These values indicate that the convection of the helium gas exists in the canister.

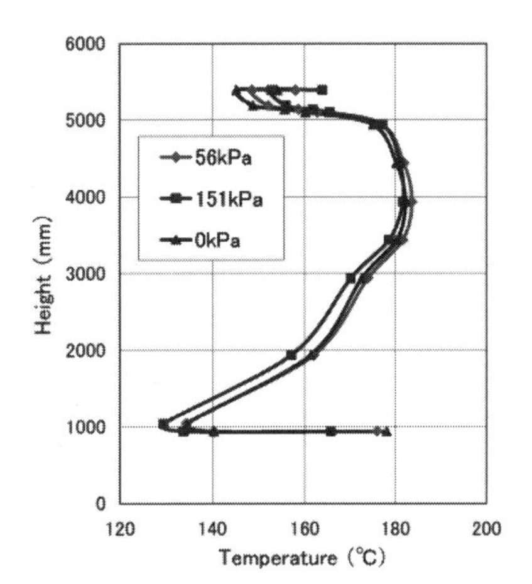

Fig. 4.7.3-12 Surface temperature distribution at various pressures

3) CASE3

The helium leak test was performed using the RC cask.

Fig. 4.7.3-13 shows the relationship between the change of ΔT_{BT} and that of the inner pressure. Fig. 4.7.3-14 shows the relationship between the change of ΔT_{BT} and that of the inlet air temperature.

In the case of the RC cask, cooling air impinges on the bottom of the canister directly. Then, the temperature of the bottom was lower than that of the CFS cask.

Thus, the values of ΔT_{BT} were minus. However, ΔT_{BT} increased after the helium leak started as well as in the tests using the CFS cask.

In this case, the inner pressure dropped from 56 kPa to 1 kPa in 2 hours. The helium leak rate was as large as 3.60×10 (Pa · m^3/s). Thus, the canister surface temperature changed gradually after the helium leak completed. The ΔT_{BT} increased about 8 °C during 24 hours.

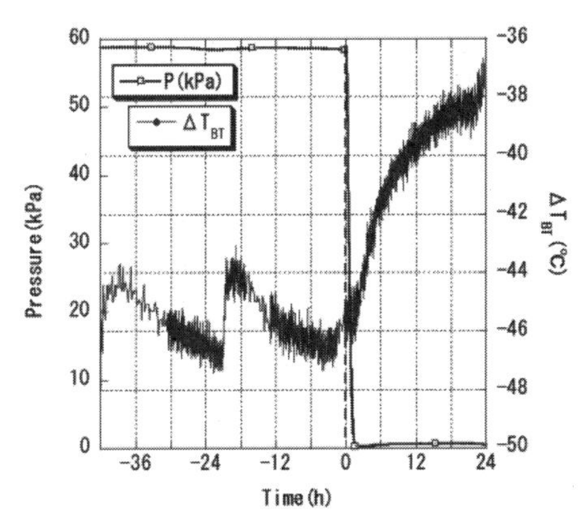

Fig. 4.7.3-13 Change of ΔT_{BT} and pressure (CASE 3)

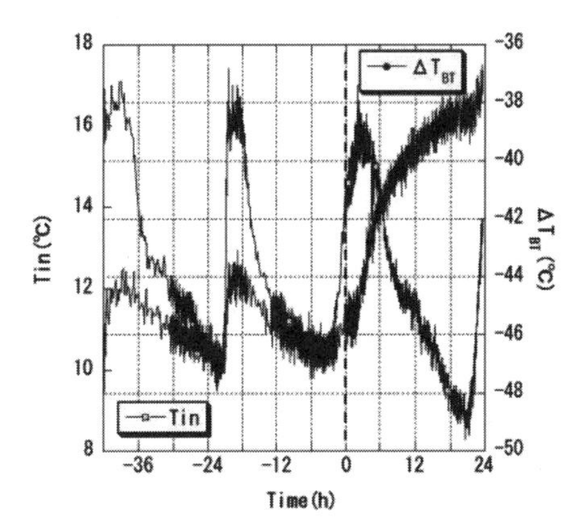

Fig. 4.7.3-14 Change of ΔT_{BT} and Tin (CASE 3)

(3) Summary

The phenomenon that the canister surface temperature changes due to the helium leak from the canister was verified. Especially, the temperatures of the bottom and the top of the canister change significantly. It was found that ΔT_{BT} increases remarkably during the helium leak. Thus, the helium leak can be detected based on the change of ΔT_{BT}.

The ΔT_{BT} increases monotonously to a fixed value during the helium leak, even if the inlet air temperature decreases. The helium leak can be detected at the early stage of the leak by observing both ΔT_{BT} and the inlet air temperature.

References

1) I. Abe, K. Matsunaga, 2002. Japan Patent Office, P2002-202400A.

2) H. Takeda, et al.,2008, "Development of the detecting method of helium gas leak from canister", Nucl. Eng. Design, Vol.238, pp.1220-1226.

4.7.4 Deterioration of concrete module

(1) Mechanism of chloride induced deterioration

In order to operate concrete module storage facilities for spent nuclear fuel in Japan, the facilities will be located in the vicinity of the coast, and the salt damage of reinforced concrete structures is concerned about. The evaluation of the salt damage to reinforced concrete structures is important problem because the corrosion of reinforcing bar and the concrete crack with the salt damage might reduce radiation shielding function and structural strength of concrete module. Particularly, the evaluation of the salt damage under high temperature is need, because when concrete modules will be stored indoors to show it in Fig. 4.7.4(1)-1, there will be an opportunity to touch the fresh air including the salt by air-cooling naturally, and it is also expected the modules is with a high temperature of around 60 °C by heat of the spent nuclear fuels[1]. However, it was not able to consider temperature of concrete in the past evaluation methods.

In this section, it is explained that the experiments which were conducted to examine influence of temperature and influence of carbonation on chloride ion diffusion phenomenon in concrete and also the experiments which were conducted to examine influence of temperature on chloride ion concentration of corrosion of reinforcing bar. Furthermore, it is explained that the discussion for developing an verification method of durability against salt attack to reinforced concrete structures under high temperature, which was improved on the past evaluation method by based on these results[2].

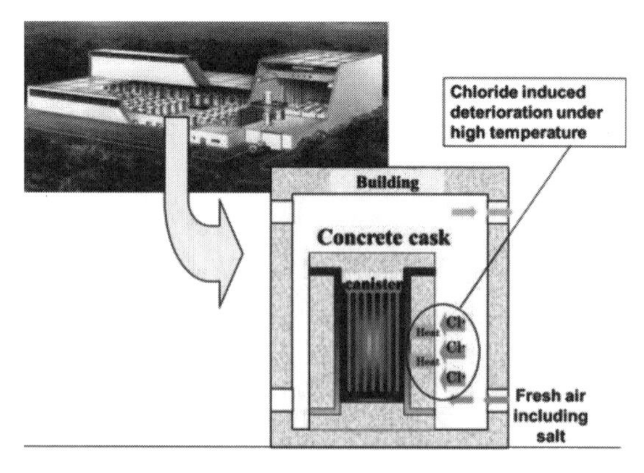

Fig. 4.7.4(1)-1 Example of concrete cask storage system and assumed part of the salt damage

1) Evaluation of influence of temperature on diffusion coefficient of chloride ion[3)]

a) Outline of immersion test

The shape of the concrete specimens for immersion tests is a column type 150 mm in diameter and 100 mm in height as shown in Fig. 4.7.4(1)-2. The concrete types had three kinds of water-cement ratio (W/C = 40, 50, and 60 %). The cement type was Ordinary Portland Cement. Immersion tests of concrete specimens using 10 % sodium chloride solution that were controlled in a constant temperature were conducted. The solution was kept at five levels of temperature, i.e. 25, 45, 65, 80, and 90 °C. Fig. 4.7.4(1)-3 shows the immersion testing set. The test periods were set as six kinds by using six specimens with the same mix proportion and the same temperature. The programming rate up to a fixed temperature was set to be 10 °C/h.

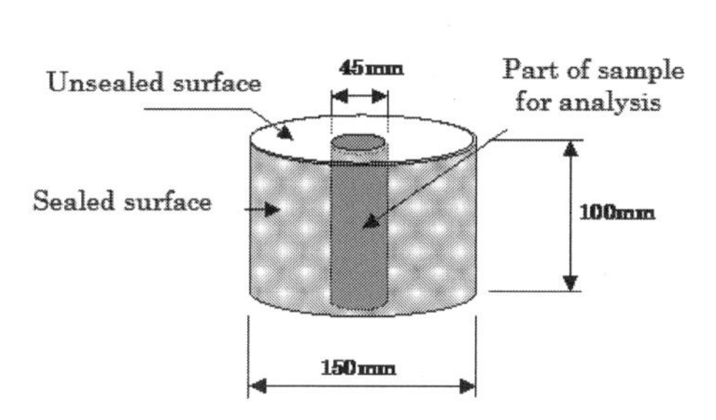

Fig. 4.7.4(1)-2 Specimen for immersion tests

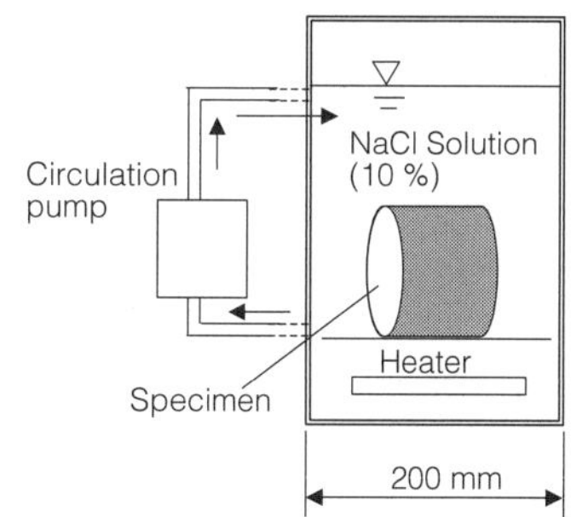

Fig. 4.7.4(1)-3 Equipment for immersion test under high temperature

b) Result of immersion test

Chloride ion is penetrated according to the Fick's second law as shown in the next expression:

$$\frac{\partial C}{\partial t} = D\frac{\partial^2 C}{\partial x^2} \quad \text{..... (1)}$$

where C is chloride ion concentration, t is time, x is distance from the surface, and D is diffusion coefficient. An analytical solution with the initial condition of $C(x, 0) = 0$ and the boundary condition of $C(0, t) = C_0$ of the above expression becomes the next expression:

$$C(x,t) = C_0\left\{1 - erf\left(\frac{x}{2\sqrt{Dt}}\right)\right\} \quad \text{..... (2)}$$

where erf(u) is Gauss's error function. The depth direction distribution of the measured ion concentrations of the chloride was approximated in accordance with the above expression by the minimum square method, and the diffusion coefficients were obtained. Fig. 4.7.4(1)-4 shows the relation between the obtained diffusion coefficient and temperature. When the temperature rises, the diffusion coefficient increases remarkably. Moreover, the concrete with a larger water-cement ratio has a tendency of having larger diffusion coefficient.

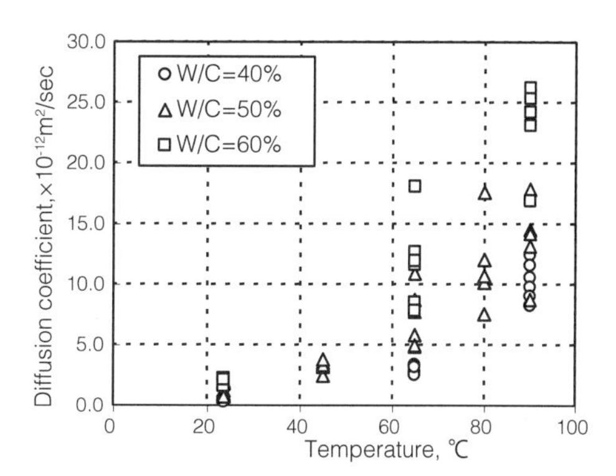

Fig. 4.7.4(1)-4 Relationship between diffusion coefficients and temperature (results of immersion tests)

c) Discussion of influence of temperature on diffusion coefficient of chloride ion

The relation between the diffusion coefficient and the temperature that is shown by an Arrhenius plot is depicted in Fig. 4.7.4(1)-5. It is understood to be able to express the temperature dependency of the diffusion coefficient roughly by an Arrhenius equation as the velocity coefficient is provided as the diffusion coefficient. Diffusion is a thermally excited reaction and its coefficient is generally given by the next expression:

$$D = A\exp(-E_a/RT) \quad ..... (3)$$

where A is the frequency factor, E_a is activation energy for diffusion, R is gas constant and T is an absolute temperature. The next expression is obtained when the diffusion coefficients at temperature T_1 and T_2 are assumed to be D_1 and D_2, respectively.

$$\ln (D_1/D_2) = -E_a (1/T_1 - 1/T_2) /R \quad ..... (4)$$

In other words, the activation energy is expressed by the next expression, and it is obtained as the value that multiplied (-2.30R) by the inclination of the graph which assumed a reciprocal number of the absolute temperature a cross axle, an logarithm of diffusivity a vertical axis.

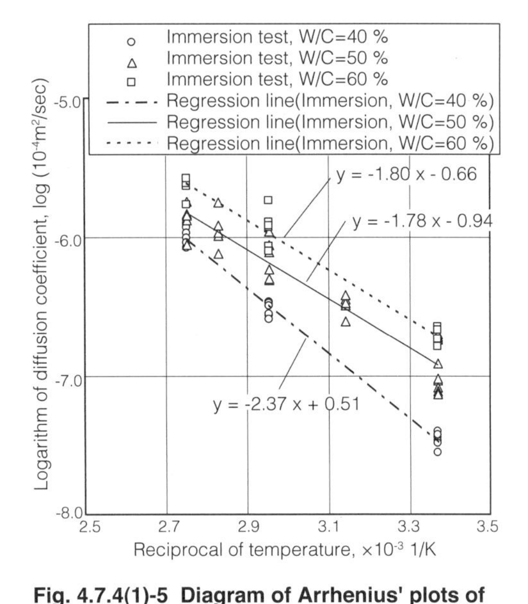

Fig. 4.7.4(1)-5 Diagram of Arrhenius' plots of all data in immersion tests and diffusion cell method tests

$$E_a = -R \times (\ln D_1 - \ln D_2) / (1/T_1 - 1/T_2)$$
$$= -2.30 \times R \times (\log D_1 - \log D_2) / (1/T_1 - 1/T_2) \quad ..... (5)$$

Activation energy obtained by expression (5) with the inclination of the regression line in Fig. 4.7.4(1)-5 is as shown in Table 4.7.4(1)-1. It is understood that the temperature dependency of W/C = 40 % is larger than that of W/C = 50 and 60 %.

Table 4.7.4(1)-1 Activation energy for diffusion of chloride ion

Water-cement ratio (%)	Activation energy (kJ/mol)
40	46.0
50	34.5
60	34.9

Though the reports of obtaining activation energy by using concrete specimen are few, it has been reported that the activation energy of cement paste of low water-cement ratio was larger than that of high water-cement ratio under the temperature of up to 45 °C[4)]. The activation energy of concrete specimen in this study has the same tendency as that of cement paste specimen. The chloride ion diffusion through concrete might depend mainly on the performance of the cement paste. Moreover, activation energy of cement paste of 50.2 kJ/mol from the experiment of the cement paste specimen (W/C = 40 %) under the temperature of up to 60 °C has been reported[5]. The reported value is larger than the result of this study. This is because of the large size pores that are produced when concrete was cured under high temperature of up to 60 °C in the experiment[6].

d) Evaluation equation

The evaluation equation of the diffusion coefficient of the chloride ion in concrete under high temperature was derived based on the test result and the following consideration:

(i) The temperature range of 65 °C or less is used as an evaluation of the safety side, because the temperature dependency tends to be small in the temperature over 65 °C.

(ii) It is predicted that diffusion coefficients in the water saturated concrete obtained by this study are larger than that of the non-saturated concrete like general structures, because the diffusion of the ion is caused in the liquid phase of concrete. Then, the temperature dependency for the saturated concrete can be conservatively used for the non-saturated concrete.

(iii) It is assumed that the diffusion coefficients at normal temperature (20 °C) are the values of the standard[7], which was based on a lot of data of the existing structures. The derived equation for evaluation of diffusion coefficient of chloride in concrete under high temperature is given by the next expression. The diffusion coefficients obtained by the equation and the results of tests are shown in Fig. 4.7.4(1)-6:

$$\log(Y) = -A \cdot X + B, \quad X = \frac{1}{T} \cdot 1000 \quad \cdots\cdots (6)$$

where Y is the diffusion coefficient of chloride ion in noncarbonated concrete (cm^2/s), T is absolute temperature (K), A is constant (2.27, 1.90 in case of W/C = 40 %, 50 %, and 1.90 in case of W/C = 60 %), and B is constant (0.05, -0.79 in case of W/C = 40 %, 50 %, and -0.28 in case of 60 %).

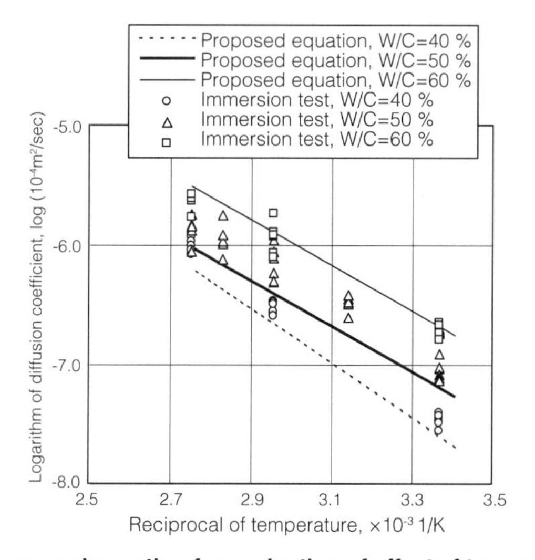

Fig. 4.7.4(1)-6 Proposed equation for evaluation of effect of temperature on chloride ion diffusion coefficient, and data of immersion tests on diagram of Arrhenius' plots

2) Evaluation of influence of temperature on the chloride ion concentration of corrosion initiation

a) Outline of corrosion test

In order to evaluate the limiting concentration of chloride for reinforcing steel corrosion initiation, corrosion tests of reinforced concrete specimens under high temperatures were conducted. The relative humidity was set to be 95 %. The shape of specimen was column type 100 mm in diameter and 200 mm in height, in which a 180mm long reinforcing steel was embedded as shown in Fig. 4.7.4(1)-7. The test section of the reinforcing steel was 8 cm, and the other parts were covered with the waterproof tape. The durations of the tests were 0, 2.5, 5.0, and 11 months.

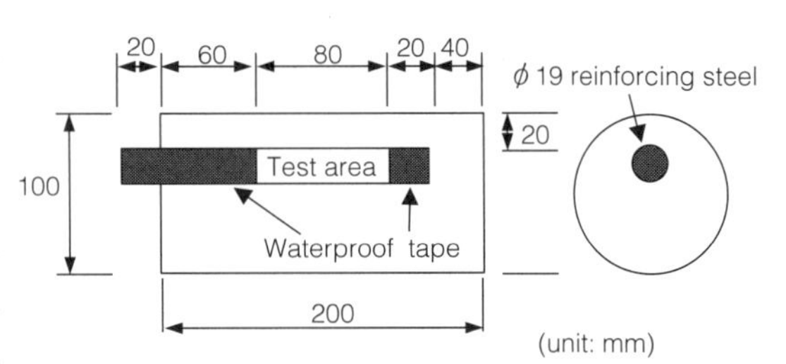

Fig. 4.7.4(1)-7 Specimen of corrosion test

b) Result of corrosion test

In general, the amount of limiting concentration of chloride for reinforcing steel corrosion initiation at normal temperature is often set as 1.2 kg/m^3 [3,7]. Then, the relation between the concentration of chloride ion and the corrosion incidence ratio under normal temperature was first examined. The corrosion incidence ratio was defined as a ratio of the number of specimens with corroded reinforcing steels to the number of all specimens. Fig. 4.7.4(1)-8 shows the relation between the corrosion incidence ratios and the concentrations of chloride ion in the concrete at room temperature. Moreover, using the following logistic function we approximated the relation between the corrosion incidence ratio and the concentrations of chloride ion in the concrete:

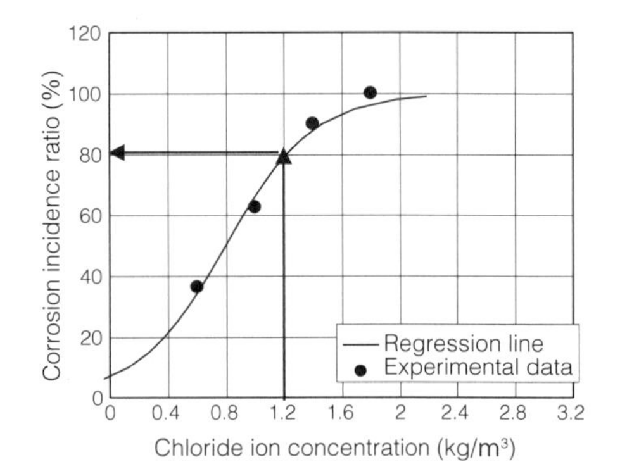

Fig. 4.7.4(1)-8 Relation between corrosion incidence ratio and chloride ion concentration at normal temperature

$$y = \frac{100}{1 + b\exp(-k \cdot x)} \quad \text{..... (7)}$$

where y is corrosion incidence ratio (%), x is concentration of chloride ion (kg/m^3), and b and k are constants. It is understood that the corrosion incidence ratio at the chloride ion concentration of 1.2 kg/m^3 is about 80 %. Then, it can be judged that the chloride ion concentration of the corrosion initiation at normal temperature is a value at the corrosion incidence ratio of about 80 %. Therefore, the chloride ion concentration at the corrosion incidence ratio of 80 % was defined in the following discussion as the chloride ion concentration of the corrosion initiation. The influence of test duration was small, as the increase or the decrease of the chloride ion concentration of corrosion initiation due to the increase of test duration was not recognized. Therefore, the distinction of duration of test was considered in the following discussions.

Fig. 4.7.4(1)-9 shows the relation between the corrosion incidence ratio and the chloride ion concentration at

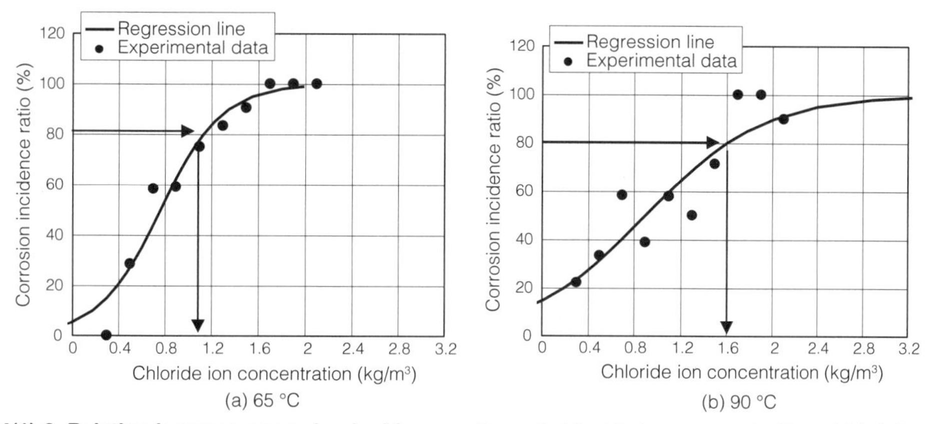

Fig. 4.7.4(1)-9 Relation between corrosion incidence ratio and chloride ion concentration at high temperature

temperatures of 65 and 90 °C. The chloride ion concentration of the corrosion initiation at 90 °C was unexpectedly large for 65 °C. The measurements at 90 °C scattered widely along the regression line. The possibility that the chloride ion concentration of the corrosion initiation decreases with the temperature that is small.

The relation between the water-cement ratios and the chloride ion concentrations of the corrosion initiation at 65 and 90 °C are shown in Fig. 4.7.4(1)-10. When the water-cement ratio becomes large, the chloride ion concentration of corrosion initiation decreases greatly, and is especially remarkable for the water-cement ratio of 60 %. In general, it is known that chloride ion is fixed in concrete up to 0.4 % of cement weight. The fixed chloride ion concentrations in concrete were calculated from the mix proportion of concrete as shown in Fig. 4.7.4(1)-10. The calculated concentrations of 0.4 % of the cement weight are almost equal to the chloride ion concentration of corrosion initiation, except for the water-cement ratio of 60 %. Then, it can be said that chloride ions are immobilized with the same rate as in the normal temperature.

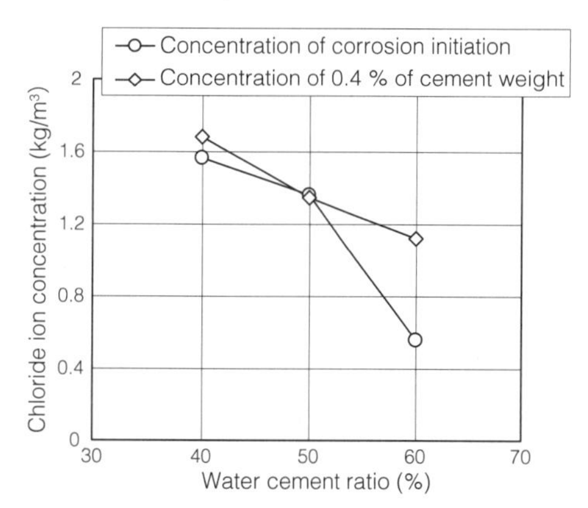

Fig. 4.7.4(1)-10 Relation between water cement ratio and chloride ion concentration for corrosion initiation

The value of 1.2 kg/m^3 at normal temperature can be considered as the chloride ion concentration of corrosion initiation at high temperature up to 90 °C, when reviewing the overall results.

3) Salt damage evaluation method

In general, the salt damage of reinforced concrete structures undergoes the deterioration process of penetration of chloride ion into concrete, corrosion of reinforcing steel, occurrence of crack of concrete due to reinforcing steel corrosion, and remarkable corrosion in cracked concrete. It is not clarified enough whether the structural performance such as load capacity, ability of deformation, etc. of reinforced concrete structures are damaged by salt attack. It is known that the structural performance hardly decreases until the crack due to corrosion is generated[8]. Any decrease of the performance of concrete structures can be conservatively found by checking the initiation of the reinforcing steel corrosion. A verification method of assuming the initiation of reinforcing steel corrosion to be the critical state is proposed as follows. The chloride ion penetration through concrete is estimated by the next equation. The chloride ion concentration at the depth of the position of reinforcing steel at the end of the use period is evaluated (as shown in Fig. 4.7.4(1)-11):

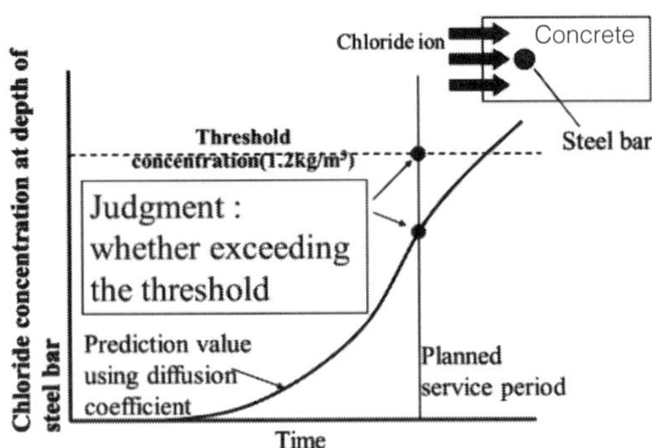

Fig. 4.7.4(1)-11 Schematic of verification method in initiation stage

$$C_d = C_0\left(1 - erf\left(\frac{0.1 \cdot c}{2\sqrt{D_d \cdot t}}\right)\right) \quad \ldots\ldots (8)$$

where C_d is design concentration of chloride ion at the reinforcing steel position (kg/m^3), C_0 is concentration of chloride ion at concrete surface (kg/m^3)[9], c depth of reinforcing steel (mm), t design life (year) and D_d is design diffusion coefficient of chloride ion (cm^2/year), which is given by the next equation :

$$\log D_d = -A \times (1/T) + B \quad \ldots\ldots (9)$$

where D_d is design diffusion coefficient of chloride ion (cm^2/year), T is absolute temperature (K), A is constant (2.27, 1.90 in case of W/C = under 40 %, 41~50 %, and 1.90 in case of W/C = 51~60 %), and B is constant (7.55, 6.71 in case of W/C = under 40 %, 41~50 %, and 7.22 in case of W/C =51~60 %).

Then, it is verified that the chloride ion concentration is less than the concentration of corrosion initiation by next equation:

$$\frac{C_d}{C_{\lim}} \leqq 1.0 \quad \ldots\ldots (10)$$

where $C_{\lim}$ is limiting chloride ion concentration of corrosion initiation (1.2 kg/m^3).

Fig. 4.7.4(1)-12 shows a calculation example of the verification method. In the case of 1.3 kg/m^3 of concentration of chloride ion at concrete surface and 40 % of water cement ratios of concrete, when concrete cover thickness (depth of the reinforcement steel) is 55, 65 mm in case of 40, 50 °C of concrete temperature, and 85mm in case of 60 °C of concrete temperature, the soundness for 40 years is guaranteed during a service period.

This verification method can be applied to the reinforced concrete structure of water-cement ratio 40~60 % used in temperature up to 90 °C due to the applicability of the composed evaluation equations.

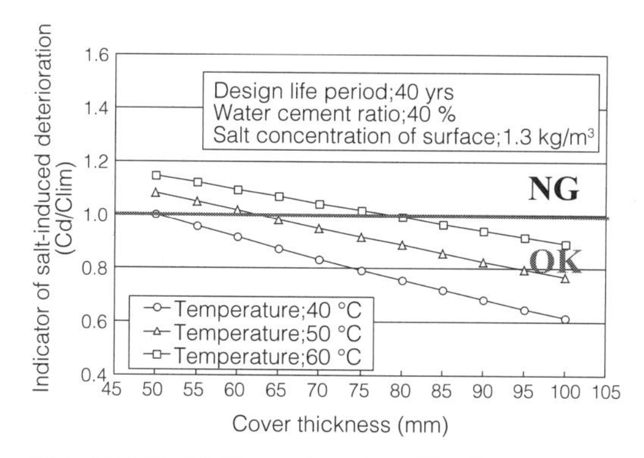

Fig. 4.7.4(1)-12 Examples of verification results

References

1) Koji Shirai, Concrete Cask Storage System for Spent Nuclear Fuel Facility, Concrete Journal, Vol.42, No.9, pp.28-31, 2004. (in Japanese)

2) Takuro Matsumura, Koji Shirai, and Toshiari Saegusa, Study on concrete cask for practical use - Development of evaluation method of salt-induced deterioration of reinforced concrete -, CRIEPI Reports, N04032, 2005. (in Japanese)

3) Takuro Matsumura, Koji Shirai, and Toshiari Saegusa, Influence of Temperature on Diffusion Coefficient of Chloride Ion in Concrete, Journal of the Society of Materials Science Japan, Vol.52, No.12, pp. 1478-1483, 2003. (in Japanese)

4) C. L. Page, N. R. Short and A. El Tarras, Diffusion of Chloride Ions in Hardened Cement Pastes, Cement and Concrete Research, Vol. 11, No. 3, pp.395-406, 1981.

5) Seishi Goto and Della M. Roy, Diffusion of Ions Through Hardened Cement Pastes, Cement and Concrete Research, Vol. 11, pp. 751-757, 1981.

6) R. J. Detwiler, K. O. Kjellsen and O. E. Gjorv, Resistance to Chloride Intrusion of Corrosion Cured at Different Temperatures, ACI Materials Journal, Vol. 88, No. 1, pp. 19-24, 1991.

7) Japan Society of Civil Engineering, Concrete standard specifications "Chapter of construction", 1999. (in Japanese)

8) Japan Concrete Institute, Rehabilitation Research Committee report, 1998. (in Japanese)

9) Japan Society of Civil Engineering, Concrete standard specifications "Chapter of construction", 2002.(in Japanese)

(2) Mechanism and evaluation of combined degradation due to salt damage and carbonation

For the evaluation of concrete casks, it is necessary to design the casks by evaluating and understanding the influence of the degradation due to the following[1)].

· Salt damage specific to the location

· Heat generation of used fuels

· Carbonation due to CO_2 in the air

Concrete materials are usually highly alkaline materials of pH 13.0 to 12.5, which is controlled by $Ca(OH)_2$, one of the major components of the concrete. This high pH causes a passivity film on the surface of reinforcing bars in the concrete structure and suppresses the progress of the corrosion of the bars. However if CO_2 in the air contacts the concrete, it reacts with $Ca(OH)_2$ to produce calcium carbonate, lowering pH. This is carbonation.

In this section, as a part of study on the establishment of concrete cask storage technologies, we show how various properties of the concrete are affected by combined degradation due to heat, salt damage, and carbonation, based on experiments and chemical analysis. For details, see Reference[2].

1) Test method of combined degradation due to salt damage and carbonation

The following was conducted to study the combined degradation due to heat, salt damage, and carbonation.

- Carbonation after spraying salt water
- Carbonation of test pieces containing salt (concentration 0, 2, 4 kg/m^3: The test pieces additionally made in Sec. 4.7.4(1) was used.)
- Salt water spraying after carbonation or heating

The test procedures of the above are shown respectively in Figs. 4.7.4(2)-1, -2, and -3.

Among the analysis items and measurement items of the above, the salt concentration analysis and carbonation measurement were already conducted in Sec. 4.7.4(1). In the present section, chemical analysis was mostly made. We studied a quality change mechanism by analyzing a test pieces before and after combined degradation occurred and observing the degradation conditions.

All the test pieces were cementitious paste test pieces.

The following analyses were made.

(i) X ray diffraction

This was conducted to identify crystalline cement hydrate produced due to the combination of heat and carbonation, that of heat and salt damage, and that of salt damage and carbonation, and check the influence of the combined degradation to the crystalline phase of the cement hydrate.

(ii) Differential thermal analysis

This was conducted to study and qualitatively identify, based on the heat analyses, the weight change of a hydrate produced by the combined degradation due to the combination of heat and carbonation, that of heat and salt damage, and that of salt damage and carbonation.

(iii) X ray macro analyzer (hereafter referred to as EPMA)

This was conducted to study the concentration and penetration depth of Cl-ion from CO_2 contact surface or salt penetration surface, where salt penetrated due to the combined degradation due to heat-carbonation, heat-salt damage, and salt damage-carbonation, by using element distribution results.

(iv) Pore size distribution

This was studied to find the influence of the heat, carbonation, and salt damage to the void diameter of the test pieces.

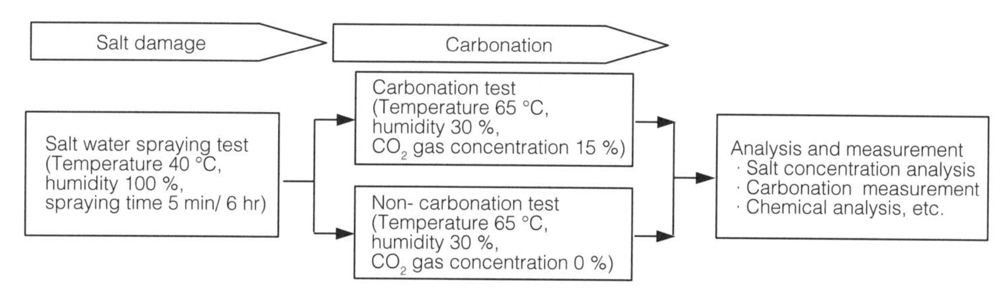

Fig. 4.7.4(2)-1 Procedure of carbonation after spraying salt water

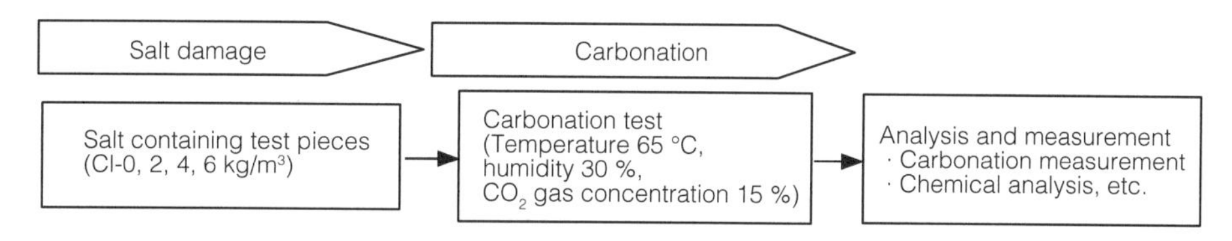

Fig. 4.7.4(2)-2 Procedure of carbonation of test body containing salt (0, 2, 4 kg/m³)

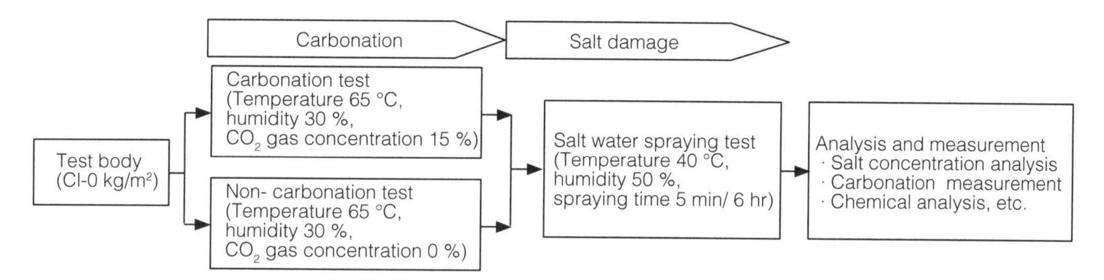

Fig. 4.7.4(2)-3 Procedure of spraying salt water after carbonation or heating

2) Various analysis results

(i) Analysis result of test pieces neutralized after salt water spraying

A paste test pieces to which salt water sprayed for 8 weeks at 40 °C and with humidity of 100 % was placed in a thermal environment at 65 °C and with humidity of 30 %. The analysis result is shown below.

(a) Analysis result of the test pieces after salt water spraying (40 °C, humidity of 100 %, for 8 weeks)

The result of EMPA showed that chlorine ions penetrated several millimeters from the surface. Generation of Friedel salt was observed in the zone where the chlorine ions invaded but not in the zone with no chlorine ions. The void diameter distribution was almost the same as that in the initial state but the total void ratio decreased by a small percent irrespective of the invasion of the chlorine ions.

Fig. 4.7.4(2)-4 shows pore size distribution measurement result.

(b) Carbonation test result after salt water spraying (temperature 65 °C, humidity 30 %, 4 weeks)

It was confirmed that the carbonation progressed from the surface of the test pieces to the inner part. The results of EPMA and differential thermal analysis show that the Friedel salt observed after the salt water spraying was separated and there was a chlorine ion concentrated area at a deeper position than the depth of the carbonation. Aragonite and Vaterite, which were not observed in the salt water sprayed test pieces, were found.

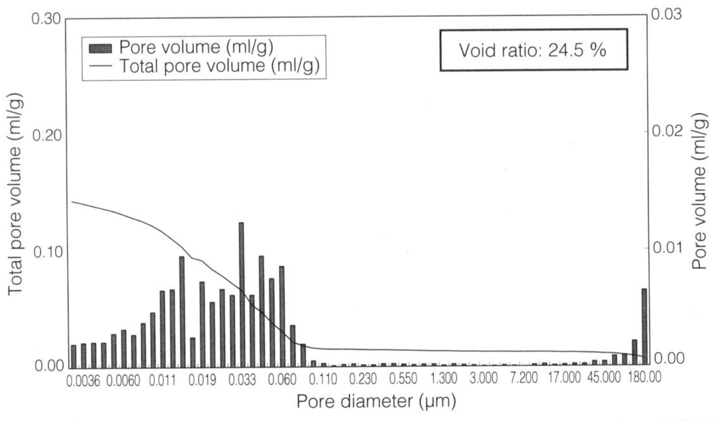

Fig. 4.7.4(2)-4(1/2) Pore diameter distribution before salt water spraying (initial pieces)

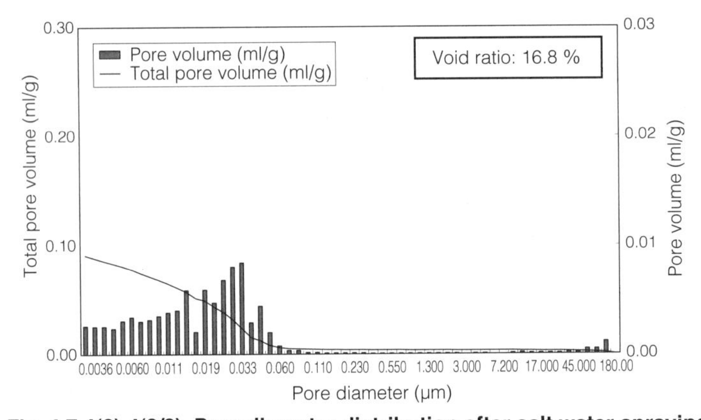

Fig. 4.7.4(2)-4(2/2) Pore diameter distribution after salt water spraying

There was no significant change in the hydrate products or void ratio even after 4 to 8 weeks of carbonation. Fig. 4.7.4(2)-5 shows the pore diameter distribution result of the test pieces which was neutralized for 4 weeks after the salt water spraying.

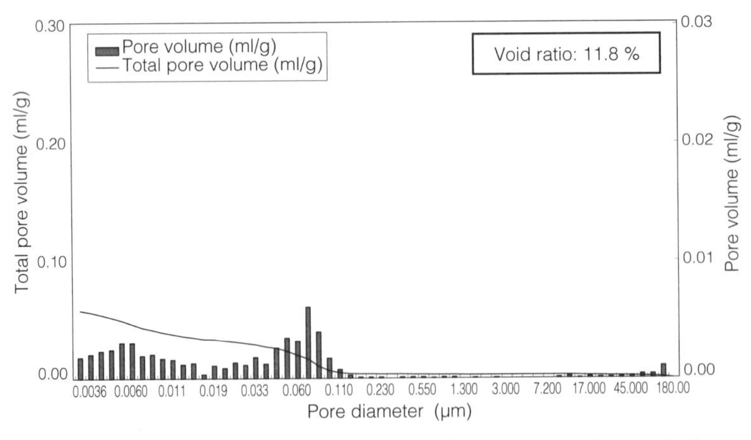

Fig. 4.7.4(2)-5 Pore diameter distribution of degraded part of test pieces which was neutralized for 4 weeks after salt water spraying

Also, the test pieces neutralized for 8 weeks after the salt water spraying was used but the analysis result was almost the same as that of the test pieces neutralized for 4 weeks.

(ii) Analysis result of carbonation of the test pieces containing salt

The analysis result of the test pieces containing salt of 0, 2, 4 kg/m^3 kept for 8 weeks at 65 °C and with humidity of 30 % is shown below.

(a) Analysis result of test pieces containing salt of 0 kg/m^3

It was confirmed that the carbonation-produced $CaCO_3$ near the surface. The hydrate of the test pieces that received only the influence of the heat had the same composition as the initial state. The void ratio of the neutralized part was around half of that of the initial state and the overall frequency of the void occurrence of every diameter decreased. This was the same behavior observed in (i). On the other hand, the void ratio of the test pieces that received only the influence of heat remained unchanged, but the void diameter distribution largely changed. As in the zone where the chlorine ions invaded mentioned in (i), the number of voids of about 3 nm, where ordinary C-S-H gels were considered to distribute, significantly decreased.

(b) Analysis result of test pieces containing salt of 2 kg/m^3

It was confirmed that the carbonation-produced $CaCO_3$ near the surface. While $Ca(OH)_2$ was slightly observed in the test pieces with no salt contained, $Ca(OH)_2$ was completely missing in the present case. The hydrate composition of non-neutralized parts was almost the same as in the initial state. The void ratio and void diameter distribution of the neutralized parts were almost the same as those of the test pieces containing no salt. The void ratio and void diameter distribution of the non-neutralized parts were almost the same as in the initial state. On the other hand, the surface of the test pieces that received only the influence of the heat was almost the same as that of the test pieces containing no salt. However the EPMA analysis showed that the void diameter distribution on the inner side of the surface zone where the chlorine ions moved into the test pieces was almost the same as that in the initial state except that the voids of relatively larger diameter increased.

Fig. 4.7.4(2)-6 shows the pore size distribution measurement result of the test pieces containing no salt and Fig. 4.7.4(2)-7 shows that of the test pieces containing salt of 2 kg/m^3.

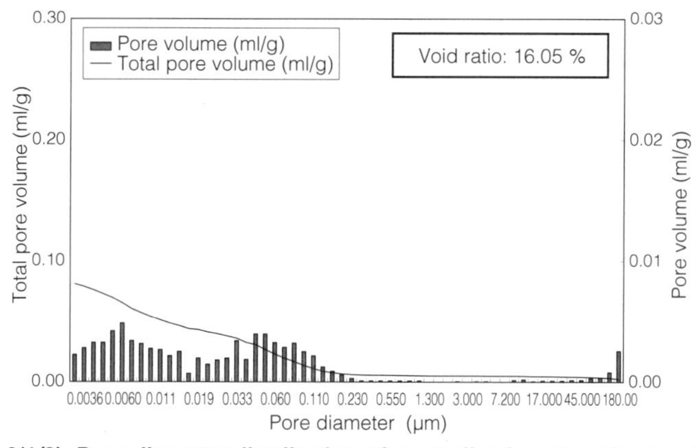

Fig. 4.7.4(2)-6(1/2) Pore diameter distribution of neutralized parts with no salt contained

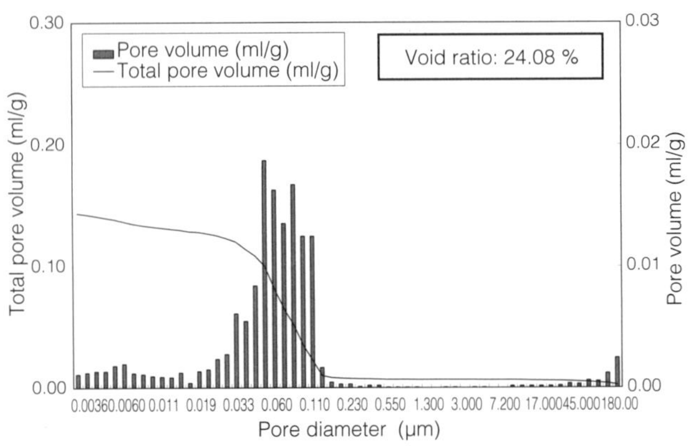

Fig. 4.7.4(2)-6(2/2) Pore diameter distribution of non-carbonated parts with no salt contained

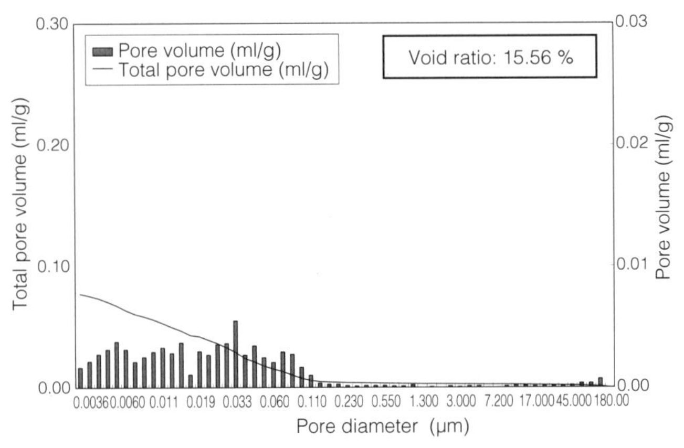

Fig. 4.7.4(2)-7(1/2) Pore diameter distribution of carbonated parts containing salt of 2 kg/m³

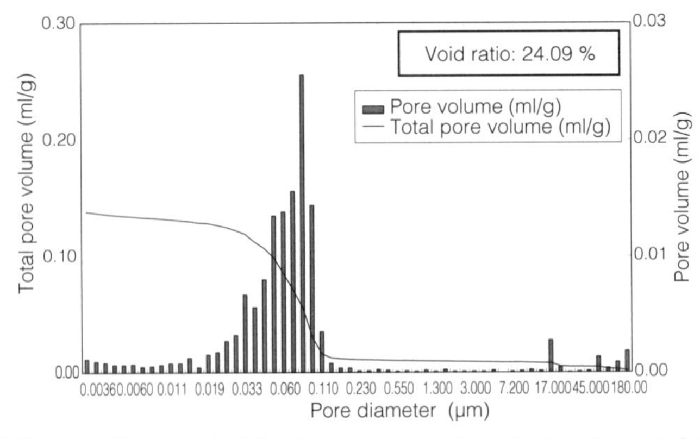

Fig. 4.7.4(2)-7(2/2) Pore diameter distribution of non- carbonated parts containing salt of 2 kg/m³

The analysis results of the test pieces containing salt of 4 kg/m^3 were almost the same as those of the test pieces containing salt of 2 kg/m^3.

(iii) Analysis result of the test pieces to which salt water was sprayed after carbonation or heating

Here we show the analysis result of the test pieces to which salt water was sprayed at 40 °C and with humidity of 100 % after it was neutralized at 65 °C and with humidity of 30 % and kept in a heat environment for 9 weeks and the analysis result of the test pieces to which salt water was sprayed at 65 °C and with humidity of 100 % after it was neutralized at 65 °C and with humidity of 30 % and kept in a heat environment for 9 weeks.

(a) Analysis result of the test pieces neutralized at 65 °C and with humidity of 30 % and kept in a heat environment

Aragonite and Vaterite were observed in the neutralized parts. These hydrates were observed under any conditions if the combined degradation due to the heat and carbonation was caused. The hydrate composition of the test pieces to which only the heat was applied was almost the same as that in the initial state. The void diameter of the neutralized part was around half of that of the initial state and the overall frequency of the void occurrence of every diameter decreased. The void ratio and the frequency of the void occurrence of the non-neutralized parts were almost the same as those in the initial state. This tendency was observed under any test conditions.

On the other hand, the void ratio of the test pieces that received only the influence of heat remained unchanged, but the void diameter distribution largely changed. The number of voids of about 3 nm significantly decreased. As in the zone where the chlorine ions invaded mentioned in (ii), the number of voids of about 3 nm, where ordinary C-S-H gels were considered to distribute, significantly decreased. According to the existing knowledge, the influence of the heat on the C-S-H gels occurs at around 60 °C or higher. Therefore we could consider that the same phenomenon occurred in the present case.

(b) Analysis result of the test pieces to which salt water was sprayed at 40 °C and with humidity of 50 %

It was confirmed that the carbonation-produced $CaCO_3$ near the surface. While $Ca(OH)_2$ was slightly observed in the test pieces after the test (a), $Ca(OH)_2$ was completely missing in the present case. Friedel salt was not observed. The hydrate composition of the non-neutralized parts was almost the same as that in the initial state.

On the other hand, for the test pieces that received only the influence of the heat, generation of Friedel salt was observed in the zone where the chlorine ions invaded. However Friedel salt was not observed in the zone with no chlorine ions and the hydrate composition was almost the same as that in the initial state. The void ratio of the neutralized part was around half of that of the initial state and the overall frequency of the void occurrence of every diameter decreased. This tendency was observed under any test conditions. The void ratio and void diameter distribution of the non-neutralized parts were almost the same as those in the initial state.

On the other hand, the void ratio of the test pieces that received only the influence of heat remained unchanged, but the void diameter distribution changed largely. As in the zone where the chlorine ions invaded mentioned in (a), the number of voids of about 3 nm, where ordinary C-S-H gels were considered to distribute, significantly decreased. According to the existing knowledge, the influence of the heat on the C-S-H gels occurs at around 60 °C or higher. However since the temperature of the present test was 40 °C, we considered that the influence of the chlorine ions was combined with the influence of the temperature.

(c) Analysis result of the test pieces to which salt water was sprayed at 60 °C and with humidity of 50 %

The hydrate composition was the same as that of the test pieces of (b). Also, the void ratio and void diameter distribution were almost the same as those of the test pieces (b), although the ratio of relatively large void increased.

Figs. 4.7.4(2)-8 to -11 show the void diameter distributions in the tests.

One can see from the figures that the void ratio in the neutralized parts decreased in every test result. On the contrary, there was no significant change in the void ratio in the other parts.

On the other hand, the difference is clear if we look at individual void diameter distribution, as shown above. Namely, the heat influence especially at 65 °C was large even to the C-S-H gels. Since the voids of relatively large diameters increased, it would significantly affect material transfer although the void ratio is unchanged. The present test also showed that the void diameter distribution of the test pieces containing salt placed in a 40 °C environment was the same as that of the test pieces placed in a 65 °C environment. This could be because the soluble salt separated Ca in the cement causing an influence to the C-S-H gels.

From these results we should study the progress of the combined degradation by focusing on the neutralized parts as one of the evaluation indices.

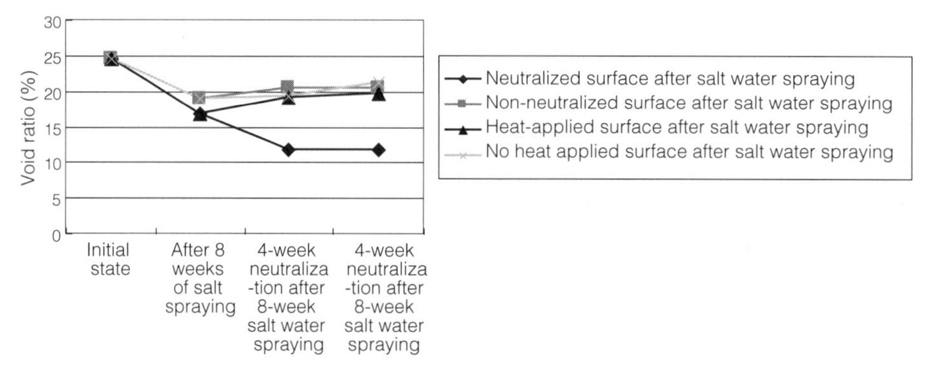

Fig. 4.7.4(2)-8 Void ratio change of the test pieces neutralized (65 °C) after salt water spraying (40 °C)

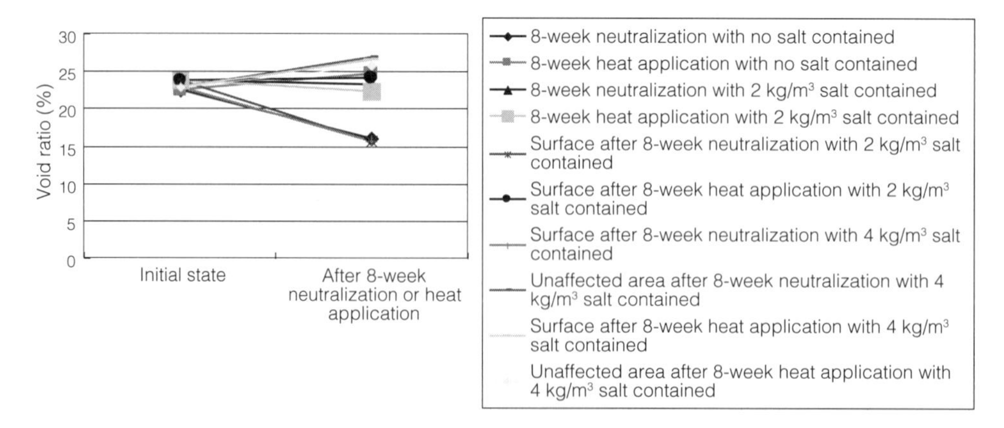

Fig. 4.7.4(2)-9 Void ratio change of the test pieces containing salt (0, 2, 4 kg/m³) neutralized at 65 °C and placed in the thermal environment for 8 weeks

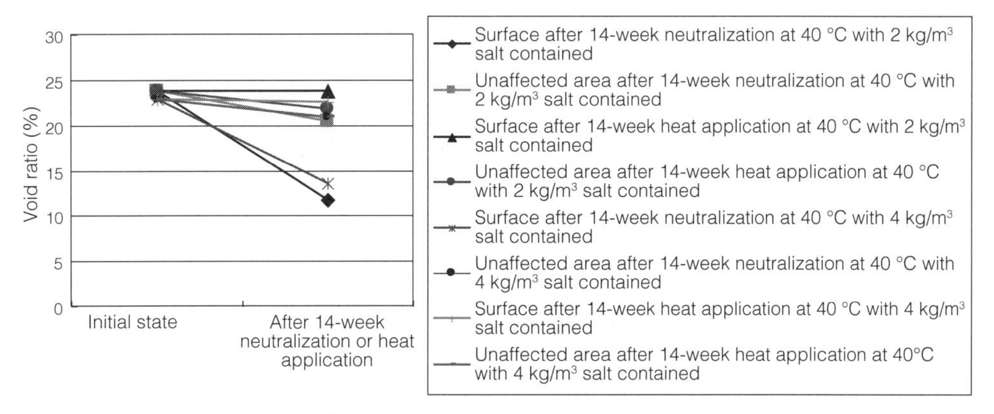

Fig. 4.7.4(2)-10 Void ratio change of the test pieces containing salt (0, 2, 4 kg/m³) carbonated at 40 °C and placed in the thermal environment for 14 weeks

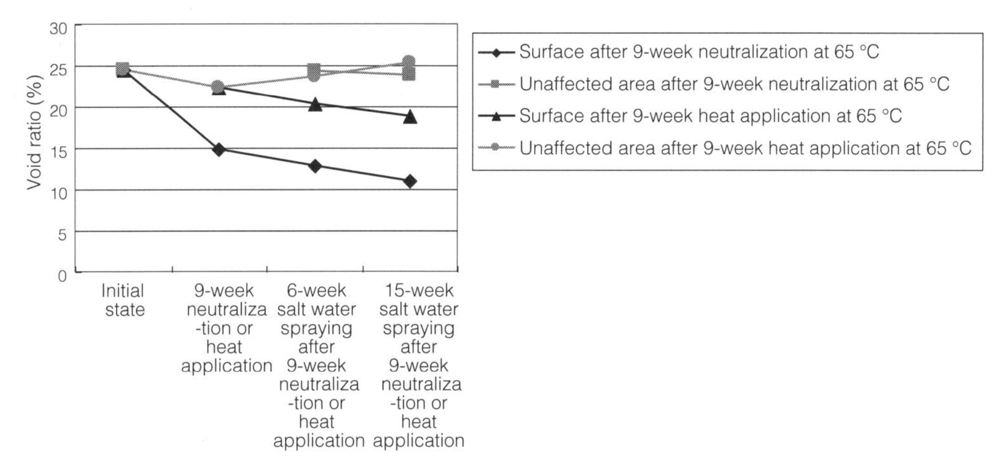

Fig. 4.7.4(2)-11 Void ratio change of the test pieces salt-water sprayed after being carbonated at 65 °C and heated

3) Summary

The carbonation of concrete lowers pH and degrades the bar corrosion suppression function. It was found that one of the major causes for the progress of the combined degradation due to heat, salt damage, and carbonation was a change in the void diameter distribution. The influence from the heat changed the distribution to the one with voids of relatively larger diameter. It was also found that chlorine ion invasion to the neutralized parts affected the hydrate of the cement materials and its structure. On the other hand, since a large change in the hydrate composition, structure, and void diameter distribution was observed under any conditions in the neutralized parts where chlorine ions invaded, the depth of the carbonation could be used as an index for the evaluation of the combined degradation.

References

1) K. Shirai: "Concrete cask type used fuel storage facilities," Concrete Engineering, Vol. 42, No.9, pp. 28-31, 2004.
2) T. Saegusa, K. Shirai, H. Takeda, et al.: "Development and evaluation of used fuel storage technology using concrete casks," Central Research Institute of Electric Power Industry, General Report N09, May 2010.

(3) Shielding performance inspection

A VSC-17 concrete cask stored since 1990 at US Idaho National Laboratory was investigated to study the ageing degradation of the shielding performance[1].

1) Influence of crack on the cask surface

In Phase I inspection for the evaluation of ageing effects in 2003, the surface dose and crack width of the cask were measured to study the influence of the cask surface cracks on the shielding performance.

Fig. 4.7.4(3)-1 shows an example of the pictures of the cask surface. Since it was difficult to visually estimate the crack width from the image data, total length of the cracks that we could recognize in the images was measured.

Fig. 4.7.4(3)-2 shows the relation between the measured gamma ray dose rate and the total crack length. The cracks were found at 35 of 96 measurement points and the maximum total crack length was 305 mm. There was no clear correlation between the measured gamma ray dose rate and the total crack length. The average gamma ray dose rate at the 35 measurement points where the cracks were found was 4.80 mR/h. It was not significantly different from 4.97 mR/h, the average gamma ray dose rate over the whole area. Therefore, it could be considered that the cracks had little affect on the dose on the surface.

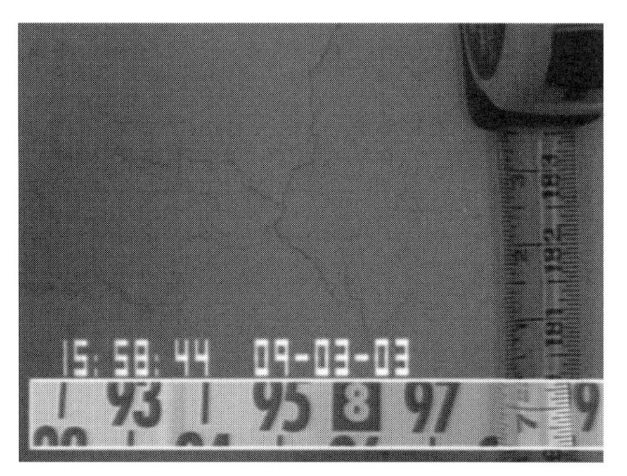

Fig. 4.7.4(3)-1 Cask surface image taken with a video camera

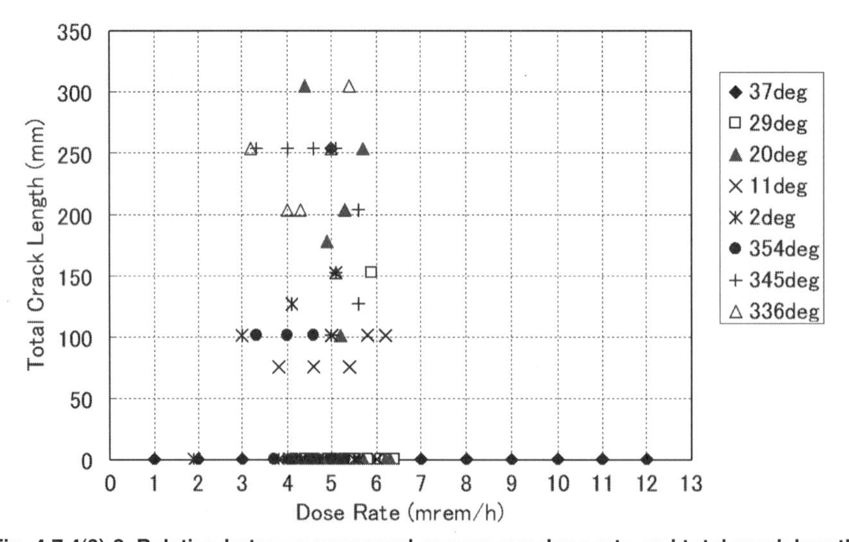

Fig. 4.7.4(3)-2 Relation between measured gamma ray dose rate and total crack length

2) Dose on the surface of and inside the cask

In a test conducted for the evaluation of aging effects in 2004, the measurement method and measured level of the radiation on the cask surface and annulus part were verified and the shielding performance was analyzed and evaluated. Fig. 4.7.4(3)-3 shows the measurement work of the radiation inside the annulus.

Fig. 4.7.4(3)-4 shows the gamma ray and neutron distributions on the cask surface and inside the annulus. The gamma ray contribution was large and exceeded the AMP-100 measurement limit 1000 R/h inside the annulus. Compared to the result obtained in October 1990 (180° and 225°), the gamma ray on the cask surface (38°) measured after 13 years (in September 2003) was lower by more than 50 %.

The gamma ray generation rate from actinide and FP calculated with ORIGEN by taking account of the burning and the cooling period, and then the dose rate was calculated from the obtained gamma ray generation rate with the gamma ray shielding effect of the above-mentioned areas being taken account of.

Table 4.7.4(3)-1 shows the calculation result. The gamma ray intensity decreased by 30 % due to the decay during the 15-year storage period from 1990 to 2004.

The gamma ray dose decrease (to 0.3) during the storage period from 1990 to 2004 was multiplied by the measured gamma ray dose rate of the year 1990 (225° and 180°) in Fig. 4.7.4(3)-4. The result is compared with the measured value in 2004 in Fig. 4.7.4(3)-5.

The calculated dose rate of the year 2004 is in good agreement with the measured dose rate in 2004, which suggests that the measured values

Fig. 4.7.4(3)-3 Radiation measurement inside annulus

Table 4.7.4(3)-1 Gamma ray intensity decrease and decayed amount

Item	Calculation result
Decrease in dose rate from 1990 to 2004	1.00 → 0.30

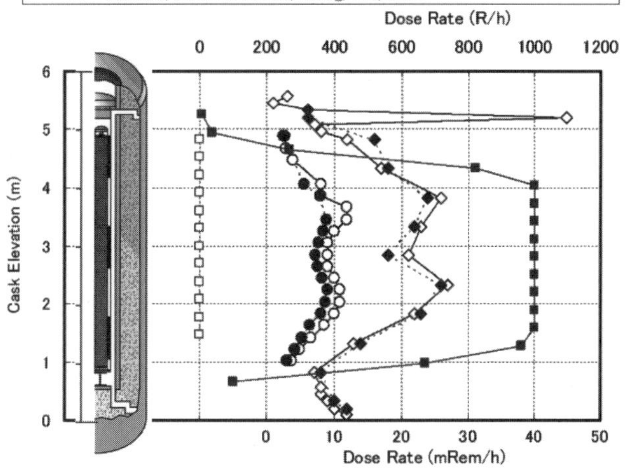

Fig. 4.7.4(3)-4 Gamma ray and neutron distributions on the cask surface and inside the annulus

in 1990 and 2004 were adequate. Let us compare the gamma ray intensity measured in 1990 and 2004. In case of the gamma ray from actinide, the gamma ray intensity does not change much because the gamma ray is also emitted from a nuclide produced in the nuclear decay. However, the overall gamma ray from fission products decays. In particular, it is affected largely by the decay of ^{134}Cs (half-life: 2.06 years) and ^{154}Eu (half-life: 8.6 years) in the energy range from 0.7 to 1.0 MeV. In the energy range from 2.0 to 4.0 MeV, the influence is large from the β decay of ^{106}Ru (half-life: 368 days), subsequent decay of ^{106}Rh (half-life: 29.9 seconds), β decay of ^{144}Ce (half-life: 284 days), and subsequent decay of ^{144}Pr (half-life: 17.3 minutes). The decay of these nuclides mainly contributes to the decrease of the dose rate.

In addition, although the measurement position (angle) was different between in 1990 and in 2004, their measured value calculated values were in good agreement. We could therefore consider that the shielding performance degradation would be small.

The gamma ray dose rate distribution in Fig. 4.7.4(3)-5 shows that the rate is about 3.5-4 mrem/h. However at the position 1.5 m lower than the exhaust duct in the 45° direction, the dose rate is locally high, i.e. 5-6 mrem/h. We therefore calculated a change of the concrete surface dose rate with a change of the density of the shielding concrete parts. In the calculation, the concrete density was assumed to decrease by 5, 8, or 10 %.

Fig. 4.7.4(3)-6 shows a change of the surface dose rate relative to the case where there is no density decrease. The concrete density decrease causes 1.44 to 2.1 times higher surface dose rate. Therefore the concrete density could be low at the position 1.5m lower than the exhaust duct.

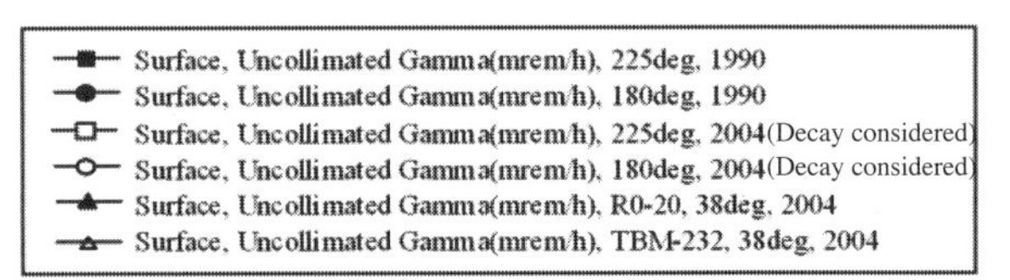

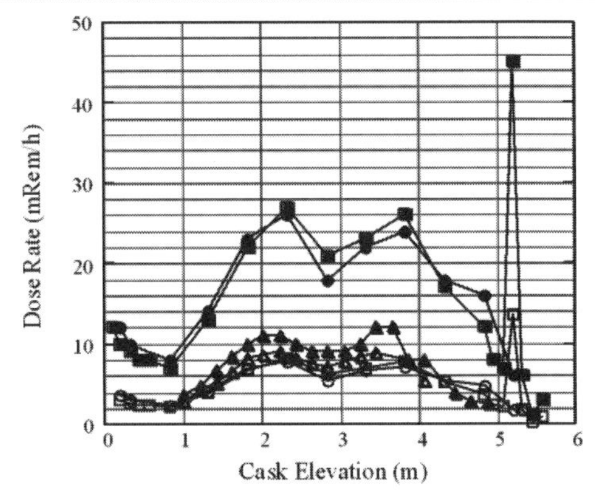

Fig. 4.7.4(3)-5 Comparison of gamma ray dose rate decrease during 15-year storage period

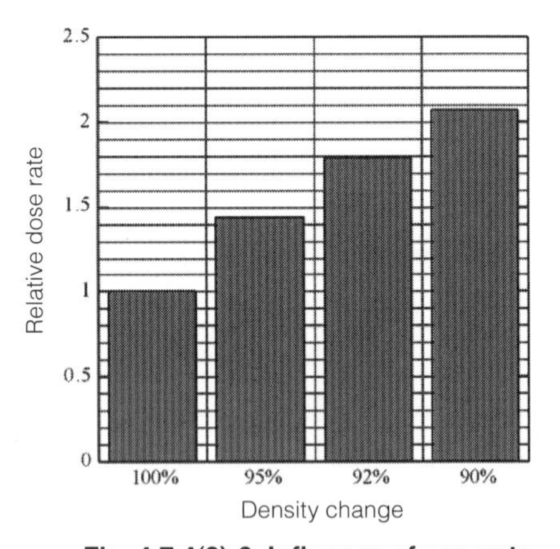

Fig. 4.7.4(3)-6 Influence of concrete density on dose rate

3) Summary of shielding test

(i) Surface dose distribution

After 15 years of the storage, the cask surface dose decreased to about one third of the initial dose. However in the area 2 m lower than the exhaust duct, a high dose rate caused by the local density decrease was observed. Even when the radiation intensity of the used fuels decreased after the 15-year storage period, the streaming from the gap between the lid and body of the cask was large. Therefore it is desirable from a viewpoint of worker's exposure control to prevent works on the lid if possible.

(ii) Influence of surface crack on shielding performance

The pictures of the cask surface area below the exhaust dust where the cracks occurred were taken with a camera and the total crack length per picture image area was measured. It was found that there was no strong correlation between the total crack length and the measured surface dose rate and the influence of the cracks on the radiation dose could be ignored.

(iii) Influence of temperature load on shielding performance

A vibration sensor with a piezoelectric element was used to measure the elastic wave velocity distribution on the concrete surface up to 2 m lower than the exhaust duct, and we found that there was an area where the elastic wave velocity decreased locally. This could be due to the concrete density decrease. The breeding water appeared in the concrete construction remained below the exhaust duct and was scattered by the temperature load for a long period of time, which could cause the density decrease.

In future concrete cask manufacturing, it will be important to optimize the selection of concrete materials and the construction method to take sufficient care of the small area between the concrete surface below the exhaust duct and the reinforcing bars.

Reference

1) K. Shirai, T. Saegusa, et al.: "Aging test of VSC-17 concrete casks over 15 years," Central Research Institute of Electric Power Industry, Research Report N08057, August 2009.

(4) Structural strength inspection (Schmidt Hammer method)

The concrete near the exhaust outlet of the concrete casks is exposed to high temperature for a long period of time. On the other hand, the surface dose measurement conducted in 2004 showed that the surface dose of the gamma rays was relatively high at the position 1.5 m lower than the exhaust duct. In the present section, for the evaluation of the concrete degradation of the cask with non-destructive measurement methods, the result of the concrete strength measurement with Schmidt Hammer method (in October 2005) and the result of the elastic wave velocity measurement with AE method (in July 2005) are shown.

1) Concrete strength measurement with Schmidt Hammer method

In the Schmidt Hammer method, the concrete surface was hammered with a heavy bob and the rebound was measured to find the concrete strength. The test was conducted in accordance with ASTM C805[1).]

Fig. 4.7.4(4)-1 shows scleroscope hardness measurement points and measurement status. There are 70 measurement points at height intervals of 2 feet (61 cm) and at angle intervals of 45° in the range from 90° to 315°. In particular, in the direction from 0° to 45° where many cracks were observed, the measurements were conducted at height intervals of 4 inches (10 cm) in the vicinity of the exhaust outlet.

Fig. 4.7.4(4)-2 shows the concrete strength distribution obtained by converting the scleroscope hardness. A low concrete strength area in the height direction of 90° might be a path through which breeding water ran when the concrete structure was constructed.

The breeding water could propagate to the area under the exhaust outlet steel pipe and remain there, which could increase the local water-cement ratio in the area. This could be the state of the concrete after the concrete structure construction finished. Since the temperature of the area under the exhaust outlet is relatively high, the long-term temperature load could accelerate the water scattering in the area and hence the local density decrease over the years.

The breeding water tended to remain in a narrow area near the exhaust duct and was kept at more than 70 °C for a long period time by the decay heat from the used fuel. Therefore the selection and prior examination of concrete materials is important to make the concrete quality uniform.

Fig. 4.7.4(4)-3 shows the relation between the concrete strength and the gamma ray measurement result. The average concrete strength over all measurement points was 47.2 MPa and the concrete strength distributed within the range of about ±10 %. In the area where the concrete strength was lower than the average 47.2 MPa, the gamma ray measurement value tended to be slightly higher, which indicated a

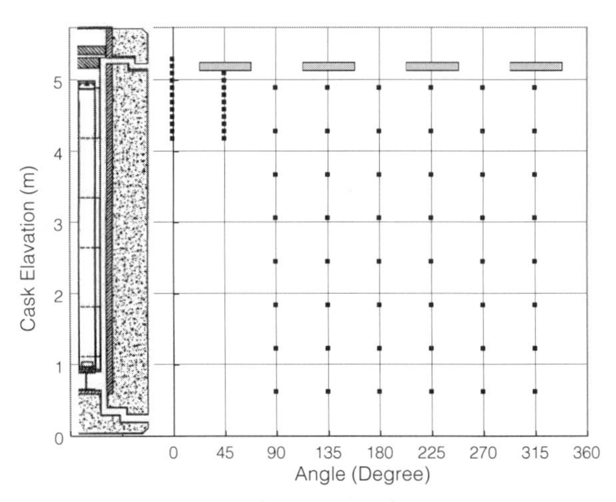

Fig. 4.7.4(4)-1 Overview of Schmidt Hammer method test

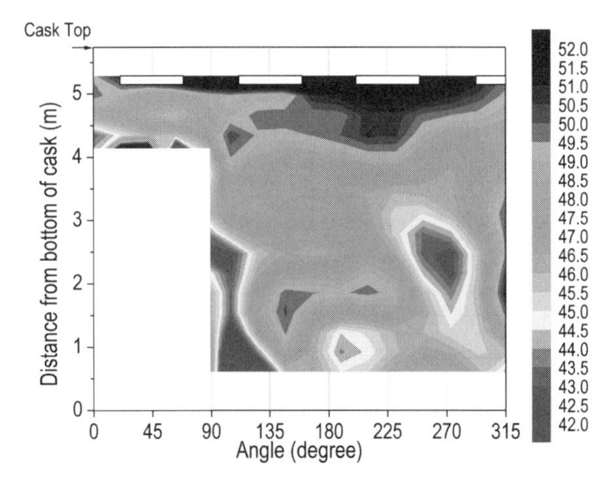

Fig. 4.7.4(4)-2 Concrete strength distribution converted from scleroscope hardness

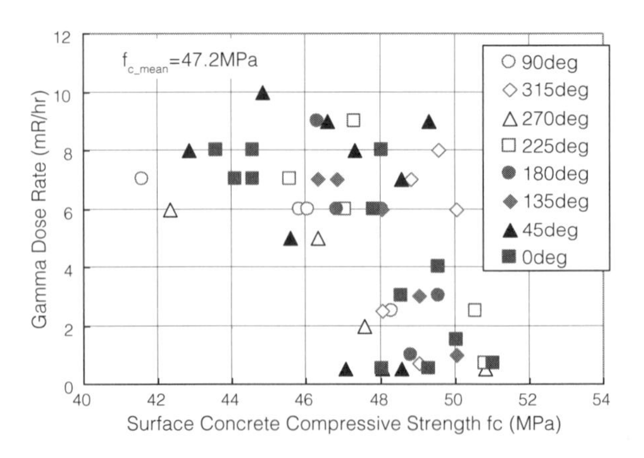

Fig. 4.7.4(4)-3 Relation between concrete strength and measured gamma ray dose rate

possibility of estimating the shielding performance from the surface strength (concrete density).

2) Measurement of elastic wave velocity distribution

As shown in Fig. 4.7.4(3)-5, the dose rate was locally high at the position 1.5 m lower than the exhaust duct in the 45° direction. At this position, a significant shielding performance decrease was expected. Since this decrease could be caused also by a defect inside the concrete due to construction failure, we measured a velocity distribution of the elastic waves on the concrete cask surface in August 2005.

Fig. 4.7.4(4)-4 shows the system and status of the elastic wave velocity measurement.

Fig. 4.7.4(4)-5 shows the elastic wave velocity distribution at the depth of 10 cm and 20 cm. A low elastic wave velocity zone exists in the 45° angle and at the depth of 10 to 20 cm. Namely, there was an area under the exhaust

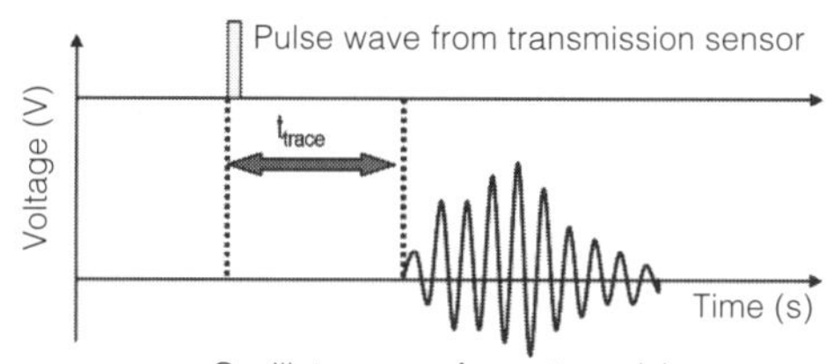

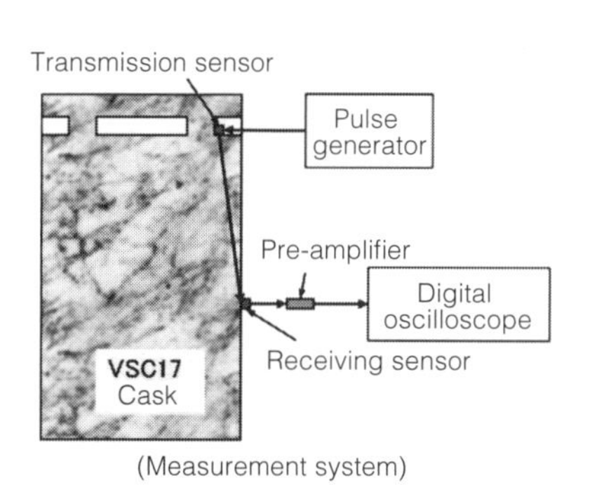

(Measurement system)

(Work for elastic wave velocity measurement)

Fig. 4.7.4(4)-4 System and work for measurement of elastic wave velocity

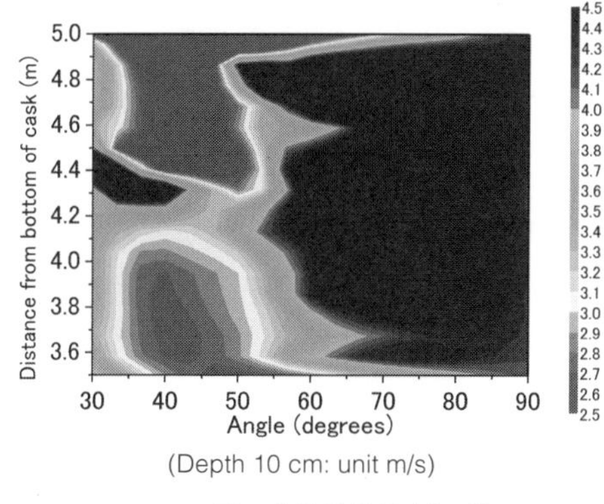

(Depth 10 cm: unit m/s)

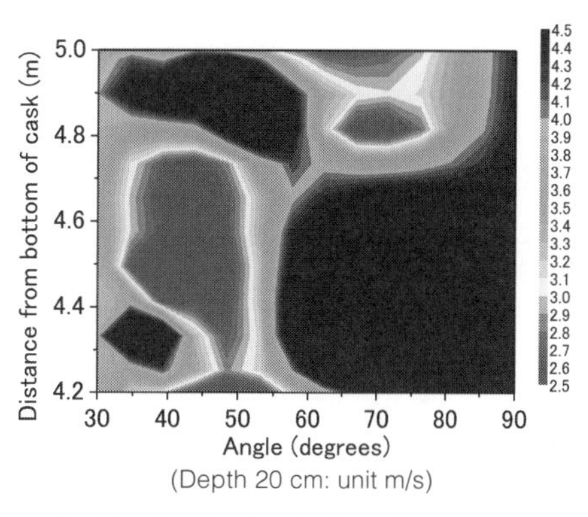

(Depth 20 cm: unit m/s)

Fig. 4.7.4(4)-5 Elastic wave velocity distribution in depth direction

outlet where the elastic wave velocity was locally low.

Reference

1) ASTM C805-02, Standard Test Method for Rebound Number of Hardened Concrete

4.7.5 Development of low activation, high performance concrete

If used fuels are stored in a concrete cask for several tens of years, a small amount of elements (e.g. Co or Eu) contained in the concrete and steels used for the cask materials are radio-activated by neutron radiation. Fig. 4.7.5-1 shows an example of the radiation level after the end of the concrete cask storage period. After the storage period, the cask body has induced radioactivity higher than the clearance level. Therefore if the storage vessel is treated as ordinary industrial waste after the storage period, it is necessary to put it under control for a certain period of time.

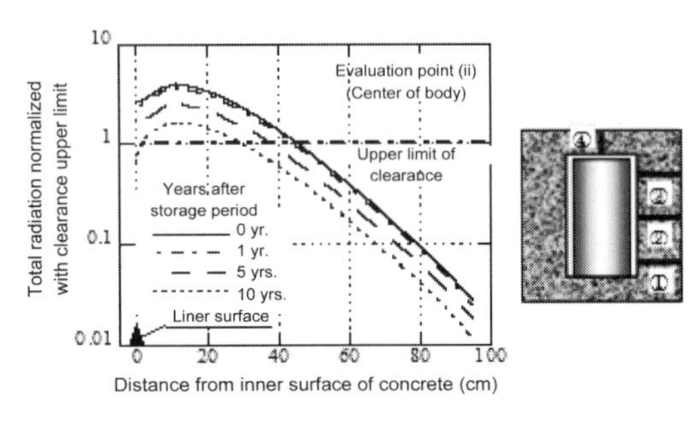

Fig. 4.7.5-1 Concrete cask radiation level after storage

So, we developed a low activation, high performance material to use for concrete casks. Fig. 4.7.5-2 shows the definition of the low activation, high performance concrete. It should contain less Eu or Co that could cause radio-activation and be highly heat resistant.

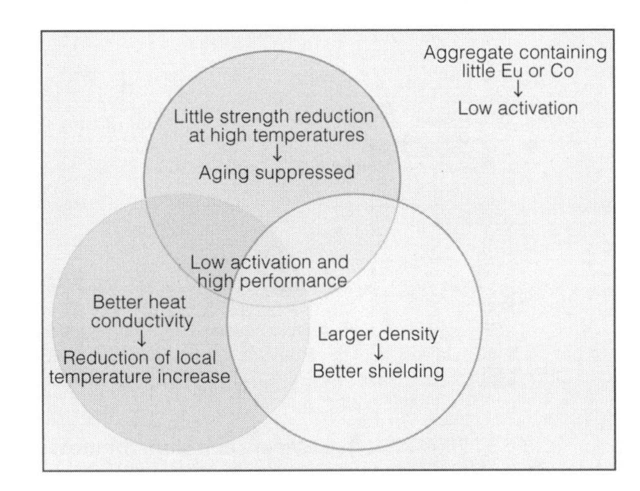

Fig. 4.7.5-2 Definition of low activation, high performance concrete

(1) Evaluation of activation

We first determined the material composition of reinforced concrete including minor elements. Then activation calculation (computational code: THIDA-2) was performed for concrete casks for used fuels of (maximum burnup of 55 GWD/tU, 10-year cooling, 21 casks) with the decay of the radiation source taken into consideration in order to forecast an induced radiation concentration distribution in 40 years.

As a result it was found that the concrete cask used for 40 years had radioactivity higher than the clearance level of the radioactive waste and that it was caused by nuclides such as Eu-152 and Co-60. From the inverse analysis of this result, we clarified the allowable content of these nuclides for low activation materials of the casks.

(2) Trial production of low activation, high performance concrete

For the activation test and analysis, we chose limestone as coarse aggregate and relatively-inexpensive alumina ceramic material (alumina aggregate) as fine aggregate. As a result we found that this combination of the aggregates was appropriate for low activation concrete since the content of Eu and Co was in the allowable range to be used for low activation materials. In addition, fly ash was added to this concrete composition to adjust the granularity of the fine aggregate and the new composition was found to have good workability.

Next we put the concrete material of this composition in an environment at 65-105 °C for 10 months and evaluated the influence of the compressive strength and the initial elastic modulus. Our developed concrete was placed in the air at 20 °C and then kept at 65 °C for 6 months and at 80 °C for a month but the concrete strength did not decrease. Therefore, the concrete is expected to have long-term durability even in a high-temperature environment.

(3) Performance check of low activation, high performance concrete

Fig. 4.7.5-3 shows thermal properties of the low activation, high performance materials that we produced.

The concrete that we developed was highly thermally-conductive and effective for removing heat from the cask. Owing to its small thermal expansion coefficient, the concrete was less subject to cracks due to the thermal expansion. Also since the concrete density was larger by about 10 % than the previous type of concrete where general natural aggregate was used, the shielding thickness of the cask could be reduced.

Cylindrical concrete test bodies (inner diameter ϕ 590 mm × outer diameter ϕ 1200 mm × height 300 mm) were made with ordinary concrete and our developed concrete and a temperature crack test was conducted to compare their properties. Compared to the test body made with ordinary concrete, the cylindrical test body with the developed concrete had less temperature cracks and was more durable. This was because the test body with the developed concrete had lower maximum temperature and smaller thermal expansion due to its higher thermal conductivity.

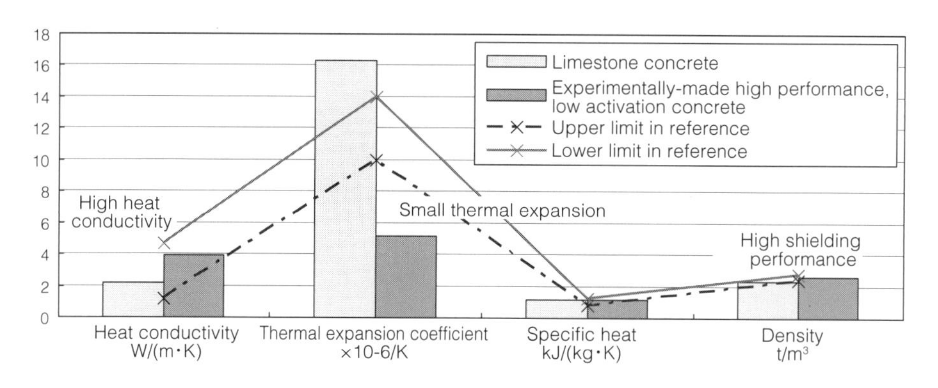

Fig. 4.7.5-3 Thermal properties of experimentally-produced low activation, high performance material

VAULT STORAGE, etc.

5.1 Vault storage

The "vault" originally means a room with thick walls and a strong door used for keeping valuable things safe. The vault storage is a method to store encapsulated spent fuel in canisters vertically in a large room of a concrete building. Heat from the spent fuel is removed naturally by air. There are two types of vault storage depending on the air flow pattern. One type is to make the storage tubes double and let the air flow in-between the tubes (parallel flow type) and the other is to let the air flow transversally among the tubes (cross flow type). Spent fuel is encapsulated in a canister with helium gas. Fig. 5.1-1 shows an example of design concept of vault storage method. The characteristic of this method is economical for a large amount of storage. In Hungary, spent fuel of the former Soviet Union's VVER reactors has been stored since 1997[1]. Fig. 5.1-2 shows an example of vault storage design concept that is being planned to store both spent fuel and vitrified wastes in Spain[2].

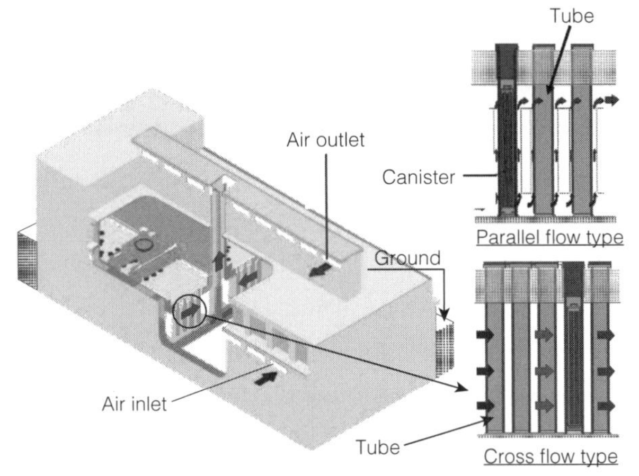

Fig. 5.1-1 Example of design of vault storage method

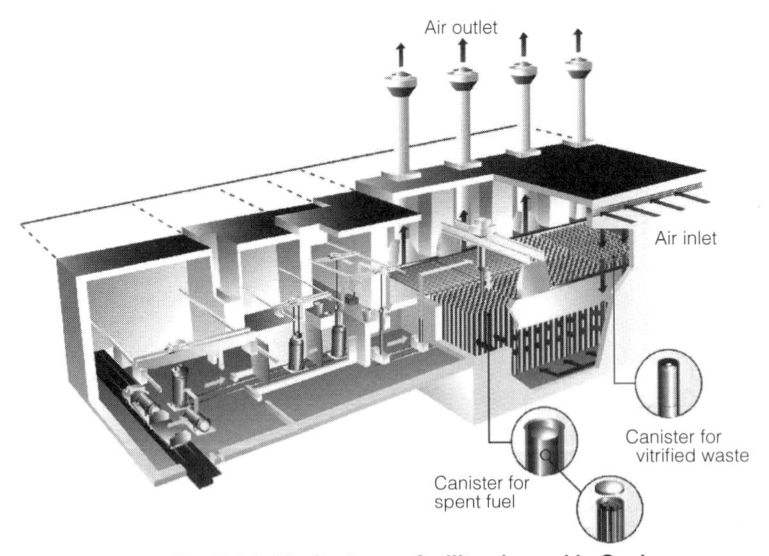

Fig. 5.1-2 Vault storage facility planned in Spain

5.2 Above ground vault storage

Vault storage method is a system to store spent fuel efficiently by allocating canister in high density, where heat removal function is important. Because the vault storage system is a natural cooling system, it is important to verify the evaluation method by heat removal test. We conducted heat removal test of the vault storage system with natural cooling phenomenon with a 1/5 scale model in order to establish the method that evaluates the natural cooling performance of the vault storage facility.

5.2.1 Cross flow type vault heat removal test

In this section, an overview of heat removal test of cross flow type vault storage facilities is shown.

(1) Idea of vault storage system of cross flow type

Cross flow type vault storage is a method of confining spent fuels in a canister, confining the canister in a storage tube and placing the storage tubes in a storage module where cooling air introduced by natural convection flows perpendicularly to the tubes to remove decay heat produced during the storage period (Fig. 5.2.1-1). A storage module contains several tens of the storage tubes (canisters) and the heat removal safety shall be ensured for each of the storage modules. Modules can be added to increase the storage capacity.

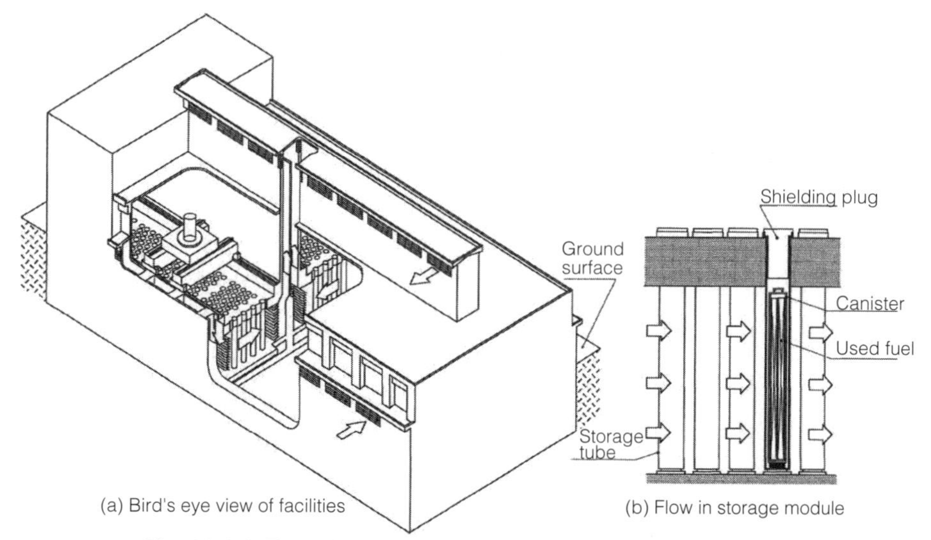

Fig. 5.2.1-1 Example of cross flow type vault storage facilities

(2) Test equipment

Since larger scale equipment is preferable to examine convection phenomena of whole storage facilities, 1/5-scale model test equipment, i.e. one-fifth size of expected actual test equipment, was used in the present study. Table 5.2.1-1 compares the major specifications of the test equipment and those of actual facilities, Fig. 5.2.1-2 shows an overview diagram, and Fig. 5.2.1-3 shows the layout of heat generator inside the storage unit.

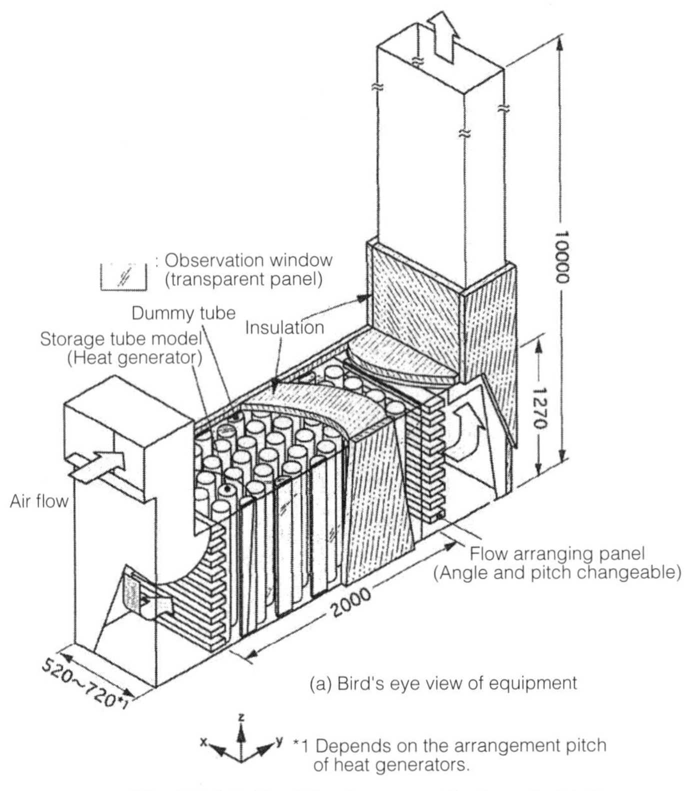

Fig. 5.2.1-2 Vault heat removal test equipment

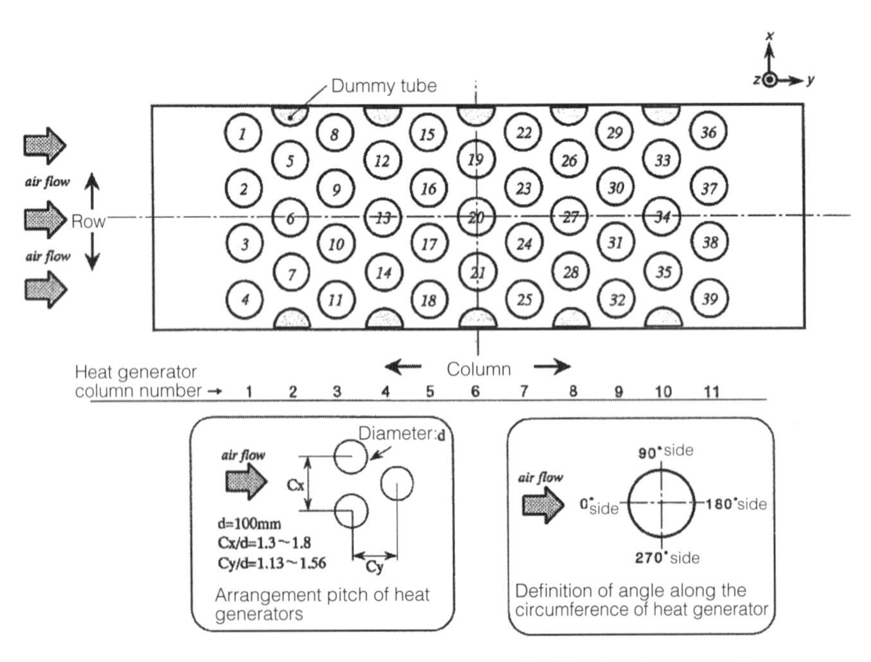

Fig. 5.2.1-3 Layout of heat generators inside the storage unit

Table 5.2.1-1 Major specifications of vault heat removal test equipment

Item	Expected actual facilities	Vault heat removal performance test equipment
Effective dimensions of storage unit	6,000 mm D × 9,500 mm W × 6,300 mm H *1	600 mm D × 2,000 mm W × 1,270 mm H *3
Storage tube (heat generator)	ϕ 508 mm OD × 7,805 mm L · Four BWR fuels or one PWR fuel can be stored. · Including the penetration length of the shielding plug	ϕ 100 mm OD × 1,301 mm L · The shielding plug not considered
Number of storagetubes	83 (staggered arrangement of 11 × 7 or 8)	39 (staggered arrangement of 11 × 3 or 4)*4
Heat generation	1,188 W/tube*2	Around 240 W/heat generator*5

*1 For 1 module
*2 Assuming 4 BWR tubes of large heat generation
*3 For arrangement pitch 150 mm of heat generators
*4 Symmetry in row direction taken into consideration
*5 Heat generators are divided into six in the axis direction, each of which produces 40 W heat.

(3) Test method

From a phenomenological viewpoint, the flow condition in the storage unit of the cross flow type vault storage system is determined by the relation between the inertia force inside the unit caused by the stacking effect and the buoyancy caused by the heat from the heat generators. This relation is determined by geometrical conditions such as diameter and configuration of the heat generators and operational conditions such as heat generation amount and flow-in air volume. If the inertia force is larger than the buoyancy a cross flow is generated inside the storage unit, and if the buoyancy is larger than the inertia force, a parallel flow (which flows upward vertically by the buoyancy caused by the heat from the heat generators) dominates.

With focus on the storage unit, Ri number can be defined as the dimensionless ratio of inertia force to buoyancy by Eq. (1).

$$Ri = \frac{g\,\beta \Delta TL}{U^2} = \frac{Gr_b}{Re_{0,b}^2} \cdot \frac{d^2}{L^2} \quad \cdots\cdots (1)$$

This definition used in the present study was obtained by applying Boussinesq approximation to the Ri number shown by Prandtl (Schlichting, 1979a).

As shown in Eq. (1), the Ri number can be correlated with the flow condition of the facilities and defined by Gr and Re numbers. To accurately simulate the actual facilities, these dimensionless numbers should be equal to actual values. However it is not realistic to match the Gr number in the test since the heat generation has to be set extremely large in the test. Therefore we set an appropriate Ri number by controlling the heat generation and flow volume individually.

In the test we first studied the influence of the Ri number on the flow condition by using it as parameter. Then by using the Re and Pe numbers as parameters under the condition where a cross flow is produced, we quantitatively examined the heat-transfer coefficient.

(4) Test results

1) Flow condition inside storage unit

We studied a method of estimating flow conditions inside the storage unit by changing the heat amount from the heat generators and the air flow volume from the fan. A visualization test was conducted to observe the flow and determine whether it is a cross flow, parallel flow, or a combination of them inside the storage unit. Then the results were analyzed with the Ri number defined by Eq. (1). Upward flows were observed inside the whole unit. When a stagnation area that might affect the heat removal performance was observed in the storage unit, it was interpreted as parallel flow. When a cross flow was observed on the upstream side (inlet side) of the storage unit and a parallel flow was observed on the downstream side (outlet side), the flow condition was interpreted as combinatory flow.

Fig. 5.2.1-4 shows the results of the present test and design values of actual facilities. For 7≦Ri, parallel flows were observed in entire area inside the storage area, which indicates that the buoyancy dominated. For 3≦Ri<7, both cross flow and parallel flow were observed. For Ri<3, cross flows were observed in the entire area inside the storage unit.

Fig. 5.2.1-5 shows the flow distribution on a two-dimensional vertical cross section that we observed from a window on the side of the storage unit for the cross flow case (Ri = 2.2) and parallel flow dominant case (Ri = 10.4). In case of the parallel flows, there is a stagnation area appearing on the upper side of the storage unit and the air temperature was much higher in the upper area than in the lower area. Therefore, for designing a vault storage system, it is important to create a cross flow of the cooling air inside the storage unit. For this purpose, we need to have Ri<3.

Since the design examples of actual facilities shown in Fig. 5.2.1-4 have Ri<3, we could consider that cross flows are dominate inside the storage unit. However if the Ri number is sufficiently small, satisfying Ri<3, there could be a margin in the heat removal design and the design could be rationalized by, for example, making stacks lower.

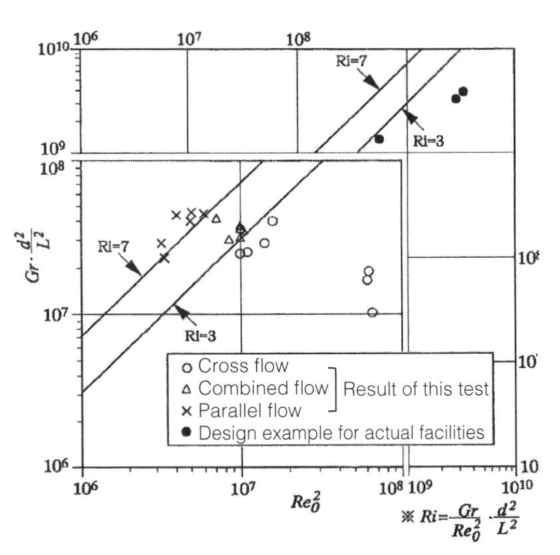

Fig. 5.2.1-4 Flow map inside storage

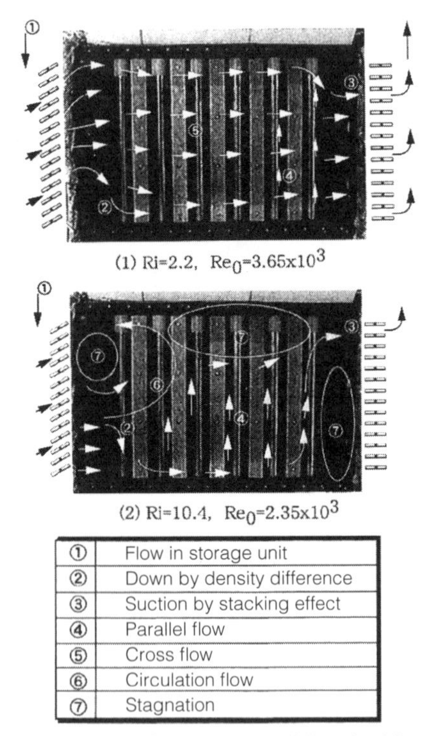

Fig. 5.2.1-5 Observation of flow inside storage (test condition: Cx/d = 1.5, 102 W/heat generator)

2) Flow speed and temperature distributions in storage unit

We studied the flow speed distribution and the temperature distribution in detail over a wide range of the Ri number.

Fig. 5.2.1-6 shows the spatial temperature distribution on the column-direction cross section with different Ri numbers for the canister location pitch of Cx/d = 1.5.

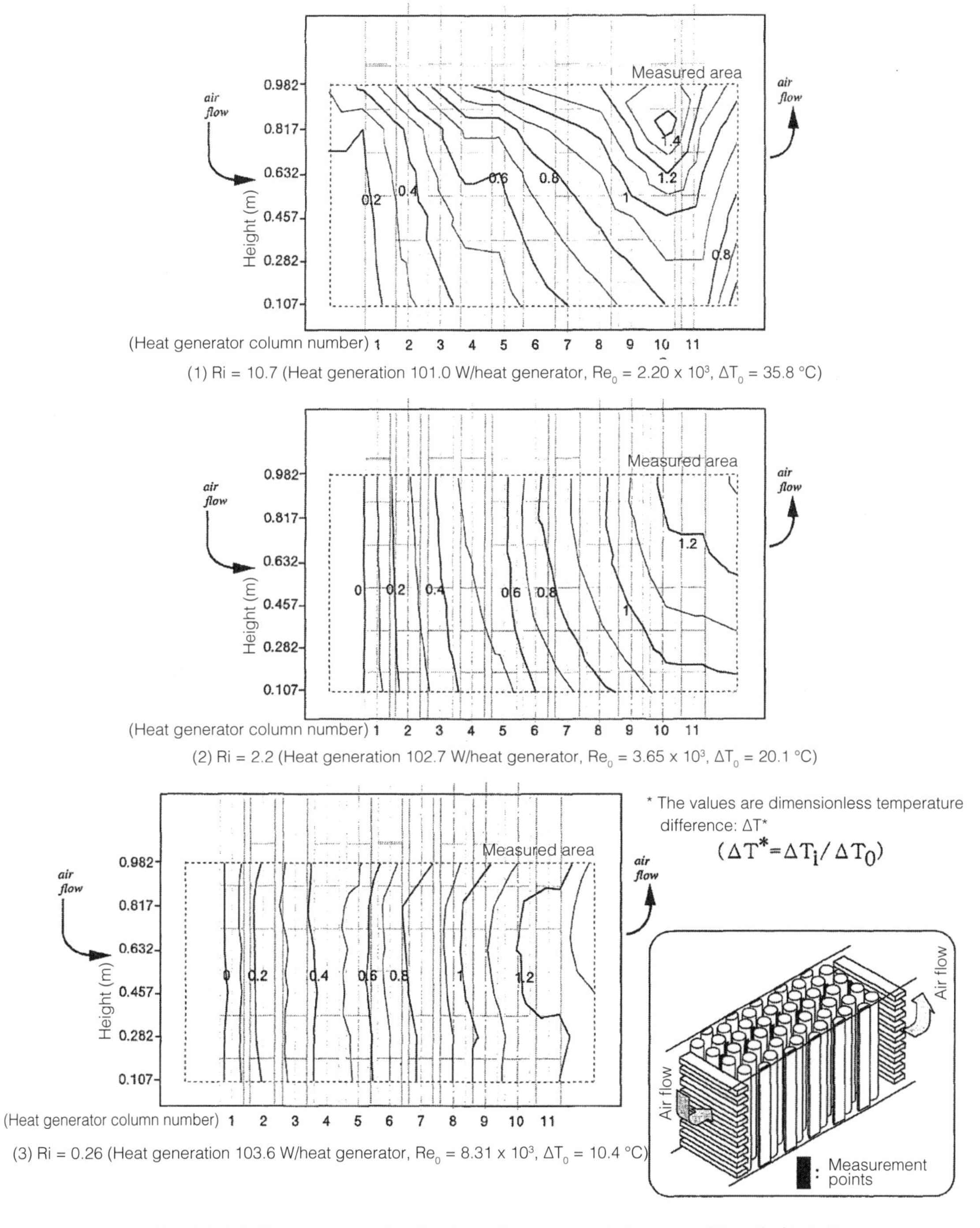

Fig. 5.2.1-6 Temperature distribution of storage unit (test condition:Cx/d=1 .5)

When Ri = 10.7, the parallel flow seems to dominate since there is a high temperature region on the top of the storage unit. When Ri = 2.2, a cross flow was observed but influence of the buoyancy was found near the last column. When Ri = 0.26, the temperature difference in the height direction is small which is a characteristic of the cross flow.

[Symbols]

Ri : Richardson number ($= g\beta\Delta TL/U^2$)(-)

Re : Reynolds number ($= dU/\nu$)(-)

Gr : Grashof number ($= g\beta\Delta TL^3/\nu^2$)(-)

Pe : Peclet number (=Pr Re)(-)

x : Row direction

y : Column direction

z : Height direction

L : Storage unit height (m)

D : Diameter of heat generator (m)

Cx : Row-direction pitch of heat generators (m)

Cy : Column-direction pitch of heat generators (m)

U : Air flow speed ($m\ s^{-1}$)

ΔT : Temperature rise (K)

g : Gravitational acceleration ($m\ s^{-2}$)

β : Volume expansion coefficient (K^{-1})

[subscripts]

* : Dimensionless value

y : y component or y direction

z : z component or z direction

0 : Corresponding to the case where air flows in the minimum flow passage area in the heat generator configuration (in the pipe assembly)

b : Main flow

5.2.2 Heat removal test of vault storage with parallel flow

The present subsection describes the outline of our heat removal tests[5)] of a vault storage facility with parallel flows. The tests aim at developing a method to confirm the heat removal capacity of facility, which must lead to suitable thermal conditions of spent fuels with generated decay heat through heat transfer to cooling air.

The vault storage facilities are divided into two broad categories according to the directions of cooling air flow: one of these is the cross flow type as described in the previous subsection and the other is the parallel flow type. The decay heat from spent fuels is mainly removed by a cooling air flow. The air flow is developed along annulus, which is composed of vertically placed a storage tube (container) and a canister in a facility. The driving force of air flow is buoyancy caused by temperature difference between the canister surface and cooling air. Thus, the cooling air flow parallels the canister surface and also buoyancy (gravity).

Understanding of the heat removal capacity is vital for estimating the surface temperature of canister, which is an important parameter to confirm the soundness of the facility. The surface temperature depends on the calorific value of spent fuel and heat transfer rate: the surface temperature decreases with decreasing calorific value or with increasing heat transfer rate. The heat transfer of decay heat of spent fuel to the cooling air is carried out three-kinds of paths, “convection”, “conduction” and (thermal) “radiation”. The contribution of heat removal capacity due to convection with cooling air flow is generally large compared with those of conduction and radiation, and also the estimation of this contribution must be quite difficult.

An approach for estimating the heat removal capacity of convection with air flow (heat transfer rate) is heat removal test by using scale models. Numerical simulation with computational fluid dynamics (CFD) codes, which will become a powerful tool with rapid increasing the performance of computers as discussing the next subsection, also requires the experimental data for the verification and validation of used CFD code.

Two parameter must be focused on understanding the heat transfer rate: one is the flow rate of cooling air, which is generated by buoyancy with temperature difference between canister surface and ambient air, in the annulus and the other is convective heat transfer rate from the canister surface heated by decay heat of spent fuel to the cooling air in the annulus. The total amount of heat removal is estimate by the product of the flow rate and the heat transfer rate. The flow rate in the annulus depends on the flow in the whole of the facility, while the heat transfer rate depends on the flow in the annulus, especially in the vicinity of the canister surface. The scale of dominant flow among two processes is much different, and thus, the combination of two kinds of experiments must be successful; one is for examining the flow rate in the annulus with the scale model of the whole of the facility and the other is for examining the heat transfer rate with the scale model only of the annulus.

Table 5.2.2-1 presents the main specifications of the experimental apparatus for the scale model of the whole of the facility used in our experiments and also a real facility. A 1/5 scale model based on the Euler number similarity, which gives agreement of the geometry and drag of flow path, was used. The model had 39 simulated storage tubes with staggered arrangement, the number of which was reduced by the symmetry of the facility. The temperature and velocity of cooling air and the surface temperature of canister and container were measured by using T-type thermocouples and hot-wires. The measurements on flow patterns and temperature distributions in the whole of the facility and also the flow rate in each annulus were accumulated under various conditions on spent fuel and facility. The measurements have indicated that no stagnation of air occurs in the facility and the cooling air uniformly flow into each annulus.

Table 5.2.2-1 Main specifications of scale model and a real facility

Item	Scale model	A real facility
storage area dimensions	820 mm (D), 2000 mm (W), 1270 mm (H)	8300 mm (D), 9500 mm (W), 6300 mm (H)
canister dimensions	100 mm (ϕ), 1301 mm (L)	508 mm (ϕ), 7805 mm (L)
number of canister	39	83
heat flux of a canister	240 W (max.)	1188 W
dimension of container	139.8 mm (ϕ), 850 mm (L), 2.8 mm (t)	720 mm (ϕ), 4250 mm (L), 6.0 mm (t)

The experiments focusing on the heat transfer process from the container surface to cooling air employed the real-scale model of a container and a canister to confirm the Rayleigh number and Reynolds number similarity, which gives agreement of the heat transfer rate of the heated surface with buoyant and forced flows. The velocity and temperature distributions of cooling air in the annulus including near the canister surface, were measured. Fig. 5.2.2-1 depicts the schematics of a special thermocouple to measure the air temperature in the vicinity of the canister surface; the very fine probe, the diameter of which is 0.05 mm, the spacing between probe and prong, and arrangement of the prong reduce the measurement errors due to the thermal conduction of the thermocouple. Fig. 5.2.2-2 illustrates a measured temperature profiles of cooling air in the vicinity of the canister surface. The horizontal and vertical axes refer to the normal distance from the canister surface and normalized cooling air temperature, respectively. The air temperatures near the surface (in the viscous layer) clearly show the linear profiles, corresponding to theoretical results. The slopes of the profiles give heat transfer rates and suggest that the heat transfer rate at the height of

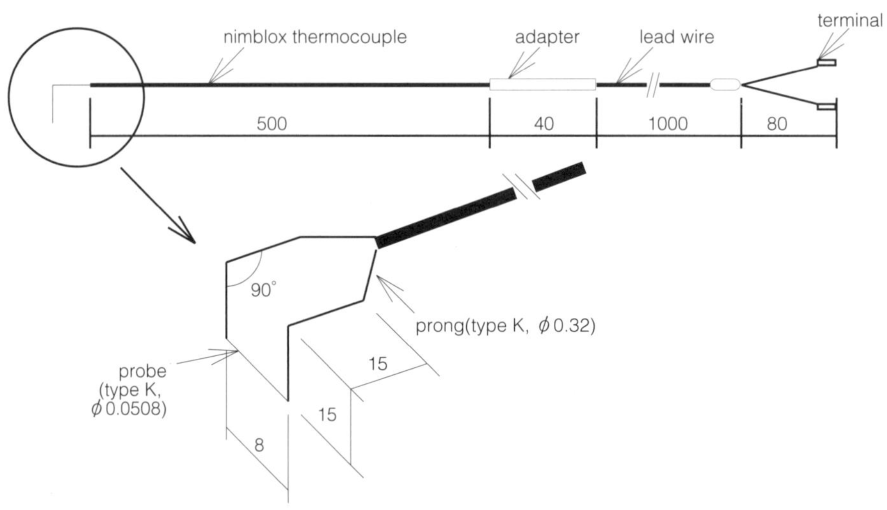

Fig. 5.2.2-1 Schematics of special thermocouple to measure air temperature near canister surface

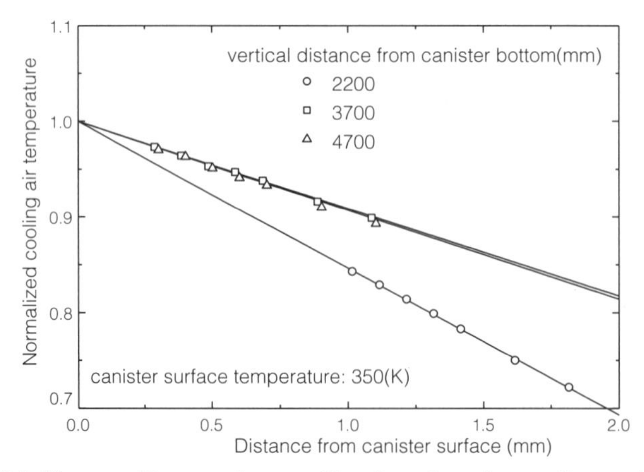

Fig. 5.2.2-2 Measured temperature profile of cooling air near the canister surface

3700 mm agrees with that at the height of 4700 mm. Such estimated heat transfer rates under various heat and fluid conditions are shown in Fig. 5.2.2-3. The heat transfer rates are normalized with that of pure buoyant (natural convection) flow and that of pure forced driven (forced convection) flow. These heat transfer rates reveal that the heat transfer rate in the annulus of vault facility with parallel flow is much large compared with that of natural and forced convection. Fig. 5.2.2-4 shows a measured distributions of upward velocity in the annulus for various heat and fluid conditions. All profiles are skew and takes maximum values near the canister surface, implying that the heat transfer is carried out by combination of buoyant and cooling air flows.

Through these experiments, the insight into heat transfer process in the vault facility, which yields establishment of evaluation methods for heat removal capacity, have been expanded and deepened. Further discussion on effects of various cooling air and canister heated conditions on heat transfer with buoyant flow have been carried out[6)-8)].

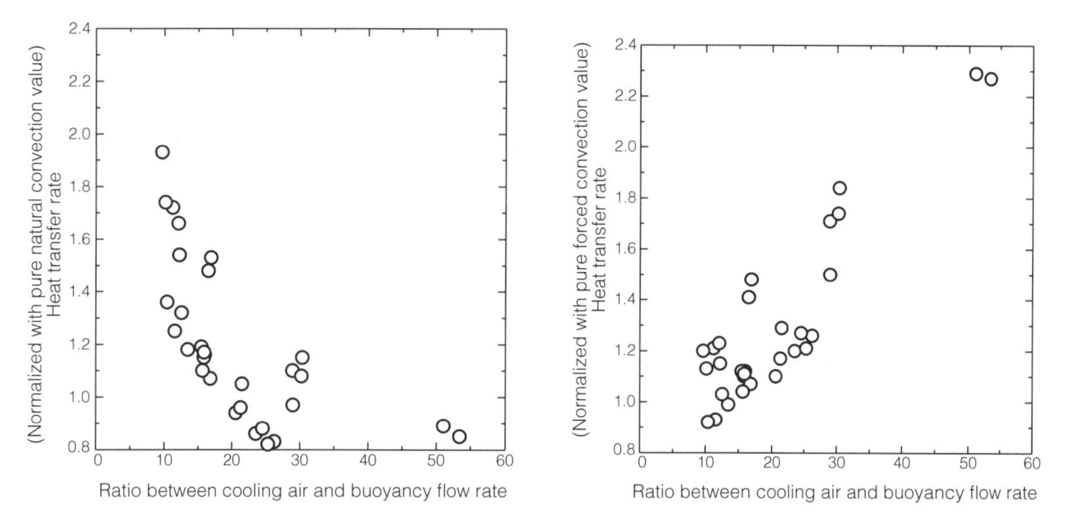

Fig. 5.2.2-3 Heat transfer rate normalized with that of pure natural and forced convection

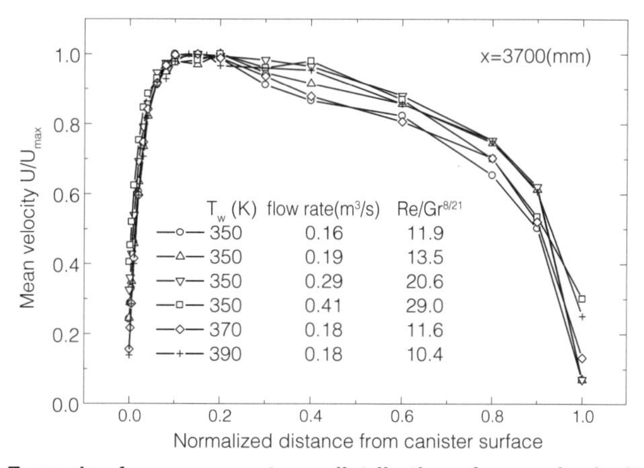

Fig. 5.2.2-4 Example of measurements on distribution of upward velocity in annulus between canister and container

5.2.3 Numerical simulation for thermal analysis of vault facility

The present subsection describes the outline of our numerical simulation[9] to evaluate heat removal characteristics of a vault storage facility. As mentioned in previous subsections, the experimental approach is a strong tool for estimating the heat removal capacity of the vault facility. On the other hand, expectations for numerical simulation with CFD codes have increased recently with rapid progress of performance of computing system. A purpose of the simulations presented here is to check the reproducibility for the heat removal tests.

The combination of two kinds of numerical simulations must be successful, which is similar to heat removal tests shown in 5.2.2; one is for simulating the flow rate in the annulus with the scale model of the whole of the facility and the other is for simulating the heat transfer rate with the scale model only of the annulus. Notice that, even the latest computing systems have some difficulties to perform the simulation to firmly grasp the flows of two kinds of scale.

The numerical simulations for discussing the flow pattern in the whole of the facility should properly represent flow fields over complex geometries, which consists a building, storage tubes, containers. The present simulations attempted a porous body model, which is a scheme to represent complex geometry by adding some physical models of governing equations in the regular arrangement coordinate, while the boundary fit coordinates (BFC) must give a solution. The comparison with experimental data shows that such simulations have capabilities to correctly capture the complex flow patterns and air temperature distributions observed in heat removal test for the vault facility with cross flow.

The numerical simulations for discussing the heat removal characteristics from the container surface to cooling should properly represent turbulence flows developed along the walls. The turbulence flows is generated by the non-linearity of the governing equations of fluid motions (Navier-Stokes equations), and this non-linearity yields the activation of thermal diffusion (turbulence diffusion). Thus, the accurate estimation of this turbulence diffusion is essential for discussing the heat transfer processes. In our simulations, the improvement of physical model to consider the buoyancy effects on the turbulence diffusion gives agreement of heat transfer rate with the experiments. Fig. 5.2.3-1 shows

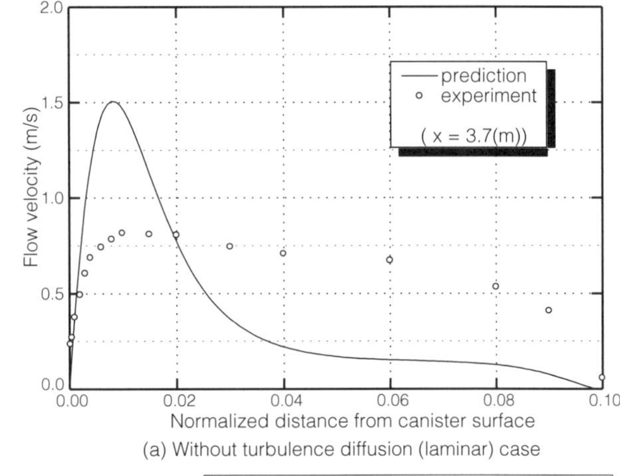

(a) Without turbulence diffusion (laminar) case

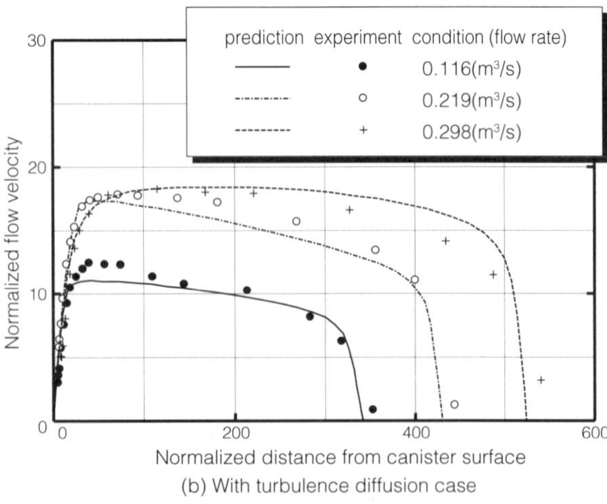

(b) With turbulence diffusion case

Fig. 5.2.3-1 Example of numerical simulation results on distribution of upward velocity in annulus between canister and container

comparison of the profiles of upward velocity in the annulus between numerical simulations and experiments. The simulations with turbulence diffusion clearly improve the reproducibility: the simulations with turbulence diffusion gives good agreement of these profiles with experiments, while the simulations without turbulence diffusion (laminar flow analysis) overestimate the upward velocity near the canister.

Moreover, the numerical simulation techniques presented here have been already applied the facilities under real conditions and the numerical results shows the soundness of thermal conditions, such as temperatures at building concrete surfaces, canister surfaces.

5.3 Shallow underground vault storage [10)]

While the surface vault storage described in section 5.2 uses buildings on the surface, the shallow underground vault storage uses underground space or tunnel. The word “shallow” is added because the rock or soil around the tunnel is not required of the function of confinement of radionuclide, different from the fact that the rock around the radioactive waste disposal facility is required of the confinement function. And also because tunnel type storage facilities utilizing the slope topography of the sites of existing nuclear power stations are mainly considered here to increase options for the siting of spent fuel storage facility. In this section, the concept and feasibility of the shallow underground vault storage is described. Please refer to the literature[10)] for details.

5.3.1 Design concept

(1) Basic premises

The specifications of the canister to be stored, which have been used for the conceptual design and feasibility study, are shown in Table 5.3-1. The scale of the storage is approximately 2000 tU of spent fuel or about 200 canisters in the present specifications.

Table 5.3-1 Specifications of the canister for the present study

Item	Specification
Number of fuel assembly / Canister	56(BWR)
	21(PWR)
Uranium weight (kg / Canister)	9744
External diameter (mm)	1607
Total length (mm)	4840
Thickness of the body (mm)	16
Material	
Body portion	SUS(*) * basically SUS304L
Bottom part	SUS(*)
Shield lid	SUS304L
Primary lid	SUS(*)
Secondary lid	SUS(*)
Basket	
Grid plate	SUS304L
Neutron absorber	Boral (Al+B)
Total weight with the fuel (ton)	about 33
Heat generation (kW / Canister)	about 22

Heat removal shall be based on natural convection method. In this method, buoyancy of the heated air with smaller density is used. The method requires stacks or shafts to obtain the necessary buoyancy of the heated air for the natural ventilation, but does not require supplying power or measures in case of power failure. Therefore it is more robust than the active ventilation system.

Lower part of the space of the storage facility shall be the storage area and upper part over the intermediate slab floor shall be the transporting area. Canisters will be conveyed by a special transporting equipment to specified positions to be put in the storage area. The taken in air by the natural convection ventilation goes through the storage area cooling down the canisters and goes out through the exhaust stacks.

The siting of this shallow underground vault storage is not restricted to the inside of the site of nuclear power stations, but it is considered to be a likely option.

Taking into consideration such siting scenario, the following two cases of topography have been set as the representatives in Japan.

i) Topography of steep slope:

The geology of steep slopes is usually hard rock with favorable properties.

ii) Topography of gentle slope:

The geology of gentle slopes would often be soft rock or condensed sandy soil.

Figs. 5.3-1 and 5.3-2 show the vertical sections of the geological model for the topography of steep slope and gentle one. Table 5.3-2 gives the mechanical properties of the supposed two geologies. B, CH, CM, CL and D in the figures are the rock grades shown in the table.

Intermediate type between the two topographies may be added, but in that case, the method of the construction of the shallow underground vault would be tunnel excavation just as in the case of the topography of steep slope. Therefore the intermediate type is represented by the topography of steep slope. On the other hand, in the case of the topography of gentle slope, cut and cover method is presupposed for the construction of the shallow underground vault.

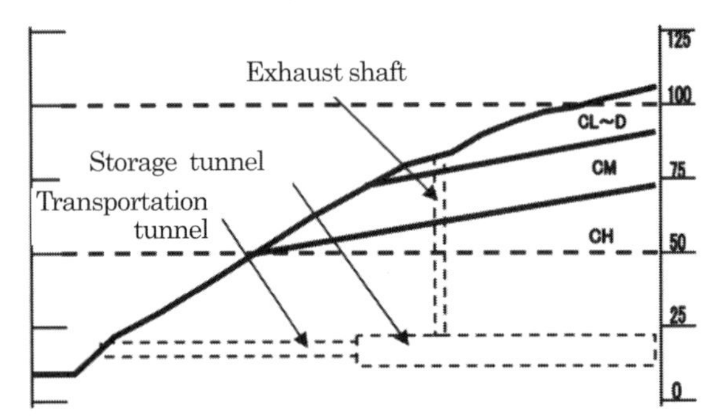

Fig. 5.3-1 Steep slope topography case (height: m)

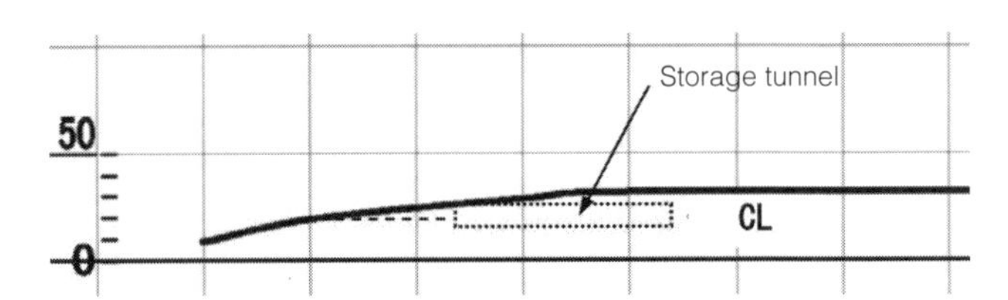

Fig. 5.3-2 Gentle slope topography case (height: m)

Table 5.3-2 Mechanical properties of the two geologies

Properties	Rock classification	Steep slope, Hard rock	Gentle slope, Soft rock
Cohesion C (MPa)	B	—	—
	CH	1.2	—
	CM	1.0	—
	CL	0.4	0.12
	D	0.1	—
Internal friction angle ϕ (deg.)	B	—	—
	CH	50	—
	CM	45	—
	CL	40	40
	D	35	—

※ Representative properties of hard and soft rock are presumed.

(2) Design concept

The basic design concepts of shallow underground vault storage for the two cases of topography are shown in the following.

1) Shallow underground storage for steep slope topography

As stated in (1), about 200 canisters are stored in a storage tunnel. The width of the tunnel is supposed to be about 20 m, considering the size of the existing tunnels and underground spaces. Fig. 5.3-3 shows the vertical and plan views of the storage tunnel.

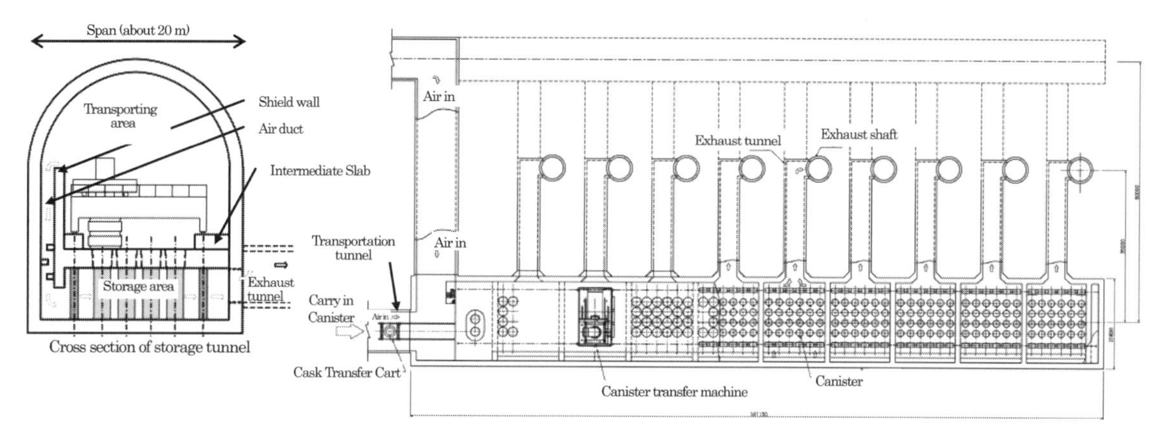

Fig. 5.3-3 Storage facility in the case of steep slope topography (vertical and plan views)

The storage facility requires transportation tunnel, ventilation tunnel for air intake, ventilation shaft for exhaust, and so on besides the storage tunnel. The transportation tunnel can serve as the ventilation tunnel but other ventilation tunnel(s) is required to maintain robustness of the heat removal system. Also, plural ventilation shafts are required for the robustness and flexibility of the heat removal operation.

At a side of the storage tunnel there is a shield wall to protect the transportation area from the radiation, and the wall also serves as a part of air duct to the storage area. The shield wall's thickness may need to be about 1m. Three labyrinth plates are installed in the air duct and a hanging wall is built under the shield wall. These are for the suppression of streaming radiation in the duct.

2) Shallow underground storage for gentle slope topography

In the case of gentle slope topography, we cannot expect to construct a self-supporting large size underground space needed for the shallow underground vault storage because strength of the rock and soil is small. In this case of topography, cut-and-cover tunnel would be feasible.

After the construction of the concrete structure of the storage tunnel, it is to be covered with soil for shielding of radiation and scenic point of view.

Fig. 5.3-4 shows the vertical and plan views of the shallow underground vault storage facility for the gentle slope topography.

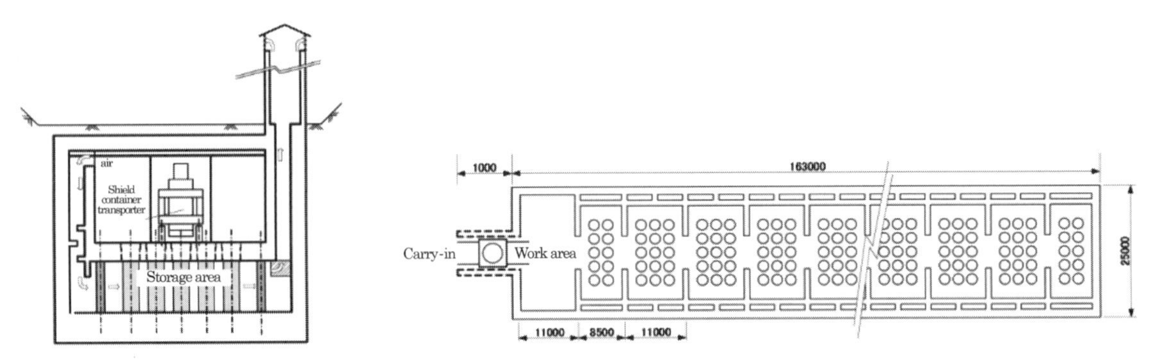

Fig. 5.3-4 Storage facility in the case of gentle slope topography (vertical and plan views)

5.3.2 Technical feasibility

Among the four basic safety functions of interim storage facility, namely, confinement, criticality prevention, radiation shielding and heat removal, the former two are quite the same as those for the surface vault storage, therefore the latter two, radiation shielding and heat removal are briefly described here.

In the case of the surface vault storage, the function of radiation shielding is obtained by the sufficient thickness of concrete wall and ceiling, and by an adequate design of labyrinth duct. In the case of the shallow underground storage, the storage facility is surrounded by rock. Rock has the similar shielding ability as concrete because the density of both materials is similar. Since the thickness of the surrounding rock is far larger than that of concrete wall (about 1 m), radiation permeation through rock is negligible. Radiation through air supply and exhaust openings can be coped with by design method of labyrinth duct just as in the case of surface vault storage. From these considerations, there is no anxiety about the radiation shielding in the case of shallow underground storage facility.

The feasibility of heat removal by natural convection can be judged by whether or not the buoyancy of heated air is larger than the resistance to flow necessary for stipulated heat removal capacity. From the maximum heat generation 22 kW/canister specified in Table 5.3-1, number of canisters 200, air temperature 30 °C at inlet and 50 °C at outlet, the required flow rate is calculated to be at least 22.3 kg/sec by a heat balance calculation. The maximum temperature 50 °C at the outlet is premised considering the durability of concrete. The obtained flow rate and the flow paths for the two topography cases lead to the evaluation of pressure loss (flow resistance) of about 40Pa for the steep slope topography and about 24 Pa for the gentle slope topography. The combined height of shaft and stack is calculated to be more than 65 m for the steep slope topography and 40 m for the gentle slope topography so that the buoyancy of heated air is larger than the pressure loss by air flow. Detailed layouts of the storage facilities for the both topographies are shown in Figs. 5.3-3 and 5.3-4. Therefore there is no anxiety about the heat removal function of the shallow underground storage facility. It can be concluded that, concerning the basic safety functions of spent fuel interim storage facility, the shallow underground vault storage is technically feasible.

A rough economic feasibility assessment has also been carried out. The total costs of construction for surface storage facility and shallow underground storage facility in the two slope topography cases were compared. The total cost of construction consists of costs for canisters, temporary utilities, construction work and civil engineering. The comparison shows that the shallow underground storage method is considered economically feasible.

5.3.3 Issues to be resolved

For the metal cask storage of spent fuel on the surface, items to be examined for government licensing have been already prescribed in Safety Requirements. Items which may be newly required for the governmental examination when the storage facility goes underground will include, stability of rock cavern and structural integrity under earthquake, safety under fire and safety under water inrush. Among these items, the aseismic item is considered to be especially important and briefly explained about its present status.

Since Japan has no experience of constructing an underground nuclear related facility yet, we cannot presume what the governmental requirements will be like, but for the cut-and-cover type facility for gentle slope topography, description in Safety Requirements about aseismicity of underground reinforced concrete structures in nuclear power stations will be of use for reference.

For the excavated tunnel type storage facility for steep slope topography, basic principle about standard earthquake ground motion and structural design is considered to be similar to that of existing Safety Requirements. Although Regulatory Guide for Reviewing Seismic Design of Nuclear Power Reactor Facilities[11] stipulates the combination of loads and threshold limit for structural design to be considered for building, concrete structure, piping and equipments, there is no stipulation about the rock around a cavern or its reinforcement. Therefore a revision of the Regulatory Guide or other standards[12),13] by JEA:Japan electric association may be necessary. The literature[10] shows an example of the stipulations concerning the underground structure based on a survey of previous work about underground nuclear power station. The outline of the concept is to evaluate the safety factor of shear slip for the surrounding rock under earthquake and to evaluate the ultimate strength for reinforced concrete which is fixed to rock.

5.4 Horizontal silo storage

The horizontal silo is a method like horizontal concrete cask, where an encapsulated spent fuel is stored horizontally in a concrete module. Fig. 5.4-1 shows an example of design concept. The structure of the canister is almost the same to that of concrete cask. It is a characteristic of this method that the canister with spent fuel does not have to be lifted high when it is transferred from a transfer cask horizontally into the concrete module. The other characteristics and safety design consideration are similar to those of concrete cask. Fig. 5.4-2 is an example of the horizontal silo (NUHOMS) used in USA, which shows a system to install encapsulated spent fuel into the concrete module.

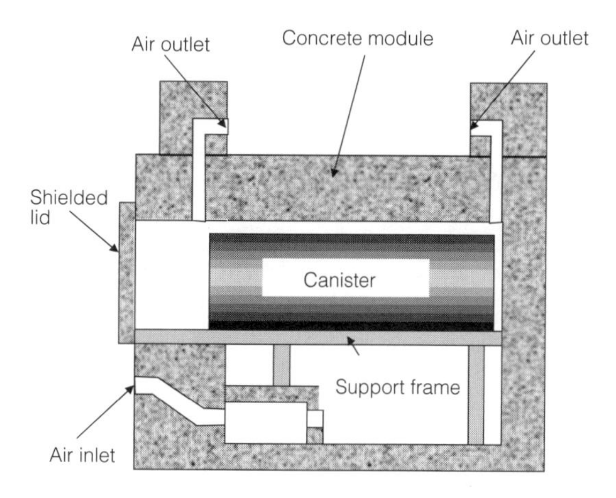

Fig. 5.4-1 Cross section of horizontal silo (example)

Fig. 5.4-2 Design concept and implementation of horizontal silo (NUHOMS), USA (Copyright TN)

References

1) Republic of Hungary: "National Report, Third Report prepared in the framework of the Joint Convention on the Safety of Spent Fuel Management and on the Safety of Radioactive Waste Management", (2008). http://www-ns.iaea.org/conventions/results-meetings.asp˜s=6&l=40.

2) Spain: "Fourth Spanish National Report for Joint Convention on the Safety of Spent Fuel Management and on the Safety of Radioactive Waste Management", October 2011. ibid.

3) K.Sakamoto, T.Koga, M.Wataru, and Y.Hattori, : "Development of Evaluation Method for Heat Removal Design of Dry Storage Facilities (Part 1) - Heat Removal Test on Vault Storage System of Cross Flow Type -", CRIEPI Report U97047, November 1997.

4) K.Sakamoto, T.Koga, M.Wataru, Y.Hattori, Heat Removal Characteristics of Vault Storage System of Cross Flow Type for Spent Fuel, Nucl. Eng. Des. Vol. 195, 57-68 (2000)

5) Y. Hattori, K. Sakamoto, E. Kashiwagi, T. Koga, and M. Wataru: "Development of Evaluation Method for Heat Removal Design of Dry Storage Facilities (Part 2) - Heat Removal Test on Vault Storage System of Parallel Flow Type -", CRIEPI Report U98005, July 1998.

6) Y. Hattori, T. Tsuji, Y. Nagano, N. Tanaka: "Characteristics of turbulent combined-convection boundary layer along a vertical heated plate", Int J Heat and Fluid Flow 21 (2000) 520-525.

7) Y. Hattori, T. Tsuji, Y. Nagano, N. Tanaka: "Effects of freestream on turbulent combined-convection boundary layer along a vertical heated plate", Int J Heat and Fluid Flow 22 (2001) 315-322.

8) Y. Hattori, T. Tsuji, Y. Nagano, N. Tanaka: "Turbulence characteristics of natural-convection boundary layer in air along a vertical plate heated at high temperatures", Int J Heat and Fluid Flow 22 (2001) 315-322.

9) K. Sakamoto, Y. Hattori, T. Koga, and M. Wataru: "Development of Evaluation Method for Heat Removal Design of Dry Storage Facilities (Part 4) - Numerical Analysis on Vault Storage System of Cross Flow Type -", CRIEPI Report U98031, March 1999.

10) K. Shin, T. Saegusa, T. Koga, M. Teramura, Y. Kaga, E. Yoshimura, K. Shirahama, K. Takeuchi and S. Sato:

"Feasibility study on underground vault storage of spent nuclear fuel", CRIEPI Report N07015, 2007.
11) Nuclear Safety Commission: "Regulatory Guide for Reviewing Seismic Design of Nuclear Power Reactor Facilities", September 19, 2006.
12) Japan Electric Association: "Guideline for Seismic Design Technology of Nuclear Power Plant", JEAG 4601-1987.
13) Japan Electric Association: "Addendum of Guideline for Seismic Design Technology of Nuclear Power Plant", JEAG 4601-1991, JEAG4601-2008.

CHAPTER 6 SPENT FUEL INTEGRITY

6.1 What is spent fuel integrity ?

For interim storage, spent fuel assemblies to be stored in metal casks are required to be confirmed for their integrity by reactor records, sipping inspection of fuel assemblies, etc. if necessary. Additionally, during dry storage, spent fuel cladding integrity must be maintained over the design storage period. Thus, it is required to maintain the integrity of spent fuel cladding in the interim storage. Definition and criteria of spent fuel integrity may vary among countries[1)]. In Japan, “integrity of spent fuel” means that spent fuel cladding is not damaged (pin hole or hairline cracking on the cladding that may yield incidentally is not categorized as damage) and that spent fuel integrity at loading into metal casks is properly maintained (i.e. there are no excessive deformation or no degradation of material characteristics)[2)]. Namely, “integrity of spent fuel” means that the spent fuel assemblies have no leakage due to pinhole or hairline cracking, and there are no fuel rod bowing and no deformation of the assembly components during reactor operation.

References

1) Spent Fuel Performance Assessment and Research: Final Report of a Coordinated Research Project (SPAR II), IAEA-TECDOC-1680.

2) Nuclear and Industrial Safety Subcommittee of the Advisory Committee for Natural Resources and Energy, Nuclear Fuel Cycle Safety Subcommittee, Interim Storage Working Group and Transport Working Group: “Long-term Integrity of the Dry Metallic Casks and their Contents in the Spent Fuel Interim Storage Facilities”, (2009.6.25)

6.2 Nuclide composition of high burnup spent fuel

6.2.1 Evaluation of source term in spent fuel[1),2)]

Calculation accuracy of shielding, criticality and heat removal for spent fuel cask or interim storage facility depends directly on the calculation accuracy of the amount of actinides and fission products in the spent fuels. As developing of higher burnup of UO_2 fuel and utilization of mixed oxide fuel (MOX fuel), the nuclide composition of the actinides and the fission products in these spent fuels after discharge will change, and consequently the source intensity and heat generation of these spent fuels will increase more than those of the conventional burnup UO_2 fuel (about 40 MWd/kgHM). Thus, improvement of the calculation accuracy of the nuclide composition in the spent fuels will contribute to improving the evaluation accuracy of various nuclear energy field that includes from the front-end such as high performance reactor management to the back-end such as reprocessing.

In this section, experimental nuclide composition data are compared with the calculated one for high burn-up PWR-UO_2 and PWR-MOX spent fuel, and improving the calculation accuracy for the nuclide composition is discussed.

6.2.2 Specification of fuel and calculation method

In this study, chemical isotopic analyses of actinides and fission products were carried out on high burn-up PWR-UO_2 and high burn-up MOX spent fuel as shown in Table 6.2-1. The high burn-up PWR-UO_2 spent fuel rod used in this study (3.8 wt%^{235}U, 60.2 MWd/kgHM declared average burn-up) was loaded in 15 x 15 fuel assembly and irradiated for five cycles in an European commercial nuclear power reactor. A chemical isotopic analysis of samples A, B, C and D, extracted from four different axial positions of the high burn-up PWR-UO_2 fuel, was carried out. Concerning PWR-MOX fuels, two segments with declared burn-ups of about 46.0 MWd/kgHM were used in this study. The segments (MOX1, MOX2) were extracted from the central part along the axial direction of two different PWR-MOX spent fuel pins (Pu enrichment: 5.07 wt.%) that were irradiated in the same fuel assembly for four cycles. The initial plutonium vector in the MOX segments was ^{238}Pu/^{239}Pu/^{240}Pu/^{241}Pu/^{242}Pu/^{241}Am=1.50/59.00/24.38/9.34/4.85/0.94. The fissile plutonium content was 68.3 wt.%.

Table 6.2-1 Specification of PWR-UO_2 and PWR-MOX spent fuel

Fuel	Initial enrichment of ^{235}U/Pu (wt%)	Burnup (MWd/kgHM)
PWR-UO_2	3.8	52.8, 60.0, 63.5, 64.7
PWR-MOX	5.07	46.0, 46.6

6.2.3 Comparison of calculation with measurement for nuclide composition in spent fuels

(1) Chemical isotopic analysis and calculation

In the chemical isotopic analysis, the total dissolution of the samples that are four samples for PWR-UO_2 and two samples for PWR-MOX was performed in nitric acid of 10-15 N/hydrogen fluoride solution using an autoclave. Hence, even insoluble isotopes such as ruthenium completely dissolved in the solution. Various analysis methods based on mass- and energy-based spectrometry techniques were applied to the samples to determine the nuclide composition of 17 actinides and 40 fission products. Table 6.2-2 shows number densities per gram solution of the actinides and fission products obtained experimentally.

Table 6.2-2 Example of chemical isotopic analysis for actinide and fission products

	PWR-UO_2 spent fuel				PWR-MOX spent fuel	
	Numbe /1 gram solution (× 10^{16})					
Nuclide	A	B	C	D	MOX1	MOX2
^{235}U	16.4	29.6	21.4	16.0	23.5	11.5
^{238}U	6.2×10^3	7.0×10^3	7.1×10^3	6.8×10^3	8.6×10^3	4.1×10^3
^{239}Pu	36.0	36.8	40.0	34.1	1.1×10^2	53.2
^{240}Pu	21.9	21.5	24.5	21.8	1.0×10^2	48.1
^{241}Pu	11.0	9.58	11.2	9.57	46.8	20.4
^{242}Pu	11.1	8.73	11.9	11.4	40.2	18.7
^{244}Cm	1.97	0.93	1.62	1.64	6.30	2.90
^{133}Cs	24.8	28.3	33.8	22.0	27.5	15.1
^{134}Cs	1.29	0.68	0.94	0.90	0.81	0.20
^{135}Cs	9.03	8.54	10.7	7.61	16.1	8.42
^{137}Cs	26.5	23.2	27.5	26.2	29.0	11.3
Burnup (MWd/kgHM)	64.7	52.8	60.0	63.5	46.0	46.6

The nuclide composition of actinides and fission products was computed using the SWAT code that is an integrated burn-up code developed at the Japan Atomic Energy Agency (JAEA). In this study, the neutron spectrum and the effective cross sections were calculated using the ultra-fine resonance absorption calculation (PEACO) routine of SRAC in the SWAT code, and the effective cross sections for all of the fission products evaluating their C/E ratios were calculated using PEACO. The libraries JENDL-3.2, JENDL-3.3, ENDF/B-VI.5, ENDF/B-VI.8, JEF-2.2 and JEFF-3.0, and the fission yields of JNDC-V2 were used in the SWAT calculation.

The local burn-ups for the samples determined by the chemical isotopic analysis were used in the calculation. A square cell model was applied, considering an equivalent volume ratio of fuel to moderator with the whole fuel assembly. The geometry used in the calculations was composed of three regions: fuel pellet, cladding and moderator.

(2) Comparison of experimental and computational compositions

The difference between the amounts of actinides and fission products obtained by chemical isotopic analysis and calculated by SWAT was evaluated as the C/E ratios. The C/E ratios were normalized by the residual amounts of ^{238}U. As an example, Fig. 6.2-1 shows the C/E values that are averaged of those of samples A, B, C and D for each nuclide for the PWR-UO_2 fuel and those of MOX1 and MOX2 for the PWR-MOX fuel.

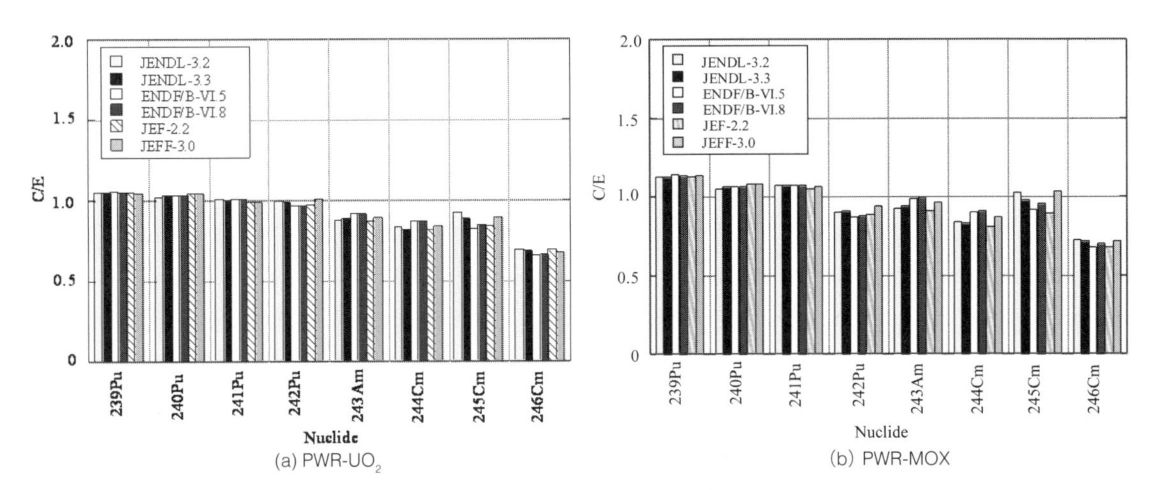

Fig. 6.2-1 Measured and calculated amounts of actinides as C/E ratio

Curium-244 is the major neutron emission source in spent fuels after 3 to 4 year-cooling time. The C/E ratio for ^{244}Cm calculated with any library was underestimated by 10 to 20 %. The fission products were also evaluated by the C/E ratios.

Actinide ^{244}Cm and fission products ^{90}Sr, ^{106}Ru, ^{133}Cs and ^{135}Cs that are important nuclides for the shielding, criticality calculation and burnup indicator, were further investigated to improve their C/E values.

6.2.4 Improvement of calclation accuracy by sensitive analysis

Simplified burnup chains for actinide ^{244}Cm and fission products ^{90}Sr, ^{106}Ru, ^{133}Cs and ^{135}Cs were developed, and production paths with large sensitivities for concerned nuclides were investigated.

(1) Determination of main production path by sensitive analysis

In simplified burn-up chains of the production path for ^{244}Cm, ^{90}Sr, ^{106}Ru, ^{133}Cs and ^{135}Cs, the fission yields for short-lived fission product nuclides were treated as cumulative fission yields. As a result of the comparison of nuclide compositions between the simple depletion using simplified burnup chains and SWAT calculations, in the production path for ^{244}Cm, the amounts of uranium and plutonium isotopes obtained by the two calculations agreed within 4 %. The ^{244}Cm amounts agreed within 8 %. For fission products that are discussed in the following sections, the two calculations agreed within about 3 %. Fig. 6.2-2 shows an example of simplified burn-up chains for ^{131}Xe-^{139}La. The similar simplified burnup chains were developed for ^{244}Cm, ^{90}Sr and ^{106}Ru.

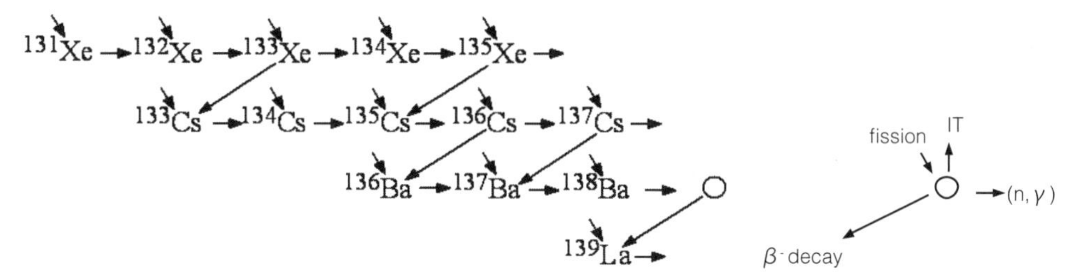

Fig. 6.2-2 Example of depletion chains of fission products

Hence, the sensitivity analysis and the correction for the fission yield or capture cross section were carried out using these simple depletion calculations.

(2) Correction of fission yield or capture cross section based on sensitivity analysis

From the sensitive analysis for the actinides, the main sensitive production path of ^{244}Cm was ^{238}U→^{239}Np→^{239}Pu →^{240}Pu→^{241}Pu→^{242}Pu→ ^{243}Pu/^{243m}Pu→^{243}Am→^{244}Cm. Thus, the capture cross section of ^{243}Am incresed by 16.3 % and those of ^{240}Pu and ^{241}Pu increased by 5 % based on the literature[3] and the C/E ratio in Fig. 6.2-1.

From the sensitive analysis for the fission products, the C/E ratios for ^{90}Sr and ^{106}Ru were improved by the correction for their own fission yield. The correction for the ^{133}Xe fission yield improved the C/E ratio for ^{133}Cs. Concerning ^{135}Cs, its C/E ratio was improved by the correction for the ^{135}Xe fission yield or capture cross section.

Table 6.2-3 shows the C/E ratios for ^{244}Cm and ^{135}Cs resulted from the correction, as an example. Other fission products that have low accuracy for their calculated amounts were also improved by the corrections of capture cross sections or fission yields of related nuclides on the production path.

Table 6.2-3 C/E ratios for ^{244}Cm and ^{135}Cs with correction

(a) ^{244}Cm

Actinide	C/E ratio before correction	C/E ratio after correction for capture cross section of ^{243}Am, ^{240}Pu, ^{241}Pu
^{240}Pu	1.06	1.02
^{241}Pu	1.07	1.07
^{242}Pu	0.91	0.96
^{243}Am	0.94	0.92
^{244}Cm	**0.83**	**0.96**

(b) ^{135}Cs

Fission product	C/E ratio before correction	C/E ratio after correction for fission yield of ^{135}Xe	C/E ratio after correction for capture cross section of ^{135}Xe
^{133}Cs	0.87	0.87	0.87
^{134}Cs	0.83	0.83	0.83
^{135}Cs	**0.88**	**1.00**	**1.03**
^{137}Cs	0.95	0.95	0.95

References

1) A. Sasahara, et al., J. Nucl. Sci. Technol., Vol. 45, No. 4, p.313-327 (2008).

2) A. Sasahara, et al., J. Nucl. Sci. Technol., Vol. 45, No. 5, p.390-401 (2008).

3) M. Ohta, S. Nakamura, H. Harada, "Measurement of Effective Capture Cross Section of Americium-243 for Thermal Neutrons," J. Nucl., Sci., Technol., 43, 1441 (2006).

6.3 Spent fuel integrity during normal storage condition

During the dry storage of spent nuclear fuel in casks or vault facilities, it is very important to deal properly with the decay heat of the spent fuel from the view point of environmental safety. For example, the number of fuel assemblies to be contained in a cask and the cooling time of the spent fuel before storage will be determined by the consideration into the thermal behavior. The temperature limitation of fuel claddings is believed to be most important among the temperature limitations of each part of storage system. Blackburn, et al.[1)] extracted the following 6 phenomena potentially affecting the integrity of fuel cladding based on the experience in the operation of light water reactors and evaluated the possibility leading to the factor determining the temperature limitation, the maximum allowable temperature.

(1) Failure due to the mechanical overload under rapid loading condition,

(2) creep rupture,

(3) stress corrosion cracking,

(4) rapid fracture due to internal or surface defects,

(5) oxidation (inner or outer surfaces),

(6) hydride formation and others.

After the evaluation, they pointed out the importance of (2) creep rupture as the determining factor and proposed the maximum allowable temperature of 653 K based on the concept of creep damage.

Most of the maximum allowable temperatures were proposed after considering creep deformation or creep rupture[2)] including the example of Blackburn, et al. mentioned above. An example of the maximum allowable temperature of fuel cladding was proposed in Germany based on the creep deformation, calculating the accumulated creep strain according to the temperature profile during dry storage and comparing the accumulated creep strain with a creep strain criterion of 1 %[2)]. In Japan, according to the standard of Atomic Energy Society of Japan, AESJ-SC-F002:2010: Standard for Safety Design and Inspection of Metal Casks for Spent Fuel Interim Storage Facility [3)], the maximum allowable temperature of fuel cladding must be determined as the lowest temperature among those obtained from consideration into the following three phenomena;

(1) The initial temperature of the fuel cladding which leads to the accumulated creep strain of 1 %, when the creep

calculation was made considering the temperature profile,

(2) The temperature of the fuel cladding which induces annealing of irradiation hardening to the strength level of fuel cladding corresponding to the design criterion,

(3) The temperature of the fuel cladding which prevent degradation of mechanical property due to reorientation of hydrides.

In the following section, the method for evaluating the maximum allowable temperature will be described according to the standard of Atomic Energy Society of Japan.

6.3.1 Temperature limit determined by creep behavior

In the standard of Atomic Energy Society of Japan, "b. Design Criterion in 4.2.5 Design of Heat Removal" requests that "the spent fuel cladding temperature must not exceed the temperature limit to be determined to maintain the integrity of the cladding". And "c. Design Method" provides that "the temperature limitation based on the creep behavior should be determined as the initial temperature of fuel cladding which leads to the accumulated creep strain of 1 % when the creep calculation is made by experimental or theoretical creep equations considering the temperature profile and the resultant decrease in the internal pressure of the cladding". The standard may consider the results of the following previous studies to provide the above mentioned concept to determine the temperature limitation.

CRIEPI has been conducting extensive researches and developments from various fields to realize the dry storage of spent fuel using casks. In the course of the research and development, the integrity of the spent fuel was also evaluated considering creep deformation and rupture as the most influencing factor[2)]. CRIEPI made a series of creep tests as shown below on fuel claddings to develop and propose a method to determine the temperature limit based on creep deformation and a strain criterion.

(1) Creep deformation tests on Zircaloy-4 under low applied stress conditions to formulate the deformation behavior[4)].

(2) Variable temperature creep tests to confirm the applicability of the strain hardening rule[5)].

(3) Creep deformation tests on Zircaloy-2 to formulate the deformation behavior and to evaluate the effect of neutron irradiation on creep deformation[6)].

(4) Creep deformation/rupture test on the actual PWR spent nuclear fuel claddings[7)].

(5) Development and proposal of a method to determine the temperature limit based on the test results shown in (1) through (4)[8),9)].

Fig. 6.3-1 shows a flow chart of the method to determine the temperature limit of fuel claddings. This method determines the temperature limit by comparing the accumulated creep strain calculated by a creep equation for un-irradiated Zircaloy cladding with the strain criterion of 1 % just the same value as that of Germany[2)], based on the facts; 1) creep deformation of spent fuel cladding or irradiated cladding is hard to occur in comparison with that of un-irradiated claddings[6)7)], 2) creep rupture strain of spent fuel claddings exceeds 3 % at the temperatures of 633 K and higher. Hence, the temperature limit of fuel claddings changes depending on the specification of the fuel and the heat removal efficiency of a cask.

Japan Nuclear Energy Safety Organization(JNES) also conducted creep tests on the spent nuclear fuel claddings,

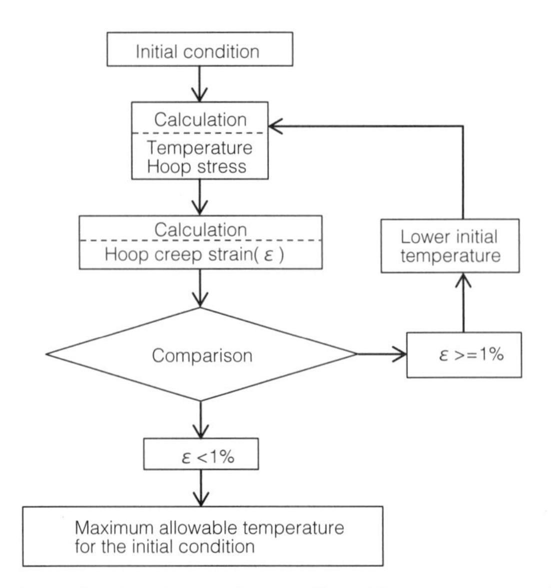

Fig. 6.3-1 Method for evaluating the maximum allowable temperature of fuel cladding[8), 9)]

and proposed two creep equations for the burn-up regions of 0 to 30 GWd/t and 30 to 53 GWd/t, respectively[10)]. To formulate the two creep equations, extensive creep tests were conducted on both the PWR spent fuel claddings and un-irradiated claddings in the temperature region of 603 K through 723 K with applied hoop stress level of 30 to 250 MPa. The saturated primary creep strain and the secondary creep rate of the spent fuel claddings were smaller than those of the un-irradiated claddings in all the temperature and the applied stress regions. This result suggested that creep deformation was retarded by the neutron irradiation in service.

In reference to the previous studies as mentioned above, the standard of Atomic Energy Society of Japan provides a method to limit the spent fuel cladding temperature as the accumulated creep strain does not exceed the criterion of 1 % during the evaluation period to maintain the fuel integrity, in "N.1 Limitation by Creep", of appendix N: Limitation of Spent Fuel Claddings[3)]. More precisely, the accumulated creep strains for BWR fuel and PWR fuel should be calculated by the experimental creep equation of CRIEPI[6)] and that proposed by JNES[10)], respectively, and then should be compared with the strain criterion of 1 % after multiplying by the uncertainty coefficient of the data to the accumulated creep strains. Some examples of the cladding temperature evaluation were provided in appendix P(reference). The temperature limit of BWR new 8x8 fuel with zirconium liner claddings is around 633 K under the assumption of the maximum burn-up of 40 GWd/t, the cooling time of 18 years, the evaluation time of 60years with a temperature profile. And the temperature limit of PWR 17x17 fuel is 593 K under the assumption of the maximum burn-up of 48 GWd/t, the cooling time of 15 years, the evaluation time of 60 years with a temperature profile.

References

1) L.D.Blackburn, et al., HEDL-TME 78-37, 1978.

2) M.Mayuzumi and H.Tanaka, A review of Temperature Limit of Spent Fuel Cladding under Dry Storage Conditions(in Japanese), CRIEPI Report 285063, CRIEPI, 1986.

3) AESJ-SC-F002:2010: Standard for Safety Design and Inspection of Metal Casks for Spent Fuel Interim Storage Facility, Atomic Energy Society of Japan, 2010.
4) M.Mayuzumi, T.Onchi, Creep Deformation of an Unirradiated Zircaloy Nuclear Fuel Cladding Tube under Dry Storage Condition, J. of Nuclear Materials, 381-388, 1989.
5) M.Mayuzumi, T.Onchi, The Applicability of the Strain-hardening Rule to Creep Deformation of Zircaloy Fuel Cladding Tube under Dry Storage Condition, J. of Nuclear Materials, 73-79, 1990.
6) M.Mayuzumi, N.Yoshiki, T,Yasuda and M.Nakatsuka, The maximum Allowable Temperature of Zry-2 fuel cladding under dry storage condition, CRIEPI Report T88068, CRIEPI, 1989.
7) M.Mayuzumi, K.Murai, Post Irradiation Creep and Rupture of Irradiated PWR Fuel Cladding, Proc. of the Conf. on Nuclear Waste Management and Environmental Remediation, Vol.1, pp.607-612, Prague, 1993.
8) M.Mayuzumi, T.Yoshiki, T.Yasuda, M.Nakatsuka, A Method for Evaluating Maximum Allowable Temperature of Spent Fuel during Dry Storage, Proc. Inter. Seminar on Spent Fuel Storage -Safety, Engineering and Environmental Aspects, IAEA/OECD, Vienna, 1990.
9) T.Saegusa, M.Mayuzumi, C.Itoh, K.Shirai, Experimental Studies on Safety of Dry Cask Storage Technology of Spent Fuel — Allowable Temperature of Cladding and Integrity of Cask under Accidents, J. of Nuclear Science and Technology, Vol.33, No.3, pp.250-258.
10) Japan Nuclear Energy Safety Organization, Improvement of the Safety Evaluation Code for the Storage facility of Recyclable Spent Nuclear Fuel - The Final Report on the Long term Safety of Spent Nuclear Fuel, JNES, 2004.

6.3.2 Hydrogen redistribution in axial direction of fuel cladding

Given the pickup of hydrogen by the cladding (Zircaloy) during irradiation, hydrogen can migrate from the high temperature to the low temperature region of the cladding. Then, hydrogen may precipitate in the colder regions in the form of hydrides, and consequently, lead to increased brittleness, which may affect fuel integrity[1)-3)]. In this section, the hydrogen redistribution in the axial direction of the fuel cladding will be described.

Hydrogen migration mainly results from the axial temperature profile formed by axial burnup profile. In this study, the hydrogen redistribution experiments in the axial direction of cladding tube were carried out and the data related to hydrogen migration such as heat of transport, hydrogen diffusion coefficient and solubility limit were obtained using a twenty-year dry-stored (i.e. in air) spent PWR-UO_2 fuel rods.

(1) Spent fuels used in the hydrogen redistribution experiment

Two spent fuels rods obtained from PWR-UO_2 that were irradiated in a European commercial PWR were used in the hydrogen redistribution experiments. These fuel rods had been stored in an air-filled small metal cask for twenty years after discharged from reactor. The fuel rods had been stored as the bundle composed of about twenty fuel rods. The initial temperature at dry storage is therefore expected to be lower than that in a real storage condition. The specification of the spent fuel rods used in the experiments is shown in Table 6.3-1. The cladding materials of the spent fuel rods stored twenty years were standard Sn Zircaloy-4. As the reference cladding materials, the unirradiated standard Zircaloy-4 (1.5 wt.% Sn) and the low Sn Zircaloy-4 (1.3 wt.% Sn) that are in service today in Japan were also used in the experiments.

Table 6.3-1 Specification of the spent fuel rod used in the experiments

Fuel rod	Burnup (MWd/kgHM)	Enrichment of ^{235}U (wt%)	Year of discharged from reactor
Fuel-1	31.2	2.9	1972
Fuel-2	58.2	3.6	1976

(2) Apparatus of hydrogen redistribution experiment

A schematic apparatus of hydrogen redistribution experiment is shown in Fig. 6.3-2. Samples of 30-mm length were mounted between heaters, and surrounded with thermal insulators. The temperature on one end of the sample was kept at 380 °C (653 K) and that on the other end was kept at 260 °C (533 K). Thermocouples were mounted at the positions of 7, 11, 15, 19 and 23 mm from the higher temperature end (653 K). The duration of heating was ten days when the hydrogen redistribution has not been stable yet. The atmosphere during the experiment was air in order to keep an oxide layer on the cladding surface to prevent hydrogen from escaping out of the sample.

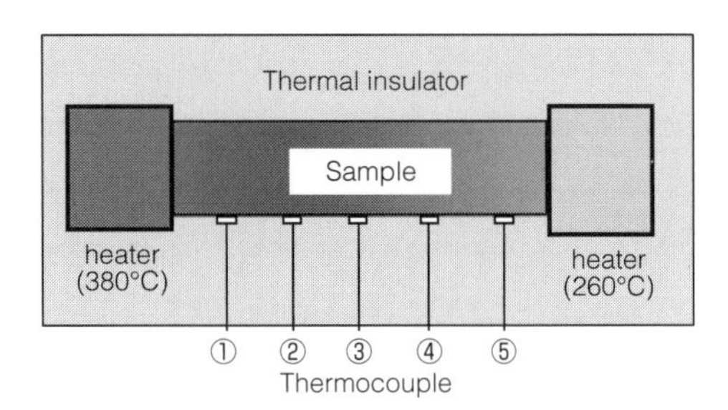

Fig. 6.3-2 Apparatus used for the hydrogen redistribution experiment

(3) Samples used in hydrogen redistribution experiments

The hydrogen redistribution experiment was carried out twice. Sample-1AR (30 mm length) taken from Fuel-2 at the third span (max. oxide thickness span) and Sample-1BR (30 mm length) taken from Fuel-1 at the forth span (max. burnup span) were used in the first experiment. Additionally, Sample-4AR (30 mm length) taken from Fuel-2 at the third span (max. oxide thickness span) and Sample-4BR (30 mm length) taken from Fuel-1 at the forth span (max. burnup span) were used in the second experiment.

The hydrogen concentration was measured on the specimens that just faced both ends of the sample. The results showed that hydrogen profiles in the samples were regarded as almost flat. The results of the measured hydrogen concentration in the specimens are shown in Table 6.3-2 (a) and (b).

Table 6.3-2 Hydrogen concentration in specimen

(a) First experiment (Irradiated sample)

	High level H_2 sample		Low level H_2 sample	
	Sample-1AR (58.2 MWd/kgHM)		Sample-1BR (31.2 MWd/kgHM)	
H_2 in both ends (ppm)	119.0	119.0	57.8	56.3
Average (ppm)	119		57	

(b) Second experiment (Irradiated sample)

	High level H_2 sample		Low H_2 level sample	
	Sample-4AR (58.2 MWd/kgHM)		Sample-4BR (31.2 MWd/kgHM)	
H_2 in both ends (ppm)	119.0	109.0	56.3	51.2
Average (ppm)	114		54	

As a reference cladding material, unirradiated samples that were charged almost the same hydrogen concentration as that in irradiated ones were used in the hydrogen redistribution experiments. The unirradiated samples were a standard Zircaloy-4 (1.5 wt% Sn) and a low Sn Zircaloy-4 (1.3 wt% Sn). Using the uniradiated samples, the effect of Sn content in the cladding samples on heat of transport was also investigated. The hydrogen concentration in the unirradiated standard Zircaloy-4 (1.5 wt% Sn) samples were 100 ppm for Sample-2AU and 65 ppm for Sample-2BU. Concerning the unirradiated low Sn Zircaloy-4 (1.3 wt% Sn), hydrogen concentration was 96 ppm for Sample-3AU and 61 ppm for Sample-3BU.

(4) Results of hydrogen redistribution experiments

After the hydrogen redistribution experiment, the samples of 30 mm length were dismounted and cut to six segments of which each length was 5 mm and hydrogen analysis was carried out to determine the hydrogen concentration on each segment. The heat of transport, diffusion coefficient and solubility limit of hydrogen were calculated by best fitting using a time-dependent hydrogen diffusion equation. Fig. 6.3-3 shows the diffusion coefficients and the solubility limits obtained by the experiments. The diffusion coefficients of the unirradiated samples are in the range of the results of Sawatzky[1] and Kearns[4]. In Fig. 6.3-3 (a), the diffusion coefficients on the irradiated Zircaloy claddings were approximately the same as or slightly smaller than that on the unirradiated

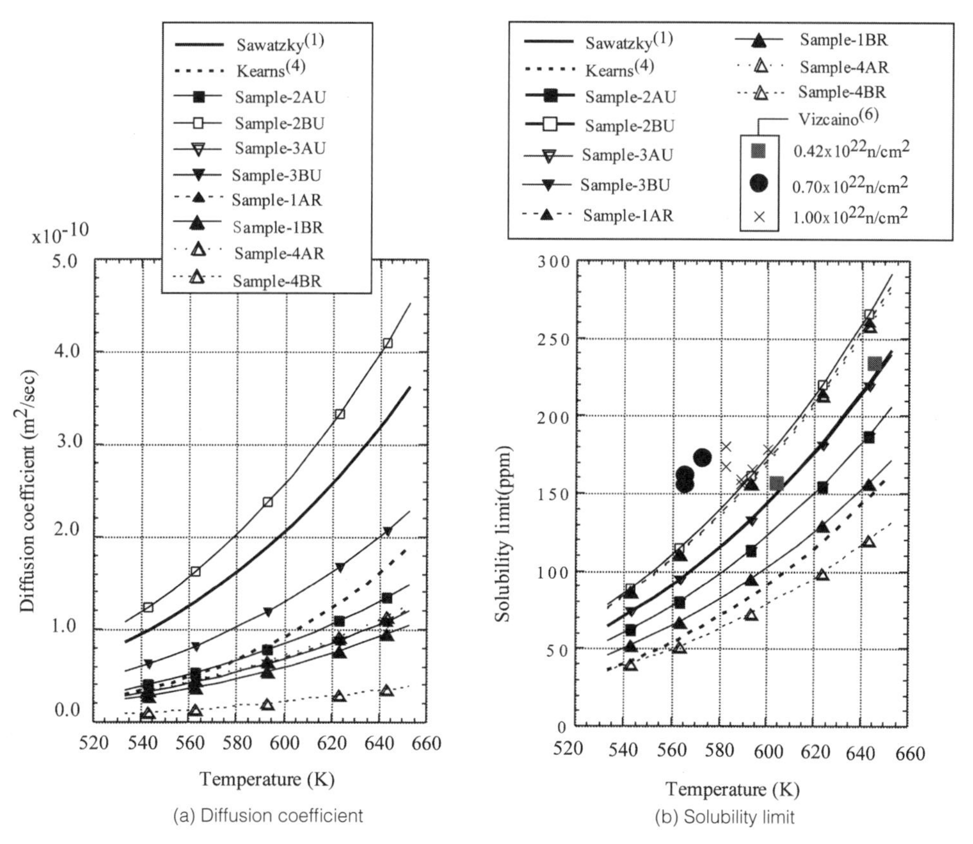

(a) Diffusion coefficient (b) Solubility limit

Fig. 6.3-3 Diffusion coefficient and the solubility limit of Zircaloy-4 cladding

cladding. Additionally, the difference in the irradiation fluence on the samples less affected the diffusion coefficients. Thus, the effect of radiation damage on the diffusion coefficient may be saturated as well as heat of transport when a certain amount of radiation damage accumulates in the cladding.

It was reported that the solubility limit of irradiated cladding tended larger than that of unirradiated cladding and hypothesis was proposed that in the irradiated cladding, the defects produced by radiation damage can trap hydrogen[5),6)]. The report described that the solubility limit of hydrogen in the cladding material was measured by differential scanning calorimetry (DSC) and the temperature of the solubility limit rose by degree with repeat of the measurement and consequently, the temperature of the solubility limit of the irradiated cladding was close to that of the unirradiated one.

In this hydrogen redistribution experiment, the results showed that there was not significantly difference in the solubility limit between the irradiated cladding and unirrdadiated one.

(5) Calculated hydrogen axial distribution after a forty-year dry storage

Hydrogen redistribution in the PWR-UO_2 rod after forty years of dry storage was estimated by one-dimensional diffusion calculations, using the measured heat of transport. Fig. 6.3-4(a) shows the initial temperature [7),8)] for spent fuels obtained from real dry stored metal cask and the literature data of hydrogen axial profile[9)] used in the calculation as an initial profile. Fig. 6.3-4 (b) shows the calculated hydrogen profile in the fuel rod after forty years of dry storage. Although under rather high temperature condition (320 °C), the axial hydrogen migration would not be significant after forty years of dry storage. As a result, it is concluded that the effect of hydrogen migration on cladding integrity is small.

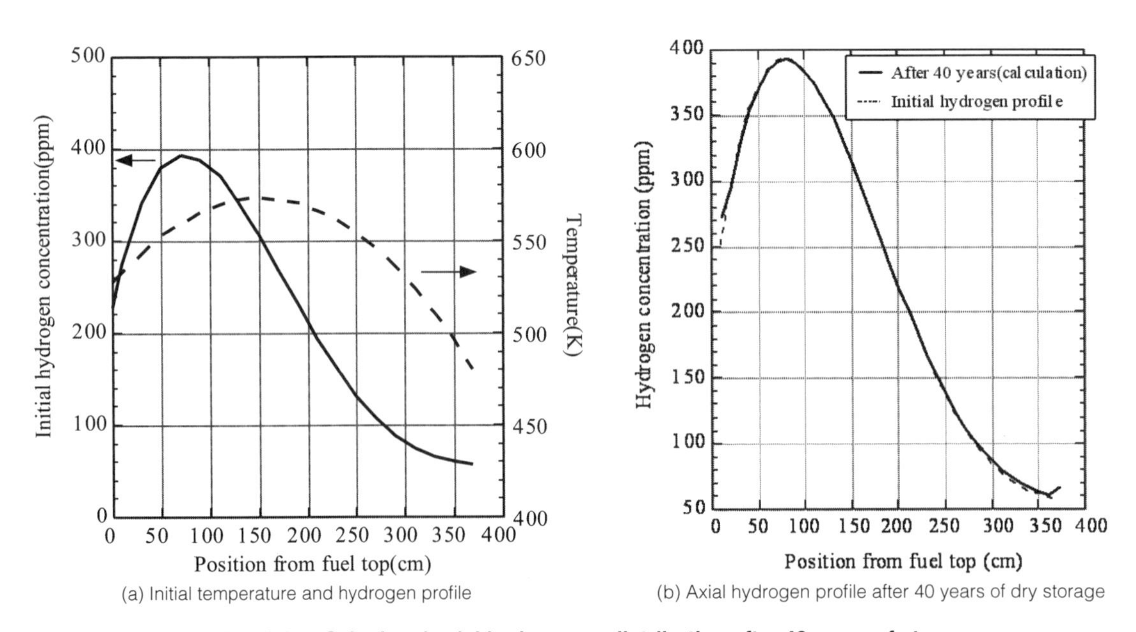

(a) Initial temperature and hydrogen profile

(b) Axial hydrogen profile after 40 years of dry storage

Fig. 6.3-4 Calculated axial hydrogen redistribution after 40 years of storage

References

1) A.Sawatzky, "Hydrogen in Zircaloy-2 : Its Distribuion and Heat of Transport", J.of Nucl. Mat. 2, No.4(1960)321-328.
2) H.S.Hong, S.J.Kim, K.S.Lee, "Thermotransport of Hydrogen in Zircaloy-4 and modified Zircaloy-4", J.of Nucl. Mat. 257(1998)15-20.
3) K.Forsberg, A.R.Massih, "Redistribution of Hydrogen in Zircaloy", J.of Nucl. Mat. 172(1990)130-134
4) J.J.Kears, "Thermal Solubility and Partitioning of Hydrogen in the Alpha Phase of Zirconium, Aircaloy-2 and Zircaloy-4", J.of Nucl. Mat. 22(1967)292-303
5) A.Mcminn, E.C.Darby, J.S.Schofield, "The Terminal Solid Solubility of Hydrogen in Zirconium Alloys", Zirconium in the Nuclear Industry : Twelfth International Symposium, Toronto, 15-18 June 1998, ASTM STP1354 American Society for Testing and Materials, West Conshohocken, PA, 2000, 173-195.
6) P.Vizcaino, A.D.Banchik, J.P.Abriata, "Solubility of hydrogen in Zircaloy-4:irradiation induced increase and thermal recovery", J.of Nucl.Mat.,No304(2002)96-106.
7) M.A.Mckinnon, J.M.Creer, C.L.Wheeler, J.E.Tanner, E.R.Gilbert, R.L.Goodman,"The MC-10 PWR Spent-Fuel Storage Cask : Testing and Analysis", EPRI Report, NP-5268, (1987)
8) D.Dziadosz, E.V.Moore, "The Castor-V/21 PWR Spent-Fuel Storage Cask : Testing and Analyses", EPRI Report, NP-4887, (1986)
9) K. Hashizume, Y. Hatano, R. Seki, M. Sugisaki, "Redistribution of Hydrogen in Fuel Cladding and Its Influence on Mechanical Properties of Cladding under Dry Storage Conditions", Kyushu university engineering science report, Vol. 21, No. 3, 281-288, 1999.

6.3.3 Hydride reorientation in radial direction

Hydrogen absorbed in fuel cladding circumferentially precipitates as hydrides. These hydrides dissolve into the fuel cladding at the elevated temperatures during drying process, and precipitate again in the fuel cladding as hydrides as the temperature drops after drying process or during dry storage. However, the hydrides have the tendency to change their direction (reorientation) vertical to the tensile direction, when large tensile stress is applied (Fig. 6.3-5). Because the inner pressure of fuel rods is higher than the outer pressure during storage, hoop stress is generated in the fuel cladding. Therefore, some hydrides change their direction radially, which leads to degradation of mechanical properties of the fuel cladding.

Hydride reorientation test and ring compression test were conducted on Zry-2 fuel cladding of 40 GWd/t BWR fuel (without Zr-liner) and 50 GWd/t and 55 GWd/t BWR fuel (with Zr-liner) and Zry-4 fuel cladding of 39 GWd/t and 48 GWd/t PWR fuel that were irradiated in LWRs in Japan. As a result, the conditions for the fuel cladding temperature and hoop stress that will not degrade mechanical properties of the fuel cladding (mechanical properties do not degrade against the post irradiation condition) were identified for each fuel type (Table 6.3-3). Thus, degradation of mechanical properties can be prevented by controlling the fuel cladding temperatures and hoop stress.

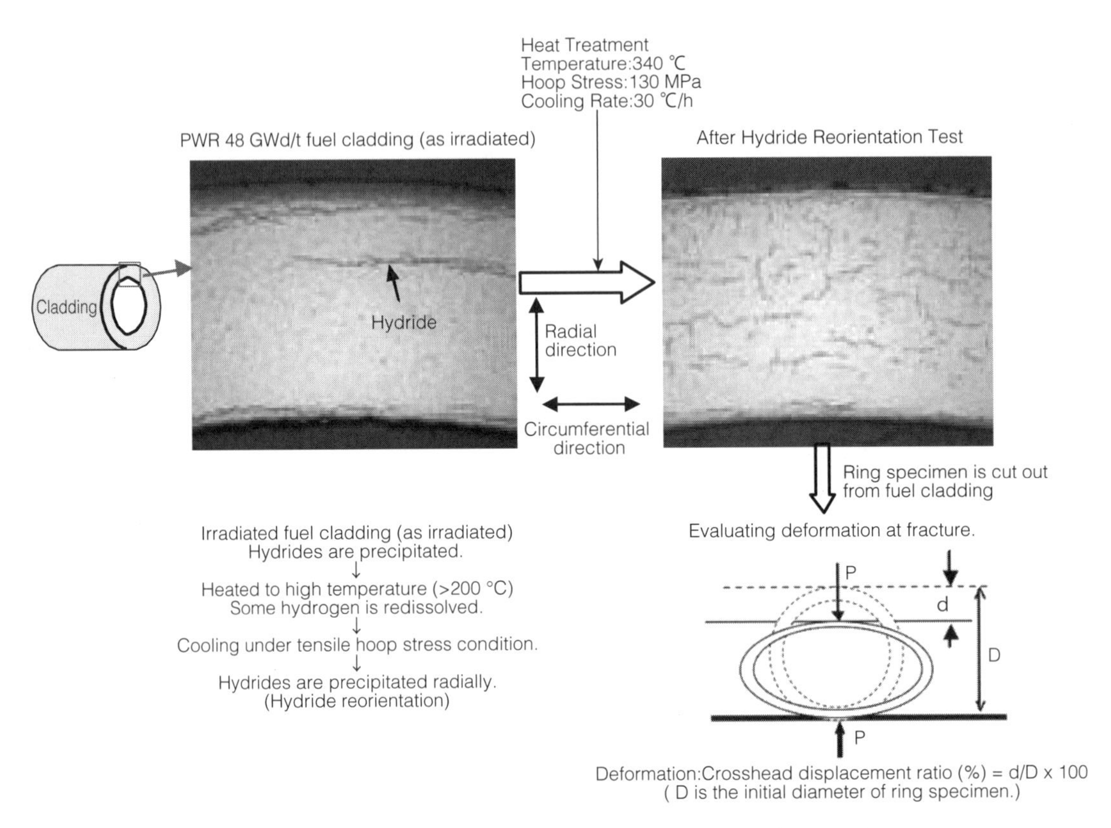

Fig. 6.3-5 Hydride reorientation test and ring compression test[1]

Table 6.3-3 Conditions that will not degrade mechanical properties of the fuel cladding test[1]

Type of fuel cladding		Conditions	
		Temperature	Hoop Stress
BWR	40 GWd/t Type (without Zr-liner)	200 °C or under	70 MPa or under
	50 GWd/t Type (with Zr-liner)	300 °C or under	70 MPa or under
	55 GWd/t Type (with Zr-liner)	300 °C or under	70 MPa or under
PWR	39 GWd/t Type	275 °C or under	100 MPa or under
	48 GWd/t Type	275 °C or under	100 MPa or under

Reference

1) Nuclear and Industrial Safety Subcommittee of the Advisory Committee for Natural Resources and Energy, Nuclear Fuel Cycle Safety Subcommittee, Interim Storage Working Group and Transport Working Group: "Long-term Integrity of the Dry Metallic Casks and their Contents in the Spent Fuel Interim Storage Facilities", (2009.6.25)

6.3.4 Hydride embrittlement and irradiation-hardening recovery

Amount of hydrogen absorbed in fuel claddings is about 400 ppm at maximum during irradiation in BWR except a special case, as shown in Fig. 6.3-6. With this amount of hydrogen absorption, the tensile strength of the fuel claddings will be maintained as shown in Fig. 6.3-7.

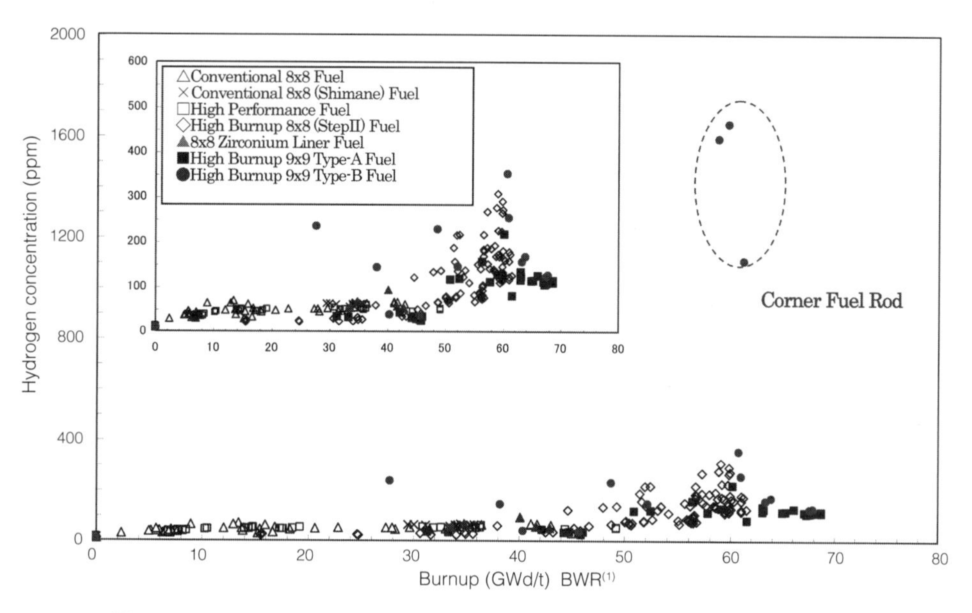

Fig. 6.3-6 Burnup dependence of hydrogen concentration of fuel cladding

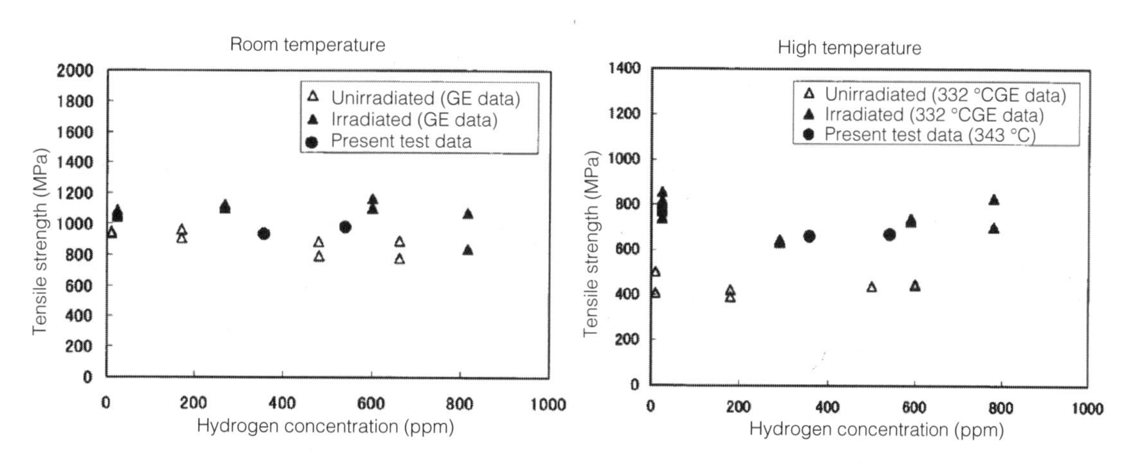

Fig. 6.3-7 Hydrogen concentration and tensile strength of Zircaloy 2 (BWR)[1)]

Amount of hydrogen absorption in standard Sn Zry-4 and low Sn Zry-4 with burnup of 39 and 48 MWd/kgHM in PWR is about 400 ppm at maximum as in BWR. The strength of the fuel claddings will be also maintained as for BWR.

The strength of fuel cladding increases with accumulation of radiation damage such as defect when irradiated in the reactor core, but its ductility decreases. If it is held at a high temperature of more than about 300 °C for a long period of time, irradiation hardening gradually recovers. Results of the irradiation hardening recovery test, using Zry-2 cladding of 50 MWd/kgHM BWR and Zry-4 cladding of 48 MWd/kgHM PWR irradiated in the LWRs in Japan (Fig. 6.3-8) showed that the possibility of recovery of irradiation hardening is small at 270 °C and 300 °C, respectively. It is necessary to consider these results appropriately in the fuel cladding integrity evaluation.

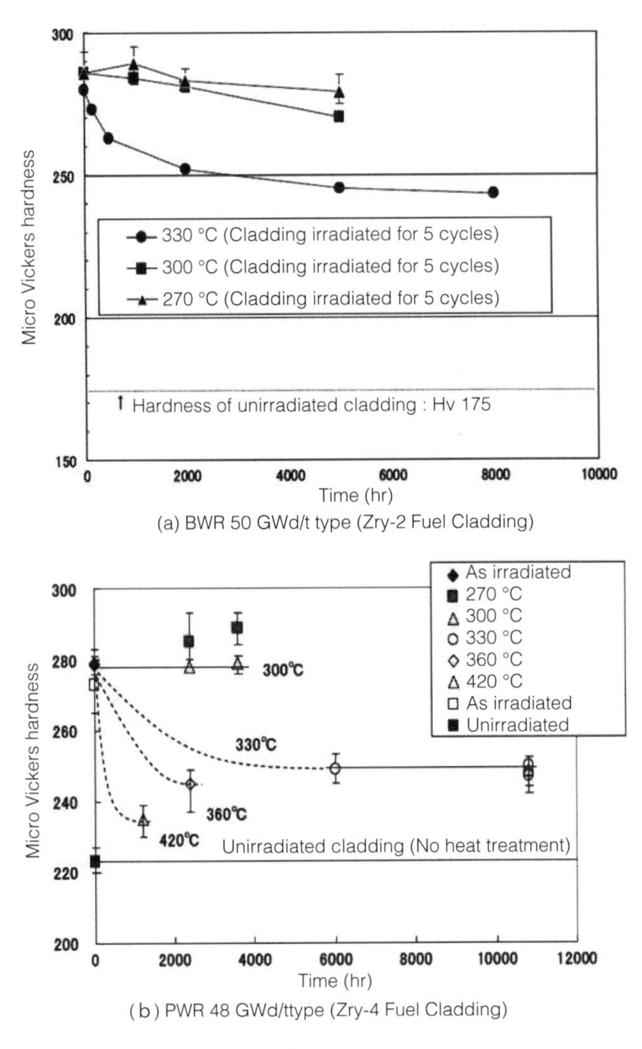

(a) BWR 50 GWd/t type (Zry-2 Fuel Cladding)

(b) PWR 48 GWd/ttype (Zry-4 Fuel Cladding)

Fig. 6.3-8 Irradiation-hardening recovery[1)]

Reference

1) Atomic Energy Society of Japan: "Standard for Safety Design and Inspection of Metal Casks for Spent Fuel Interim Storage Facility", AESJ-SCF002: 2010 (in Japanese).

6.4 Spent fuel integrity during postulated accident condition

Several kinds of accident will be considered during dry storage depending on the types of the storage systems. Here, we postulate the accident as a case that the storage building collapses over the casks and the heat removal system does not function. If we postulate a cask is in an adiabatic condition, as an extreme case, the spent fuel temperature will increase gradually with time. Under this postulated condition, creep deformation/rupture is the most probable factor leading to failure of the spent fuel cladding when the cask itself maintains the integrity. If the sealing of the cask is lost, then oxidization of the fuel cladding also should be considered carefully. Hence, the methods of integrity evaluation considering creep and oxidization of fuel cladding will be discussed in the following section.

6.4.1 Temperature limit determined by creep behavior

Einziger, et al.[1] conducted high temperature creep test on the actual fuel rods irradiated in a PWR during 2 to 3 cycles. The test fuel rods were WH 15x15 type fuel rods irradiated in Turky Point Unit-3 until burn-up of 27000 to 31000 MWd/t, and then cooled for around 3.6years. Although three test temperatures were set with estimated rupture times of 167 to 860 h, based on the creep rupture data of irradiated Zircaloy, no rods ruptured until 4662, 7680 and 1740 h at the temperatures of 755, 783 and 844 K, respectively. They pointed out the cause of no failure of the fuel rods as the decrease in the internal pressure(hence, hoop stress) due to increase in the free volume of the fuel rods in response to the progress of creep deformation. They also measured hardness of the test rods before and after the test, and showed that irradiation hardening had been fully recovered during the high temperature holding. Porsch, et al. of ISPRA also made creep test using actual spent fuel rods[2]. They confirmed that creep rupture of the fuel rods did not occur even though the test rod subjected the step like temperature history of 673, 703 and 723 K with the holding times of 7, 5 and 4.5 months, respectively. These test results suggest us that the spent fuel rod does not fail even at the temperature of around 873 K in the postulated accident condition since the hoop stress of the fuel rods decrease with the increase of the free volume with creep deformation.

CRIEPI conducted a series of creep tests shown in below to develop a method to determine the temperature limit based on the creep deformation of fuel rods.

(1) Development of a creep equation of un-irradiated Zircaloy-4 applicable to the temperature ranges of 727 to 855 K[3].

(2) Evaluation of effect of the free volume of a fuel rod on the creep deformation by the creep tests using simulated spent fuel rod specimens[4].

(3) Development and proposal of a method to evaluate the temperature limit in the postulated accident condition[5].

In the course of the creep test, it was shown that creep rupture did not occur in the simulated fuel rods having the [free volume] to [pellet volume] ratios of 0.1 to 0.3 with the internal pressures of 40 to 80 bar even after the holding of 2000 h at temperatures of 774 to 868 K. The effect of irradiation hardening was not considered in the test since it was expected that the irradiation hardening would have been fully recovered in the temperature region as Einziger, et al.[1] pointed out.

Fig. 6.4-1 shows the method to determine the temperature limit considering the creep deformation of spent fuel rods[5]. In the evaluation method, the storage temperature, the internal pressure of the fuel rod and the temperature increase rate during the accident are set as the initial condition. The creep strain of spent fuel rods is calculated by the creep equation[3] considering the internal pressure increase with the increased temperature and then the accumulated creep strain is compared with the strain criterion which is confirmed not to lead to the cladding failure. If the accumulated creep strain does not reach to the strain criterion, the internal pressure is calculated considering the increased free volume and the increased temperature, and then the accumulated creep strain is calculated and compared with the strain criterion again. The process is repeated until the accumulated creep strain reaches to the

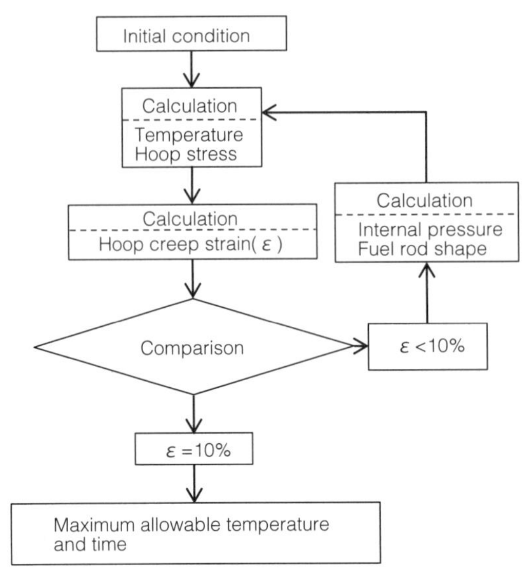

Fig. 6.4-1 Method to evaluate the temperature limit of fuel cladding in accident condition[5)]

strain criterion and the temperature limit is obtained. In the following example, the temperature limit is calculated under the assumed initial condition of the spent fuel; the temperature in the normal storage condition: 673 K, the internal pressure of fuel rods: 55 bar at 298 K, the [free volume] to [pellet volume] ratio: 0.05. If we assume the strain criterion of 10 %, and the linear temperature increase rates of 0.1 K/h and 10 K/h, the temperature limits are 821 K and 928 K, respectively. Thus, the temperature limit becomes higher for the larger temperature increased rate. However, the time until reaching to the temperature limit becomes shorter for the larger temperature increase rate, as 1480 h and 25 h for 0.1 K/h and 10 K/h, respectively, in this particular case. These times are important factors showing the allowable time permitted for recovery activity from the accident.

Here, we described a method for evaluating the temperature limit during a postulated accident condition. Although a method to use creep rupture as the criterion also can be considered, a method using a strain limit will be more realistic when we consider the previous test results on the actual spent fuel rods. The above mentioned method calculates the temperature limit of spent fuel during a postulated accident condition under simple assumptions of fuel condition. The following efforts will be necessary to advance and practically use the developed method to evaluate the temperature limit of spent fuel during the accident condition. (1) To set a realistic strain criterion based on the various test results on the actual spent fuel reds, (2) consideration of the temperature profile along the axial direction of fuel rods, (3) usage of a realistic temperature increase curve, and (4) evaluation of the creep behavior of fuel rods under oxidizing condition by assuming the loss of sealing of the cask.

References

1) R.E.Einziger, S.D.Atkin, et al., High Temperature Postirradiation Materials Performance of Spent Pressurized Water Reactor Fuel Rods under Dry Storage Conditions, Nuclear Technology, Vol.57, p.65, 1982.

2) G.Porsch, J.Fleisch, B.Heits, Accelerated High-temperature Tests with Spent PWR and BWR Fuel Rods under Dry Storage Conditions, Nuclear Technology, Vol.74, p.287, 1986.
3) M.Mayuzumi, T.Onchi, Creep Deformation and Rupture Properties of Unirradiated Zircaloy-4 Nuclear Fuel Cladding Tube at Temperatures of 727K and 857K, J. of Nuclear Materials, 135-142, 1990.
4) M.Mayuzumi, Effect of Free to Pellet Volume Ratio on Creep Deformation of Nuclear Fuel Rods, CRIEPI Report T90061, CRIEPI, 1991.
5) M.Mayuzumi, T.Onchi, A Method to Evaluate the Maximum Allowable Temperature of Spent Fuel in Dry Storage Condition During a Postulated Accident Condition, Nuclear Technology, Vol.93, p.382, 1991.

6.4.2 Oxidation of fuel and cladding by air

In dry storage, spent fuels are stored in the casks under inert atmosphere such as helium gas. Thus, oxidation reaction of fuel cladding with oxygen will not occur in the cask. If 10 wt.% of the cover gas in the cask were residual water and all oxygen of the residual water reacts with the cladding materials during storage, oxide layer thickness formed on the cladding surface with oxidation reaction would be less than 1 μm and the oxide layer would not affect integrity of the fuel cladding[1].

On the other hand, if containment of the cask or canister were lost and the air flew in and if there were through-wall defects such as pinhole or hairline cracking on the cladding that have been loaded into the central region of the cask, the air would come in contact with the fuel inside the rod. Then, oxidation reaction of fuel may proceed as $UO_2 \rightarrow U_4O_9 \rightarrow U_3O_8$. The transition from UO_2 to U_3O_8 results in a volume expansion of greater than 30 %[2]. Small initial defect may expand to result in failure of cladding[3]. To investigate the effect of burnup and temperature on oxidation behavior of a pellet, oxidation experiments were carried out on the irradiated UO_2 fuels.

(1) Pellet oxidation experiment

The UO_2 fuels with burnups of 50 MWd/kgHM and 65 MWd/kgHM were used in the pellet oxidation experiment. In the experiments, fuel samples were exposed to dry air at temperatures of 300 and 350 °C. The weight gain of the fuel sample was often measured to investigate evolution of oxidation[3]. Table 6.4-1 shows the relation between weight gain and oxidation state of the UO_2 fuel.

Table 6.4-1 Relation between weight gain and oxidation state

Oxide	Weight gain (wt.%)	Oxygen/Uranium ratio
UO_2	0	2.0
U_4O_9	1.5	2.25
U_3O_8	4.0	2.67

(2) Results

Fig. 6.4-2 shows the weight gain curves at 300 °C and 350 °C. A weight gain of 4 % corresponds to 100 % formation of U_3O_8 from UO_2.

The oxidation reaction from UO_2 to U_4O_9 will start by oxygen diffusion from the grain boundaries into grains and U_3O_8 will nucleate from U_4O_9 phase. At 300 °C, there was no significant burnup difference in the oxidation behavior between 50 and 65 MWd/kgHM. On the other hand, at 350 °C, the oxidation rate of the sample with a burnup of 65 MWd/kgHM became higher than that with a burnup of 50 MWd/kgHM beyond the weight gain of 2.4 %. It has been reported that oxidation rate decreases with higher burnup fuel because of accumulation of fission products in the pellet. However, from the experiments, it is speculated that change in the microstructure of the pellet also affect the oxidation behavior.

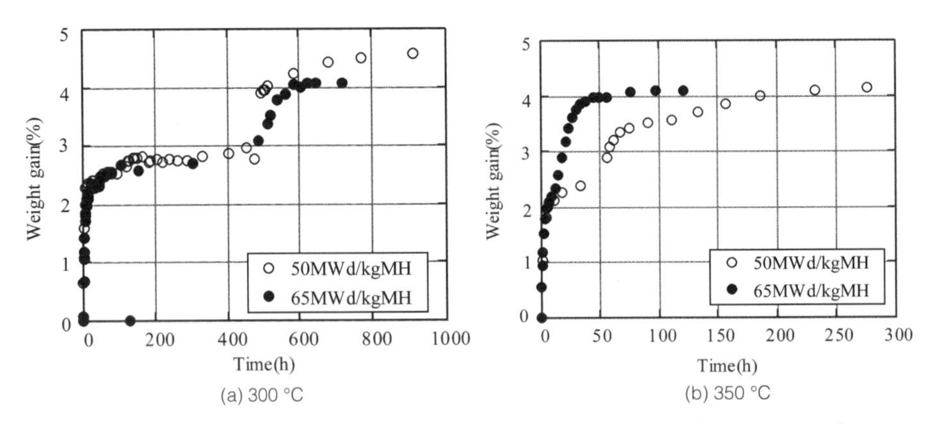

Fig. 6.4-2 Oxidation behavior dependence on burnup and temperature[4)]

References

1) Nuclear and Industrial Safety Subcommittee of the Advisory Committee for Natural Resources and Energy, Nuclear Fuel Cycle Safety Subcommittee, Interim Storage Working Group and Transport Working Group: "Long-term Integrity of the Dry Metallic Casks and their Contents in the Spent Fuel Interim Storage Facilities", (2009.6.25).
2) TU Annual Report 1995 (EUR 16368 EN).
3) A. Sasahara, T. Matsumura, D. Papaioannou: "Examinations of Spent Fuels during Interim Storage (5) - Oxidation Behavior of Irradiated UO2 Fuel in Air-", H3, Autumn meeting of the Atomic Energy Society of Jaoan, 2003.
4) Spent Fuel Performance Assessment and Research: Final Report of a Coordinated Research Project (SPAR II), IAEA-TECDOC-1680.

6.5 Inspection method for ageing

6.5.1 Non-destructive analysis of spent fuel in canister

When carrying out dry storage of spent fuels with a storage canister, it is difficult to observe condition of spent fuels visually. As fission products (FP) are accumulated in a spent fuel, rare gas (FP gas) among FP such as Kr and Xe, is expected to spread over the storage canister, when the spent fuel is damaged.

A part of FP gas accumulated in a fuel pellet diffuses the inside of crystal of the fuel pellet (UO_2), and reaches the grain boundary of the crystal. A part of FP gas trapped in the grain boundary is emitted to inner side free space of the spent fuel cladding by a crack or deformation of the fuel pellet. It is also known that material structure change of a fuel pellet with burn-up promote gas diffusion and emission in the fuel pellet.

In the post irradiation examination (PIE) of a spent fuel, FP gas release is actually measured by digging a hole in a fuel cladding and collecting FP gas.

It is shown that about several percent of FP gas generated in the fuel pellet moves from a fuel pellet, and is accumulated in inner side free space of the fuel cladding in such a PIE. When a spent fuel is damaged, FP gas accumulated in the fuel cladding spread over a storage container.

The nuclide with one year or more of half-life is only ^{85}Kr (half-life: 10.7 years) among FP gas shown in Table 6.5-1. If the gamma ray emitted from ^{85}Kr spread over the storage canister is detectable from the outside of the storage canister, it will become possible to confirm the spent fuel soundness without breaking the sealing performance of the storage canister, when damage of the spent fuel is suspected.

Table 6.5-1 Evaluation of the main FP gas amount in one ton of spent fuel

nuclides	half-life	After Burn-up of 47 GWd/t (T Bq)	After cooling of 60 days (T Bq)
^{85}Kr	10.7 years	48	47.5
^{85m}Kr	4.5 hours	769	0
^{88}Kr	2.8 hours	2020	0
^{133}Xe	5.2 days	6940	3.07
^{133}Xe	9.1 hours	1400	0

Note) As FP gas amount changes with a fuel design and condition of burn-up, etc., above FP gas amount is an example of evaluation

The possibility of detection of the gamma ray emitted from ^{85}Kr of a storage canister from the outside of the storage canister is investigated with a reference design of canister structure (loaded capacity: 21 PWR spent fuels) of a concrete cask (Fig. 6.5-1) [1)].

In order to collect more gamma ray of ^{85}Kr in free volume of the canister and reduce the gamma ray from the spent fuel, un-penetrated gamma ray detection hole is set in the lid part of the canister. Most of gamma rays from the spent fuel are shielded in lid part of the canister. The un-penetrated gamma ray detection hole keeps the sealing performance of the storage canister.

Since the energy of gamma rays of ^{85}Kr is 514 keV, and closes to energy of the gamma ray of positron annihilation (511 keV), discrimination with this positron annihilation gamma ray becomes important. Discrimination with the positron annihilation gamma ray which generated in the canister bottom plate should be considered in the detection design of the gamma ray of ^{85}Kr. For this reason, discrimination limit curve of the gamma ray of ^{85}Kr is derived experimentally with a small mock-up canister.

Fig. 6.5-2 shows an example of gamma ray measurement with a high resolution germanium semiconductor detector (FWHM: 1.84 keV for 1333 keV). Germanium semiconductor detector is a radiation detector using a germanium semiconductor, and is excellent in energy resolution. FWHM (full width at half maximum) means the energy resolution of the gamma ray. 1.84 keV for 1333 keV is a standard value as a germanium semiconductor detector with high resolution.

Evaluation error of ^{85}Kr amount was acquired from comparison between set up values and estimated value from

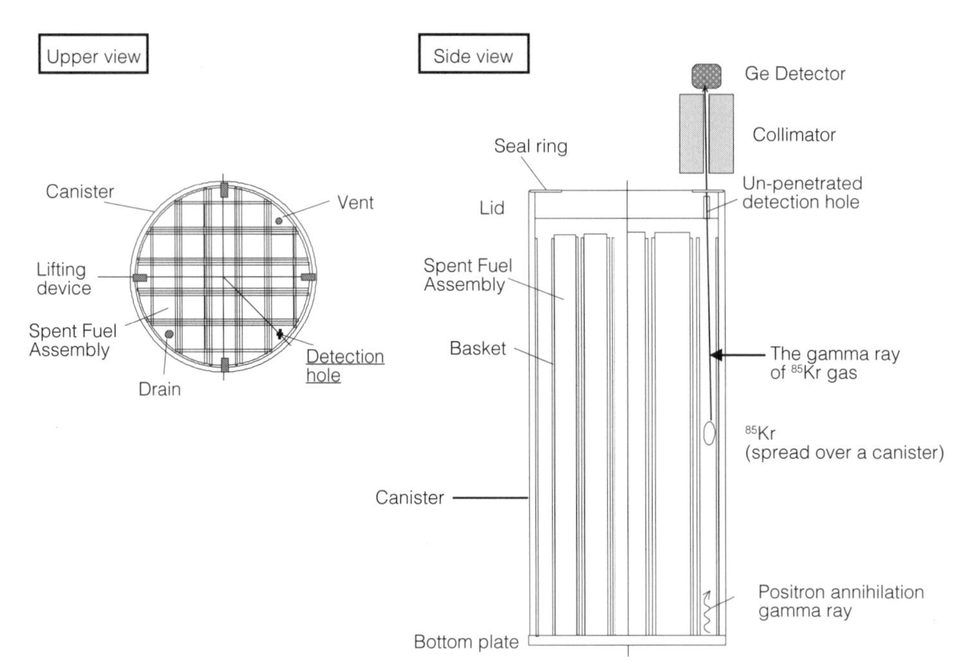

Fig. 6.5-1 Conceptual design of spent fuel monitoring

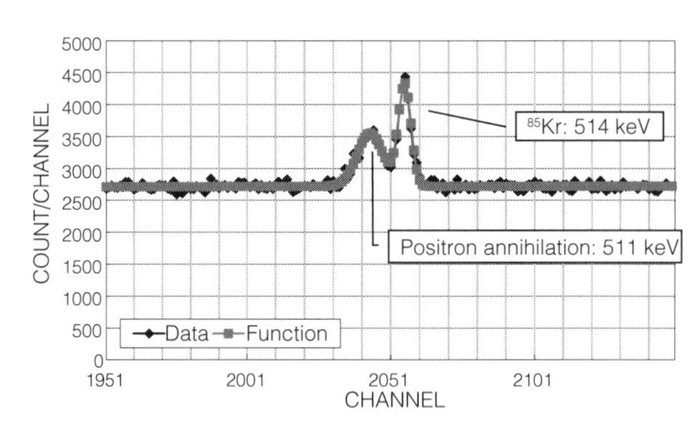

Fig. 6.5-2 Example of gamma ray measurement of ^{85}Kr (514 keV) in experiment with small mock-up canister

experiments in various conditions. Using parameters of "peak count of 514 keV (^{85}Kr)", "peak ratio (gamma ray ratio of 511 keV / 514 keV)", "base ratio (Compton scattering / ^{85}Kr gamma ray at 514 keV)", and consideration the error of experiments data and statistics calculation, the evaluation error function was derived by several experiments as;

$$D_t = \sqrt{\frac{(1+2B)}{S} + (P \cdot g \cdot u)^2} \quad ..... (1)$$

D_t : evaluation error (statistical error)

S : peak count of 514 keV (^{85}Kr)

B : base ratio (Compton scattering / ^{85}Kr gamma ray at 514 keV)

P : peak ratio (gamma ray ratio of 511 keV / 514 keV)

u : correction factor of peak ratio (0.0116)

g : relative error of correction factor of peak ratio (25 %)

This evaluation error function can be used as the Discrimination limit curve. Fig. 6.5-3 shows the result of calculating a 514 keV peak discrete value required to make one sigma error of 20 % using the above-mentioned evaluation error function.

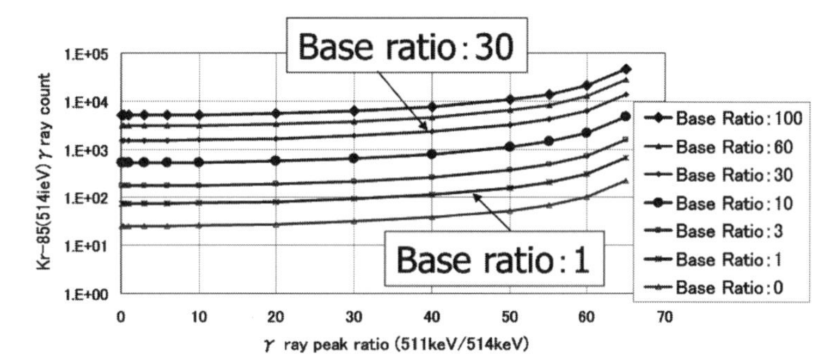

Fig. 6.5-3 Discrimination limit curve of ^{85}Kr gamma ray

The detection performance of gamma rays was examined analytically using the Discrimination limit curve. Gamma rays scattered around the un-penetrated hole of the lid and annihilation gamma rays generated in the canister bottom plate were dominant as noise gamma rays which interfere with the detection of signal gamma rays emitted from ^{85}Kr gas (Fig. 6.5 -1).

When 10 % (7.09×10^{10} Bq) of ^{85}Kr gas of one fuel rod is released, a 514 keV peak measurement rate is estimated to be 2.98×10^{-2} cps (counts per second) in consideration of the volume of the storage container, the diameter of un-penetrated gamma ray detection hole, etc. Peak ratio (P: gamma ray ratio of 511 keV / 514 keV) is calculated as 25.2 with the value of the positron annihilation gamma ray from a canister bottom plate. And by Base ratio (B) of 0.001 with scattered gamma ray from spent fuels, 514 keV peak counts required to be 29 to keep less than 20 % (one sigma) of measurement error. This value fulfills the Discrimination limit curve with 970 seconds of measurement time.

On the other hand, in case of 1 % (7.09×10^{9} Bq) of release of ^{85}Kr gas of one fuel rod, Peak ratio with a positron annihilation gamma ray increases to be 252, and make it impossible to get 20 % (one sigma) of measurement error with finite measurement time due to strong positron annihilation gamma ray. For this reason, this monitoring method is inapplicable for small amount release of ^{85}Kr gas without additional technique of reducing positron annihilation gamma ray from the canister bottom plate.

From these evaluations, ^{85}Kr gas release of about 10^{11} Bq could be detectable under the noise gamma rays by using the detection system with a collimator, which is about 10 % of ^{85}Kr inventory in a fuel rod. It turned out that detection of damage of fuel rods is possible when there is detectable release of ^{85}Kr gas (Table 6.5-2). Since the major part of ^{85}Kr gas accumulated in a fuel is held in a fuel pellet, and several percent of ^{85}Kr gas accumulated in a fuel rod will be released into the canister when the fuel cladding is damaged as mentioned above, it is thought that damage of several fuel rods causes detectable release of ^{85}Kr gas.

Table 6.5-2 Estimation of detectable ^{85}Kr gas release

^{85}Kr gas Release (Bq)	7.09×10^{9}	7.09×10^{10}	7.09×10^{11}
Ratio (%) of ^{85}Kr in a fuel rod	1 %	10 %	100 %
514 keV Peak (cps)	2.98×10^{-3}	2.98×10^{-2}	2.98×10^{-1}
Measuring time (sec.) required to keep less than 20 % (one sigma) of measurement error.	×	970	97

As soundness of spent fuels (no deformation, no defect etc.) is checked at the loading time into the storage canister, it is judged that the soundness of spent fuel under storage is kept. The soundness of the spent fuel by dry storage was actually checked for about 20 years [2)]. However, when the storage container falls or damaged under storage, and the soundness of spent fuels needs to be checked, it may be thought that the necessity of pre-evaluation of the soundness of the spent fuel comes out without braking the sealing performance of the storage canister, before transporting the storage container to an inspection facility. This monitoring methodology enables to inspect whether the spent fuel was damaged bulkily with keeping the sealing performance of the storage canister.

When the spent fuels store for long period, as ^{85}Kr gas decay with half-life of 10.7 years, improvement of measurement is require, which raise measurement sensitivity, such as increasing the diameter and the number of the detection hole in the lid.

References

1) T. Matsumura, A. Sasahara, Y. Nauchi and T. Saegusa "Development of monitoring technique for the confirmation of spent fuel integrity during storage", Nuclear Engineering and Design 238, 1260-1263 (2008),

2) A. Sasahara, T. Saegusa: Evaluation of long-term integrity of spent fuels stored in Idaho National Laboratory -Cover gas sampling inside metal cask and its analysis-, CRIEPI Report L10017 (2010) [in Japanese]

CHAPTER 7 UTILIZATION TECHNOLOGIES OF WASTE HEAT AND RADIATION

Heat and radiation are emitted from stored spent fuel over long periods of time. Until now, there was not such a concept that these kinds of energy are utilized at spent fuel storage facilities. We examined a system of utilizing the heat and radiation of spent fuel in terms of effective utilization of energy.

A dry storage system was chosen as a storage system to be examined on the assumption of large-capacity storage away from reactors.

7.1 Waste heat utilization technology

When waste heat is recovered at dry storage facilities for spent fuel, a method to recover thermal energy from air which has cooled containers storing spent fuel is safe and efficient. Waste heat recovery from a low grade heat source was not performed due to high costs in past days. However, the waste heat recovery has increased because the efficiency of waste heat recovery was improved due to technical improvement and the costs decreased. If this technology is applied, the waste heat recovery from spent fuel storage facilities is also possible and the effective utilization of energy can be achieved[1].

(1) Facility concept

There are several storage systems for dry storage facilities, such as cask systems, vault systems, etc., and the vault storage system was employed on the examination. In the vault storage system, the thermal energy emitted from spent fuel concentrates at stack parts and is removed by cooling air, so that the waste heat is easy to recover. Specifications of spent fuel and a spent fuel storage facility for the examination are shown in Table 7.1. One module is constituted by 77 storage pipes and a heat removal function is ensured at each module. Eight modules constitute one storage building. Each module is separated by walls and cooling air does not flow between the modules.

Table 7.1 Specifications of spent fuel and storage facility

Item	Specifications
Storage system, Capacity	Closed cycle vault storage system, PWR fuel (17×17), 1130 MTU
Fuel assemblies stored in 1 canister	4 (PWR fuel)(Canister : Outer diameter 698×Height 5173×Thickness 14 mm)
Number of canisters in 1 storage pipe	1 (Storage pipe: Outer diameter 750×Height 7750×Thickness 15 mm)
Arrangement of storage pipes in 1 module, Number of modules	Staggered arrangement of 9 rows×8,9 columns (Total: 77)
Fuel	Average burnup: 40000 MWD/tU, Cooling: 5 years, Calorific value: 2.158 kW/tU
Facility scale	107 mW x 27 mD x 35 mH (Above-ground part: 25 mH)

The storage facility has the following system: a heat exchanger for waste heat recovery is installed to a stack part; water circulating in the heat exchanger is warmed; and the warmed water is utilized, or warmed further by a heat pump so as to be utilized.

As requirements for a heat recovery device, it is required in light of heat recovery from the spent fuel storage facility that its mechanism is simple and the maintenance is required as little as possible. As the most important requirement, safety as the spent fuel storage facility can be secured even in the case of maintenance and failure of the heat recovery device (fail safe). In a concept shown in Fig. 7.1, heat removal operation is performed in such a closed

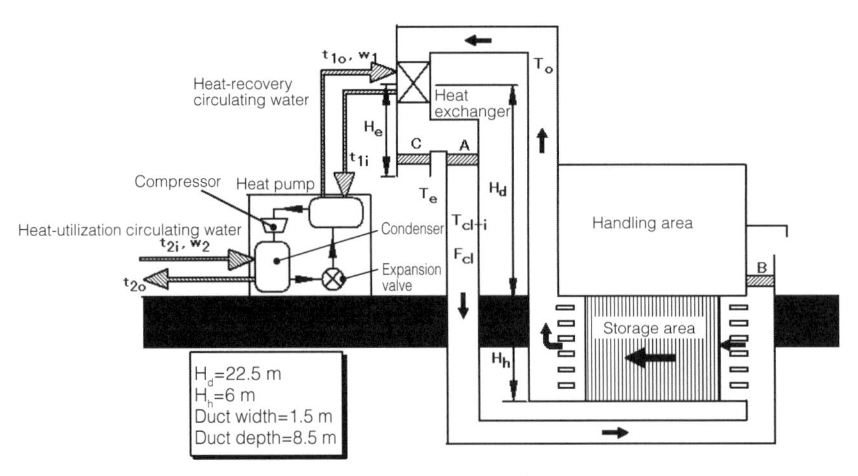

Fig. 7.1 Conceptual diagram of storage facility with waste heat recovery device

cycle that B and C are closed and A is opened at normal times, and is performed in such an open cycle that B and C are opened and A is closed at the time of failure of a heat exchanger, so that the safety is secured. An advantage of the closed cycle operation is that the heat recovery can be performed at a constant temperature without being affected by an ambient temperature.

(2) Heat removal safety evaluation

In the closed cycle method, the flow rate and outlet temperature of cooling air relating to heat removal performance is obtained by the following formulas: Formula (1) expressing draft power (ΔD), Formula(2) expressing a pressure loss (ΔP) of a storage module and a heat exchanger, and Formula (3) as a balance equation of a calorific value (Q) and recovered heat quantity. Here, the calculation was performed based on an assumption that the all calorific value of spent fuel is recovered. The heat removal function is ensured at each module, so that the calculation was performed on one module.

$$\Delta D = g\left(\rho(T_{cl-i}) - \rho(T_0)\right)H_d + g\left(\rho(T_{cl-i}) - \rho\left(\frac{T_{cl-i} + T_0}{2}\right)\right)H_h \quad \ldots\ldots (1)$$

$$\Delta P_{total} = \Sigma(\Delta P_n) + \Delta P_e \qquad \Delta P_n = \xi_n \rho \left(\frac{F_{cl}}{\rho A_n}\right)^2 / 2 \quad \ldots\ldots (2)$$

$$Q = C_p F_{cl}\left(T_0 - T_{cl-i}\right) \quad \ldots\ldots (3)$$

g: Gravitational acceleration   ρ: Air density   Tcl-i,0: Air temperature of inlet and outlet of storage part

Hd,h: Distance of heat exchanger and floor face of storage part from earth surface

ΔPn: Pressure loss of each part of cooling air channels   ΔPe: Pressure loss of heat exchanger

ξn: Pressure loss coefficient of each part of cooling air channels   Fcl: Cooling air flow rate

An: Cross section of each part of cooling air channels   Cp: Specific heat of air

The flow rate and outlet temperature of cooling air flowing in a storage facility were obtained by using the above formulas (1) to (3) while setting the temperature efficiency ratio of a heat exchanger to a specific value and inlet and outlet temperatures on the water side to 10 °C and 15 °C respectively. Furthermore, the temperature of the inside of storage pipes was obtained by a thermal conductivity analysis of a radial two-dimensional cross section based on the obtained cooling air temperature as a boundary condition, and the maximum temperature of fuel claddings was evaluated. The result is shown in Table 7.2. An example of temperature evaluation of a conventional vault storage facility which does not take into account the heat recovery is also shown in Table 7.2 for comparison.

The result clearly shows that the waste heat recovery system which does not impair heat removal safety essentially held by the storage facility can be established by performing such heat removal design that the heat emitted by spent fuel is sufficiently removed and the temperature of fuel claddings, etc. is below the restrictive temperature value.

(3) Future issues

The examination of utilization methods appropriate for locations of spent fuel storage facilities is required in order to realize the utilization of the waste heat of spent fuel. In simple cost evaluation in terms of hot water provision, the waste heat recovery system has more cost-competitive strength than an electric heater system which requires running cost when the cost of heat transportation from a storage facility to a place to use the waste heat is excluded; however, it is considerably more expensive than a boiler system. Purposes of utilizing the waste heat recovered from spent fuel storage facilities are the promotion of locating the facilities and regional development other than the effective utilization of energy, and the use application of the waste heat and location environments (e.g. seashores, mountainous areas, city suburbs, depopulated area, etc.) are also required to be closely examined.

Table 7.2 Result of heat removal evaluation

Item	Storage facility with waste heat recovery system	Closed cycle without recovering waste heat	Restrictive temperature
Maximum temperature of fuel claddings	299 °C	301 °C	380 °C
Inlet air temperature of storage part	26.2 °C	38 °C	
Outlet air temperature of storage part	39.2 °C	48.9 °C	65 °C
Flow rate of cooling air (1 module)	23.5 kg/s	27.9 kg/s	
Flow rate of water of 45 °C warmed by heat pump	150 kg/s (Total of 8 modules)	—	

7.2 Radiation utilization technology

Radiation affects a human body and the environment if it is mishandled; however, it can be utilized for various purposes if it is safely controlled and applied. In fact, radiation utilization is proceeding in various areas such as the manufacturing of high polymer material by radiation polymerization in the industrial world. If spent fuel can be utilized as a radiation source, stored spent fuel is effectively utilized [2)].

(1) Facility concept

There are cask, vault, and silo systems as dry storage systems. Here, the vault storage system was chosen to be an examination object because a space for conveying an object to be irradiated to the vicinity of radiation sources is structurally easy to secure and a relatively high dose rate can be obtained in the system. In this regard, however, irradiation is performed not in a storage pit but in an irradiation facility which is additionally built as an annex of the storage facility, in consideration of the maintenance of equipment and safety. The irradiation facility should meet the following requirements.

- The storage and irradiation facilities are built side by side and have such structure that canisters are easy to move by sharing a crane.
- When canisters in the irradiation facility decrease in source strength, they are replaced with the ones in the storage pit.
- When a conveying device, etc. in the irradiation facility is maintained, once canisters in the facility are moved to the storage pit and workers perform operations inside.
- The irradiation sources are fixed and irradiated objects are moved by the conveying device, so that irradiation is performed.
- The irradiation facility has shielding, criticality prevention, containment, and heat removal functions as well as the storage facility.
- Monitoring is performed on a constant basis for nuclear safeguard.

A conceptual diagram of an irradiation facility is shown in Fig. 7.2. We examined the facility on the assumption of implementing the general irradiation service of accepting anything received from companies as long as problems such as dimensional limits and activation are not caused. Specifications of spent fuel and a canister for the examination are shown in Table 7.3.

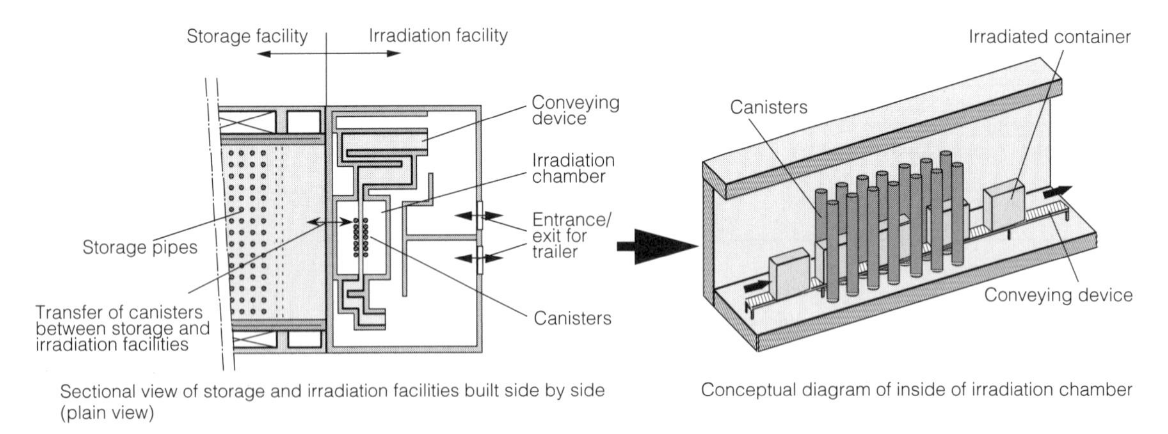

Fig. 7.2 Conceptual diagram of irradiation facility using spent fuel as irradiation source

Table 7.3 Specifications of spent fuel and canister

Item	Specifications
Fuel	PWR fuel(17×17), Average burnup: 40000MWD/tU, Cooling: 5 years, Calorific value: 2.158 kW/tU
Fuel assemblies stored in 1 canister	6 (Canister: Outer diameter 890×Height 5200×Thickness 15 mm)

(2) Radiation source evaluation

Most of γ-ray sources used as radiation sources at usual irradiation facilities are Co-60 (half-life: 5.27 years, γ-ray energy: 1.173 MeV and 1.332 MeV) and Cs-137 (half-life: 30.17 years, γ-ray energy: 0.662 MeV). In contrast, spent fuel includes various radionuclides, so that the γ-ray energy varies in distribution. Cs-137/Ba-137m series which are fission products have large contribution as γ-ray sources, and their ratios to the total increase with the lapse of time. The peak value of the γ-ray energy of Ba-137m T is 0.662 MeV. In radiation irradiation application, an irradiation effect is determined by the total absorbed dose of the object to be irradiated. A calculation result by the ORIGEN-2 is shown in Fig. 7.3. Contrary to general γ-ray sources, neutron rays of spontaneous fissionable nuclides such as Pu-238, 240, 242, and Cm-242, 244, 246 are emitted from spent fuel. Cm-244 accounts for 90 % or more of the total, and its neutron emission rate per assembly is approximately 2.27×10^{8}(n/s) in total.

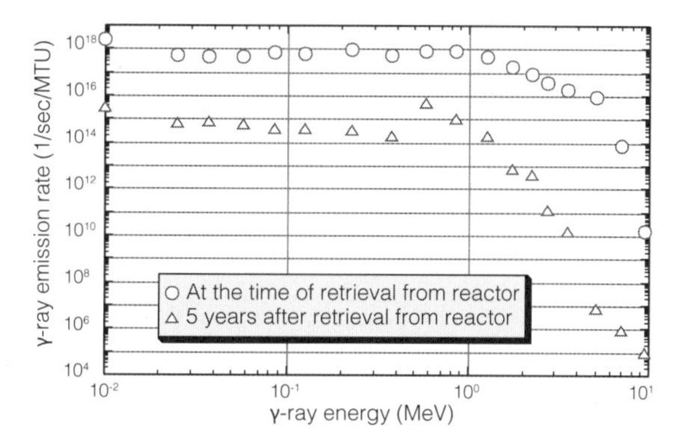

Fig. 7.3 Relation between energy and emission rate of γ ray from spent fuel

When canisters each of which stores six PWR spent fuel assemblies are set as radiation sources, they cannot achieve the intended purpose if the dose rate obtained in a space adjacent to them is lower than a practical level. Accordingly, the performance of spent fuel as a radiation source was calculated. The calculation was performed on the condition that the canisters were arranged in two rows and an irradiated object was passed in between. In this structure, as high a dose rate as possible can be obtained and the conveyance of the irradiated object is possible. We calculated the dose rate on the movement axis of the irradiated object in the case where the number of arranged canisters were from 2×2 to 2×7. A result of the calculation relative to the arrangements of the canisters by the QAD-CGGP2 code is shown in Fig. 7.4. Here, the calculation was performed on the condition that the canisters were not put in store pipes because of the obtainment of the highest dose rate and there was nothing between the canisters and the irradiated object. In the safety design of actual irradiation facilities, another containment boundary other than canisters is considered to be necessary on the assumption that radioactive substances are leaked from the canisters by any possibility. For example, if a conveyance path of irradiation objects is isolated by stainless plates and an irradiation chamber is made into an enclosed space, the dose rate decreases by 10 %.

It was clarified from literature research and investigation at general irradiation facilities, that in irradiation business, Co-60 etc. is used as a radiation source, irradiated objects are irradiated after being moved close to the radiation source, the dose rate in an irradiation field is about 10 kGy/h, and about 1 kGy/h is required at least. The maximum value of the dose rate increases as the number of rows of canisters increases; however, its increasing rate decreases when the number of rows reaches seven. The highest dose rate in the case of seven rows is about 1.1 kGy/h, and it is confirmed that a dose rate of 1 kGy/h which is required as an irradiation facility can be obtained.

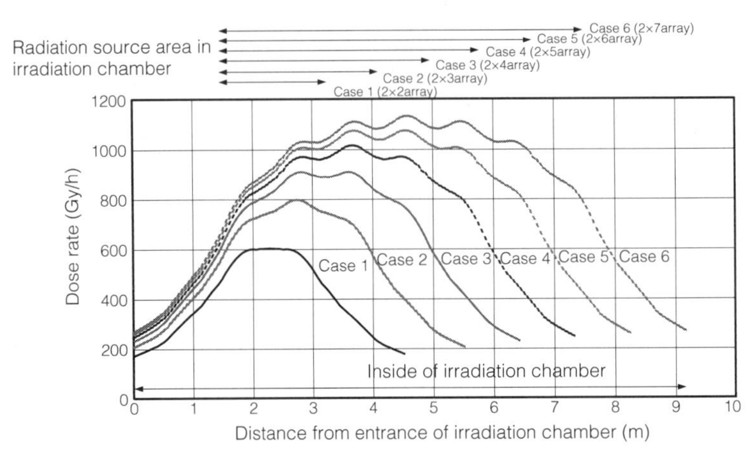

Fig. 7.4 Calculation result of dose rate in irradiation chamber

(3) Safety evaluation

Contrary to Co-60 and Cs-137 as radiation sources, the influence of activation due to neutrons emitted from spent fuel is required to be evaluated in the case of the use of spent fuel as a radiation source. The neutron emission rate of spent fuel is approximately 2.27×10^{8}(n/s) as mentioned above, and a calculation by a simple formula was performed relative to an injection needle and natural rubber as examples of an irradiation object by using this value. According to the calculation performed on the assumption that irradiation time was 30 hours, the absorbed doses of both objects were below 74 Bq/g which is the density value of an isotope emitting radiation mentioned in Science and Technology Agency Notification No. 15. However, a legal reference value for activated products has not decided yet at this time, so that revaluation is required to be performed when the reference value is decided.

Canisters are needed to be arranged closely in order to make the dose rate in the irradiation chamber as high as possible. However, if they are arranged closely, the temperature of fuel claddings might exceed the restrictive temperature because decay heat is generated from spent fuel in them and an internal temperature rises. Therefore, the thermal conductivity analysis of a radial two-dimensional cross section was performed by using the ABAQUS code, and a temperature inside the canisters installed in the irradiation chamber was obtained.

As a result, in the case where heat was forcibly removed by applying a wind of 3.6 m/s from the side of the canisters at an ambient temperature of 38 °C, the surface temperature of the canisters was 190 °C and the maximum temperature of the fuel claddings was 276 °C. Thus, it was confirmed that the restrictive temperature value is satisfied by installing a canister forced-air cooling system. The heat exchanger system which maintains the inside of the irradiation chamber at a constant temperature by releasing decay heat to the outside is also required in the irradiation chamber together with the canister forced-air cooling system.

(4) Future issues

Considering a qualitative cost comparison, because the irradiation facility using spent fuel has a large facility scale and the accessory equipment of the irradiation chamber is needed, construction costs of the facility are high. However, there is an advantage that no cost of radiation sources is required, so that it is possible to make this radiation utilization method competitive depending on optimization of facility design. In the future, it is required to consider how to return profits to local areas (e.g. cooperation with local industries, creation of new business and jobs, etc.) by providing what kind of irradiation service, together with to perform safety evaluation at accidents, specific equipment design, and detailed cost evaluation. In the case of utilizing spent fuel as a radiation source, there is an advantage that spent fuel has a long half-life and is in abundant supply. Thus, such facility design that irradiation is performed on a large number of irradiated objects is beneficial if this radiation utilization method becomes widely used.

References

1) M. Wataru, K. Sakamoto, T. Saegusa, E. Kashiwagi, & Y. Sasaki, (August, 1997) "Feasibility Study on Utilization of Heat from Spent Fuel in Storage Facility" CRIEPI Report U97023

2) M. Wataru, K. Sakamoto, T. Saegusa, T. Sakaya, & H. Fujiwara, (August, 1997) "Feasibility Study on Utilization of Radiation from Spent Fuel in Storage Facility" CRIEPI Report U97024

CHAPTER 8 INTERNATIONAL TRENDS

8.1 IAEA

(1) SPAR (Spent Fuel Performance Assessment and Research)

IAEA has implemented a project of collecting and organizing information about the storage status of spent fuel and the status of research and development on the behavior of stored fuel in each country and of providing member countries with the information. Regulatory authorities, research institutes, operators, etc. of the member countries participate in the project (Table 8.1-1). Japan has participated since the project was started, and has taken part in the creation of IAEA technical documents (TECDOC) while reporting the status of spent fuel storage in Japan and research results. At present, the project is at the stage of Phase III, and a draft of TECDOC on the behavior of spent fuel and the research and development status in international major countries is supposed to be drawn up by the end of the fiscal year of 2014.

Table 8.1-1 Main participants in SPAR

Country	Organization
France	TN Int. (container manufacturer)
Korea	KAERI (research institute)
Hungary	TS ENERCOM (consultancy firm)
US	NRC (regulator), EPRI (research institute), PNNL (national research institute)
England	Sellafield (operator)
Germany	GNS (operator)
Japan	JNES, CRIEPI

(2) Coordinated research project on demonstrating performance of spent fuel and related storage system components during very long term storage

Some countries have a policy to directly dispose of spent fuel as waste (e.g. Finland), while others have a policy to recycle it as a resource after reprocessing (e.g. France). However, many countries have not yet decided whether to reprocess or directly dispose of spent fuel at this time, and took a course to store it for the time being. Thus, the amount of stored spent fuel has increased worldwide (Fig. 8.1-1)[1] and also the storage period tends to be prolonged, so that it is hoped that the ageing of fuel and related equipment in long-term storage is early clarified.

From this background, IAEA started a research project on long-term storage of spent fuel, which was scheduled to be conducted for five years at the moment, in August 2012. In the project, the accumulation of experimental data and the development on evaluation tools on the basis of the information provided by participants (Table 8.1-2) are supposed to be performed.

Japan has participated in the project and provided the results of the durability evaluation of a gasket of a lid-sealing part of a metal cask and the SCC evaluation of a weld of a canister of a concrete cask and the results obtained from the PWR fuel storage test.

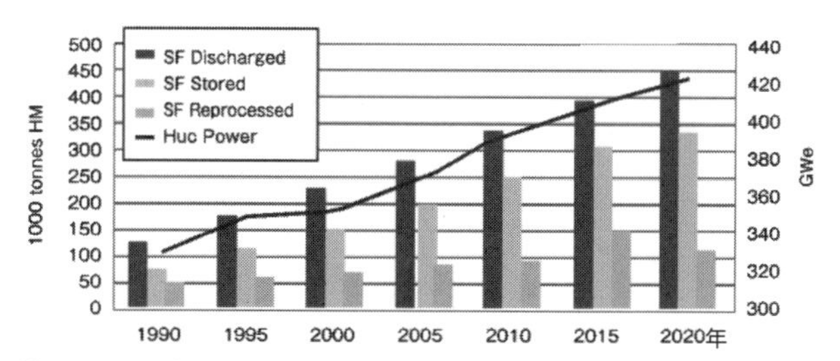

Fig. 8.1-1 Globally accumulated amount of discharged, stored, and reprocessed spent fuel

Table 8.1-2 Main participants in long-term storage project

Country	Organization
France	TN Int. (container manufacturer)
US	NRC (regulator), EPRI (research institute)
Germany	BAM (national research institute)
England	NNL (national research institute)
Spain	CSN (government organization)
Poland	NCNR (national research institute)
Japan	CRIEPI, JAPC

(3) Contribution to establishment of IAEA guides

IAEA has established guides related to transport and storage of spent fuel, and Japan has contributed to this. The latest examples are shown in Table 8.1-3.

Table 8.1-3 Examples of recent IAEA guides, etc. on transport/storage of spent fuel

Guide/Report
No.SSG-15 Specific Safety Guide Storage of Spent Nuclear Fuel (2012) (under consideration of revision)
No.NF-X.X Technical report Potential Interface Issues in Spent Fuel Management (in press)
No.NS-X.X Draft Safety Guide Guide on Dual Purpose Cask Safety Case for Transport/Storage Casks Containing Spent Fuel (in press)
No.NF-G-3.1 Guide Extending Spent Fuel Storage until Reprocessing or Disposal (in preparation)

8.2 USA

A spent fuel disposal project in Yucca Mountain was canceled, so that power companies have increased storage of spent fuel in their power plant sites. As of September 2011, the amount of pool-stored spent fuel in the sites was 49,067MTHM in 68 sites, and the amount of dry-stored spent fuel was 15,357MTHM in 52 sites [2]. As of March 2013, the number of generally-authorized storage sites was 54, and the number of site-authorized storage sites was 15. Furthermore, eight sites are expected to apply for general authorization between 2013 and 2014 (ref. Fig. 8.2-1). Recent application is all for concrete cast storage (including horizontal silo storage). It is because the Department of the Environment (DOE) is in charge of transport and disposal while the power companies are in charge of at-reactor storage, and DOE hasn't decided a transport destination/disposal site yet. In Japan, electric utilities play a role in both

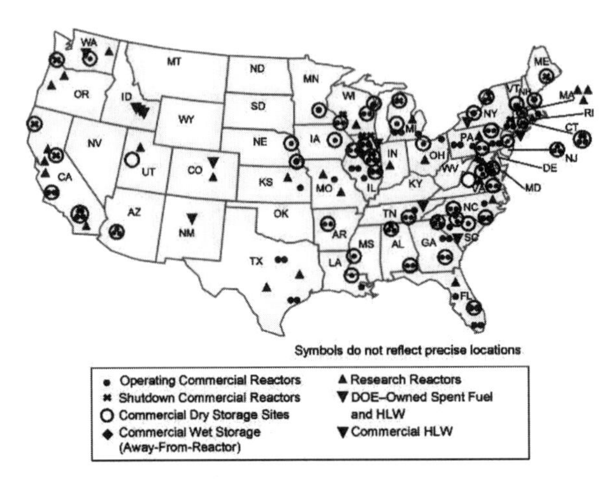

Fig. 8.2-1 Locations of spent fuel storage facilities in the USA

transport and storage of spent fuel, and a metal cask for both transport and storage is adopted because of its high cost efficiency in transport and storage.

Also, because spent fuel storage measures by DOE have been delayed as discussed below, the storage period tends to be prolonged. The storage period was initially set to 20 years, so that some facilities (power plants of Surry, Robinson, and Oconee) have already renewed the license.

(1) Department Of the Energy (DOE)

The DOE was supposed to start accepting spent fuel from the power companies in 1998 under the Nuclear Waste Policy Act (NWPA) of 1982; however, the power companies have stored it due to the cancellation of the spent fuel disposal project in Yucca Mountain. The DOE established a Blue Ribbon Committee on America's Nuclear Future by request from President Barak Obama. The committee reviewed the whole policy on the backend of an energy cycle, and produced a report counseling a new strategy [3]. The DOE decided the following policy in reaction to the report[4].

- By 2021, an interim storage facility for spent fuel from inactive nuclear power plants will be located, designed and constructed, and will start operation.
- By 2025, a large-capacity interim storage facility will come into operation through location and authorization.
- By 2048, a spent fuel disposal facility will come into operation through location and site characteristics inspection.

The DOE has promoted research and development through energy savings performance contracts (ESPC) with universities (ref. Nuclear Energy University Programs), after-mentioned EPRI, electric utilities, and research institutes.

(2) Nuclear Regulatory Commission (NRC)

1) Waste Confidence Decision and Rule

The NRC has a history of refusing to refrain from licensing nuclear reactors even though being required to perform the refrainment until disposal of spent fuel was decided in the suit (42 FR 34391) in 1977[5]. In the suit, The NRC stated that the licensing of nuclear reactors would not be continued if safe disposal of spent fuel was not assured. Since then, the federal government has intended deep geological disposal as a spent fuel disposal policy (Radioactive Waste Policy Act of 1982). The following is changes in the license period of nuclear reactors and the storage period until the present in the Waste Confidence Decision and Rule.

1984: Nuclear reactor license/40 years + Renewal/30 years

1990: Nuclear reactor license/40 years + Renewal/30 years + Storage alone/30 years

2010: Nuclear reactor license/40 years + Renewal/20 years + Storage alone/60 years

Now, the NRC plans to examine the Generic Environmental Impact Statement (GEIS) for supporting the renewal of the Waste Confidence Decision and Rule while continuing all licensing safety review operations. Three scenarios for storage are examined in the Waste Confidence Generic Environmental Impact Statement (Draft Report for Comment) issued in 2013.

a. Short-term storage within 60 years after shutdown of nuclear reactors (This is the likeliest scenario and accords with the DOE's plan of making a disposal site come into operation by 2048.)

b. Long-term storage within 160 years after shutdown of nuclear reactors (Storage facilities will be relocated every 100 years.)

c. At-reactor storage or away-from-reactor storage for an indefinite period (This is a scenario for the case that a disposal site cannot be used, and has the lowest possibility.)

The NRA promotes an environmental impact assessment according to Fig. 8.2-2.

The spent fuel disposal policy of the US is in a transition period. There is a possibility of long -term storage, and an alternative plan on the spent fuel measures could be presented. In regard to long-term storage, technological information with high degree of necessity in status quo includes the following [6].

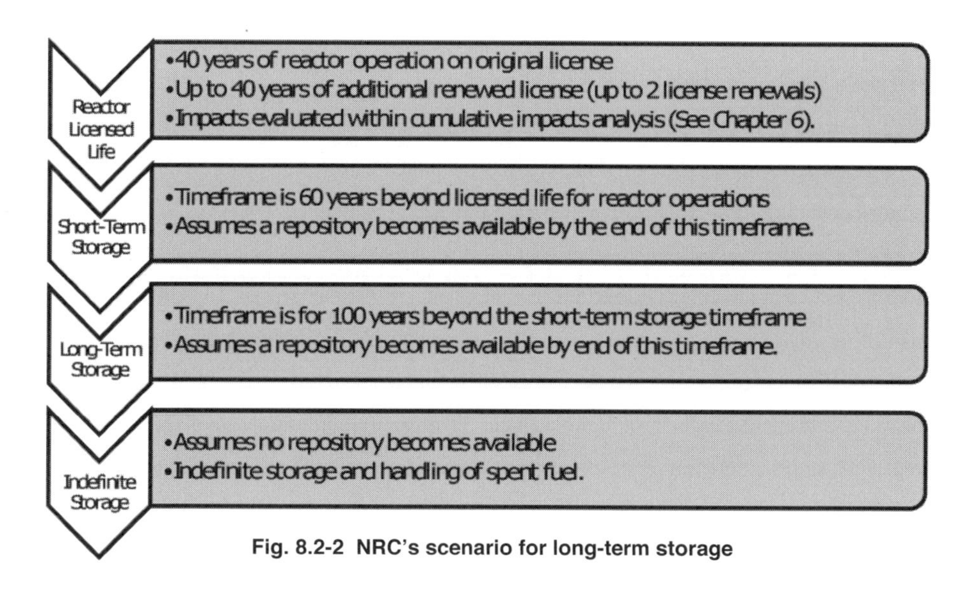

Fig. 8.2-2 NRC's scenario for long-term storage

· Stress corrosion cracking of a stainless canister

· Swelling and fragmentation of fuel

· More realistic thermal analysis models during long-term storage

· Influence of residual water after drying of fuel

· Monitoring method in service

(3) Idaho National Laboratory (INL) −Ageing evaluation of concrete−

The INL has conducted a demonstration experiment of storing real spent fuel in metal casks or concrete casks since the 1980s (Fig. 8.2-3). The CRIEPI acquired test data relating to the ageing of storage casks storing real spent fuel, which are difficult to obtain in Japan, in collaboration with INL[7].

And now, the collaborative research has already been completed. However, in regard to the handing of spent fuel in association with the accident in Tokyo Electric Power Company's Fukushima No.1 Nuclear Power Plant due to the Great East Japan Earthquake, valuable information has been obtained. The information is based on the experience of transport and storage of failed fuel generated at the accident in the Three Mile Island Nuclear Generating Station (TMI).

Fig. 8.2-3 Demonstration test on cask storage by INL

(4) Extended Storage Collaboration Program

The Extended Storage Collaboration Program (ESCP) led by Electric Power Research Institute (EPRI) was undertook for the purpose of surely promoting the long-term storage measures of spent fuel in the fiscal year of 2010. Before now, actions on a long-term storage demonstration test which is planned to start in 2020 have been performed mainly by an ESCP steering committee consisting of representatives of the government organization of the US, national institutions, power companies, nuclear power industries, etc. (Table. 8.2-1). The CRIEPI has participated as a joint chairman of an international cooperation sub-WG under the ESCP steering committee and also as a canister SCC task member of a demonstration test sub-WG since 2010, and has worked so as to make results of the long-term storage demonstration test planed in the program beneficial for Japan in light of a trend of spent fuel storage.

Recently, the DOE and Sandia National Laboratories reported a gap analysis (insufficient data predicted when the storage period is prolonged, extraction of evaluation methods and their prioritization) especially on high burnup fuel on the assumption that the storage period is 120 years or more (an evaluation target period is 300 years) and

Table 8.2-1 Main participants in ESCP International Cooperation WG

Country	Organization
Germany	BAM (national research institute)
Hungary	Som Sys. (consultancy firm)
Korea	KAERI (national research institute)
Spain	ENRESA (public corporation)
England	NNL (national research institute)
Japan	CRIEPI
US	SNL (national research institute)

the safety of retrieving after storage and multiple times of transport after storage is ensured. The following is much-needed technical information which the US considers[8].

· Hydrogen brittleness and hydride reorientation of spent fuel claddings
· Delayed hydride cracking of spent fuel claddings
· Atmospheric corrosion of canisters under ocean environment
· Wet corrosion of canisters
· Wet corrosion of metal casks
· Nondestructive monitoring technique
· Temporal temperature distribution analysis technique
· Quantitation of residual water after drying of casks and canisters storing spent fuel
· Burnup credit
· With-lid- opened inspection on casks (CASTOR V/21 and REA) stored in Idaho National Laboratory
· Influence of drying and submergence of spent fuel during retrieving

8.3 Germany

Germany has implemented a policy to store spent fuel in transport/storage dry casks until direct disposal[9]. As of May 2012, nine light water reactors were in operation, and uranium fuel and MOX fuel were used. The license of eight nuclear reactors lapsed due to the revision of an atomic energy act in August 2011. More 12 nuclear reactors were shut down. Under the atomic energy act revised in 2002, the spent fuel other than that which was entrusted to France or England for reprocessing by 2005 shall be stored in a form of at-reactor storage. As of December 2010, the spent fuel of 3,448 tHM was pool-stored in power plant sites, the spent fuel of 3,261tHM was dry-stored in the sites, and the spent fuel of 92 tHM was dry-stored away from the sites (Figs. 8.3-1, and 8.3-2).

In 2010, periodic safety review was decided to be conducted every 10 years with the object of maintaining the safety of storage relative to prolongation of the spent fuel storage period[10]. As of 2013, the review was conducted on two plants as a trial. In 2012, Entsorgungskommission (ESK) established a proposed measure on guidelines for monitoring the ageing of spent fuel in long-term storage. For example, an exterior inspection shall be conducted randomly on transport/storage casks every 10 years, periodic measurement shall be performed on storage buildings, and a state inspection shall be conducted on storage buildings and equipment needed for storage by a field survey and site measurement.

Fig. 8.3-1 Transport cask storage facility in Gorleben
TBL -G: Spent fuel storage facility
PKA: Spent fuel processing facility
ALG: Waste storage facility
(Copyright: GNS)

Fig. 8.3-2 Dry cask s for transport /storage in facility (Copyright: GNS)

References

1) IAEA: "Costing of Spent Nuclear Fuel Storage", No. NF-T-3.5 (2009)

2) U.S. Department of Energy, "United States of America Fourth National Report for the Joint Convention on the Safety of Spent Fuel management and on the Safety of Radioactive Waste Management", September 2011.

3) Blue Ribbon Commission on America's Nuclear Future: Report to the Secretary of Energy, January 2012.

4) Department of Energy, USA: Strategy for the Management and Disposal of Used Nuclear Fuel and High-Level Radioactive Waste, January 2013.

5) US NRC: Waste Confidence Generic Environmental Impact Statement (Draft Report for Comment), NUREG-2157, September 2013.

6) M. Lombard: Status Update: Extended Storage and Transportation Waste Confidence, 2013 NEI used Fuel Management Conference, May 7-9, 2013, St. Petersburg, Florida.

7) A.Sasahara, T.Saegusa, "Evaluation of Long-term Integrity of Spent Fuels Stored in Idaho National Laboratory -Cover Gas Sampling inside Metal Cask and Its Analysis-", CRIEPI Report L10017, July 2011.

8) EPRI: Extended Storage Collaboration Program International Subcommittee Report - International Perspectives on Technical Data Gaps Associated With Extended Storage and Transportation of Used Nuclear Fuel -, 2012 Technical Report, EPRI

9) Federal Ministry for the Environment, Nature Conservation and Nuclear Safety, "Report of the Federal Republic of Germany for the Fourth Review Meeting in May 2012 for Joint Convention of Safety of Spent Fuel Management and on the Safety of Radioactive Waste Management."

10) G.Arens, Ch.Gotz, S.Geupel, B.Gmal, W.Mester, "Interim Storage of Spent Nuclear Fuel before Final Disposal in germany- Regulator's View", Proc. OECD/NEA Int'l Workshop on Safety of Long Term Interim Storage Facilities, Munich, Germany, 21-23 May 2013.

INDEX

List of Figures

4.5.1(2)

4.5.2

4.5.3

4.5.4

4.6.1

4.6.2

4.7.1(1)

4.7.1(2)

4.7.1(3)

4.7.1(4)

List of Tables

CHAPTER 8

8.1

8.2

Authors

Chapter 1 : Toshiari Saegusa

Chapter 2 : Toshiari Saegusa

Chapter 3 : Toshiari Saegusa, Chihiro Ito, Hidetsugu Yamakawa, Hirofumi Takeda, Masumi Wataru, Hidetoshi Tamura, Akio Kosaki, Akihiro Sasahara, Norihiro Kageyama, Koji Shirai, Kosuke Namba

Chapter 4 : Toshiari Saegusa, Koji Shirai, Hirofumi Takeda, Masumi Wataru, Akio Kosaki, Taku Arai, Masanori Gotoh, Junichi Tani, Shuzo Eto, Takashi Fujii, Takuro Matsumura, Michihiko Hironaga

Chapter 5 : Toshiari Saegusa, Masumi Wataru, Yasuo Hattori, Koichi Shin

Chapter 6 : Akihiro Sasahara, Masami Mayuzumi, Tetsuo Matsumura

Chapter 7 : Masumi Wataru

Chapter 8 : Toshiari Saegusa

Toshiari Saegusa

Executive Research Scientist / Civil Engineering Research Laboratory

Masumi Wataru

Senior Research Scientist / Nuclear Fuel Cycle Backend Research Center / Civil Engineering Research Laboratory

Norihiro Kageyama

Research Scientist / Nuclear Fuel Cycle Backend Research Center / Civil Engineering Research Laboratory (Now in Transnuclear, LTD.)

Chihiro Ito

Associate Vice President / Director, Nuclear Fuel Cycle Backend Research Center / Civil Engineering Research Laboratory

Hidetoshi Tamura

Research Scientist / Fluid Dynamics Sector / Civil Engineering Research Laboratory

Koji Shirai

Deputy Associate Vice President / Nuclear Fuel Cycle Backend Research Center / Civil Engineering Research Laboratory

Hidetsugu Yamakawa

Consultant / Nuclear Fuel Cycle Backend Research Center / Civil Engineering Research Laboratory

Akio Kosaki

Consultant / Structural Engineering Sector / Civil Engineering Research Laboratory

Kosuke Namba

Research Scientist / Nuclear Fuel Cycle Backend Research Center / Civil Engineering Research Laboratory

Hirofumi Takeda

Senior Research Scientist / Nuclear Fuel Cycle Backend Research Center / Civil Engineering Research Laboratory

Akihiro Sasahara

Senior Research Scientist / Nuclear Technology Research Laboratory

Taku Arai

Senior Research Scientist / Structural Materials Sector / Materials Science Research Laboratory

Masanori Gotoh

Research Scientist / Nuclear Fuel Cycle Backend Research Center / Civil Engineering Research Laboratory (Now in Hitachi Zosen, Corp.)

Takuro Matsumura

Senior Research Scientist / Leader, Structural Engineering Sector / Civil Engineering Research Laboratory

Masami Mayuzumi

Associate Vice President / Director, Nuclear Power Plant Maintenance Research Team / Materials Science Research Laboratory

Junichi Tani

Senior Research Scientist / Structural Materials Sector / Materials Science Research Laboratory

Michihiko Hironaga

Senior Research Scientist / Nuclear Fuel Cycle Backend Research Center / Civil Engineering Research Laboratory

Tetsuo Matsumura

IERE Secretary General / Research Advisor / Central Research Institute of Electric Power Industry

Shuzo Eto

Research Scientist / Sector, Applied High Energy Physics / Electric Power Engineering Research Laboratory

Yasuo Hattori

Senior Research Scientist / Fluid Dynamics Sector / Civil Engineering Research Laboratory

Takashi Fujii

Senior Research Scientist / Sector, Applied High Energy Physics / Electric Power Engineering Research Laboratory

Koichi Shin

Senior Research Scientist / Nuclear Fuel Cycle Backend Research Center / Civil Engineering Research Laboratory

Basis of Spent Nuclear Fuel Storage

2015年 8月30日　　初版　第1刷発行

著　　者　一般財団法人電力中央研究所

発 行 人　長田　高

発 行 所　株式会社 ERC 出版
〒107-0062　東京都港区南青山 3-13-1　小林ビル 2F
電話　03-3479-2150　振替　00110-7-553669

印刷製本　芝サン陽印刷株式会社　東京都中央区新川 1-22-13
電話　03-5543-0161

ISBN978-4-900622-55-5

Basis of Spent Nuclear Fuel Storage

First Printing　August 30, 2015

Author: Central Research Institute of Electric Power Industry

Published by　ERC PUBLISHING CO., LTD
3-13-1 Minamiaoyama Minato-ku Tokyo Japan 107-0062

ISBN978-4-900622-55-5